Logic in Asia: Studia Logica Library

Logic in Asia: Studia Logica Library

This book series promotes the advance of scientific research within the field of logic in Asian countries. It strengthens the collaboration between researchers based in Asia with researchers across the international scientific community and offers a platform for presenting the results of their collaborations. One of the most prominent features of contemporary logic is its interdisciplinary character, combining mathematics, philosophy, modern computer science, and even the cognitive and social sciences. The aim of this book series is to provide a forum for current logic research, reflecting this trend in the field's development.

The series accepts books on any topic concerning logic in the broadest sense, i.e., books on contemporary formal logic, its applications and its relations to other disciplines. It accepts monographs and thematically coherent volumes addressing important developments in logic and presenting significant contributions to logical research. In addition, research works on the history of logical ideas, especially on the traditions in China and India, are welcome contributions.

The scope of the book series includes but is not limited to the following:

- Monographs written by researchers in Asian countries.
- Proceedings of conferences held in Asia, or edited by Asian researchers.
- Anthologies edited by researchers in Asia.
- Research works by scholars from other regions of the world, which fit the goal of "Logic in Asia".

The series discourages the submission of manuscripts that contain reprints of previously published material and/or manuscripts that are less than 165 pages/90,000 words in length.

Please also visit our webpage: http://tsinghualogic.net/logic-in-asia/background/

Relation with Studia Logica Library

This series is part of the Studia Logica Library, and is also connected to the journal Studia Logica. This connection does not imply any dependence on the Editorial Office of Studia Logica in terms of editorial operations, though the series maintains cooperative ties to the journal.

This book series is also a sister series to Trends in Logic and Outstanding Contributions to Logic.

For inquiries and to submit proposals, authors can contact the editors-in-chief Fenrong Liu at fenrong@tsinghua.edu.cn or Hiroakira Ono at ono@jaist.ac.jp.

Katsuhiko Sano · Ryo Hatano · Hiroakira Ono
Editors

Exploring Negation, Modality and Proof

6th Asian Workshop on Philosophical Logic

Springer

Editors
Katsuhiko Sano
Faculty of Humanities and Human Sciences
Hokkaido University
Sapporo, Hokkaido, Japan

Hiroakira Ono
Japan Advanced Institute of Science
and Technology
Nomi, Ishikawa, Japan

Ryo Hatano
Department of Industrial and Systems
Engineering
Tokyo University of Science
Noda, Chiba, Japan

ISSN 2364-4613 ISSN 2364-4621 (electronic)
Logic in Asia: Studia Logica Library
ISBN 978-981-96-8565-3 ISBN 978-981-96-8566-0 (eBook)
https://doi.org/10.1007/978-981-96-8566-0

This Springer imprint is published by the registered company Springer Nature Singapore Pte Ltd.
The registered company address is: 152 Beach Road, #21-01/04 Gateway East, Singapore 189721, Singapore

Organization

Program Committee Chairs

Ryo Hatano, Tokyo University of Science, Japan
Katsuhiko Sano, Hokkaido University, Japan

Steering Committee

Phan Minh Dung, Asian Institute of Technology, Thailand
Hu Liu, Sun Yat-sen University, China
Beishui Liao, Zhejiang University, China
Fenrong Liu, Tsinghua University, China
Hiroakira Ono, Japan Advanced Institute of Science and Technology, Japan
Ram Ramanujam, Institute of Mathematical Sciences, India
Hsing-chien Tsai, National Chung Cheng University, Taiwan

Program Committee

Thomas Ågotnes, University of Bergen, Norway
Sankha S. Basu, Indraprastha Institute of Information Technology Delhi, India
Hans van Ditmarsch, University of Toulouse, CNRS, IRIT, France
Sujata Ghosh, Indian Statistical Institute, India
Yuichiro Hosokawa, Gunma Prefectural Women's University, Japan
Ryo Ito, Waseda University, Japan
Raheleh Jalali, Utrecht University, Netherlands
Fengkui Ju, Beijing Normal University, China
Makoto Kanazawa, Hosei University, Japan

Md. Aquil Khan, Indian Institute of Technology Indore, India
Daisuke Kimura, Toho University, Japan
Kohei Kishida, University of Illinois Urbana-Champaign, USA
Hidenori Kurokawa, Kanazawa University, Japan
Kok Yong Lee, National Chung Cheng University, Taiwan
Dazhu Li, Chinese Academy of Sciences, China
Hanti Lin, University of California, Davis, USA
Tadeusz Litak, FAU Erlangen-Nurnberg, Germany
Fenrong Liu, Tsinghua University, China
Hu Liu, Sun Yat-Sen University, China
Minghui Ma, Sun Yat-sen University, China
Koji Mineshima, Keio University, Japan
Hitoshi Omori, Tohoku University, Japan
Takuro Onishi, Kyoto University, Japan
Elise Perrotin, Computer science Research Institute of Lens, France
Teeradaj Racharak, Japan Advanced Institute of Science and Technology, Japan
Ram Ramanujam, Institute of Mathematical Sciences, India
Takahiro Sawasaki, National Institute of Informatics, Japan
Satoshi Tojo, Asia University, Japan
Jonni Virtema, University of Sheffield, UK
Ren-June Wang, National Chung-Cheng University, Taiwan
Yì Nicholas Wáng, Sun Yat-sen University, China
Tomoyuki Yamada, Hokkaido University, Japan
Byeong-Uk Yi, University of Toronto, Canada

Additional Reviewers

Jonas Becker Arenhart
Patrick Blackburn
Mike Cruchten
Asta From
Rustam Galimullin
Iris van der Giessen
Marianna Girlando
Reijo Jaakkola
Yusuke Kubota
Gavin St. John
Xiaolong Liang
John Lindqvist
Jun Otsuka
Vít Punčochář
Umberto Rivieccio
Takahiro Seki

Ian Shillito
Shawn Standefer
Ryo Takemura
Masanobu Toyooka
Fernando R. Velazquez-Quesada

Preface

We are pleased to present the Proceedings of the Sixth Asian Workshop on Philosophical Logic (AWPL 2024), held at Hokkaido University, Japan, from March 4 to 6, 2024. The Asian Workshop on Philosophical Logic (AWPL) is a prestigious series of events initiated by a group of Asian logicians. Since its inception at the Japan Advanced Institute of Science and Technology (JAIST) in 2012, AWPL has been dedicated to promoting awareness, understanding, and collaboration among researchers in philosophical logic and related fields. With a special emphasis on the interplay between philosophical ideas and formal theories, the workshop covers a wide range of topics, including non-classical logic, philosophical logic, algebraic logic, and their applications in disciplines such as computer science, cognitive science, and social science. Notably, this workshop is the first AWPL event to be held in Sapporo, Hokkaido, Japan, since the disruptions caused by the global COVID-19 pandemic.

We set the length of submitted papers at 12 pages and sent out the first call for papers in July 2023. In the end, we received 42 submissions for presentations at the workshop, with authors from various countries not limited to Asia, each of which was then reviewed by two reviewers determined by the international program committee. Of the 42 submissions, 15 were accepted for regular presentation and 9 for short presentation. In addition, there were 4 invited talks by Patrick Blackburn, Ryo Kashima, Shawn Standefer, and Fan Yang. After the workshop, we invited the authors of 18 presentations, including the invited speakers, to submit extended versions of their papers for inclusion in the proceedings. These 18 papers were again peer-reviewed and finally included in the proceedings after careful revision based on the reviewers' suggestions and comments.

This collection of accepted papers explores various areas within philosophical logic, including different approaches to negation, the formalization of modal and epistemic concepts, the development of proof systems, the study of non-monotonic reasoning, and the investigation of the interplay between metaphysical concepts and logical frameworks. These papers collectively inspire new ideas under the title "Exploring Negation, Modality and Proof."

The first part of this volume starts with the topics of negation. Fan Yang explores propositional team-based logics with full intuitionistic negation, whereas the languages of logics based on team semantics typically allow only atomic negation or restricted forms of negation. Takuro Onishi proposes a new approach to negation in the operational semantics for relevant logic to define a logic that validates contraposition axiom at the expense of contraction, reflecting their incompatibility in operational semantics. Shawn Standefer presents two approaches to epistemic logic within the framework of relevant logics. The first approach uses equivalence relations to model indistinguishability, while the second approach, more common in relevant logic research, eschews the use of equivalence relations. In order to systematically construct models of mereology, Santiago Jockwich Martinez proposes a method called mereological forcing. Pedro Teixeira Yago studies arbitrary objects within the framework of the propositional bilateral system called weak rejectivist logic to avoid objections against the principle of generic attribution, which states that an arbitrary object has the properties common to all individual objects in its domain.

The second part is related to modality and modal logic. Satoru Suzuki offers a simplified semantic completeness proof for Modal-Qualitative-Probability Logic (MQPL), improving upon Segerberg's proof for qualitative probability logic, where MQPL overcomes limitations in existing models of comparative epistemic modalities, like Kratzer's and Yalcin's, by employing qualitative-probability orderings, aligning with Kratzer's understanding of epistemic modality. Jialiang Yan, Maria Aloni, and Fenrong Liu propose a multi-agent model that combines team-based semantics with a two-dimensional semantics to analyze epistemic might. Patrick Blackburn, Torben Braüner, and Julie Lundbak Kofod discuss how Arthur Prior placed propositional quantifiers in the context of his early work and his creation of hybrid tense logic and egocentric logic, and review what they call Prior's Ideal Language (PIL), a hybrid modal logic enriched with propositional quantifiers. Yuki Nishimura and Tsubasa Takagi proposes a tableau method of the hybrid logic for undirected graphs and shows the completeness theorem and the termination property of their tableau method, which leads us to prove the decidability.

The third part deals with various papers on proofs in a broader sense. Yuecong Zhang and Xuefeng Wen provide a new generalization of strict-tolerant logic in terms of generalized matrices in abstract algebraic logic to reflect the main idea that strict truth is the standard of truth accepted by everyone, and tolerant truth is the standard accepted by someone. Takahiro Seki studies variations of the implicational axioms K and S from a syntactical perspective within the framework of substructural logics, without assuming both exchange and associativity. Andrzej Indrzejczak establishes the Craig interpolation theorem for elementary ontology of Leśniewski (LO), a comprehensive calculus of names and a theoretical basis for mereology, an alternative foundation of mathematics. To characterize cut-eliminability, Ryo Takemura proposes a sequent calculus mG3 for multi-sequents, where a multi-sequent is a sequence of ordinary sequents representing successive applications of the cut rule and proof-reduction rules for eliminating the cut-rule are represented as proof-construction rules in mG3. Rohan French studies from intuitionistic logical viewpoints nonreflexive and nontransitive logics that reject structural principles, like the reflexivity or transitivity of the entailment relation, and that provide a solution to the semantic paradoxes. Yukiko Yana, Koji Mineshima, and Daisuke Bekki use dependent type semantics to propose an alternative approach to Sudo's work to the binding problem of two-dimensional theory in natural language semantics, which distinguishes between assertion and presupposition. The binding problem means that when analyzing sentences with indefinite subjects, it cannot always be ensured that the subject in the presupposition corresponds to the same person as in the assertion.

The remaining part of this volume introduces various aspects of philosophical logic. Hayato Ito investigates a variant of David Lewis' imaging and explores their relationship with conditionalization, which is a basic notion of belief change in Bayesianism, in terms of AGM-style postulates. Hanti Lin develops a Peircean alternative to the traditional Carnapian conception of inductive logic and use it to unify formal learning theory, statistics, and a significant part of machine learning, i.e., supervised learning. Byeong-Uk Yi argues that the thesis of composition as identity (CAI) is incoherent because it collapses into the triviality thesis, which states that everything is a part of itself and nothing else, regardless of how the composition relation is analyzed.

Hiroakira Ono played a key role in advising on the overall structure and outline of the volume, as well as providing supportive recommendations for this volume to the Logic in Asia (LIAA). We are grateful to Zhang He, Kruthika Srinivasan and Shalini Monica Clement Selvam for their support and patience in the process of publication of this volume.

Finally, the Sixth Asian Workshop on Philosophical Logic (AWPL 2024) is hosted by the Laboratory of Philosophy and Ethics, Hokkaido University, and co-sponsored by JSPS Kakenhi Grant-in-Aid for Scientific Research (C) Grant No. JP19K12113, and under the auspices of CAEP (Center for Applied Ethics and Philosophy), Faculty of Humanities and Human Sciences, Hokkaido University.

Sapporo, Hokkaido, Japan — Katsuhiko Sano
Noda, Chiba, Japan — Ryo Hatano
Nomi, Japan — Hiroakira Ono
January 2025

Contents

Chapter 1
There Are (Other) Ways to Negate in Propositional Team Semantics

Fan Yang

Abstract The languages of logics based on team semantics typically only allow atomic negation or restricted negation. In this paper, we explore propositional team-based logics with full (intuitionistic) negation. We demonstrate that including full intuitionistic negation does not complicate the axiomatisation of propositional team-based logics with the downward closure property. We also review known expressive completeness results for these logics, highlighting how relevant complemented properties are expressed in propositional dependence logic without directly using negation. Building on these insights, we also prove a new result: propositional logic extended with both dependence and inclusion atoms is expressively complete.

Keywords Team semantics · Dependence logic · Negation

1.1 Introduction

Hodges [20, 21] introduced *team semantics* as a compositional semantics for Hintikka and Sandu's independence-friendly logic [18, 19]. Team semantics was then further developed in Väänänen's dependence logic [34] with atoms of the type $=(x, y)$ expressing that the value of y is determined by the value of x. Team semantics characterises such dependency formulas by evaluating them over sets of assignments (called *teams*). This framework has since then been extended significantly. Many logics based on team semantics have been introduced, such as independence logic (with independence atoms) [12], inclusion logic (with inclusion atoms) [10, 11] and their propositional and modal variants. The idea of team semantics has also been independently adopted in inquisitive logic [8] to characterise questions in natural language.

F. Yang (✉)
Department of Philosophy and Religious Studies, Utrecht University, Utrecht, The Netherlands
e-mail: f.yang@uu.nl

K. Sano et al. (eds.), *Exploring Negation, Modality and Proof*,
Logic in Asia: Studia Logica Library,
https://doi.org/10.1007/978-981-96-8566-0_1

Many of these team-based logics do not include full negation in their syntax. Indeed, negation is often considered a tricky connective in team semantics. The most commonly known negation in team semantics, the dual negation, originates from the game theoretic interpretation of team semantics. In the evaluation game of a formula, when a position associated with a negated formula $-\alpha$ is reached, the Falsifier and Verifier swap their roles. Such a negation is, however, not a semantic operator, in the sense that preservation of equivalence under replacement fails under the scope of negation. For instance, we have $- =\!\!(x, y) \equiv \bot$, whereas $- - =\!\!(x, y) \equiv =\!\!(x, y) \not\equiv \top \equiv - \bot$. In fact, the failure of dual negation to function as a semantic operator is even more extreme, as shown first by Burgess [5] in the related context of Henkin quantifiers [17] and later by other authors in team-based logics [2, 26]. This problem can, nevertheless, be avoided by considering both the team that validates the formula and the team that falsifies the formula (known also as "co-team"), as done in Hodges' [20, 21] and in Väänänen's monograph [34].

Apart from these technical concerns, there is also a conceptual challenge with the dual negation. When defining the semantics of a negated dependence atom $- =\!\!(x, y)$, in order to preserve the desirable downward closure property that all the other formulas in dependence logic have, one has to define $- =\!\!(x, y)$ to be true only on the empty team (as $=\!\!(x, y)$ is always trivially true on any singleton team). In other words, $- =\!\!(x, y)$ is defined to be equivalent to the constant falsum $\bot$. This technical solution makes it difficult to provide a meaningful interpretation of the negated dependence atoms, and thus has often been criticised.

In order to avoid all these complications, most of the literature on team-based logics restricts the occurrence of negation in front of arbitrary formulas. This is done by assuming negation normal form and excluding negated dependence atoms from the class of well-formed formulas. However, negation as an important connective cannot be entirely avoided, particularly in the context of proof theory. For instance, in the partially complete deduction systems for first-order dependence, independence and inclusion logics [13, 27, 39],[1] the scope of negation was extended in an ad hoc manner to cover all classical first-order formulas (i.e., those without dependence, independence or inclusion atoms). Similarly, in propositional and modal team-based logics, negation is often allowed to occur in front of classical formulas (see, e.g., [36, 40]). However, full negation is still often excluded in these frameworks. In this paper, we study team-based propositional logics with full negation.

Negation can, of course, also be defined in terms of an appropriate implication. In particular, in inquisitive logic, the formula $\phi \to \bot$ (with the intuitionistic implication $\to$) is (as usual) treated as the definition of the (intuitionistic) negation $\neg\phi$ of ϕ. In this paper, we define the full negation $\neg\phi$ exactly through the semantic clause of the intuitionistic negation $\phi \to \bot$. This negation is conservative over the dual negation, as for formulas α in the original syntax of propositional logic, $\neg\alpha$ (i.e., $\alpha \to \bot$) is equivalent to the dual negation $-\alpha$. The semantic clause of this negation actually coincides with that of the *flattening* of the dual negation, as considered by Hodges

[1] Note that these logics cannot be fully axiomatised, due to their strong expressive power (see the cited papers for details).

[20, 21] when he introduced team semantics. Since such an intuitionistic negation is defined compositionally in the usual manner, without referring to co-teams, failure of replacement does not any more occur; in particular, we have $\neg =\!\!(x, y) \equiv \bot$ and $\neg\neg =\!\!(x, y) \equiv \neg\bot \equiv \top$. In the literature on team semantics, this compositional full negation has been touched upon in some instances (such as in [22, 29, 30]). In this paper, we aim to present a new perspective on this negation.

In the context of dependence logic, while the intuitionistic implication (and thus intuitionistic negation) has also been introduced [1], they are usually not included in the standard syntax of first-order dependence logic and its variants. This is because, for instance, first-order dependence logic is expressively equivalent to existential second-order logic [25, 34], but its extension with intuitionistic implication is expressively too strong—equivalent to full second-order logic [35]—and consequently many desirable properties, such as compactness and others, are lost. In fact, already first-order dependence logic extended with full intuitionistic negation alone is able to express finiteness of the model (via the intuitionistic negation $\neg\phi_\infty$ of the dependence logic sentence ϕ_∞ that expresses infinity[2]), and thus compactness fails for the logic.

However, the situation changes significantly when we restrict our attention to propositional and modal team-based logics. Many noteworthy propositional and modal team-based logics are shown to be expressively complete with respect to certain closure properties (such as downward and union closure) [15, 16, 40–42]. The intuitionistic negation preserves most of these closure properties (as $\neg\phi$ is always flat). Consequently, extending these logics with full intuitionistic negation does not increase their expressive power. It is thus both safe and reasonable to consider full negation in propositional and modal settings.

Moreover, incorporating this full negation does not complicate the axiomatisation of propositional team-based logics with downward closure property, as we will demonstrate in this paper. We show in Sect. 1.4 that for classical propositional logic extended with global disjunction $\mathbin{\rotatebox[origin=c]{0}{$\vee$}\!\!\!\vee}$ (**PL**($\mathbin{\vee\!\!\!\vee}$)) with full negation, a complete natural deduction system can be obtained by adapting the existing system of **PL**($\mathbin{\vee\!\!\!\vee}$) with restricted negation in a natural way. In Sect. 1.4 we also revisit the known systems for **PL**($\mathbin{\vee\!\!\!\vee}$) [6, 7, 41] in the literature and present a simpler, folklore-based alternative. In addition, we document several folklore observations regarding the system, including the fact the Split axiom of inquisitive logic is derivable from a distributive law in the system of **PL**($\mathbin{\vee\!\!\!\vee}$).

Closely related to the lack of appropriate full negation in the typical formulation of team-based logics is the challenge of defining the complement of a given property. This is particularly relevant in the proof of the expressive completeness of propositional dependence logic (**PL**($=\!\!(\cdot)$)) as given in [41]. Similar expressive completeness result for **PL**($\mathbin{\vee\!\!\!\vee}$) is proved in [41] by expressing a relevant team property using a formula $\bigvee\!\!\!\bigvee_{t \in T} \chi_t$ in disjunctive normal form, where each disjunct χ_t defines

[2] To be more precise, the formula ϕ_∞ is defined as $\exists z \forall x \exists y (=\!\!(x, y) \wedge =\!\!(y, x) \wedge y \neq z)$, which is true in a first-order model M iff there is an injective function $f : M \to M$ that is not onto, meaning that M is infinite. See e.g., [34, Sect. 4.2] for details.

a specific (positive) property that is easily expressible in classical logic. In contrast, for $\mathbf{PL}(=\!\!(\cdot))$ without the global disjunction, a conjunctive formula $\bigwedge_{t\in T}\rho_t$ is used as the defining formula in the proof, where each conjunct ρ_t defines the complement of a certain property P (i.e., $\overline{P}$). One might want to define $\overline{P}$ using the negation of a formula ϕ that defines P, but the appropriate negation in this context is the classical negation $\sim$, which is lacking in the logic's language. Adding such a classical negation $\sim$ would break the downward closure of $\mathbf{PL}(=\!\!(\cdot))$ and resulting logic has quite different properties (see [28, 31] for studies on this and related logics). Let us also remark that negated formulas $\neg\phi$ with our newly introduced full intuitionistic negation do not generally contribute to expressing complemented properties, as $\neg\phi$ is always flat and can thus only define flat properties. Nevertheless, the property $\overline{P}$ can be expressed in $\mathbf{PL}(=\!\!(\cdot))$ by other means.

In Sect. 1.3, we review these results from the literature and highlight how complemented properties are expressed without directly using negation in the logic $\mathbf{PL}(=\!\!(\cdot))$. This further explains why full negation has often been excluded from the typical formulations of propositional dependence logic and its variants—the versions without full negation are already sufficiently expressive, after all.

It is worth noting that in the first-order setting, as expected, incorporating classical negation $\sim$ into first-order dependence logic dramatically increases the expressive power of the logic to full second-order logic too [24]. Alternative ways to express complemented properties without increasing the expressive power in the first-order setting have been explored in [38].

In Sect. 1.3, we also revisit the disjunctive and conjunctive normal forms for propositional team-based logics. We illustrate how the normal form approach for obtaining expressive completeness results for these logics aligns with the standard normal form technique used in the usual (single valuation-based) classical logic. Building on these insights, we prove a new result: classical propositional logic extended with both dependence and inclusion atoms ($\mathbf{PL}(=\!\!(\cdot), \subseteq)$) is expressively complete for the class of all team properties that contain the empty team. This logic has so far been largely overlooked in the literature. Our result corresponds well with the known result for the first-order variant of $\mathbf{PL}(=\!\!(\cdot), \subseteq)$, which is also expressively complete, in the sense that it captures all properties definable in existential second-order logic [10]. Our proof also offers yet another way to express complemented properties without directly using negation in propositional team-based logics.

Overall, in this paper we aim to demonstrate that adding full (intuitionistic) negation to propositional team-based logics does not introduce significant technical complications, especially for downward closed logics. The obstacle to incorporating full negation is, in a sense, more conceptual than technical. For instance, one remaining challenge is how to meaningfully interpret negated dependence or inclusion atoms. Nevertheless, there are interesting well-behaved propositional team-based logics that do not contain these dependency atoms in the syntax, such as $\mathbf{PL}(\vee\!\!\!\vee)$, which can accommodate full negation well. We leave the exploration of these and other issues to future research, which is discussed in Sect. 1.5.

1.2 Preliminaries

We first fix our notations for the language of classical (propositional) logic, denoted as **PL**. Let Prop be a (countable) set of propositional variables. We define formulas of **PL**, also called *classical formulas*, recursively as

$$\alpha ::= p \mid \bot \mid \neg\alpha \mid \alpha \wedge \alpha \mid \alpha \vee \alpha$$

where $p \in \mathsf{Prop}$. The standard classical logic evaluates formulas over *valuations*, which are functions $v : A \to \{0, 1\}$ from a set $A \subseteq \mathsf{Prop}$ of propositional variables to the set $\{0, 1\}$ of truth values. The satisfaction relation $v \models \alpha$ is defined inductively as usual, namely, e.g., $v \models p$ iff $v(p) = 1$, $v \models \neg\alpha$ iff $v \not\models \alpha$, etc.

In this paper we study propositional logics with *team semantics*. One such logic, denoted as **PL**($\mathbin{\vee\!\!\!\vee}$), has the language **PL** extended with *global disjunction* $\mathbin{\vee\!\!\!\vee}$[3] (in contrast, the disjunction of **PL** is referred to as *local disjunction*). To be more precise, formulas of **PL**($\mathbin{\vee\!\!\!\vee}$) are defined recursively as

$$\phi ::= p \mid \bot \mid \neg\phi \mid \phi \wedge \psi \mid \phi \vee \psi \mid \phi \mathbin{\vee\!\!\!\vee} \psi$$

where $p \in \mathsf{Prop}$. Unlike most literature on team semantics where negation is restricted to atomic or classical formulas only, in this paper we allow negation to occur in front of arbitrary formulas.

Formulas of **PL**($\mathbin{\vee\!\!\!\vee}$) are evaluated over (propositional) teams. A *team* is a set t of valuations $v : A \to \{0, 1\}$ over the same domain $A \subseteq \mathsf{Prop}$. For any formula ϕ and team t whose domain contains all the propositional variables occurring in ϕ, we define the satisfaction relation $t \models \phi$ inductively as:

- $t \models p$ iff $v \models p$ for all $v \in t$;
- $t \models \bot$ iff $t = \emptyset$
- $t \models \neg\phi$ iff $\{v\} \not\models \phi$ for all $v \in t$;
- $t \models \phi \wedge \psi$ iff $t \models \phi$ and $t \models \psi$
- $t \models \phi \vee \psi$ iff there exist $r, s \subseteq t$ such that $t = r \cup s$, $r \models \phi$ and $s \models \psi$;
- $t \models \phi \mathbin{\vee\!\!\!\vee} \psi$ iff $t \models \phi$ or $t \models \psi$

We write $\phi \models \psi$ if $t \models \phi$ implies $t \models \psi$ holds for all teams t. If both $\phi \models \psi$ and $\psi \models \phi$, we write $\phi \equiv \psi$ and say that ϕ and ψ are (semantically) equivalent.

The above semantic clause for the full negation $\neg\phi$ is clearly compositional. We will make further remarks about the full negation at the end of this section. Let us now focus on recalling basic properties of the logic.

It is easy to verify by induction that arbitrary formulas ϕ in **PL**($\mathbin{\vee\!\!\!\vee}$) satisfy the empty team and downward closure property, defined as:

---

[3] The disjunction $\mathbin{\vee\!\!\!\vee}$ is also known in the literature as *intuitionistic disjunction*, *inquisitive disjunction*, and sometimes also *Boolean disjunction* or *classical disjunction*. In this paper we choose to adopt a more neutral term: *global disjunction*.

Empty team property: $\emptyset \models \phi$ always holds.
Downward closure: If $t \models \phi$, then $s \models \phi$ for all $s \subseteq t$.

Formulas in the classical language α satisfy, additionally, union closure and flatness property:

Union closure: If $t_i \models \alpha$ for all $i \in I \neq \emptyset$, then $\bigcup_{i \in I} t_i \models \alpha$.
Flatness: $t \models \alpha$ iff $\{v\} \models \alpha$ for all $v \in t$.

One can easily show that a formula ϕ is flat iff it satisfies downward closure, union closure and the empty team property.[4]

An easy inductive proof shows that for any classical formula α,

$$\{v\} \models \alpha \iff v \models \alpha. \tag{1.1}$$

We will thus use the above two types of satisfaction relations for classical formulas interchangeably.

Non-classical formulas are in general not flat. A typical example of a non-flat non-classical formula is $p \vee\!\!\!\vee \neg p$, which is trivially true on all singleton teams (i.e., teams of the form $\{v\}$) but not true, for example, on the team $\{v_0, v_1\}$ with $v_0(p) = 0$ and $v_1(p) = 1$. Yet, the negation $\neg(p \vee\!\!\!\vee \neg p)$ (being still non-classical) is flat. In fact, it follows immediately from the definition that all negated formulas $\neg\phi$ are flat.

We also consider dependence and inclusion atoms, which are strings of the form $=\!\!(p_1 \dots p_n, q)$ and $a_1 \dots a_n \subseteq b_1 \dots b_n$ (where $a_i, b_i \in \mathsf{Prop} \cup \{\bot, \top\}$), respectively. Their semantics are defined as

- $t \models =\!\!(p_1 \dots p_n, q)$ iff $v(p_i) = u(p_i)$ for all $1 \leq i \leq n$ implies $v(q) = u(q)$.
- $t \models a_1 \dots a_n \subseteq b_1 \dots b_n$ iff for all $v \in t$, there exists $u \in t$ such that for all $1 \leq i \leq n$, $v(a_i) = u(b_i)$.

Two simple types of these atoms will play a central role in our discussions: Dependence atoms $=\!\!(\langle\rangle, p)$ with the first component the empty sequence, which we also denote as $=\!\!(p)$ and refer to as *constancy atoms*; inclusion atoms of the form $x_1 \dots x_n \subseteq p_1 \dots p_n$ with $x_1, \dots, x_n \in \{\top, \bot\}$, which we refer to as *primitive inclusion atoms*. The semantic clauses of these simple atoms reduce to

- $t \models =\!\!(p)$ iff for all $v, u \in t$, $v(p) = u(p)$ (namely, p has a constant value in the team t).
- $t \models x_1 \dots x_n \subseteq p_1 \dots p_n$ iff there exists $v \in t$ such that for all $1 \leq i \leq n$, $v(p_i) = \underline{x_i}$, where $\underline{\top} = 1$ and $\underline{\bot} = 0$ (namely, the sequence $\underline{x_1} \dots \underline{x_n}$ of values for $p_1 \dots p_n$ is present in the team t).

---

[4] Even though this fact is immediate, in some earlier publications on team semantics (such as the author's joint paper [42]), the empty team property was mistakenly forgotten. Neither did they include the restriction that in the definition of union closure, the index set I should be assumed nonempty, for otherwise union closure would (unintentionally) imply the empty team property. The author would like to thank Fausto Barbero for bringing this issue to her attention several years ago.

We write $\mathbf{PL}(=\!\!(\cdot))$ for $\mathbf{PL}$ extended with dependence atoms, known as *propositional dependence logic*, $\mathbf{PL}(\subseteq)$ for $\mathbf{PL}$ extended with inclusion atoms, known as *propositional inclusion logic*, $\mathbf{PL}(\subseteq, \mathbin{\mathbb{v}})$ for $\mathbf{PL}$ extended with inclusion atoms and global disjunction, and $\mathbf{PL}(=\!\!(\cdot), \subseteq)$ for $\mathbf{PL}$ extended with both dependence and inclusion atoms. To be more precise, formulas of, e.g., $\mathbf{PL}(=\!\!(\cdot))$ are defined recursively as

$$\phi ::= p \mid \bot \mid =\!\!(p_1 \dots p_n, q) \mid \neg\phi \mid \phi \wedge \psi \mid \phi \vee \psi$$

where $p \in \mathsf{Prop}$. It is easy to verify that formulas in $\mathbf{PL}(=\!\!(\cdot))$ are downward closed, formulas in $\mathbf{PL}(\subseteq)$ are union closed, and formulas in all these four new logics have the empty team property.

We conclude this section with a few additional remarks on the full negation we defined. Our semantic clause for the full negation $\neg\phi$ actually corresponds to that of the familiar intuitionistic negation $\phi \to \bot$, where $\to$ denotes the *intuitionistic implication* from the literature (see e.g., [1, 8]) and is defined as

- $t \models \phi \to \psi$ iff for any $s \subseteq t$, $s \models \phi$ implies $s \models \psi$.

While we, in this paper, do not include the intuitionistic implication in our syntax, adding it to, e.g., $\mathbf{PL}(\mathbin{\mathbb{v}})$ does not increase the expressive power of the logic, as was proved essentially in [41] and as we will review in Theorem 3.2. (The intuitionistic implication is, however, not uniformly definable in $\mathbf{PL}(\mathbin{\mathbb{v}})$, by the argument in [37].[5])

As expected for intuitionistic negation, the excluded middle law fails with respect to local or global disjunction in the downward closed logics $\mathbf{PL}(\mathbin{\mathbb{v}})$ and $\mathbf{PL}(=\!\!(\cdot))$, namely neither $\phi \vee \neg\phi$ nor $\phi \mathbin{\mathbb{v}} \neg\phi$ is valid in general for these logics. To see why, for the logic $\mathbf{PL}(\mathbin{\mathbb{v}})$, the two formulas $(p \mathbin{\mathbb{v}} \neg p) \vee \neg(p \mathbin{\mathbb{v}} \neg p)$ and $p \mathbin{\mathbb{v}} \neg p$ are both invalid. For $\mathbf{PL}(=\!\!(\cdot))$, our counter-example is $=\!\!(p) \vee \neg =\!\!(p)$, where note that $\neg =\!\!(p) \equiv \bot$.

Over union closed team-based logics, $\phi \vee \neg\phi$ is actually valid. To see why, consider any team t, and let $r = \{v \in t \mid \{v\} \models \phi\}$ and $s = \{v \in t \mid \{v\} \not\models \phi\}$. Since ϕ is closed under unions, $r \models \phi$ and $s \models \neg\phi$, which give $t \models \phi \vee \neg\phi$.

1.3 Expressive Completeness via Normal Form and Ways to Negate

All five of the team-based logics we defined are expressively complete. The expressive completeness result for the downward closed logics $\mathbf{PL}(\mathbin{\mathbb{v}})$ and $\mathbf{PL}(=\!\!(\cdot))$ with atomic negation only was proved in [41], while for $\mathbf{PL}(\subseteq)$ with restricted negation, it was proved essentially in [16] (see also [40]), and the result for $\mathbf{PL}(\subseteq, \mathbin{\mathbb{v}})$ with restricted negation follows essentially from [23]. The expressive completeness of these four logics with full negation follows immediately. In this section, we review

---

[5] Without going into too much detail, we note that any context formed by the connectives in $\mathbf{PL}(\mathbin{\mathbb{v}})$ is monotone, while the formula $\neg\phi$ is not. See [37] for the definition of a context and a monotone context.

the role of disjunctive and conjunctive normal forms in obtaining the expressive completeness results for these logics. Our goal is to highlight how this approach aligns with the standard techniques used in the usual (single valuation-based) classical logic. We also discuss how the complement of a property was expressed in **PL**(=(·)) without directly using negation, and point out that our more relaxed full negation does not readily offer new methods for capturing these complemented properties. Finally, building on the insights of this alternative way of expressing complemented properties, we prove a new result that the logic **PL**(=(·), ⊆) is expressively complete, by using a conjunctive normal form and by expressing complemented properties in yet another way.

1.3.1 Classical Logic and Normal Forms

It will be instructive for our discussion to first recall the standard expressive completeness result for classical logic, along with how it can be proved by constructing formulas in disjunctive and conjunctive normal forms.

Throughout Sect. 1.3, we fix a set $N = \{p_1, \ldots, p_n\}$ of n propositional variables. All formulas considered in this section have propositional variables only in N and all teams have the set N as their domain (referred to as N-*teams*). For any **PL**-formula α with propositional variables from N, consider the set

$$\|\alpha\| = \{v \in 2^N : v \models \alpha\}$$

of valuations that satisfy α. We shall call the set $\|\alpha\| \subseteq 2^N$ the *property characterised by* α (over N). We now review the result that classical logic is *expressively complete*, in the sense that for any property $t \subseteq 2^N$ over N, classical logic has a *characteristic formula* χ_t satisfying $\|\chi_t\| = t$, or, in other words, for any valuation $v \in 2^N$,

$$v \models \chi_t \iff v \in t. \tag{1.2}$$

Let us examine two "brute-force" methods to construct such a formula. These are based on two alternative interpretations of the equivalence (1.2), leading to a disjunctive normal form and a conjunctive normal form for classical logic.

We first observe that for an arbitrary single valuation $v \in 2^N$, the classical formula

$$\chi_v = p_1^{v(p_1)} \wedge \cdots \wedge p_n^{v(p_n)}, \quad \text{where } p_i^1 := p_i \text{ and } p_i^0 := \neg p_i,$$

defines the valuation, in the sense that for any valuation $u \in 2^N$,

$$u \models \chi_v \iff u = v. \tag{1.3}$$

For any property $t \subseteq 2^N$ (which is also a team), we now form the characteristic formula

$$\chi_t = \bigvee_{v \in t} \chi_v \tag{1.4}$$

by taking disjunctions of the χ_v formulas. Clearly, the χ_t formula in disjunctive normal form satisfies (1.2), as it expresses the property that

"the valuation v in question is equal to some valuation in t"

There is also an indirect, alternative way to interpret the equivalence in (1.2), namely to let χ_t express

"the valuation v in question is not equal to any valuation in the complement of t (and thus has to be in t)."

This idea can be captured by using conjunction and negation through defining

$$\chi_t = \bigwedge_{v \in \bar{t}} \neg \chi_v \tag{1.5}$$

where note that the set $\bar{t} = 2^N \setminus t$ is finite. The formula χ_t can be easily turned into an equivalent formula $\bigwedge_{v \in \bar{t}} (p_1^{1-v(p_1)} \vee \cdots \vee p_n^{1-v(p_n)})$ in *conjunctive normal form* by applying the De Morgan laws and double negation elimination law to each formula $\neg\chi_v$. Let us emphasise that (classical) negation plays a crucial role in the conjunctive characteristic formula χ_t in (1.5).

1.3.2 Team-Based Logics: Downward Closed Case

We now turn to reviewing the proofs given in the literature for the expressive completeness of the downward closed team-based logics $\mathbf{PL}(\mathbin{\vee\!\!\!\vee})$ and $\mathbf{PL}(=\!\!(\cdot))$.

Define a *team property* (over N) as a set $T \subseteq \wp(2^N)$ of teams. The *team property characterised by a formula* ϕ (over N) is defined as the set

$$[\![\phi]\!] = \{t \subseteq 2^N : t \models \phi\}$$

of N-teams that satisfies ϕ. As an example, the team property characterised by a classical formula α can be easily computed from the standard (single-valuation) property $\|\alpha\|$ characterised by the formula:

Lemma 3.1 $[\![\alpha]\!] = \wp(\|\alpha\|)$ *for any classical formula* α.

Proof For any $t \in [\![\alpha]\!]$, we have $t \models \alpha$. Thus, $v \models \alpha$ for any $v \in t$, by flatness of α. It then follows that $t \subseteq \|\alpha\|$, i.e., $t \in \wp(\|\alpha\|)$.

Conversely, for any nonempty $t \in \wp(\|\alpha\|)$, i.e., $\emptyset \neq t \subseteq \|\alpha\|$, we have $t \models \alpha$ by flatness of α since $v \models \alpha$ for each $v \in t$. The fact $\emptyset \models \alpha$ follows from the empty team property of α. □

The team property $\wp(\|\alpha\|)$ is clearly *flat*, meaning (naturally) that $t \in \wp(\|\alpha\|)$ iff $\{v\} \in \wp(\|\alpha\|)$ for all $v \in t$. We can define a team property T being *downward closed* or *union closed* in a similar way.

Let $\mathcal{C}$ be a class of team properties. We say that a team-based logic L is *expressively complete for* $\mathcal{C}$ if the following two conditions hold:

- for every L-formula ϕ, the team property $[\![\phi]\!]$ belongs to $\mathcal{C}$, i.e., $[\![\phi]\!] \in \mathcal{C}$;
- for every team property $T \in \mathcal{C}$, there exists an L-formula ϕ such that $[\![\phi]\!] = T$.

Theorem 3.2 *The logics* $\mathbf{PL}(\mathbin{\mathpalette\make@circled\vee})$ *and* $\mathbf{PL}(=\!\!(\cdot))$ *are both expressively complete for the class of nonempty and downward closed team properties.*

The above theorem was proved essentially already in [41] (where, however, arbitrary negation was not allowed in the syntax). We now revisit these proofs and demonstrate how the standard approach of constructing formulas in disjunctive and conjunctive normal form from the previous subsection can be extended to the context of team-based logics, with a twist regarding negation.

The first condition in the expressive completeness definition for the two logics $\mathbf{PL}(\mathbin{\mathpalette\make@circled\vee})$ and $\mathbf{PL}(=\!\!(\cdot))$ follows easily from the fact that formulas ϕ in the logics are downward closed and have the empty team property (thus $\emptyset \in [\![\phi]\!] \neq \emptyset$). For the nontrivial second condition, we need to characterise every nonempty downward closed team property T using a formula χ_T in the two logics; that is to guarantee that

$$s \models \chi_T \iff s \in T. \tag{1.6}$$

This equivalence is the team semantics counterpart to equivalence (1.2) from the previous subsection, where the classical formula χ_t was used as a characteristic formula.

As a starting point of our proof, observe that on the team level, the classical characteristic formula χ_t (given by either (1.4) or (1.5)) defines the team t up to its subteams in the following sense:

Lemma 3.3 *For any* N*-team* s*,*

$$s \models \chi_t \iff s \subseteq t.$$

In other words, $[\![\chi_t]\!] = \wp(t)$.

Proof By Lemma 3.1 and (1.2), $[\![\chi_t]\!] = \wp(\|\chi_t\|) = \wp(t)$. □

A corollary that the classical language **PL** over team semantics is also expressively complete follows easily:

Corollary 3.4 **PL** *is expressively complete for the class of flat team properties. Consequently, a formula is flat iff it is equivalent to a classical formula.*

Proof For the expressive completeness, the first condition follows from the fact that classical formulas are flat. For the second condition, for any flat team property T, observe that $T = \wp(\bigcup T)$. Thus, by Lemma 3.3, we have $[\![\chi_{\bigcup T}]\!] = \wp(\bigcup T) = T$, where $\chi_{\bigcup T}$ is a classical formula.

For the second claim in the corollary, we have already seen that classical formulas are flat. Conversely, any flat formula ϕ defines a flat property $[\![\phi]\!]$, which can then be defined by a **PL**-formula α., i.e., $[\![\phi]\!] = [\![\alpha]\!]$. Thus, $\phi \equiv \alpha$. □

Now, in the logic **PL**($\mathbin{\mathbb{V}}$) we define the characteristic formula $\chi_T := \mathop{\mathbb{V}}_{t \in T} \chi_t$ as a disjunction of the classical characteristic formulas χ_t. The reader can verify that (1.6) holds for the formula, since T is downward closed and nonempty (thus $\emptyset \in T$); see [41] for proof details. Essentially, the formula χ_T states

"the team s in question is a subteam of some team t in T (and thus in T)."

The global disjunction $\mathbin{\mathbb{V}}$, which plays a crucial role in the above argument, is however not available in the logic **PL**($=\!(\cdot)$). Yet in **PL**($=\!(\cdot)$) we can construct χ_T as a conjunctive formula $\chi_T = \bigwedge_{t \in \overline{T}} \rho_t$ by using a similar idea to the previous subsection. One naive idea would be to take $\rho_t = \neg\chi_t$, as we did in (1.5) in classical logic. However, our negation $\neg$ does not behave classically on the team level. In particular, $\neg\chi_t$ is always flat, and thus $\bigwedge_{t \in \overline{T}} \neg\chi_t$ would also be flat, which already means that it cannot be the defining formula for any downward closed team property T that is not flat.

The proof in [41] instead builds the formula ρ_t as in the lemma below:

Lemma 3.5 *For any nonempty N-team t, there is a formula ρ_t in* **PL**($=\!(\cdot)$) *such that for any N-team s,*

$$s \models \rho_t \iff t \nsubseteq s.$$

Proof The argument is due to Huuskonen. Let $|t| = k + 1$. Define $\rho_t = \mu_k \vee \chi_{\overline{t}}$, where

$$\mu_k = \bigvee_{i=1}^{k} (=\!(p_1) \wedge \cdots \wedge =\!(p_n))$$

and $\overline{t} = 2^N \setminus t$. To show that the formula ρ_t is as required, one applies Lemma 3.3 for the classical characteristic formula $\chi_{\overline{t}}$ and a crucial observation for the formula μ_k that for any N-team s,

$$s \models \mu_k \iff |s| \leq k.$$

We refer the reader to [41] for details. □

Then, the conjunctive formula $\chi_T = \bigwedge_{t \in \overline{T}} \rho_t$ would satisfy (1.6) as it states

"the team s in question is not a superteam of any team in the complement of T (and thus has to be in T)."

Note that T was assumed to be downward closed and nonempty, which implies $\emptyset \in T$. Thus, every team $t \in \overline{T}$ is nonempty, and each conjunct ρ_t in the formula χ_T is guaranteed to exist by Lemma 3.5.

1.3.3 The Team-Based Logics with Inclusion Atoms

In this section, we focus on the remaining three team-based logics, namely $\mathbf{PL}(\subseteq)$, $\mathbf{PL}(\subseteq, \vee\!\!\!\vee)$ and $\mathbf{PL}(=\!\!(\cdot), \subseteq)$ with inclusion atoms.

Recall that for any valuation $v \in 2^N$, in the usual classical logic **PL**, we have a formula χ_v that defines v in the sense of (1.3), which, on the team level, can be rendered as

$$\{u\} \models \chi_v \iff u = v.$$

This same equivalence can also be captured by the following (primitive) inclusion atom:

$$\iota_v = \underline{v(p_1)} \dots \underline{v(p_n)} \subseteq p_1 \dots p_n,$$

where $\underline{v(p_i)} = \top$ if $v(p_i) = 1$, and $\underline{v(p_i)} = \bot$ if $v(p_i) = 0$. That is,

$$\{u\} \models \iota_v \iff u = v.$$

Moreover, for any nonempty team $s \subseteq 2^N$,

$$s \models \iota_v \iff v \in s. \tag{1.7}$$

Now, as in the case for **PL**, the formula ι_v can be used as a building block to construct characteristic formulas for teams, in conjunctive and disjunctive normal forms. First, consider the formula $\iota_t = \bigwedge_{v \in t} \iota_v$ (due to [16, 40]). By (1.7), the formula defines the team t up to its super-teams modulo the empty team, in the following sense:

$$s \models \iota_t \iff t \subseteq s \text{ or } s = \emptyset. \tag{1.8}$$

Recall that the **PL**-formula χ_t defines the team t up to its subteams in the sense of Lemma 3.3. Thus, the formula $\chi_t \wedge \iota_t$ defines the team t precisely modulo the empty team, in the following sense:

$$s \models \chi_t \wedge \iota_t \iff s = t \text{ or } s = \emptyset. \tag{1.9}$$

By using this fact, it was shown in [16, 40] that propositional inclusion logic $\mathbf{PL}(\subseteq)$ is expressively complete for the class of union closed team properties that contain the empty team (the argument was due to [16], where a slightly more complex formulation of the formula ι_v was used than the one above or in [40], and thus a syntactically more complex logic was considered in [16]). The idea for the nontrivial condition is that for any such team property $\mathcal{T} \subseteq \wp(2^N)$, one proves that the formula $\bigvee_{t \in \mathcal{T}}(\chi_t \wedge \iota_t)$ in disjunctive normal form defines $\mathcal{T}$, namely, $\mathcal{T} = [\![\bigvee_{t \in \mathcal{T}}(\chi_t \wedge \iota_t)]\!]$, by using the fact that $\mathcal{T}$ is union closed and $\emptyset \in \mathcal{T}$. See [40] for the proof details.

By adopting a different disjunction—global disjunction—in the characteristic formula, we can prove that the logic $\mathbf{PL}(\subseteq, \mathbin{\mathbb{V}})$ is expressively complete for a larger class of team properties, a result that follows essentially from [23].

Theorem 3.6 *The logic* $\mathbf{PL}(\subseteq, \mathbin{\mathbb{V}})$ *is expressively complete for the class of team properties that contain the empty team.*

Proof For the nontrivial condition, for any team property $T \subseteq \wp(2^N)$ with $\emptyset \in T$, define $\chi_T = \mathbb{V}_{t\in T}(\chi_t \wedge \iota_t)$. Now, by (1.9), we have that for any N-team s,

$$s \models \chi_T \iff s \models \chi_t \wedge \iota_t \text{ for some } t \in T \iff s \in T \text{ or } s = \emptyset \iff s \in T,$$

namely $[\![\chi_T]\!] = T$. □

Finally, let us present our new result that the logic $\mathbf{PL}(=\!\!(\cdot), \subseteq)$ is also expressively complete for the class of team properties that contain the empty team. The global disjunction $\mathbin{\mathbb{V}}$ that played a crucial role in the proof above is not available in the language $\mathbf{PL}(=\!\!(\cdot), \subseteq)$. We thus have to take a different approach to construct the characteristic formula. The idea is that we shall define each team property T with $\emptyset \in T$ by using a $\mathbf{PL}(=\!\!(\cdot), \subseteq)$-formula $\bigwedge_{t\in\overline{T}} \psi_t$ in conjunctive normal form with each ψ_t satisfying

$$s \models \psi_t \iff s \neq t. \tag{1.10}$$

Then, as with the classical logic case we reviewed in Sect. 1.3.1, the conjunctive formula would then state that

"the team in question is not equal to any team t in the complement of T (and thus has to be in T)."

Let us emphasise (again) that defining ψ_t to be the negated formula $\neg(\chi_t \wedge \iota_t)$ does not give the equivalence (1.10) (or equivalently $[\![\neg(\chi_t \wedge \iota_t)]\!] \neq \wp(2^N) \setminus \{t\}$), since, for instance, the team property $[\![\neg(\chi_t \wedge \iota_t)]\!]$ is always flat, while $\wp(2^N) \setminus \{t\}$ needs not be.

Now, in order to construct the appropriate formula ψ_t, we first recall that the formula ι_t satisfying (1.8) was built by taking conjunction of all ι_v's with $v \in t$. Interestingly, the dual property of (1.8) can be defined by a dual formula $\sigma_t = \bigvee_{v\in\bar{t}} \iota_v$, formed by taking (local) disjunction of ι_v's, as we show in the next lemma:

Lemma 3.7 *Given any nonempty N-team t, we have that for any N-team s,*

$$s \models \sigma_t \iff s \nsubseteq t \text{ or } s = \emptyset.$$

Proof If $s = \emptyset$, then $s \models \sigma_t$ follows from the empty team property of σ_t. Now assume that $s \neq \emptyset$. Suppose $\emptyset \neq s \nsubseteq t$. Then there exists $v \in s$ such that $v \in \bar{t}$. By (1.7), we obtain $s \models \iota_v$, and thus $s \models \sigma_t$.

Conversely, if $\emptyset \neq s \models \sigma_t$, then $s = \bigcup_{v\in\bar{t}} s_v$ for some $s_v \subseteq s$ and each $s_v \models \iota_v$. At least one such s_v with $v \in \bar{t}$ must be nonempty and satisfies $v \in s_v$ by (1.7). It then follows that $v \in s_v \setminus t$, implying $s \nsubseteq t$. □

Now, we can build the formula ψ_t that satisfies (1.10).

Lemma 3.8 *Given any nonempty N-team t, we have that for any N-team s,*

$$s \models \rho_t \vee \sigma_t \iff s \neq t.$$

Proof Suppose $s \neq t$. Then either $t \not\subseteq s$ or $s \not\subseteq t$. In the former case, $s \models \rho_t$ by Lemma 3.5; in the latter case, $s \models \sigma_t$ by Lemma 3.7. Thus, in both cases we obtain $s \models \rho_t \vee \sigma_t$ by the empty team property.

Conversely, suppose $s \models \rho_t \vee \sigma_t$. Then $s = s_1 \cup s_2, s_1 \models \rho_t$ and $s_2 \models \sigma_t$ for some $s_1, s_2 \subseteq s$. If $s_2 \neq \emptyset$, then by Lemma 3.7 we have that $s_2 \not\subseteq t$, which implies $s \not\subseteq t$, meaning $s \neq t$. If $s_2 = \emptyset$, then $s_1 = s$. By Lemma 3.5 we have that $t \not\subseteq s_1 = s$, meaning $t \neq s$. □

Finally, we derive the expressive completeness of $\mathbf{PL}(=\!(\cdot), \subseteq)$ as an immediate corollary:

Theorem 3.9 *The logic $\mathbf{PL}(=\!(\cdot), \subseteq)$ is expressively complete for the class of team properties that contain the empty team.*

Proof For the nontrivial condition, for any team property $T \subseteq \wp(2^N)$ with $\emptyset \in T$, define $\chi_T = \bigwedge_{t \in \overline{T}} (\rho_t \vee \sigma_t)$, where note that $\emptyset \notin \overline{T}$. Thus, $[\![\chi_T]\!] = T$ follows from Lemma 3.8. □

Clearly the characteristic formulas used in this proof and the other proofs in this section only use atomic negations. Therefore the results in this section also apply to the versions of these logics with atomic negation only.

1.4 Natural Deduction System for PL($\mathbin{\rotatebox[origin=c]{-90}{$\geqslant$}}$) and Its Extensions Revisited

In this section, we discuss the axiomatisation for the logic $\mathbf{PL}(\mathbin{\rotatebox[origin=c]{-90}{$\geqslant$}})$ and highlight that the natural deduction system for $\mathbf{PL}(\mathbin{\rotatebox[origin=c]{-90}{$\geqslant$}})$ with restricted negation, as known in the literature, remains complete for our version of $\mathbf{PL}(\mathbin{\rotatebox[origin=c]{-90}{$\geqslant$}})$ with full negation. In this sense, the inclusion of full intuitionistic negation in the syntax of $\mathbf{PL}(\mathbin{\rotatebox[origin=c]{-90}{$\geqslant$}})$ does not complicate its axiomatisation. We also make some remarks about the known proof system of the extension of $\mathbf{PL}(\mathbin{\rotatebox[origin=c]{-90}{$\geqslant$}})$ with intuitionistic implication.

The first complete natural deduction system for $\mathbf{PL}(\mathbin{\rotatebox[origin=c]{-90}{$\geqslant$}})$ (with only atomic negation) was introduced in [41]. Later, [6] introduced a complete natural deduction system for $\mathbf{PL}(\mathbin{\rotatebox[origin=c]{-90}{$\geqslant$}})$ extended with intuitionistic implication $\rightarrow$ (with negation defined via the implication as $\neg\phi := \phi \rightarrow \bot$), which combines and simplifies rules in the system of [41] and the axioms in the Hilbert system of inquisitive logic [8]. Such a system was further simplified and generalised to the intuitionistic case in [7]. It has been folklore that a complete and simpler system for $\mathbf{PL}(\mathbin{\rotatebox[origin=c]{-90}{$\geqslant$}})$ (with negation for classical formulas only) can be obtained by omitting the rules for implication in the system of [7]. We now present a natural deduction system for **PL**($\mathbb{v}$) based on the system in [7] with rules for negation adapted for the (full) intuitionistic negation.

Definition 3.1 The system of natural deduction for **PL**($\mathbb{v}$) consists of the following rules, where the formula α ranges over classical formulas only:

Rule for $\bot$:

$$\frac{\bot}{\phi}\ \bot E$$

Rules for $\wedge$:

$$\frac{\phi \qquad \psi}{\phi \wedge \psi}\ \wedge I \qquad \frac{\phi \wedge \psi}{\phi}\ \wedge E_r \qquad \frac{\phi \wedge \psi}{\psi}\ \wedge E_l$$

Rules for $\neg$:

$$\frac{\begin{array}{c}[\phi]\\ \vdots \\ \bot\end{array}}{\neg\phi}\ \neg I \qquad \frac{\phi \qquad \neg\phi}{\psi}\ \neg E \qquad \frac{\begin{array}{c}[\neg\alpha]\\ \vdots \\ \bot\end{array}}{\alpha}\ RAA$$

Rules for $\mathbb{v}$:

$$\frac{\phi}{\phi \mathbb{v} \psi}\ \mathbb{v}I_r \qquad \frac{\phi}{\psi \mathbb{v} \phi}\ \mathbb{v}I_l \qquad \frac{\phi \mathbb{v} \psi \qquad \begin{array}{c}[\phi]\\ \vdots \\ \chi\end{array} \qquad \begin{array}{c}[\psi]\\ \vdots \\ \chi\end{array}}{\chi}\ \mathbb{v}E$$

Rules for $\vee$:

$$\frac{\phi}{\phi \vee \psi}\ \vee I \qquad \frac{\phi \vee \psi \qquad \begin{array}{c}[\phi]\\ \vdots \\ \alpha\end{array} \qquad \begin{array}{c}[\psi]\\ \vdots \\ \alpha\end{array}}{\alpha}\ \vee E$$

$$\frac{\phi \vee \psi}{\psi \vee \phi}\ \vee Com \qquad \frac{\phi \vee \psi \qquad \begin{array}{c}[\phi]\\ \vdots \\ \chi\end{array}}{\chi \vee \psi}\ \vee Mon$$

Rule for interaction between the two disjunctions:

$$\frac{\phi \vee (\psi \mathbin{\mathbb{V}} \chi)}{(\phi \vee \psi) \mathbin{\mathbb{V}} (\phi \vee \chi)}\, Dis \vee \mathbb{V}$$

The consequence relation $\Gamma \vdash \phi$ for the deduction system is defined as usual. In particular, we write $\phi \dashv\vdash \psi$, if both $\phi \vdash \psi$ and $\psi \vdash \phi$ hold.

Our rules for negation are more general than those in the (folklore) system in [7] in that we adopt the standard introduction and elimination rules, while in [7] the principal formula ϕ are assumed classical. The soundness of our $\neg$I and $\neg$E rules follows easily from the fact that $\neg\phi$ is flat. The soundness of the other rules is also easy to verify.

Since our logic and deduction system are both conservative over classical propositional logic, the above system is clearly complete over classical formulas:

Lemma 3.2 *For any set* $\Delta \cup \{\alpha\}$ *of classical formulas,*

$$\Delta \models \alpha \iff \Delta \vdash \alpha.$$

Proof By flatness of classical formulas and by the equivalence (1.1), it can be verified $\Delta \models \alpha \iff \Delta \models^c \alpha$, where $\models^c$ denotes the semantic entailment relation for the usual (single valuation-based) classical logic **PL**. By the completeness theorem of classical propositional logic, we have $\Delta \models^c \alpha \iff \Delta \vdash_{\mathbf{PL}} \alpha$. Finally, since when restricted to classical formulas, our system coincide with the standard system for classical logic, we have $\Delta \vdash_{\mathbf{PL}} \alpha \iff \Delta \vdash_{\mathbf{PL}(\mathbb{V})} \alpha$. □

Let $\phi[\chi/p]$ denote the formula obtained from ϕ by replacing one particular instance of p with χ. An easy induction shows that, as usual, the replacement lemma holds:

$$\text{If } \chi \dashv\vdash \chi', \text{ then } \phi[\chi/p] \vdash \phi[\chi'/p], \tag{1.11}$$

where in the negation case one may apply the following antitone rule that is clearly derivable from $\neg$E and $\neg$I:

$$\frac{\neg\phi \qquad \begin{array}{c}[\psi]\\ \vdots\\ \phi\end{array}}{\neg\psi}\,\neg\mathsf{Anton}$$

As in [7], every formula is provably equivalent in the system of $\mathbf{PL}(\mathbb{V})$ to one in disjunctive normal form (c.f. the more specific disjunctive normal form in Sect. 1.3.2):

Lemma 3.3 *For any formula* ϕ, *we have that*

$$\phi \dashv\vdash \mathbb{V}_{i\in I}\alpha_i$$

for some finite set $\{\alpha_i \mid i \in I\}$ *of classical formulas.*

Proof The lemma is proved by induction on ϕ. The proof for all cases except for negation case can be found in, e.g., [6, 7]. Notably, for the key case $\phi = \psi \vee \chi$, one applies the rule Dis$\mathbb{V}\mathbb{V}$ along with the induction hypothesis.

For the new case $\phi = \neg\psi$, by induction hypothesis, $\psi \dashv\vdash \mathbb{V}_{i\in I}\alpha_i$ for some classical formulas α_i. Thus, we have $\neg\psi \dashv\vdash \neg\mathbb{V}_{i\in I}\alpha_i$ (by (1.11)). We now show $\neg\mathbb{V}_{i\in I}\alpha_i \dashv\vdash \bigwedge_{i\in I}\neg\alpha_i$, where the formula on the right hand side is classical.

For the left to right direction, for each $j \in I$, we have $\alpha_j \vdash \mathbb{V}_{i\in I}\alpha_i$ by $\mathbb{V}$I. Thus, $\neg\mathbb{V}_{i\in I}\alpha_i, \alpha_j \vdash \bot$, which implies $\neg\mathbb{V}_{i\in I}\alpha_i \vdash \neg\alpha_j$ for each $j \in I$ by $\neg$I. Thus, we conclude $\neg\mathbb{V}_{i\in I}\alpha_i \vdash \bigwedge_{i\in I}\neg\alpha_i$ by $\wedge$I.

For the converse direction, we first derive $\mathbb{V}_{i\in I}\alpha_i, \bigwedge_{i\in I}\neg\alpha_i \vdash \bot$ by applying $\mathbb{V}$E and $\wedge$E. Then we conclude $\bigwedge_{i\in I}\neg\alpha_i \vdash \neg\mathbb{V}_{i\in I}\alpha_i$ by $\neg$I. □

The completeness theorem of our system can be proved through an argument that uses the disjunctive normal form result above:

Theorem 3.4 (Completeness) *For any set* $\Gamma \cup \{\phi\}$ *of formulas of* **PL**($\mathbb{V}$),

$$\Gamma \models \phi \iff \Gamma \vdash \phi.$$

Proof The detailed proof can be found in, e.g., [6, 7]; we only give a proof sketch. Since **PL**($\mathbb{V}$) is compact (see [41]), we may assume Γ is finite and $\psi = \bigwedge\Gamma$. Suppose $\psi \models \phi$. W.l.o.g., assume further that $\psi = \mathbb{V}_{i\in I}\alpha_i$ and $\phi = \mathbb{V}_{j\in J}\beta_j$ are in disjunctive normal form. Then for each $i \in I$, there exists $j_i \in J$ such that $\alpha_i \models \beta_{j_i}$. By Lemma 3.2, it follows that $\alpha_i \vdash \beta_{j_i}$. It then follows easily that $\mathbb{V}_{i\in I}\alpha_i \vdash \mathbb{V}_{j\in J}\beta_j$, which then gives $\psi \vdash \phi$ by Lemma 3.3. □

As we demonstrated above, both the deduction system of **PL**($\mathbb{V}$) with full (intuitionistic) negation and its completeness proof can be easily adapted from those for **PL**($\mathbb{V}$) with restricted negation in the literature. We now summarise a few further folklore observations regarding this system of **PL**($\mathbb{V}$).

First, in the RAA and $\vee$E rules, we restrict the formula α to be classical, but the two rules are sound as long as the formula α is flat. A larger syntactically defined class of flat formula is the class of *Harrop formulas*, defined as

$$\delta ::= p \mid \bot \mid \neg\phi \mid \delta \wedge \delta \mid \delta \vee \delta$$

where ϕ ranges over arbitrary **PL**($\mathbb{V}$) formula. For instance, all classical formulas, $\neg(p\mathbb{V}\neg p)$, and $\neg(\phi\mathbb{V}\psi) \vee (q \wedge r)$ are Harrop formulas, whereas $p\mathbb{V}\neg p$ is not a Harrop formula. Harrop formulas are studied originally in the literature on intuitionistic logic [14]. They are considered by Punčochář [9, 32] also in the context of inquisitive logic and team-based logics. (Both contexts consider Harrop formulas in languages with intuitionistic implication, which is not available in **PL**($\mathbb{V}$).) It is easy to verify by induction that all Harrop formulas are flat, and by Corollary 3.4, every Harrop formula δ is equivalent to some classical formula δ^*. In fact, δ^* can be obtained directly from δ by replacing every global disjunction $\mathbin{\vee\!\!\!\vee}$ in δ with the local disjunction $\vee$ (follows from an easy inductive proof by using the flatness of δ). The more general version of the (sound) rules RAA and $\vee$E with α being a Harrop formula is, obviously, derivable in our (complete) system.

Next, we remark that some earlier presentations of the system (such as those in [6, 41]) included more rules for the local disjunction, particularly the associativity rule

$$\frac{(\phi \vee \psi) \vee \chi}{\phi \vee (\psi \vee \chi)}\ \vee\mathsf{Ass}$$

and the rule $\phi \vee \bot/\phi$. These rules are actually derivable in the system, via the disjunctive normal form lemma above:

Proposition 3.5 *The following are derivable:*

(i) $(\phi \vee \psi) \vee \chi \vdash \phi \vee (\psi \vee \chi)$
(ii) $\phi \vee \bot \vdash \phi$

Proof (i) By Lemma 3.3, we have that $\phi \dashv\vdash \mathop{\vee\!\!\!\vee}_{i\in I}\alpha_i$, $\psi \dashv\vdash \mathop{\vee\!\!\!\vee}_{j\in J}\beta_j$ and $\chi \dashv\vdash \mathop{\vee\!\!\!\vee}_{k\in K}\gamma_k$ for some classical formulas α_i, β_j and γ_k. Then we have that

$$\begin{aligned}
(\phi \vee \psi) \vee \chi &\vdash \big((\mathop{\vee\!\!\!\vee}_{i\in I}\alpha_i) \vee \mathop{\vee\!\!\!\vee}_{j\in J}\beta_j\big) \vee \mathop{\vee\!\!\!\vee}_{k\in K}\gamma_k\\
&\vdash \big(\mathop{\vee\!\!\!\vee}_{i\in I, j\in J}(\alpha_i \vee \beta_j)\big) \vee \mathop{\vee\!\!\!\vee}_{k\in K}\gamma_k \qquad \text{(by } \vee\,\mathsf{Mon}, \text{Dis} \vee \mathbin{\vee\!\!\!\vee})\\
&\vdash \mathop{\vee\!\!\!\vee}_{i\in I, j\in J, k\in K}\big((\alpha_i \vee \beta_j) \vee \gamma_k\big)\\
&\vdash \mathop{\vee\!\!\!\vee}_{i\in I, j\in J, k\in K}(\alpha_i \vee (\beta_j \vee \gamma_k)) \qquad \text{by Lemma 4.2, classical rule}\\
&\vdash (\mathop{\vee\!\!\!\vee}_{i\in I}\alpha_i) \vee \mathop{\vee\!\!\!\vee}_{j\in J, k\in K}(\beta_j \vee \gamma_k)\\
&\vdash (\mathop{\vee\!\!\!\vee}_{i\in I}\alpha_i) \vee \big((\mathop{\vee\!\!\!\vee}_{j\in J}\beta_j) \vee \mathop{\vee\!\!\!\vee}_{k\in K}\gamma_k\big) \qquad \text{(by } \vee\,\mathsf{Mon}, \text{Dis} \vee \mathbin{\vee\!\!\!\vee})\\
&\vdash \phi \vee (\psi \vee \chi).
\end{aligned}$$

(ii) By Lemma 3.3, we have that $\phi \dashv\vdash \mathop{\vee\!\!\!\vee}_{i\in I}\alpha_i$ for some classical formulas α_i. Then,

$$\begin{aligned}
\phi \vee \bot &\vdash (\mathop{\vee\!\!\!\vee}_{i\in I}\alpha_i) \vee \bot \qquad (\vee\mathsf{Mon})\\
&\vdash \mathop{\vee\!\!\!\vee}_{i\in I}(\alpha_i \vee \bot) \qquad (\vee\mathsf{Com}, \mathit{Dis} \vee \mathbin{\vee\!\!\!\vee})\\
&\vdash \mathop{\vee\!\!\!\vee}_{i\in I}\alpha_i \qquad \text{By Lemma 4.2, classical rule}\\
&\vdash \phi.
\end{aligned}$$

□

These derived rules allow us to derive interesting clauses such as the following ones:

Proposition 3.6 *The following are derivable:*

(i) $\alpha \vee \phi, \neg\alpha \vdash \phi$
(ii) $\alpha \vee \phi, \neg\alpha \vee \psi \vdash \phi \vee \psi$

Proof (i) Since $\alpha, \neg\alpha \vdash \bot$ (by $\neg\mathsf{E}$), we have $\alpha \vee \phi, \neg\alpha \vdash \bot \vee \phi$ by $\vee\mathsf{Mon}$. By Proposition 3.5(i), we have $\bot \vee \phi \vdash \phi$. Putting these together, we conclude $\alpha \vee \phi, \neg\alpha \vdash \phi$.

(ii) Follows from (i), and $\vee\mathsf{Mon}$ applied to $\neg\alpha$ in the disjunction $\neg\alpha \vee \psi$. □

Let us end with some comments on the system for the language of **PL**($\mathbin{\vee\mkern-12mu\vee}$) extended with the intuitionistic implication $\to$, which can also be viewed as inquisitive logic extended with the local disjunction. The (classical) systems in [6, 7] for this language contain two additional rules for intuitionistic implication and the Split rule:

Rules for intuitionistic implication:

$$\dfrac{\begin{matrix}[\phi]\\ \vdots\\ \psi\end{matrix}}{\phi \to \psi}\to I \qquad\qquad \dfrac{\phi \to \psi \qquad \phi}{\psi}\to \mathsf{E}$$

Split rule:

$$\dfrac{\alpha \to (\phi \mathbin{\vee\mkern-12mu\vee} \psi)}{(\alpha \to \phi)\mathbin{\vee\mkern-12mu\vee}(\alpha \to \psi)}\,\mathit{Split}$$

Intuitionistic implications of the form $\alpha \to \phi$ with α classical behaves classically in the sense that it is equivalent to the local disjunction $\neg\alpha \vee \phi$. This fact can also be obtained proof-theoretically in our extended deduction system:

Proposition 3.7 $\alpha \to \phi \dashv\vdash \neg\alpha \vee \phi$.

Proof For the left to right direction, note that we have $\vdash \neg\alpha \vee \alpha$ by Lemma 3.2. Now, $\neg\alpha \vee \alpha, \alpha \to \phi \vdash \neg\alpha \vee \phi$ by $\to\mathsf{E}$ and $\vee\mathsf{Mon}$. Putting these together, we obtain $\alpha \to \phi \vdash \neg\alpha \vee \phi$.

For the right to left direction, by Proposition 3.6(i) we have $\neg\alpha \vee \phi, \neg\neg\alpha \vdash \phi$, and by Lemma 3.2 we have $\alpha \vdash \neg\neg\alpha$. Thus, we obtain $\neg\alpha \vee \phi, \alpha \vdash \phi$, which by $\to\mathsf{I}$ implies $\neg\alpha \vee \phi \vdash \alpha \to \phi$. □

Given this, the Split rule, which is often viewed as a distinguishing feature of inquisitive logic, is in fact derivable from Dis$\mathbb{V}\vee$ and other rules in the system. To see how to derive it in the system, note that by Proposition 3.7, to derive the Split rule it is sufficient to derive

$$\neg\alpha \vee (\phi \mathbb{V} \psi) \vdash (\neg\alpha \vee \phi) \mathbb{V} (\neg\alpha \vee \psi),$$

but this follows immediately from Dis$\vee\mathbb{V}$.

1.5 Concluding Remarks and Further Directions

In this paper, we have explored propositional team-based logics with full (intuitionistic) negation. We illustrated how the complement of certain properties can be characterised by other means without directly using negation, and we also showed that including the full intuitionistic negation does not substantially affect the natural deduction system of the downward closed logic **PL**($\mathbb{V}$). The original version of **PL**($\mathbb{V}$) with negation restricted to classical formulas only has been recently introduced a deep inference style calculus [4]. Generalising this calculus to cover full intuitionistic negation as well is left as future work.

Another downward closed logic with full intuitionistic negation we considered in the paper is **PL**($=\!\!(\cdot)$). The version of **PL**($=\!\!(\cdot)$) with atomic negation was axiomatised in [41] with a complete natural deduction system that contains all rules for **PL**($\mathbb{V}$) (with restricted negation) except for those that involve global disjunction $\mathbb{V}$ and some additional (deep inference style) rules for dependence atoms. It is essentially clear that a complete natural deduction system for **PL**($=\!\!(\cdot)$) with full intuitionistic negation can be obtained in the same way as we did in Definition 3.1 for **PL**($\mathbb{V}$). We leave the proof details for future research, as they are more involved.

Similarly, the modal versions of the logics we considered in this paper with restricted negation have been axiomatised in e.g., [3, 28, 33, 36]. Generalising these systems to encompass the full negation case should be reasonably achievable. Working out the details, especially the interaction between negation and modalities, is left as future work.

Much of the flexibility in these results relies on the downward closed setting. As for the non-downward closed logics we considered, it is not clear whether, e.g., a direct adaptation to the system of **PL**($\subseteq$) with restricted negation gives rise to a complete system for **PL**($\subseteq$) with full negation. In particular, the completeness proof for the system of **PL**($\mathbb{V}$) relies heavily on a syntactic normal form lemma (Lemma 3.3), which, as we showed, can be easily generalised to cover the full negation case. However, the corresponding syntactic normal form lemma for **PL**($\subseteq$) from [40] involves more complex formulas, and thus generalising this to cover the full negation case does not seem to be straightforward.

Finally, we remark that apart from the intuitionistic implication $\to$, the full negation induces two other implications defined in terms of the two different versions of disjunctions:

$$\phi \rightharpoondown \psi := \neg\phi \vee \psi \quad \text{and} \quad \phi \overline{\rightharpoondown} \psi := \neg\phi \mathbin{\vee\!\!\!\vee} \psi.$$

It is easy to verify that $\phi \rightharpoondown \psi \equiv \phi \to \psi$ if ϕ is union closed. Modus Ponens is sound for both implications, but deduction theorem fails in general for both implications. Studying the properties of these implications and possibly other connectives induced by the full negation is an interesting further direction.

Acknowledgements The author would like to thank Aleksi Anttila and Fausto Barbero for discussions related to the topic of this paper. She is also grateful to the anonymous referee for their corrections.

References

1. Abramsky, S., Väänänen, J.: From IF to BI. Synthese **167**(2), 207–230 (2009)
2. Anttila, A.: Further remarks on the dual negation in team logics (2024). https://arxiv.org/abs/2410.07067
3. Aleksi, A., Häggblom, M., Yang, F.: Axiomatizing modal inclusion logics and its variants. Arch. Math. Logic **64**, 755–793 (2025)
4. Anttila, A., Iemhoff, R., Yang, F.: A deep-inference sequent calculus for basic propositional team logic (without delving too deep) (2025). https://arxiv.org/abs/2508.07509
5. Burgess, J.P.: A remark on henkin sentences and their contraries. Notre Dame J Formal Logic **44**(3), 185–188 (2003)
6. Ciardelli, I.: Dependency as question entailment. In: Abramsky, S., Kontinen, J., Väänänen, J., Vollmer, H. (eds.) Dependence Logic: Theory and Application, Progress in Computer Science and Applied Logic, pp. 129–182. Birkhauser (2016)
7. Ciardelli, I., Iemhoff, R., Yang, F.: Questions and dependency in intuitionistic logic. Notre Dame J. Formal Logic **61**(1), 75–115 (2020)
8. Ciardelli, I., Roelofsen, F.: Inquisitive logic. J. Philos. Log. **40**(1), 55–94 (2011)
9. Ferguson, T., Punčochář, V.: Structural completeness and superintuitionistic inquisitive logics. In: Helle Hvid Hansen, A.S..R.J.G.B.D.Q. (ed.) Logic, Language, Information, and Computation: 29th International Workshop (WoLLIC), pp. 194–210 (2023)
10. Galliani, P.: Inclusion and exclusion in team semantics: On some logics of imperfect information. Ann. Pure Appl. Logic **163**(1), 68–84 (2012)
11. Galliani, P., Hella, L.: Inclusion logic and fixed point logic. In: Computer Science Logic 2013. Leibniz International Proceedings in Informatics (LIPIcs), vol. 23, pp. 281–295. Schloss Dagstuhl - Leibniz-Zentrum fuer Informatik (2013)
12. Grädel, E., Väänänen, J.: Dependence and independence. Stud. Logica. **101**(2), 399–410 (2013)
13. Hannula, M.: Axiomatizing first-order consequences in independence logic. Ann. Pure Appl. Logic **166**(1), 61–91 (2015)
14. Harrop, R.: On disjunctions and existential statements in intuitionistic systems of logic. Math. Ann. **132**(4), 347–361 (1956)
15. Hella, L., Luosto, K., Sano, K., Virtema, J.: The expressive power of modal dependence logic. In: Proceedings of Advances in Modal Logic 10, pp. 294–312. College Publications (2014)
16. Hella, L., Stumpf, J.: The expressive power of modal logic with inclusion atoms. In: Proceedings of the 6th GandALF, pp. 129–143 (2015)
17. Henkin, L.: Some remarks on infinitely long formulas. In: Infinitistic Methods. Proceedings Symposium Foundations of Mathematics, pp. 167–183. Pergamon, Warsaw (1961)
18. Hintikka, J.: The Principles of Mathematics Revisited. Cambridge University Press (1998)

19. Hintikka, J., Sandu, G.: Game-theoretical semantics. In: van Benthem, J., ter Meulen, A. (eds.) Handbook of Logic and Language. Elsevier (1996)
20. Hodges, W.: Compositional semantics for a language of imperfect information. Logic J. IGPL **5**, 539–563 (1997)
21. Hodges, W.: Some strange quantifiers. In: Mycielski, J., Rozenberg, G., Salomaa, A. (eds.) Structures in Logic and Computer Science: A Selection of Essays in Honor of A. Ehrenfeucht, Lecture Notes in Computer Science, vol. 1261, pp. 51–65. Springer, London (1997)
22. Iemhoff, R., Yang, F.: Structural completeness in propositional logics of dependence. Arch. Math. Logic **55**(7), 955–975 (2016)
23. Kontinen, J., Müller, J., Schnoor, H., Vollmer, H.: 24th EACSL Annual Conference on Computer Science Logic (CSL 2015), Stephan Kreutzer, pp 277–291 (2015)
24. Kontinen, J., Nurmi, V.: Team logic and second-order logic. Fund. Inform. **106**, 259–272 (2011)
25. Kontinen, J., Väänänen, J.: On definability in dependence logic. J Logic, Lang Inf **18**(3), 317–332 (Erratum: The Same Journal **20**(1) (2011), 133–134) (2009)
26. Kontinen, J., Väänänen, J.: A remark on negation in dependence logic. Notre Dame J. Formal Logic **52**(1), 55–65 (2011)
27. Kontinen, J., Väänänen, J.: Axiomatizing first-order consequences in dependence logic. Ann. Pure Appl. Logic **164**(11) (2013)
28. Lück, M.: Axiomatizations of team logics. Ann. Pure Appl. Logic **169**(9), 928–969 (2018)
29. Lück, M.: Team logic: axioms, expressiveness, complexity. PhD thesis, University of Hanover, Hannover, Germany (2020)
30. Lück, M.: On the complexity of linear temporal logic with team semantics. Theoret. Comput. Sci. **837**, 1–25 (2020)
31. Punčochář, V.: Weak negation in inquisitive semantics. J. Logic Lang. Inform. **24**, 323–355 (2015)
32. Punčochář, V.: A Generalization of Inquisitive Semantics. J. Philos. Logic **45**(4), 399–428 (2016)
33. Sano, K., Virtema, J.: Axiomatizing propositional dependence logics. In: 24th EACSL Annual Conference on Computer Science Logic (CSL 2015). Leibniz International Proceedings in Informatics (LIPIcs), vol. 41, pp. 292–307 (2015)
34. Väänänen, J.: Dependence Logic: A New Approach to Independence Friendly Logic. Cambridge University Press, Cambridge (2007)
35. Yang, F.: Expressing second-order sentences in intuitionistic dependence logic. Stud. Logica. **101**(2), 323–342 (2013)
36. Yang, F.: Modal dependence logics: Axiomatizations and model-theoretic properties. Logic J. IGPL **25**(5), 773–805 (2017)
37. Yang, F.: Uniform definability in propositional dependence logic. Rev. Symb. Logic (2017)
38. Yang, F.: Negation and partial axiomatizations of dependence and independence logic revisited. Ann. Pure Appl. Logic **170**(9), 1128–1149 (2019)
39. Yang, F.: Axiomatizing first-order consequences in inclusion logic. Math. Log. Q. **66**(2), 195–216 (2020)
40. Yang, F.: Propositional union closed team logics. Ann. Pure Appl. Logic **173**(6), 103102 (2022)
41. Yang, F., Väänänen, J.: Propositional logics of dependence. Ann. Pure Appl. Logic **167**(7), 557–589 (2016)
42. Yang, F., Väänänen, J.: Propositional team logics. Ann. Pure Appl. Logic **168**(7), 1406–1441 (2017)

Chapter 2
A Note on Negation in the Operational Semantics for Relevant Logic

Takuro Onishi

Abstract In this paper, we explore a new approach to negation in the operational semantics for relevant logic. Urquhart's seminal semilattice frame semantics does not validate all the negation principles of relevant logic R, notably Contraposition. We introduce a logic that validates Contraposition at the expense of Contraction, reflecting their incompatibility in operational semantics. Three different types of frames for this non-contractive yet contrapositive logic are presented. (1) Contrapositive frame, which slightly modifies Urquhart's original semilattice frame. (2) Functional multiset-multiset (FMM) frame, adapting Standefer's functional set frame which has proven equivalent to the semilattice frame for the positive fragment. (3) Bi-operational frame that incorporates dual binary operations, serving as a bridge between contrapositive frames and FMM frames. We demonstrate that these three frame types are equivalent, allowing for isomorphic transformations among them. The paper concludes by presenting a labeled sequent calculus for this logic.

Keywords Relevant logic · Operational semantics · Collection frame · Labeled sequent calculus

2.1 Introduction

In this paper, we introduce a new treatment of negation in the operational semantics for relevant logic. The logic defined by Urquhart's *semilattice frame semantics* [9–11] does not validate all the negation principles of relevant logic **R**. This occurs because the frame constraints for Contraction and Contraposition are incompatible in the sense that when combined, the logic collapses into classical logic. Urquhart chose Contraction and gave up Contraposition. We explore another path, namely, dropping Contraction and taking Contraposition.

We begin by defining the notion of *contrapositive frame* that validates Contraposition, by slightly modifying Urquhart's semantics. Following this we provide two

T. Onishi (✉)
Kyoto University, Kyoto, Japan
e-mail: takuro.onishi@gmail.com

K. Sano et al. (eds.), *Exploring Negation, Modality and Proof*,
Logic in Asia: Studia Logica Library,
https://doi.org/10.1007/978-981-96-8566-0_2

different presentations equivalent to it. *Functional multiset-multiset frame* (FMM frame for short) builds upon Restall and Standefer's idea of *collection frame* [7] and its extension, *multiset-multiset frame* [6]. Standefer [8] has already introduced *functional set frame* that is equivalent to Urquhart's semilattice frame for the positive fragment. Our FMM frame extends it in a certain way to a frame for logic including negation. Additionally, we define the concept of *bi-operational frame* equipped with two binary operations, serving as a bridge connecting contrapositive frames and FMM frames. We demonstrate that these three types of frames are all equivalent, that is, isomorphic transformations can be defined between them. The paper concludes by presenting a labeled sequent calculus for this logic, built on the previous work of Urquhart [10] and Kashima [3].

2.2 Semilattice Frame and Functional Set Frame

In this section, we briefly review Urquhart's *semilattice frame* for the positive fragment of relevant logic, and Standefer's *functional set frame* that is equivalent to semilattice frame.[1] Until the negation operator is introduced in the next section, we work with the language L of the positive fragment of relevant logic. That is, we define L as the set of propositional formulas built from a countably infinite set PV of propositional variables by the connectives t (*truth*), $\wedge$, $\vee$ (*extensional conjunction and disjunction*), $\circ$ (*intensional conjunction*), and $\rightarrow$ (*relevant implication*).

Definition 2.1 (*Semilattice frame and model* [9, 11]) A semilattice frame is a structure $\langle P, +, 0\rangle$, where P is a non-empty set, $+$ a binary operation on P, $0 \in P$ such that for any $x, y, z \in P$,

$$
\begin{aligned}
&0 + x = x, && \text{(Unit)}\\
&x + y = y + x, && \text{(Commutativity)}\\
&(x + y) + z = x + (y + z), && \text{(Associativity)}\\
&x + x = x. && \text{(Contraction)}
\end{aligned}
$$

A *valuation* v on a semilattice frame $S = \langle P, +, 0\rangle$ is a function that assigns 1 or 0 to each pair of an element of P and a propositional variable. The pair $\langle S, v\rangle$ of a semilattice frame S and a valuation v on S is called a semilattice model.

[1] We refer the reader to this paper [8] on functional set frame of Standefer's also for a comprehensive survey of the field of operational semantics for relevant logic including semilattice frame.

Given a semilattice model $M = \langle P, +, 0, v\rangle$, the forcing relation $\Vdash_M \subseteq P \times L$ is inductively defined as follows. For each $x \in P$,

$$
\begin{aligned}
x \Vdash_M t &\iff x = 0,\\
x \Vdash_M p &\iff v(p) = 1 \text{ for any } p \in PV,\\
x \Vdash_M A \wedge B &\iff x \Vdash_M A \,\&\, x \Vdash_M B,\\
x \Vdash_M A \vee B &\iff x \Vdash_M A \text{ or } x \Vdash_M B,\\
x \Vdash_M A \circ B &\iff \exists y, z : y + z = x \,\&\, y \Vdash_M A \,\&\, z \Vdash_M B,\\
x \Vdash_M A \to B &\iff \forall y : y \Vdash_M A \Longrightarrow x + y \Vdash_M B.
\end{aligned}
$$

We omit the subscript M in $\Vdash_M$ as long as no confusion is caused. We say that A is true (false) at point x if $x \Vdash A$ ($x \nVdash A$). Also we say that

- A is valid in a semilattice model $M = \langle P, +, 0, v\rangle$ if $0 \Vdash A$,
- A is valid in a semilattice frame S if A is valid in every semilattice model defined on S, and
- A is valid in **UR** if A is valid in any semilattice frame.

We call the set of formulas that are valid in **UR** the logic **UR**.

It is known that while the disjunction-free fragment of **UR** coincides with that of the relevant logic **R** without negation, **UR** properly extends it. For example,

$$
\begin{aligned}
&(A \to B \vee C) \wedge (B \to D) \to (A \to D \vee C) \text{ and}\\
&(A \to (A \to A) \vee A) \to (A \to A)
\end{aligned}
$$

are valid in **UR** but not in **R**.

We proceed to functional set frame. It is a kind of *collection frame* introduced by Restall and Standefer [7] as a generalization and simplification of ternary relation frame for relevant logic. In general terms, a collection frame is a set equipped with a relation that relates a collection of elements of the set to an element. A collection may have some kind of structure. It may be a tree, a list, a multiset, or a set. The choice of the kind of structure roughly corresponds to what structural rules the target logic assumes.

Functional set frame introduced below, as the name suggests, is a frame that relates a finite subset of a set with a single element. It assumes an additional property called Functionality to obtain a correspondence to the operation $+$ on semilattice frame.

Definition 2.2 (*Functional set frame and model* [8]) Let P be a non-empty set and $\mathcal{F}(P)$ be the set of finite subsets of P. A functional set frame (FS frame for short) is a structure $\langle P, R\rangle$ where $R \subseteq \mathcal{F}(P) \times P$ such that

For any $X \in \mathcal{F}(P)$, there is a unique y such that XRy, (Functionality)
$\{x\}Rx$, (Reflexivity)
$(X \cup W)Ry \iff \exists u : XRu \,\&\, (\{u\} \cup W)Ry$. (Compositionality)

A functional set model (FS model for short) is a pair $M = \langle F, v \rangle$ of an FS frame F and a valuation v on F. The forcing relation $\Vdash_M$ is inductively defined as follows. For any $x \in P$,

$$\begin{aligned}
x \Vdash_M t &\iff \emptyset Rx, \\
x \Vdash_M p &\iff v(p) = 1 \text{ for any } p \in PV, \\
x \Vdash_M A \wedge B &\iff x \Vdash_M A \,\&\, x \Vdash_M B, \\
x \Vdash_M A \vee B &\iff x \Vdash_M A \text{ or } x \Vdash_M B, \\
x \Vdash_M A \circ B &\iff \exists y, z : \{y, z\}Rx \,\&\, y \Vdash_M A \,\&\, z \Vdash_M B, \\
x \Vdash_M A \to B &\iff \forall y, z : \{x, y\}Rz \,\&\, y \Vdash_M A \Longrightarrow z \Vdash_M B.
\end{aligned}$$

We omit the subscript M in $\Vdash_M$ as before. We say that A is valid in an FS model if $x \Vdash A$ holds for any $x \in P$ such that $\emptyset Rx$, valid in an FS frame if it is valid in every FS model defined on the frame, and valid in **FS** if A is valid in any FS frame. We call the set of formulas that are valid in **FS** the logic **FS**.

Standefer shows that **UR** = **FS** by providing transformations between semilattice frames and functional set frames.

Theorem 2.1 (Standefer [8]) *There are transformations σ from the class of semilattice frames to the class of FS frames, and λ from the class of FS frames to the class of semilattice frames such that for any semilattice frame S and FS frame F,*

$$\sigma \circ \lambda(F) = F \text{ and } \lambda \circ \sigma(S) = S.$$

*Moreover, since σ and λ preserve the forcing relation, it immediately follows that **UR** = **FS**.*

2.3 Negation

Hereafter we include negation $\sim$ in the language L. On semilattice frames, Urquhart defines negation using a unary operation on a frame, which we denote by $*$, as

$$x \Vdash \sim A \iff x^* \nVdash A,$$

where $*$ satisfies

$$0^* = 0 \text{ and } x^{**} = x.$$

The negation so defined does not validate all principles of **R**. While the law of excluded middle (thanks to $0^* = 0$) and the double negation laws (thanks to $x^{**} = x$) are validated,

$$(A \to \sim A) \to \sim A \qquad \text{(Reduction)}$$
$$(A \to B) \to (\sim B \to \sim A) \qquad \text{(Contraposition)}$$

are not. Indeed it is impossible to validate both of these in the present setting: Contraposition corresponds to the constraint

$$x + y = z \iff x + z^* = y^*, \qquad (*)$$

in the sense that Contraposition is valid in any semilattice frame satisfying (*) and, conversely, it is invalid in any frame in which (*) fails. However, any semilattice frame with (*) is indeed a singleton, since for any $x \in P$, we have

$$x + x^* = 0^* \text{ by } (*) \text{ from } 0 + x = x, \text{ and}$$
$$x + x^* = x^* \text{ by } (*) \text{ from } x + x = x,$$

and hence $0 = x$. The logic defined by those frames would be classical logic. Dropping Contraction would prevent this collapse, but then we could not validate Reductio to which Contraction corresponds under (*). Thus, Contraction and (*) are mutually exclusive in the setting of semilattice frame for relevant logic. Urquhart chose Contraction. In this paper, we instead choose (*) giving up Contraction, and provides three different but equivalent characterizations of the logic.[2]

2.4 Two Kinds of Frames

This section introduces and demonstrates the equivalence of two distinct types of frames for our logic: *contrapositive frame* and *bi-operational frame*. Contrapositive frame modifies Urquhart's semilattice frame by dropping Contraction, thereby transforming it into a commutative monoid. This frame includes an operation $*$ that adheres to the Contraposition constraint (*). On the other hand, bi-operational frame features two distinct commutative monoid operations. These operations interact under a principle called Compositionality, which was previously introduced in the context of functional set frames. This principle is critical as it governs the interaction between the two operations, ensuring that the operation $*$ is definable within

[2] There are several works on contractionless or non-contractive logics based on operational semantics [1–4], but to the best of my knowledge, there are no studies discussing it in relation to negation or the Contraposition constraint.

a bi-operational frame. The relationship with collection frames will become clear in the next section when we introduce functional multiset frames.

Definition 2.3 (*Contrapositive frame*) A contrapositive frame is a structure $F = \langle P, +, 0, *\rangle$ where $\langle P, +, 0\rangle$ is a commutative monoid, i.e. semilattice minus Contraction, and $*$ is an operation on P that satisfies $0^* = 0$, $x^{**} = x$, and (*).

Example 2.1 Here is an example of contrapositive frame. Let $P = \{0, x, x^*\}$. $+$ and $*$ are defined as below.

$+$	0	x	x^*
0	0	x	x^*
x	x	x^*	0
x^*	x^*	0	x

$*$	
0	0
x	x^*
x^*	x

Definition 2.4 (*Bi-operational frame*) A bi-operational frame is a structure $\langle P, +, \times, 0\rangle$ where $\langle P, +, 0\rangle$ and $\langle P, \times, 0\rangle$ are commutative monoids that satisfies

$$x + w = y \times z \iff \exists u \in P : x = y \times u \;\&\; u + w = z. \qquad \text{(Compositionality)}$$

The notions of valuation and validity for these frames are defined in the same way as for semilattice frames.

The additional operation $\times$ can be used to define *intentional disjunction* $\bullet$, and *exclusion* (or *co-implication*) $\rhd\!\!-$ as

$$x \Vdash A \bullet B \iff \forall y, z : x = y \times z \;\&\; y \nVdash A \Longrightarrow z \Vdash B$$
$$x \Vdash A \rhd\!\!- B \iff \exists y : y \Vdash A \;\&\; x \times y \nVdash B.$$

They are indeed redundant. As we will see shortly, the $*$ operation is definable on any bi-operational frame, and hence negation is interpretable by the defined $*$ operation. Then through the following equations they are reduced to the existing vocabulary:

$$A \bullet B = \sim\!(\sim A \circ \sim B);\;\; A \rhd\!\!- B = \sim\!(A \to \sim B).$$

Let us check that the two kinds of frames above are equivalent.

Proposition 2.1 *There are transformations α from the class of contrapositive frames to the class of bi-operational frames, and β from the class of bi-operational frames to the class of contrapositive frames such that for any contrapositive frame F and bi-operational frame G,*

$$\alpha \circ \beta(G) = G \text{ and } \beta \circ \alpha(F) = F.$$

Proof Given a contrapositive frame $F = \langle P, +, 0, *\rangle$, define $\alpha(F) = \langle P, +, \times, 0\rangle$ by defining $\times$ by

$$x \times y = (x^* + y^*)^*.$$

It is immediate that the operation $\times$ satisfies Unit, Commutativity, and Associativity. For Compositionality, we first claim that for any $x, y, w, z \in P$

$$x + w = y \times z \iff x + y^* = w^* \times z$$

holds. Indeed, suppose that $x + w = y \times z$. Then,

$$\begin{aligned} & x + w = (y^* + z^*)^*, \\ \therefore\ & x + y^* + z^* = w^*, \\ \therefore\ & x + y^* + w = z, \\ \therefore\ & w + z^* = (x + y^*)^*, \\ \therefore\ & x + y^* = (w + z^*)^* = w^* \times z. \end{aligned}$$

Note further that from this it follows that

$$\begin{aligned} x + w = z &\iff x = w^* \times z, \text{ and} \\ x = y \times z &\iff x + y^* = z \iff y^* = x^* \times z. \end{aligned}$$

Now if $x + w = y \times z$, then $x + y^* = w^* \times z$. Put $u = w^* \times z$. Then

$$\begin{aligned} & x + y^* = u, \text{ hence } x = y \times u, \text{ and} \\ & u = w^* \times z, \text{ hence } u + w = z. \end{aligned}$$

Conversely, if $x = y \times u$ and $u + w = z$ for some $u \in P$, then $x + y^* = u = w^* \times z$, hence $x + w = y \times z$. Thus, $\alpha(P)$ is well-defined as a bi-operational frame.

We proceed to define the contrapositive frame $\beta(G)$ for any bi-operational frame $G = \langle P, +, \times, 0 \rangle$. By Compositionality, for each $x \in P$, we can argue as

$$\begin{aligned} & x = x \\ \therefore\ & 0 + x = x \times 0, \\ \therefore\ & \exists u \in P : 0 = x \times u \,\&\, u + x = 0. \qquad \text{(Compositionality)} \end{aligned}$$

And such u is unique since if there exists $u' \in P$ such that

$$0 = x \times u' \,\&\, u' + x = 0,$$

we can obtain $u = u'$ using the right-to-left direction of Compositionality. Let us define x^* to be the u that exists uniquely for x. The so-defined $*$ satisfies $0 = 0^*$, $x^{**} = x$ and (*): since x^* is the unique y that satisfies $x + y = 0$ and $0 = x \times y$,

$$\begin{aligned} & 0 + 0 = 0 \text{ and } 0 = 0 \times 0, \text{ hence } 0^* = 0, \text{ and} \\ & x^* + x = 0 \text{ and } 0 = x \times x^*, \text{ hence } x = x^{**}. \end{aligned}$$

Further, if $x + y = z$, then since $y + y^* = z + z^* = 0$,

$$x = x + y + y^* = z + y^*$$
$$\therefore \; x + z^* = y^* + z + z^* = y^*.$$

Thus, $\beta(G)$ has been defined as a contrapositive frame. The proof that α and β constitute an isomorphism is tedious but straightforward.

It should be clear that the transformations α and β preserve forcing relations over arbitrary formulas between contrapositive frames and bi-operational frames: Let $F_c = \langle P, +, *, 0\rangle$ and $F_b = \langle P, +, \times, 0\rangle$ be an equivalent pair of contrapositive frame and bi-operational frame (i.e. $\alpha(F_c) = F_b$ and $\beta(F_b) = F_a$). Then for any valuation v (and the forcing relation it induces), for any $x \in P$ and any formula $A \in L$,

$$x \Vdash_{M_c} A \iff x \Vdash_{M_b} A,$$

where on M_b negation $\sim B$ is interpreted by the $*$ operation which has been shown to be definable in bi-operational frames. So the two classes of frames define the same set of valid formulas.

2.5 Functional Multiset-Multiset Frame

We now turn to define the notion of *functional multiset-multiset* (FMM) frame and prove the equivalence with that of bi-operational frame. FMM frame is simply a functional version of multiset-multiset frame presented in [6] as an extension of Restall and Standefer's multiset frame. A multiset-multiset frame is a frame with a relation between two finite multisets of elements of a set. Compared to Standefer's functional set frame, we use multisets instead of sets, in accordance with the giving up of Contraction in the operational semantics. Moreover we dualize the relation, that is, use the (multi)set-(multi)set form instead of the (multi)set-point form. The parallelism with bi-operational frame should be evident.

The idea of dualization is derived from the notion of GS frame [5], frame that has two dual ternary relations with the constraint called Generalized Star postulate which implies the definability of the $*$ operation on the frame. Multiset-multiset frame is the counterpart of GS frame on the side of collection frame, and it turns out that the Compositionality constraint adopted to the multiset-multiset setting nicely corresponds to Generalized Star postulate.

From the above discussion, the equivalence between FMM frame and bi-operational frame may conceptually be evident. In the following, we will formally define FMM frame and demonstrate the equivalence.

Definition 2.5 (*Functional multiset-multiset frame*) A functional multiset-multiset frame is a structure $\langle P, \bowtie \rangle$ with the following constraints, where P is a non-empty set and $\bowtie$ is a binary relation on the set of finite multisets of elements from P, which we denote by $\mathcal{FM}(P)$: for any $X, W, Y, Z \in \mathcal{FM}(P)$ and $x \in P$,

there exist unique t and u s.t. $X \bowtie \{t\}$ and $\{u\} \bowtie X$	(Functionality)
$\{x\} \bowtie \{x\}$	(Reflexivity)
$\emptyset \bowtie \emptyset$	($\emptyset$)
$(X \cup W) \bowtie (Y \cup Z) \iff \exists u \in P : X \bowtie (Y \cup \{u\}) \,\&\, (\{u\} \cup W) \bowtie Z.$	(Compositionality)

The notions concerning valuation and validity are defined in the same way as in functional set frame and model.

The set theoretic notation used here, such as the operation of union, is of course for multisets. Handling multisets generally requires different considerations than sets, but it is unlikely to pose any special problems in the following.

The constraint ($\emptyset$) here is the counterpart of $0^* = 0$ in contrapositive frames and the assumption that 0 is the unit both for $+$ and $\times$ in bi-operational frame. We could consider frames without these constraints and the resultant logic would invalidate the law of Excluded Middle.

Note that the constraint of Functionality only requires the existence and uniqueness of *single* succedent (or antecedent) for a given antecedent (or succedent), and does not exclude the cases for various multiple succedents or antecedents. For example, even when $X \bowtie \{t\}$ holds, it is possible that $X \bowtie Y$ holds for some multiset Y with multiple (or zero) elements.

Example 2.2 (*Cf. Example 1 in* [6]) Take the set $\mathbb{Z}$ of integers and the addition $+$ on $\mathbb{Z}$. Define the relation $\bowtie^=$ between finite multisets X, Y of integers as

$$X \bowtie^= Y \iff \sum X = \sum Y,$$

where $\sum \emptyset = 0$. Then it is clear that $\langle \mathbb{Z}, \bowtie^= \rangle$ satisfies the constraints Functionality, Reflexivity, and $\emptyset$. The $\Longleftarrow$ direction of Compositionality is also obvious: $\sum X = \sum Y + u$ and $u + \sum W = \sum Z$ imply $\sum(X \cup W) = \sum(Y \cup Z)$. For the $\Longrightarrow$ direction, consider, for illustration, the case where we have $x + w = y + z$. If we put $u = x - y = z - w$, then $x = y + u$ and $u + w = z$, i.e. $\{x\} \bowtie^= \{y, u\}$ and $\{u, w\} \bowtie^= \{z\}$. Thus $\langle \mathbb{Z}, \bowtie^= \rangle$ is a functional multiset-multiset frame. One can see that this essentially establishes Compositionality in bi-operational frame as well. Thus, $\langle \mathbb{Z}, +, +, 0 \rangle$ is a bi-operational frame.

Proposition 2.2 *There are transformations δ from the class of bi-operational frames to the class of FMM frames, and γ from the class of FMM frames to the class of bi-operational frames such that for any bi-operational frame G and FMM frame H,*

$$\gamma \circ \delta(G) = G \text{ and } \delta \circ \gamma(H) = H.$$

Proof Given a bi-operational frame $G = \langle P, +, \times, 0\rangle$, we define the FMM frame $\delta(G) = \langle P, \bowtie\rangle$ as follows. First we define two operations Σ and Π on the set of finite multiset of P by induction on the size of multiset [3]:

$$\Sigma\emptyset = 0 \qquad \Pi\emptyset = 0$$
$$\Sigma\{x\} = x \qquad \Pi\{x\} = x$$
$$\Sigma\{x, y\} = x + y \qquad \Pi\{x, y\} = x \times y$$
$$\Sigma(\{x\} \cup X) = x + \Sigma X \qquad \Pi(\{x\} \cup X) = x \times \Pi X.$$

Next define $\bowtie$ by for any $X, Y \in \mathcal{FM}(P)$ by

$$X \bowtie Y \iff \Sigma X = \Pi Y.$$

It is clear that $\bowtie$ satisfies ($\emptyset$), Functionality and Reflexivity. Compositionality

$$(X \cup W) \bowtie (Y \cup Z) \iff \exists u : X \bowtie (Y \cup \{u\}) \,\&\, (\{u\} \cup W) \bowtie Z$$

is verified by reducing all cases to the one where X, W, Y, Z are all singletons $\{x\}, \{w\}, \{y\}, \{z\}$. The reduction is possible because for any X, W, Y, Z, we have

$$(X \cup W) \bowtie (Y \cup Z) \iff \{\Sigma X, \Sigma W\} \bowtie \{\Pi Y, \Pi Z\}.$$

For the singleton case we argue as

$$\begin{aligned} \{x, w\} \bowtie \{y, z\} &\iff x + w = y \times z \\ &\iff \exists u : x = y \times u \,\&\, u + w = z \\ &\iff \exists u : \{x\} \bowtie \{y, u\} \,\&\, \{u, w\} \bowtie \{z\}, \end{aligned}$$

appealing to Compositionality in G.

Conversely, for any FMM frame $H = \langle P, \bowtie\rangle$, we define the bi-operational frame $\gamma(H) = \langle P, +, \times, 0\rangle$ as follows. For any $x, y \in P$, by Functionality, there are unique t and u such that

$$\{x, y\} \bowtie \{t\} \text{ and } \{u\} \bowtie \{x, y\}$$

Define $x + y$ to be such t and $x \times y$ to be such u. And since $\emptyset \bowtie \emptyset$, there is unique u such that

$$\emptyset \bowtie \{u\} \text{ and } \{u\} \bowtie \emptyset$$

[3] We have to check that the operations Σ and Π are defined in a coherent way, that is, whether they are uniquely defined regardless of the representation of the input multiset. For example, $\Sigma\{x, y\}$ and $\Sigma\{y, x\}$, and $\Sigma(X \cup (Y \cup Z))$ and $\Sigma((X \cup Y) \cup Z)$ must be defined to be the same. This can be shown appealing to commutativity and associativity inherent in multisets.

by Compositionality and Functionality. Let 0 to be such u. Under this definition, Unit is evident. Commutativity and Associativity follow immediately from the structural properties of multisets. Now it should also be clear that Compositionality of $\gamma(H)$ follows from that of H:

$$\begin{aligned} x + w = y \times z &\iff \{x, w\} \bowtie \{y, z\} \\ &\iff \exists u : \{x\} \bowtie \{y, u\} \,\&\, \{u, w\} \bowtie \{z\} \\ &\iff \exists u : x = y \times u \,\&\, u + w = z. \end{aligned}$$

Now that we have defined the transformations γ and δ, let us check that they form an isomorphism between the classes in question. Consider a bi-operational frame $G = \langle P, +, \times, 0\rangle$, and write

$$\delta(G) = \left\langle P^{\delta}, \bowtie^{\delta}\right\rangle \text{ and } \gamma \circ \delta(G) = \left\langle P^{\gamma\circ\delta}, +^{\gamma\circ\delta}, \times^{\gamma\circ\delta}, 0^{\gamma\circ\delta}\right\rangle.$$

We show that $\gamma \circ \delta(G) = G$. Obviously $P = P^{\gamma\circ\delta}$ and $0 = 0^{\gamma\circ\delta}$. Moreover,

$$x + y = x +^{\gamma\circ\delta} y$$

holds for any $x, y \in P$ since

$$x +^{\gamma\circ\delta} y = u \iff \{x, y\} \bowtie^{\delta} \{u\} \iff \Sigma\{x, y\} = \Pi\{u\} \iff x + y = u.$$

The case for $\times$ is similar.

Next, given an FMM frame $H = \langle P, \bowtie\rangle$, write

$$\gamma(H) = \langle P^{\gamma}, +^{\gamma}, \times^{\gamma}, 0^{\gamma}\rangle\,, \text{ and } \delta \circ \gamma(H) = \left\langle P^{\delta\circ\gamma}, \bowtie^{\delta\circ\gamma}\right\rangle.$$

We show that $\delta \circ \gamma(H) = H$. For that it suffices to show that

$$X \bowtie Y \iff X \bowtie^{\delta\circ\gamma} Y$$

for any $X, Y \in \mathcal{FM}(P)$. First, we note that

$$X \bowtie \{u\} \iff \{\Sigma^{\gamma} X\} \bowtie \{u\} \text{ and } \{u\} \bowtie Y \iff \{u\} \bowtie \{\Pi^{\gamma} Y\},$$

where Σ^{γ} and Π^{γ} are the Σ and Π operations defined in terms of $+^{\gamma}$ and $\times^{\gamma}$. We prove the case for Σ^{γ} by induction on the size of X. The cases for $X = \emptyset$ and $X = \{x\}$ are trivial. Suppose that $X = \{x\} \cup X'$. Then

$$
\begin{aligned}
\{\Sigma^{\gamma} X\} \bowtie \{u\} &\iff \{\Sigma^{\gamma}(\{x\} \cup X')\} \bowtie \{u\} && \\
&\iff \{x +^{\gamma} \Sigma^{\gamma} X'\} \bowtie \{u\} && \text{(Definition of } \Sigma^{\gamma}\text{)} \\
&\iff x +^{\gamma} \Sigma^{\gamma} X' = u && \text{(Reflexivity and Functionality)} \\
&\iff \{x, \Sigma^{\gamma} X'\} \bowtie \{u\} && \text{(Definition of} +^{\gamma}\text{)} \\
&\iff \exists t : \{\Sigma^{\gamma} X'\} \bowtie \{t\} \,\&\, \{t, x\} \bowtie \{u\} && \text{(Compositionality)} \\
&\iff \exists t : X' \bowtie \{t\} \,\&\, \{t, x\} \bowtie \{u\} && \text{(Inductive hypothesis)} \\
&\iff \{x\} \cup X' \bowtie \{u\}. && \text{(Compositionality)}
\end{aligned}
$$

Now,

$$
\begin{aligned}
X \bowtie Y &\iff \exists u : X \bowtie \{u\} \,\&\, \{u\} \bowtie Y && \text{(Compositionality)} \\
&\iff \exists u : \{\Sigma^{\gamma} X\} \bowtie \{u\} \,\&\, \{u\} \bowtie \{\Pi^{\gamma} Y\} && \text{(The above claim)} \\
&\iff \Sigma^{\gamma} X = \Pi^{\gamma} Y && \text{(Reflexivity and Functionality)} \\
&\iff X \bowtie^{\delta \circ \gamma} Y. && \text{(Definition of } \bowtie^{\delta \circ \gamma}\text{)}
\end{aligned}
$$

This concludes the proof.

The $*$ operation is also definable on FMM frames since they are equivalent with bi-operational frames. So, again, negation is interpretable with the usual truth clause on an FMM frame. And it is also clear that forcing relations are preserved through the transformations δ and γ as in the case of α and β. Thus we have:

Proposition 2.3 *Let $F_c = \langle P, +, *, 0\rangle$, $F_b = \langle P, +, \times, 0\rangle$ and $F_f = \langle P, \bowtie\rangle$ be the equivalent trio of contrapositive frame, bi-operational frame, and FMM frame: namely $F_b = \alpha(F_c)$, $F_f = \delta(F_b)$ and $F_c = \beta \circ \gamma(F_f)$. Consider a valuation v on P and let $M_c = \langle F_c, v\rangle$, $M_b = \langle F_b, v\rangle$, $M_f = \langle F_f, v\rangle$. Then for any $x \in P$ and any formula $A \in L$,*

$$x \Vdash_{M_c} A \iff x \Vdash_{M_b} A \iff x \Vdash_{M_f} A.$$

Hence the three classes of frames define the same set of valid formulas.

2.6 Labeled Sequent Calculus

In this section we define a labeled sequent calculus for the logic of contrapositive frame. While the calculi defined in the previous work of Urquhart [10] and Kashima [3] for positive semilattice logics use finite sets of positive integers as labels, we use finite multisets of non-zero (positive and negative) integers, reflecting the non-contractive and dual structure of frames (the use of multisets for non-contractive logics is mentioned in [3]).

2.6.1 Labeled Sequent Calculus LCon

First let us introduce a bunch of notions concerning *labels*. We call a finite multiset of non-zero integers a *label*, for which we use greek letters α, β etc. A label α is *normalized* by canceling the appearances of a number in α with those of its opposite. For example, if $\alpha = \{-2, -1, -1, 1, 2, 3\}$ then it normalizes to $\{-1, 3\}$. We write $[\alpha]$ for the result of normalization of α and call it the *normalized label* of α. Formally, it is defined as follows. A label α as a multiset can be regarded as a function $\alpha : \mathbb{Z} \setminus \{0\} \to \mathbb{N}$ that assigns the number of occurrences to each non-zero integer. Then its normalized label $[\alpha]$ is defined to be a function such that

$$[\alpha](x) = \alpha(x) - \alpha(-x)$$

for each *positive* integer x. E.g. if $[\alpha](3) = -2$, it means that -3 appears twice more often than 3 in α, and hence $[\alpha]$, regarded as a multiset, contains two occurrences of -3 (and no occurrence of 3). Finally, α^* denotes the label obtained by inverting the sign of all numbers in α. As a function, α^* is defined to be such that $\alpha^*(x) = \alpha(-x)$ for each non-zero integer x.

A *term* is a pair of a label, say α, and a formula, say A, for which we write $\alpha : A$. A *sequent* is a expression of the form $X \Rightarrow Y$ where X, Y are multisets of terms. $L(X)$, $L(Y)$ and $L(X \Rightarrow Y)$ denote the union of all labels that appear in X, Y or $X \Rightarrow Y$.

Definition 2.6 We define the labeled sequent calculus **LCon** for the logic of contrapositive frames to include the following axiom and rules.

$$\text{(Id) } X, \alpha : A \Rightarrow \beta : A, Y, \text{ where } [\alpha] = [\beta]$$

$$(t_L)\ X, \alpha : t \Rightarrow Y \quad (t_R)\ X \Rightarrow \beta : t, Y, \text{ where } [\alpha] \neq \emptyset \text{ and } [\beta] = \emptyset.$$

$$\frac{X, \alpha : A \wedge B, \alpha : A, \alpha : B \Rightarrow Y}{X, \alpha : A \wedge B \Rightarrow Y}\ (\wedge_L)$$

$$\frac{X \Rightarrow \alpha : A, \alpha : A \wedge B, Y \qquad X \Rightarrow \alpha : B, \alpha : A \wedge B, Y}{X \Rightarrow \alpha : A \wedge B, Y}\ (\wedge_R)$$

$$\frac{X, \alpha : A \vee B, \alpha : A \Rightarrow Y \qquad X, \alpha : A \vee B, \alpha : B \Rightarrow Y}{X, \alpha : A \vee B \Rightarrow Y}\ (\vee_L)$$

$$\frac{X \Rightarrow \alpha : A, \alpha : B, \alpha : A \vee B, Y}{X \Rightarrow \alpha : A \vee B, Y}\ (\vee_R)$$

$$\frac{X, \alpha : A \circ B, \{x\} : A, \alpha \cup \{x\}^* : B \Rightarrow Y}{X, \alpha : A \circ B \Rightarrow Y}\ (\circ_L), \text{ where } x \notin L(X) \cup L(Y) \cup \alpha.$$

$$\frac{X \Rightarrow \beta : A, \alpha : A \circ B, Y \qquad X \Rightarrow \alpha \cup \beta^* : A, \alpha : A \circ B, Y}{X \Rightarrow \alpha : A \circ B, Y} (\circ_R)$$

$$\frac{X, \alpha : A \to B \Rightarrow Y, \beta : A \qquad X, \alpha : A \to B, \alpha \cup \beta : B \Rightarrow Y}{X, \alpha : A \to B \Rightarrow Y} (\to_L)$$

$$\frac{X, \{x\} : A \Rightarrow \{x\} \cup \alpha : B, \alpha : A \to B, Y}{X \Rightarrow \alpha : A \to B, Y} (\to_R), \text{ where } x \notin L(X) \cup L(Y) \cup \alpha.$$

$$\frac{X, \alpha : \sim A \Rightarrow Y, \alpha^* : A}{X, \alpha : \sim A \Rightarrow Y} (\sim_L) \qquad \frac{X, \alpha^* : A \Rightarrow Y, \alpha : \sim A}{X \Rightarrow \alpha : \sim A, Y} (\sim_R)$$

The notion of provability is defined as usual. We write $X \vdash Y$ if $X \Rightarrow Y$ is provable in **LCon**.

Example 2.3 Here is a proof of Excluded Middle.

$$\cfrac{\cfrac{\emptyset : A \Rightarrow \emptyset : A, \emptyset : \sim A}{\Rightarrow \emptyset : A, \emptyset : \sim A} (\sim_R)}{\Rightarrow \emptyset : A \vee \sim A} (\vee_R)$$

$(\sim_R)$ is applicable because $\emptyset : A$ on the left hand side of the first line can be regarded as $\emptyset^* : A$ since $\emptyset = \emptyset^*$, which corresponds to the constraint $0 = 0^*$. We also present a proof of contraposition in **LCon**. For simplicity the redundant occurrences of principal formulas in the upper sequents are omitted.

$$\cfrac{\cfrac{\cfrac{\cfrac{\{x, y\}^* : A \Rightarrow \{x, y\}^* : A \qquad \{x\} \cup \{x, y\}^* : B \Rightarrow \{y\}^* : B}{x : A \to B, \{x, y\}^* : A \Rightarrow \{y\}^* : B}}{x : A \to B \Rightarrow \{x, y\} : \sim A, \{y\}^* : B}}{x : A \to B, y : \sim B \Rightarrow \{x, y\} : \sim A}}{x : A \to B \Rightarrow x : \sim B \to \sim A}$$

where $[\{x\} \cup \{x, y\}^*] = [\{x, -x, -y\}] = \{-y\} = [\{y\}^*]$.

Let $F = \langle P, +, 0, * \rangle$ be a contrapositive frame. Let I be a set of non-zero integers and $Lb(I)$ the set of all labels formed from elements of I (all finite sub-multisets of I). A mapping $f : I \to P$ is called an *interpretation of I on F* if $f(-x) = f(x)^*$ for any $x \in I$. Define $\bar{f} : Lb(I) \to P$ by

$$\bar{f}(\emptyset) = 0$$
$$\bar{f}(\{x_1, \ldots, x_n\}) = f(x_1) + \cdots + f(x_n).$$

Note that $\bar{f}(\alpha) = \bar{f}([\alpha])$ for any $\alpha \in Lb(I)$. It can be proved generally by induction on the size of α, but we give an example here.

$$\bar{f}(\{x, -x\}) = f(x) + f(-x) = f(x) + f(x)^* = 0 = \bar{f}(\emptyset) = \bar{f}([\{x, -x\}]).$$

The property (*) of contrapositive frame is necessary for the part $f(x) + f(x)^* = 0$.

Definition 2.7 A counter-model of a sequent $X \Rightarrow Y$, with I the underlying set of $L(S)$, is a pair $\langle M, f \rangle$ of a contrapositive model $M = \langle F, v \rangle$ and an interpretation f of I on F such that

- $\bar{f}(\alpha) \Vdash A$ for any $\alpha : A \in X$ and
- $\bar{f}(\beta) \nVdash B$ for any $\beta : B \in Y$.

A sequent $X \Rightarrow Y$ is said to be valid (we write $X \models Y$) if there is no counter-model.

Note that $\models \emptyset : A$ if and only if A is valid in the class of contrapositive frames in the sense that $0 \Vdash A$ holds in all contrapositive models.

Theorem 2.2 (Soundness) *If* $X \vdash Y$ *then* $X \models Y$.

Proof By induction on the construction of proof of $X \Rightarrow Y$. The aforementioned fact that $\bar{f}(\alpha) = \bar{f}([\alpha])$ validates (Id). That each rule preserves the validity is almost trivial.

2.6.2 Completeness

Let us turn to the completeness proof. In the following, we ignore the extensional connectives $\wedge, \vee$ since the treatment of them is straightforward and tedious. Also the intentional conjunction $\circ$ is omitted since it is definable as $A \circ B = {\sim}(A \rightarrow {\sim} B)$ in the presence of negation.

We follow Urquhart's method [10] for the completeness proof. A proof search procedure can be defined in the obvious way: apply the rules in reverse to a given sequent until you obtain *saturated* sequent in every branch in the following sense. Let $S = X \Rightarrow Y$ be a sequent and $L(S)$ the set of labels appearing in S. S is said to be *saturated* if

- S is an instance of some axiom, otherwise
- for any $\alpha : A \rightarrow B \in X$, it holds that for all $\beta \in L(S)$, $\beta : B \in X$ or $\alpha \cup \beta : A \in Y$,
- for any $\alpha : A \rightarrow B \in Y$, there is $\{x\} \in L(S)$ such that $x : A \in X$ and $\{x\} \cup \alpha : B \in Y$,
- for any $\alpha : {\sim} A \in X$, it holds that $\alpha^* : A \in Y$, and
- for any $\alpha : {\sim} A \in Y$, it holds that $\alpha^* : A \in X$.

Then there are three possible cases:

1. The process terminates and every leaf of the obtained tree is an instance of some axiom. Then we are done. We have a proof.

2. The process terminates, but one leaf is not an instance of any axiom though it is saturated.
3. The process fails to terminate (consider a proof search for $\Rightarrow \emptyset : ((A \rightarrow B) \rightarrow A) \rightarrow A)$).

In the second case, the tree contains a branch that leads to a non-axiom. In the third case, the tree contains at least one infinite long branch by König's lemma. Let us call such branches *full* branches.

A term is said to occur *on the left* (*on the right*) of a branch if it occurs on the left (on the right) of some sequent in the branch. Let $Lb(Br)$ be the union of all labels occurring on a branch Br. Then the following is obvious.

Lemma 1 *Let Br be a full branch.*

1. *If $\alpha : A$ occurs on the left (right) of Br, any term $\beta : A$ with $[\alpha] = [\beta]$ does not occur on the right (left) of Br.*
2. *No term of the form $\alpha : t$ such that $[\alpha] \neq \emptyset$ does not occur on the left.*
3. *Not term of the form $\beta : t$ such that $[\beta] = \emptyset$ does not occur on the right.*
4. *If $\alpha : A \rightarrow B$ occurs on the left of Br, then for all $\beta \in Lb(Br)$, $\beta : A$ occurs on the right or $\alpha \cup \beta : B$ occurs on the left.*
5. *If $\alpha : A \rightarrow B$ occurs on the right of Br, then for some $\beta \in Lb(Br)$, $\beta : A$ occurs on the left and $\alpha \cup \beta : B$ occurs on the right.*
6. *If $\alpha : \sim A$ occurs on the left (right) of Br, then $\alpha^* : A$ occurs on the right (left).*

Proof For 1, 2, and 3, see (Id), (t_L) and (t_R) respectively. 4 and 5 holds in virtue of the rules $(\rightarrow_L)$ and $(\rightarrow_R)$. 6 comes from the rules for negation.

Now we define a kind of "canonical model." Let Lb be the set of all labels. We define $\mathcal{F} = \langle [Lb], +, \emptyset, * \rangle$ to be such that

$$[Lb] = \{[\alpha] : \alpha \in Lb\}$$
$$[\alpha] + [\beta] = [\alpha \cup \beta]$$
$$[\alpha]^* = [\alpha^*].$$

$[\alpha] + [\beta]$ and $[\alpha]^*$ can be equivalently defined as functions by

$$[\alpha] + [\beta](x) = [\alpha](x) + [\beta](x) \text{ and } [\alpha]^*(x) = -[\alpha](x).$$

It is easy to see that these operations are well-defined, i.e., if $[\alpha] = [\alpha']$ and $[\beta] = [\beta']$ then $[\alpha] + [\beta] = [\alpha'] + [\beta']$ and $[\alpha]^* = [\alpha']^*$. It is also straightforward to check that $\mathcal{F}$ is a contrapositive frame. Given a full branch Br, define a valuation v^{Br} on $\mathcal{F}$ by

$$v^{Br}([\alpha], p) = 1 \iff \exists\beta : [\alpha] = [\beta] \text{ and } \beta : p \text{ occurs on the left.}$$

Lemma 2 *In the contrapositive model* $\langle \mathcal{F}, v \rangle$ *just defined, for any* α *occurring in* Br *and for any formula* A,

$$[\alpha] \Vdash^{Br} A \Longleftarrow \alpha : A \text{ occurs on the left}$$
$$[\alpha] \nVdash^{Br} A \Longleftarrow \alpha : A \text{ occurs on the right.}$$

Proof By induction on the construction of the formula A. The case where A is a propositional variable is the definition of v^{Br}.

Case $A = t$. If $\alpha : t$ occurs on the left, then by Lemma 1, $[\alpha] = \emptyset$, which implies that $[\alpha] \Vdash^{Br} t$. On the other hand if $\alpha : t$ occurs on the right, then again by Lemma 1, we have $[\alpha] \neq \emptyset$. It follows that $[\alpha] \nVdash^{Br} t$.

Case $A = B \to C$. Suppose that $\alpha : B \to C$ occurs on the left. To show that $[\alpha] \Vdash^{Br} B \to C$, suppose also that $[\beta] \Vdash^{Br} B$. By IH, $\beta : B$ does not occur on the right. Then $\alpha \cup \beta : C$ occurs on the left by Lemma 1, so $[\alpha \cup \beta](= [\alpha] + [\beta]) \Vdash^{Br} C$ by IH again. This implies $[\alpha] \Vdash^{Br} B \to C$

Suppose that $\alpha : B \to C$ occurs on the right. Then for some $\beta \in Lb(B)$, $\beta : B$ occurs on the left and $\alpha \cup \beta : C$ on the right. By IH, $[\beta] \Vdash^{Br} B$ and $[\alpha \cup \beta](= [\alpha] + [\beta]) \nVdash^{Br} C$. Thus $[\alpha] \nVdash^{Br} B \to C$.

Case $A = \sim B$. Suppose that $\alpha : \sim B$ occurs on the left. Then $\alpha^* : B$ occurs on the right. By IH, $[\alpha^*] \nVdash^{Br} B$, i.e. $[\alpha]^* \nVdash^{Br} B$. Thus, $[\alpha] \Vdash^{Br} \sim B$.

Suppose that $\alpha : \sim B$ occurs on the right. Then $\alpha^* : B$ occurs on the left. By IH, $[\alpha^*] \Vdash^{Br} B$, i.e. $[\alpha]^* \Vdash^{Br} B$. Thus, $[\alpha] \nVdash^{Br} \sim B$.

Theorem 2.3 (Completeness) *If* $X \models Y$, *then* $X \vdash Y$.

Proof Suppose that $X \nvdash Y$. Then the proof search procedure for $X \Rightarrow Y$ results either in the second or third case. Take a full branch Br in the proof attempt tree. For any term $\alpha : A \in X$, since it occurs on the left of Br, we have $[\alpha] \Vdash^{Br} A$, and for any term $\beta : B \in Y$, we have $[\beta] \nVdash^{Br} B$. If we define the interpretation f on $\mathcal{F}$ by $f(x) = \{x\}$ for each integer x, then $\langle \mathcal{F}, \Vdash^{Br}, f \rangle$ is a counter-model of $X \Rightarrow Y$. Therefore $X \not\models Y$.

2.7 Conclusion

We have presented three different kinds of frames and shown that there are isomorphic transformations between them. One is contrapositive frame obtained by modifying Urquhart's semilattice frame so that Contraposition is validated and Contraction (and Reductio) is invalidated as a trade-off. Another is functional multiset-multiset (FMM) frame, a modification of Standefer's functional multiset frame by applying the idea of GS-frame. The rest, bi-operational frame, served as a bridge that connects contrapositive frame and FMM frame. A labeled sequent calculus is defined for this logic by adapting Urquhart's and Kashima's calculi.

So, what exactly is the logic that these frames model? As mentioned in the text, the logic of Contractive operational frames (semilattice frames) is strictly stronger than that of the corresponding ternary relation frames, but it is worth noting that all examples demonstrating this appeal to contraction. To my knowledge, it is not known whether the logic of non-contractive operational frames is strictly stronger compared to that of their counterparts, the non-contractive ternary relation frames. This remains a topic for future research.

Acknowledgements The author is grateful to the audience of AWPL 2024: The 6th Asian Workshop on Philosophical Logic for their valuable comments. Appreciation is also extended to the members of the informal seminar group "Unscripted" for their discussions and encouragement regarding an earlier version of this paper.

References

1. Giambrone, S., Meyer, R.K., Urquhart, A.: A contractionless semilattice semantics. J. Symb. Logic **52**(2), 526–529 (1987)
2. Giambrone, S., Urquhart, A.: Proof theories for semilattice logics. Math. Log. Q. **33**(5), 433–439 (1987)
3. Kashima, R.: On semilattice relevant logics. Math. Logic Quart.: Math. Logic Quart. **49**(4), 401–414 (2003)
4. Meyer, R.K., Martin, E.P., Giambrone, S., Urquhart, A.: Further results on proof theories for semilattice logics. Math. Log. Q. **34**(4), 301–304 (1988)
5. Onishi, T.: Understanding negation implicationally in the relevant logic R. Stud. Log. **104**(6), 1267–1285 (2016)
6. Onishi, T.: Multiset-multiset frames. J. Philos. Log. **53**, 1241–1264 (2024)
7. Restall, G., Standefer, S.: Collection frames for distributive substructural logics. Rev. Symb. Logic **16**(4), 1120–1157 (2022)
8. Standefer, S.: Revisiting semilattice semantics. In: Alasdair Urquhart on Nonclassical and Algebraic Logic and Complexity of Proofs, pp. 243–259 (2022)
9. Urquhart, A.: Semantics for relevant logics. J. Symb. Logic **37**(1), 159–169 (1972)
10. Urquhart, A.: The Semantics of Entailment. Ph.D thesis, University of Pittsburgh (1973)
11. Urquhart, A.: Semilattice semantics for relevance logics. In: Anderson, A.R., Belnap, N.D., Dunn, J.M. (eds.), Entailment: The Logic of Relevance and Necessity, Volume II. Princeton University Press (1992)

Chapter 3
Ignorance and the Possibility of Error in Relevant Epistemic Logic

Shawn Standefer

Abstract In this paper, I will present two approaches to epistemic logic in the setting of relevant logics. One uses the framework of equivalence relations representing indistinguishability, found in much work in epistemic logics. The other, does not use equivalence relations, but is more common in the area of relevant logics. I will argue that the former has many advantages over the latter while avoiding some of the standard criticisms leveled against the use of equivalence relations in classically based epistemic logic.

Keywords Relevant logic · Epistemic logic · Equivalence relations · Ignorance

3.1 Introduction

One conception of knowledge in epistemic logics is that something is known if it is true in all epistemic alternatives. The epistemic alternatives at a given world, in turn, are the worlds that the agent cannot distinguish, given their background information. This idea of indistinguishability is formalized using equivalence relations in Kripke models. This modeling is intuitive and powerful, and it is arguably the default modeling epistemic logic.[1] Nonetheless, the modeling and its logic have faced criticisms. In this paper, I will examine this conception of knowledge against the backdrop of relevant logics, in particular the logic R. I will argue that in the context of R, this conception of knowledge can evade many of the criticisms leveled against it. I further identify three related, plausible epistemic logics and comment on their relations. To begin, in Sect. 3.2, I provide background on a standard approach to epistemic logics using equivalence relations. In Sect. 3.4, I present the frames and axioms for R. In

[1] See Meyer and van der Hoek [37], van Ditmarsch et al. [67], or Rendsvig et al. [43].

S. Standefer (✉)
Department of Philosophy and Religious Studies, North Carolina State University, Raleigh, NC, USA
e-mail: smstande@ncsu.edu

K. Sano et al. (eds.), *Exploring Negation, Modality and Proof*,
Logic in Asia: Studia Logica Library,
https://doi.org/10.1007/978-981-96-8566-0_3

Sect. 3.3, I present three plausible epistemic extensions of R, focusing on two defined in terms of equivalence classes. Finally, in Sect. 3.5, I summarize the results of the earlier sections and highlight ways in which the logics relate to issues of logical omniscience.

3.2 Epistemic Logics

Epistemic logics are most often studied with classical logic as the base logic.[2] The main results of this paper will concern an alternative base logic, but it will be useful to have the backdrop of standard epistemic logics in place before proceeding. We will throughout work in a propositional language $\mathscr{L}$ with the connectives $\{\neg, \rightarrow, \vee, \wedge, \Box\}$ and a countably infinite set of atoms $\mathsf{At} = \{p, q, r, \ldots\}$. The singulary operator $\Box$ will be the knowledge operator of the epistemic logics. We will define the possibility operator, $\Diamond$, as $\neg\Box\neg$.

The equivalence conception of knowledge is formalized using equivalence relations in Kripke frames.[3] Kripke frames with equivalence relations are the only sorts of Kripke frames used in this paper, so I will call them *classical partition frames*.

Definition 1 (*Equivalence relation*) A binary relation S on a set X is an *equivalence relation* iff it satisfies the three conditions

- $\forall x \in X$, Sxx (reflexivity),
- $\forall x, y \in X$, if Sxy, then Syx (symmetry), and
- $\forall x, y, z \in X$, if Sxy and Syz, then Sxz (transitivity).

For an equivalence relation S on a set X, we will define the notation $[w] = \{x \in X : Swx\}$. The sets $[w]$ are *equivalence classes*.

Definition 2 (*Classical partition frames*) A *classical partition frame* is a pair $\langle W, S\rangle$ where $W \neq \emptyset$ and S is an equivalence relation on W.

The relation S represents indiscernibility from the agent's point of view. Given the agent's background information, the agent cannot distinguish any of the worlds in a given equivalence class. These worlds are all the epistemic alternatives from any world in that class.

To define models, we add a valuation function.

Definition 3 (*Partition models*) Let $\langle W, S\rangle$ be a classical partition frame. $\langle W, S, V\rangle$ is a partition model where $V : \mathsf{At} \mapsto \wp W$.

---

[2] See van Ditmarsch et al. [68] and Humberstone [28, Chap. 5] for some examples and references.

[3] In Kripke frames, it is more common to use 'R' for the binary relation, but we will reserve this for the ternary relation of the frames for relevant logics.

The verification relation, $\Vdash$, for the language is defined inductively as follows.

- $w \Vdash p$ iff $w \in V(p)$
- $w \Vdash \neg B$ iff $w \nVdash B$
- $w \Vdash B \wedge C$ iff $w \Vdash B$ and $w \Vdash C$
- $w \Vdash B \vee C$ iff $w \Vdash B$ or $w \Vdash C$
- $w \Vdash B \rightarrow C$ iff $w \nVdash B$ or $w \Vdash C$
- $w \Vdash \Box B$ iff for all $x \in [w]$, then $x \Vdash B$

Given the definition of models, we can define validity.

Definition 4 (*Validity*) A formula B is valid on the class classical partition frames iff for all models built on a classical partition frame, for all $w \in W$, $w \Vdash B$.

The logic that results from this definition is S5. An axiomatization of S5 can be had by adding the following axioms and rules to classical logic, including the rule *modus ponens*. I will use '$A_1, \ldots, A_n \Rightarrow B$' for the rule from the formulas $A_1, \ldots, A_n$ to the conclusion B.[4]

- (K) $\Box(A \rightarrow B) \rightarrow (\Box A \rightarrow \Box B)$
- (Agg) $(\Box A \wedge \Box B) \rightarrow \Box(A \wedge B)$
- (T) $\Box A \rightarrow A$
- (4) $\Box A \rightarrow \Box\Box A$
- (5) $\neg\Box A \rightarrow \Box\neg\Box A$
- (B) $A \rightarrow \Box\neg\Box\neg A$
- (Nec) $A \Rightarrow \Box A$
- (Mono) $A \rightarrow B \Rightarrow \Box A \rightarrow \Box B$

This axiomatization is redundant, but the redundancies will facilitate the later discussion. So, I will leave them in. It is worth noting that given the other axioms and rules, (B) and (5) are interderivable. In terms of frames, the axioms (T), (4), (B), and (5) all correspond to S-based conditions on a frame $\langle W, S\rangle$, in the sense that for a given frame $\langle W, S\rangle$, S obeys the condition if and only if the axiom is valid on the frame. (T) corresponds to reflexivity, (4) to transitivity, and (B) to symmetry. (5) corresponds to a condition known as euclideanness, which is equivalent to symmetry given reflexivity and transitivity.

The epistemic logic S5 represents an idealization, since the knowers whose knowledge it codifies have unlimited memory and inferential capacities. While the logic S5 and its frames are perhaps the starting point for epistemic logics, it does not have many defenders as it seems to represent overly idealized agents. Stalnaker [55] has

[4] The two rules in this list are, in Smiley's terminology, rules of proof, taking one from theorems to theorems. Rules of proof are contrasted with rules of inference, which take one from truths to truths. See Humberstone [26] for discussion of this distinction, and see Brady [11] for some discussion in the context of relevant logics. The logics in this paper are in the framework FMLA of Humberstone [27], so the distinction between types of rules does not arise, but it would arise if one considered consequence relations, or SET- FMLA logics. Thanks to Rohan French for discussion of this point.

offered a limited defense of this idealization, although he thinks the modeling has serious flaws. Yap [71] has defended idealizations in epistemic logic, generally and not just for the specific case of interest.

The logic S5 has been criticized in many different ways as an epistemic logic.[5] I will focus on criticisms directed at the (B) and (5) axioms, as these two principles are interderivable given the background principles. Hintikka has argued that the idealizations inherent in an S5 approach to knowledge are implausible because "[i]t is not true that everybody could come to know the possibility of any fact whatsoever simply by following the consequences of what he already knows."[6] Stalnaker [55, 173] regards this objection as clear and decisive. As Hintikka notes, the mere fact that p is true need not imply that one knows that p is consistent with one's knowledge, where '$\neg\Box\neg p$' formalizes the claim that p is consistent with one's knowledge. There is an apparent gap between matters of non-epistemic fact and knowledge about one's epistemic states, so the (B) axiom must go, and along with it (5).

In the context of epistemic logics, (5) is often known as the *negative introspection* axiom, since not knowing something implies that one knows it is not known. It has come under much criticism as well. Hintikka [24, 106] rejects it as implausible, as does Humberstone [25, 187]. The reason is that one might, when p is false, mistakenly take oneself to know p, on the basis of misleading though persuasive evidence. The falsity of p would imply that one does not know p, and by (5), one knows that one does not know p. Given this knowledge, the agent should be able to see that they have made a mistake, even though they take their evidence to be conclusive. As Stalnaker [55, 177] puts it, the (5) principle "require[s] that rational agents be immune to error." The possibility of error is important in epistemology, and plausibly epistemic logics should allow for that possibility.

The (5) axiom is inconsistent with so-called *Rumsfeld ignorance*, that there are unknown unknowns.[7] The inconsistency is immediate, since (5) says that if a claim is unknown, then one knows that it is unknown. It is plausible that the category of unknown unknowns is not incoherent, since it does seem that there are many things that are unknown and we are ignorant of that fact as well. Echoing Hintikka [24, 106], "you may fail—unless you happen to be as sagacious as Socrates—to know your ignorance." The plausibility of the category of unknown unknowns provides additional reason to be suspicious of the (5) principle.

Further objections to the (5) principle arise in epistemic doxastic logics, which combine an operator for belief, $\mathcal{B}$, with the epistemic operator for knowledge, $\Box$. In such logics, one adopts plausible bridge principles relating belief and knowledge, such as $\Box A \to \mathcal{B}A$ and $\mathcal{B}A \to \mathcal{B}\Box A$. Humberstone [28, 373ff.] illustrates the problems that arise from (5) by presenting an argument due to Wolfgang Lenzen. Starting from the assumption that the agent has a false belief, represented as $\neg p \wedge \mathcal{B}p$, one can derive that the agent has contradictory beliefs, namely $\mathcal{B}\Box p \wedge \mathcal{B}\neg\Box p$. Depending

[5] See Humberstone [28, Chap. 5] for an overview of objections in the context of an epistemic logic of knowledge and belief.

[6] Hintikka [24, 54.]

[7] See Humberstone [28, 371], Fine [22], and Fang [18] for more on Rumsfeld ignorance.

on the logic of belief, this result may be disastrous or merely extremely unsavory. Of the principles involved in the derivation, (5) appears to be the least plausible. Given the other problems with (5), it seems to be the natural place to pin the blame.

Williamson [69, 166–167] argues that there are important epistemic asymmetries that arise in knowledge. In certain skeptical scenarios there is a good case, where one isn't being deceived, and a bad case where one is. If one is in the good case, the bad case is not epistemically accessible, while if one is in the bad case, then the good case is epistemically accessible. This sort of consideration provides a reason for rejecting the symmetry condition on S. Rejecting the symmetry of S brings with it a rejection of axioms (B) and (5). I will note that Williamson offers a more general argument, the anti-luminosity argument, against principles of the form $\_ \rightarrow \Box\_$. Such an argument cuts against the (5) axiom, as well as the (4) axiom, but I will not have anything further to say on the luminosity issue.

The epistemic logics discussed so far have all used *classical* logic. There is increasing interest in the use of *non-classical* logics in epistemic logics. There are two ways one might go about using non-classical logics. One is to keep the base logic classical while using a non-classical logic within the scope of the epistemic operators. This approach is adopted by Fagin and Halpern [17], Levesque [29], Ferenz [19], Sedlár and Vigiani [54], and Standefer et al. [64], among others. This approach attempts to combine the features of classical logic in the base language with non-classical logic providing closure conditions on the epistemic concepts. On this approach, within the scope of epistemic modal operators, one is restricted to using the non-classical logic so that the consequences of a formula within the scope of an epistemic modal may differ from the consequences outside the scope of any epistemic modal. The second approach is to use a non-classical base logic instead of classical logic and add an epistemic extension. This approach is adopted by Bílková et al. [8], Bílková et al. [7], Sedlár [51, 52], and Punčochář [41], among others. The approach pursued below falls in this second category. I will use the relevant logic R as the base logic and study some epistemic extensions of it. With that, let us turn to the background on R.

3.3 Relevant Logics

Relevant logics are logics whose implication connective enforces a strong connection between antecedent and consequent.[8] The most famous relevant logic is perhaps the logic R of Anderson and Belnap [1], and that is the logic on which I will focus for this paper.[9] Focusing on R will facilitate comparison with the results of Standefer [61], but most of the results will carry over to other relevant logics as well. Relevant

[8] See Read [42], Dunn and Restall [16], Bimbó [9], Mares [35], or Logan [30] for overviews of relevant logics.

[9] See Mares [34] for an extended defense of R.

logics have been studied axiomatically, algebraically, and frame-theoretically, and in this section I will present axioms for R as well as the ternary relational frames for it.

Definition 5 (*Ternary relational frames*) A basic ternary relational frame F is a quadruple $\langle K, N, R,^* \rangle$, where $K \neq \emptyset$, $N \subseteq K$, $R \subseteq K^3$, $^* : K \mapsto K$, where

(B1) $a \leq b =_{Df} \exists x \in N,\ Rxab$,
(B2) $\leq$ is a partial order,
(B3) $a^{**} = a$,
(B4) $a \leq b$ only if $b^* \leq a^*$, and
(B5) if $d \leq a$, $e \leq b$, $c \leq f$, and $Rabc$, then $Rdef$.

Let $Rabcd$ and $Ra(bc)d$ be defined as

- $Rabcd =_{Df} \exists x \in K(Rabx \text{ and } Rxcd)$, and
- $Ra(bc)d =_{Df} \exists x \in K(Raxd \text{ and } Rbcx)$.

A basic ternary relational frame is an R-*frame*, or a *ternary relational frame*, iff it obeys the following conditions.

(C1) $Rabc \Rightarrow Rac^*b^*$
(C2) $Rabc \Rightarrow Rbac$
(C3) $Rabcd \Rightarrow Rb(ac)d$
(C4) $Rabc \Rightarrow Rabbc$

Separately listing the conditions on the basic frames and on the R-frames can be useful for seeing how seldom the additional features of R-frames are invoked. Since we are only considering R-frames in this paper, we will not refer to any frames as 'basic ternary relational frames'.

Before proceeding, it will be worth commenting on a few aspects of the frames. First, the points in the set K are often viewed as being situations, which are partial and potentially contradictory, as opposed to worlds, which are maximally determinate and consistent. Worlds, as maximal situations, will play no role in the discussion of this paper, although they are used by some relevant logicians.[10] Second, given that understanding of the points, the relation $\leq$ is naturally understood as a kind of mereological containment relation. If $a \leq b$, then the situation b contains the situation a. There is no requirement that there be a greatest situation containing any pair of situations.

The Routley star, *, and its interpretation will play an important role in the epistemic logics, and I will return to it and its interpretation in the next section. The ternary relation, while important in the models, will not play a major role in the philosophical interpretation of the epistemic logics. Therefore, I will not take a stand on its interpretation here.[11]

Next, we define models and the verification relation.

---

[10] See Meyer and Mares [38] for an example.

[11] See Beall et al. [2], Restall [46], Mares et al. [36], and Tedder [65, 66] for discussion of interpretations of the ternary relation. For critical commentary on it, see Brady [12].

Definition 6 (*Model*) A model M is a pair $\langle F, V\rangle$ of an R-frame F, $\langle K, N, R, ^*\rangle$, and a valuation function $V : \mathsf{At} \mapsto \wp K$ such that if $a \in V(p)$ and $a \leq b$, then $b \in V(p)$. Such a model is built on F.

The valuation function is extended to a verification relation on the whole language as follows.

- $a \Vdash p$ iff $a \in V(p)$
- $a \Vdash \neg B$ iff $a^* \nVdash B$
- $a \Vdash B \wedge C$ iff $a \Vdash B$ and $a \Vdash C$
- $a \Vdash B \vee C$ iff $a \Vdash B$ or $a \Vdash C$
- $a \Vdash B \to C$ iff for all $b, c \in K$, if $Rabc$ and $b \Vdash B$, then $c \Vdash C$

Given this definition of the verification relation, we can provide the standard definitions of counterexample and validity.

Definition 7 (*Counterexample, validity*) A model M is a counterexample to a formula A iff for some $a \in N$, $a \nVdash A$. The point a is a counterexample point.

A formula A is valid on a frame F iff no model built on F is a counterexample to A.

A formula A is valid over a class of frames $\mathfrak{C}$ iff A is valid on each frame $F \in \mathfrak{C}$.

Write $\models_{\mathfrak{C}} A$ when A is valid over the class of frames $\mathfrak{C}$ and write $\models_{\mathsf{R}} A$ when A is valid over the class of R-frames.

I will note, without proof, two important lemmas appealed to without comment.

Lemma 1 (Heredity) *For all A in the language, for any model M, if $a \Vdash A$ and $a \leq b$, then $b \Vdash A$*

Lemma 2 (Verification) *The following are equivalent for any model M.*

- *For all $a \in N$, $a \Vdash B \to C$.*
- *For all $b \in K$, if $b \Vdash B$, then $b \Vdash C$.*

Both of these lemmas extend to the language of the next section, once knowledge operators are introduced.

Next, I will provide an axiomatization of R. The logic R is the least set of formulas containing axioms (R1)–(R11) and closed under the rules (R12) and (R13).

(R1) $A \to A$

(R2) $A \wedge B \to A$, $A \wedge B \to B$

(R3) $((A \to B) \wedge (A \to C)) \to (A \to (B \wedge C))$

(R4) $A \to A \vee B$, $B \to A \vee B$

(R5) $((A \to C) \wedge (B \to C)) \to ((A \vee B) \to C)$

(R6) $A \wedge (B \vee C) \to (A \vee B) \wedge (A \vee C)$

(R7) $(A \to \neg B) \to (B \to \neg A)$

(R8) $\neg\neg A \to A$

(R9) $(A \to (A \to B)) \to (A \to B)$

(R10) $A \to ((A \to B) \to B)$

(R11) $(A \to B) \to ((B \to C) \to (A \to C))$

(R12) $A, A \to B \Rightarrow B$

(R13) $A, B \Rightarrow A \wedge B$

Proof is defined in the usual way, as a sequence of formulas each of which is an axiom or follows from earlier members of the sequence by rules. We will write $\vdash_R A$ to mean that A has a proof and so is a theorem of R.

Belnap's variable-sharing property is an important feature of relevant logics, and it is arguably what makes them *relevant* logics.

Definition 8 (*Variable-sharing Property*) A logic L has the *variable-sharing property* iff for all formulas A and B, if $\vdash_{\mathsf{L}} A \to B$, then A and B share a propositional atom.

The variable-sharing property is preserved downward to sublogics, a feature that will become important later. One of the crucial features of R is that it enjoys Belnap's variable-sharing property.

Theorem 1 R *has the variable-sharing property, i.e. for all formulas A and B, if* $\vdash_{\mathsf{R}} A \to B$, *then A and B share a propositional atom.*

Proof See Anderson and Belnap [1, 252–254] or Restall [45, 184-185] for a proof using matrix methods.

The variable-sharing property provides a minimal level of formal relevance for valid implications and is typically taken as a necessary condition on being a relevant logic.[12] If a logic does not enjoy the variable-sharing property, then it is, consequently, not a relevant logic. Classical logic is a paradigm example of a non-relevant logic, with theorems such as $p \to (q \to q)$. The addition of new connectives to a relevant logic may lead to violations of variable-sharing, as is the case with the universal modality considered by Standefer [61].[13] It is, therefore, important for the relevant logician to verify that the addition of a connective to a relevant logic does not lead to violations of variable-sharing. This issue will be taken up again in the next section.

The axiom system above is adequate for the frame semantics of this section in the sense that it is sound and complete.

Theorem 1 *For all formulas A,* $\vdash_R A$ *iff* $\models_{\mathsf{R}} A$.

Proof See Routley et al. [49, Chap. 4] for a proof.

That is enough background on the base logic, R. In the next section, I will define some epistemic extensions of R. Some of these will be defined axiomatically and some will be defined by classes of frames. Both approaches, however, build on the presentations of R in this section.

---

[12] See Standefer [63] for a discussion of variable-sharing as a characterization of relevant logics.

[13] See Standefer [59] for a discussion of the role of variable-sharing in adding new connectives to relevant logics.

3.4 Relevant Epistemic Logics

Having defined the base relevant logic and its frames, we can turn to the epistemic extensions. The epistemic logic of classical partition frames in Sect. 3.2 is S5, and that would be the natural neighborhood to look for epistemic extensions of R. As shown by Standefer [61], there are distinct options for an S5-ish logic extending R, but I will initially focus on two. There is the logic RS5 obtained axiomatically and the logic Eq obtained by expanding the ternary relational frames with an equivalence relation. While the use of equivalence relations is common in epistemic logics over classical logic, they have been studied much less in the context of relevant logics.

To define RS5, we add to R the following axioms and rules, adjusting the definition of proof and theorem accordingly.

(Agg)	$(\Box A \wedge \Box B) \to \Box(A \wedge B)$	(B)	$A \to \Box\neg\Box\neg A$
(K)	$\Box(A \to B) \to (\Box A \to \Box B)$	(5)	$\neg\Box A \to \Box\neg\Box A$
(T)	$\Box A \to A$	(Nec)	$A \Rightarrow \Box A$
(4)	$\Box A \to \Box\Box A$	(Mono)	$A \to B \Rightarrow \Box A \to \Box B$

The theorems of RS5 are close to those of S5, but the former are properly contained in the latter. Apart from the base logic R not being as strong as classical logic, there are properly modal theorems of S5 that are not theorems of RS5. Two of which that will appear later are $\Box(A \vee B) \to (\Box A \vee \neg\Box\neg B)$ and $\Box(\Box A \vee B) \to (\Box A \vee \Box B)$.

To provide frames for the modal logic RS5, we add a binary accessibility relation, S, to ternary relational frames.

Definition 9 (RS5-*frame*) An RS5-frame is a quintuple $\langle K, N, R, S, ^* \rangle$, where $\langle K, N, R, ^* \rangle$ a ternary relational frame and $S \subseteq K^2$, obeying the following conditions.

(S1) If Sbc and $a \leq b$, then Sac.
(S2) If $a \in N$ and Sab, then $b \in N$. (Nec)
(S3) Saa. (T)
(S4) If Sab and Sbc, then Sac. (4)
(S5) If Sab, then Sb^*a^*. (B)
(S6) If $\exists z(Rabz$ and $Szc)$, then $\exists x \exists y(Sax, Sby$, and $Rxyc)$. (K)

The definitions of model, counterexample, holding, and validity are all adapted in a straightforward way. The verification condition for $\Box$ is the following.

- $a \Vdash \Box B$ iff for all b such that Sab, $b \Vdash B$

In the context of RS5-frames, the (B) axiom corresponds to the condition that if Sab then Sb^*a^*. This means that the S relation is not generally an equivalence relation. Given the other conditions, it turns out that there are stringent limits on what sort of equivalence relations are available for RS5-frames, because some equivalence relations invalidate the (Nec) principle, as we shall see.

Generally, the S relation in RS5 models does not represent indiscernibility, as it is not an equivalence relation. It does not have as neat an epistemic interpretation as one finds with the equivalence relations. Punčochář et al. [41] say, in effect, that $\{b \in K : Sab\}$ is the epistemic state of the agent at the situation a, the set $\{b \in K : Sab\}$ is a particular body of information for the agent at that situation. This is a slightly different conception of knowledge, knowledge as what is supported by one's information state, than one finds with the equivalence relations. Despite this difference, RS5 has many pleasant features, and it need not totally undermine the appeal of RS5 as a logic for knowledge. Rather, the RS5-models are modeling a different sort of epistemic scenario than the classical partition models. It is unfortunate to give up on the intuitive picture offered by the use of equivalence relations to represent indistinguishability, so I will turn to the logic of the equivalence conception of knowledge, Eq, to be defined below.

Definition 10 An equivalence frame is a quintuple $\langle K, N, R, \approx, ^* \rangle$, where $\langle K, N, R, ^* \rangle$ is a ternary relational frame and $\approx$ is an equivalence relation on K, obeying the condition that if $a \leq b$ and $b \approx c$, then there is $x \in K$ such that $x \leq c$ and $a \approx x$.

The class of equivalence frames will be denoted $\mathfrak{Eq}$.

Validity is adapted in the obvious way, and we will write $\models_{\mathfrak{Eq}} A$ when A is valid over $\mathfrak{Eq}$.

This definition of equivalence frame differs from that of Standefer [61] in the interaction condition between $\leq$ and $\approx$. The condition of Standefer [61] is that if $a \leq b$ and $b \approx c$, then $a \approx c$. This is the interaction condition used by RS5-frames. The condition adopted in the definition is weaker than the condition for RS5-frames, and it is also more plausible from an epistemic point of view.[14] If two points, b and c are indistinguishable, namely $b \approx c$, and b contains a, $a \leq b$, generally one would not expect a and c to be indistinguishable. Perhaps c contains a lot of things that are entirely absent from a. For example, suppose that situation b involves an apartment, α_1, with two rooms, a bedroom and a kitchen, while situation c involves a distinct but indistinguishable apartment, α_2, with two rooms, a bedroom and a kitchen. Then, situation a will involve just the kitchen of α_1. While b and c are indistinguishable, it does not follow that a and c are, since c contains a bed while a does not. Rather, there is a part of c, which we can call d, that involves only the bedroom of α_2, and, plausibly, a and d are indistinguishable. Therefore, I take the interaction condition to be motivated.

The verification condition for $\Box$, presented using $\approx$, is the following.

- $a \Vdash \Box B$ iff for all b such that $a \approx b$, $b \Vdash B$

With this in hand, we can define the logic Eq.

Definition 11 (*The logic* Eq) The logic Eq is the set of formulas valid on all equivalence frames, i.e. the set $\{A \in \mathscr{L} : \models_{\mathfrak{Eq}} A\}$.

---

[14] We could correspondingly weaken the condition on RS5-frames without changing the logic.

Table 3.1 Equivalence frame F_1

R	0	a	b	*	$\approx$
0	0	a	b	0	$0, b$
a	a	a	$0ab$	b	a
b	b	$0ab$	b	a	$0, b$

As shown by Standefer [61], the following principles are invalid on the class of equivalence frames.

- (K) $\Box(A \to B) \to (\Box A \to \Box B)$
- (B) $A \to \Box\neg\Box\neg A$
- (5) $\neg\Box A \to \Box\neg\Box A$
- (Nec) $A \Rightarrow \Box A$

I will demonstrate the invalidity of (B), (5), and (Nec) here. I will begin by demonstrating that (5) entails (B), given some additional background. While this is familiar from classical S5, it is worth reproducing in this non-classical context.[15]

Lemma 3 *In any logic extending* R *with* (T), (5) *implies* (B).

Proof From (5), we get $\neg\Box\neg\Box\neg A \to \Box\neg A$ via contraposition and double negation elimination. Using (T) and transitivity, we have $\neg\Box\neg\Box\neg A \to \neg A$. Contraposing again gets $A \to \Box\neg\Box\neg A$, as desired.

To show the invalidity of (B), I will use the frame in Table 3.1, where $K = \{0, a, b\}$ and $N = \{0\}$. To read the table, the first column is the first position of the ternary relation, the top row is the second, and the third position is represented by each point in the middle cells. For example, the upper cell containing '$0ab$' should be understood as saying that $Rab0$, $Raba$, and $Rabb$. Similarly, the $\approx$ row for 0 means that $0 \approx 0$ and $0 \approx b$.

Lemma 4 *The frame F_1 without the equivalence relation in Table 3.1 is a ternary relational frame. Further, F_1 with the equivalence relation is an equivalence frame.*

Proof The proof is by inspection of all the conditions, which will be omitted here.

Lemma 5 (B) *is invalid over the class* $\mathfrak{Eq}$.

Proof Let $v(p) = \{a, 0\}$. Then, $a \Vdash \neg p$, so $a \Vdash \Box\neg p$. It then follows that $b \nVdash \neg\Box\neg p$. Therefore, $0 \nVdash \Box\neg\Box\neg p$, As $0 \Vdash p$, this is sufficient to invalidate $p \to \Box\neg\Box\neg p$.

Corollary 1 (5) *is invalid over the class* $\mathfrak{Eq}$.

[15] Since R is the only relevant logic discussed in this paper, the lemma concerns R. Inspection of the proof reveals that it works for much weaker logics. Additionally, the converse entailment, from (B) to (5) holds as well, but it is not needed for the following results.

Table 3.2 Equivalence frame F_2

R	0	a	b	*	$\approx$
0	0	a	b	0	0
a	a	a	$0ab$	b	a
b	b	$0ab$	b	a	b

Proof This follows from the previous lemma and the fact that (T) is valid.

Next we turn to (Nec).

Lemma 6 *The rule* (Nec) *fails for* Eq.

Proof Let M be a ternary relational model where for some point $a \in K, a \nVdash p \to p$. We know there is such a model since $q \to (p \to p)$ is not a theorem of R. Add to this model an equivalence relation $\approx$ where for all $a, b \in K$, $a \approx b$. Therefore, for some $b \in N$, $b \nVdash \Box(p \to p)$, as desired.

While many of the typical modal principles of S5 fail for Eq, some are valid, namely (4), (T), and (Mono).

An important question was left open by Standefer [61]: Does Eq enjoy the variable-sharing property? This is an important question because being a proper relevant logic requires an affirmative answer. I can provide an affirmative answer to the former, namely that for all A and B, if $A \to B$ is a theorem of Eq, then A and B share an atom. To prove this, I will use the frame in Table 3.2, where $K = \{0, a, b\}$ and $N = \{0\}$.

Lemma 7 *The frame F_2 without the equivalence relation in Table 3.2 is a ternary relational frame. Further, F_2 with the equivalence relation is an equivalence frame.*

Proof The underlying ternary relational frame is the same as the frame in Table 3.2.

Lemma 8 *Let A and B be formulas with no variables in common. Over the frame F_2, define a model as follows. For all atoms p occurring in A, let $V(p) = \{a\}$. For all atoms q occurring in B, let $V(q) = \{b\}$. The following are true.*

(i) For all formulas C only containing atoms occurring in A, $a \Vdash C$.
(ii) For all formulas C only containing atoms occurring in A, $b \nVdash C$.
(iii) For all formulas D only containing atoms occurring in B, $b \Vdash D$.
(iv) For all formulas D only containing atoms occurring in B, $a \nVdash D$.

Proof For this proof, C will be used for formulas only containing atoms occurring in A and D will be used for formulas only containing atoms occurring in B. The proof is by simultaneous induction on C and D. For atoms, the four cases are immediate.

The cases for conjunction and disjunction are immediate by the inductive hypothesis.

Suppose C is $\neg E$. By the inductive hypothesis, $b \not\Vdash E$, so $a \Vdash \neg E$. By the inductive hypothesis again, $a \Vdash E$, so $b \not\Vdash \neg E$. The cases for D are similar.

Suppose C is $E \rightarrow F$. Suppose that $a \not\Vdash E \rightarrow F$, so for some $x, y \in K$, $Raxy$, $x \Vdash E$ and $y \not\Vdash F$. It follows from the inductive hypothesis that $x \in \{a, 0\}$. If $x = 0$, then $y = a$. In that case, by the inductive hypothesis, $y \Vdash F$, which is a contradiction. If $x = a$, then $y = a$, so there is a contradiction by the inductive hypothesis.

Next, suppose that $b \Vdash E \rightarrow F$. By the inductive hypothesis, $a \Vdash E$. Since $Rbab$, $b \Vdash F$. By the inductive hypothesis, however, $b \not\Vdash F$, which is a contradiction. Therefore, $b \not\Vdash E \rightarrow F$.

The remaining conditional cases are handled similarly.

The cases where C is $\Box E$ are immediate from the inductive hypothesis and the fact that each equivalence class is a singleton.

Lemma 9 *If A and B share no atoms, then $A \rightarrow B$ has a counterexample on F_2.*

Proof Over the frame F_2, let V be a valuation such that

- for all atoms p occurring in A, $V(p) = \{a\}$, and
- for all atoms q occurring in B, $V(q) = \{b\}$.

By Lemma 8, $a \Vdash A$ but $a \not\Vdash B$. Therefore $A \rightarrow B$ does not hold in this model.

With those pieces in hand, I can prove the variable-sharing theorem.[16]

Theorem 2 (Variable-sharing for Eq) *Suppose that $A \rightarrow B$ is a theorem of* Eq. *Then $A \rightarrow B$ share an atomic formula.*

Proof Suppose A and B do not share a variable. By Lemma 9, it has a counterexample on an equivalence frame. Therefore, $A \rightarrow B$ is not a theorem of Eq.

Theorem 2 settles one of the questions left open by Standefer [61]. The logic Eq enjoys variable-sharing and is a reasonable candidate for an epistemic relevant logic. The question of an axiomatization is still left open, but an axiomatization is not needed to demonstrate variable-sharing using the technique above.

The logic Eq has some nice features that make it appealing as an epistemic logic. First, its frames use equivalence relations, so they appeal to the same sort of intuitions about indiscernibility that we find with the classical partition frames. They have a clear intuitive interpretation, as opposed to the situation with the accessibility relation S in the RS5 frames. The intuitions behind the equivalence conception of knowledge are worth preserving, at least as a starting point, since they are so robust and so useful, as demonstrate by the logical work exploring extensions of it.

Second, many of the criticisms of S5 as an epistemic logic focus on the (B) and (5) axioms. Even in very weak relevant logics, these axioms are interderivable, given

[16] It is worth noting that Brady et al. [10, 100ff.] demonstrate the correspondence between the frame F_2 Belnap's 8-valued matrix M_0 used by Anderson and Belnap [1] to show that R and E enjoy variable-sharing. The idea to use this result was due to an observation of John Slaney, as communicated to me by Shay Logan.

the other modal principles, so they stand or fall together here as well. Neither of these axioms is valid over Eq. Since relevant logics are paraconsistent, one could have unknown unknowns along with the (5) axiom without the logic collapsing into triviality. That would not be ideal, since the concept of unknown unknowns is not on its face incoherent. The logic Eq allows there to be unknown unknowns consistently. Similarly, the mere truth of a claim does not on its own imply that an agent knows the truth of that claim is consistent with their knowledge. The logic Eq does not build in unrestricted negative introspection in the way that S5 and RS5 do. Thus, the equivalence frames, and the logic they generate, avoid some of the major criticisms leveled against their classical counterparts.

The preceding considerations prompt the question of what sort of epistemic agent Eq is supposed to model. The agents of Eq are highly idealized, since they have no limits on memory or computational resources. They have unlimited positive introspection, as the (4) axiom is valid. Their powers of introspection are, nonetheless, limited in certain ways. The failure of the (5) axiom means that they are unaware of their own ignorance, and one can consistently model situations in which they do not know what they do not know. Similarly, the failure of the (B) axiom means that they are limited in what consequences they draw from their own knowledge. The fact that p is true need not imply that an agent know that p is consistent with their knowledge, since they may be mistaken and their knowledge is already inconsistent. The agents modeled by the equivalence frames of Eq are in some respects highly idealized, but in other respects they are rather limited. This combination is, I think, potentially fruitful for further exploration of epistemic concepts.

The source of the failure of the (B) and (5) axioms is due to the Routley star, the part of the frame that interprets negation.[17] While we may have two points, a and b, in the same equivalence class, there is no guarantee that their stars, a^* and b^*, will be in the same class. The star does not respect the equivalence classes.[18] If, following

[17] Rather than using the Routley star, one could use a binary compatibility relation to interpret negation in the ternary relational models, following Dunn [14] and Restall [44], among others. Compatibility generalizes the Routley star, and many of the same points made below will carry over. Therefore, I will not further consider the use of compatibility relations.

One might also follow Dunn [15] in wanting to consider what happens when one has a primitive possibility operator, $\Diamond$, not defined in terms of negation and necessity. The short answer is that it will depend on whether $\Box$ and $\Diamond$ are interpreted using distinct accessibility relations or not. If they are, then the resulting logic will be similar to that of the coordinated equivalence frames defined below. If they are not, which is the standard default option in this area, then the logic will be similar to that of the equivalence frames, with the same axioms and being invalidated. I thank Hitoshi Omori for raising this question.

[18] This issue is arising because I am following the Australian plan for negation in relevant logics, as defended by Berto [3] and Berto and Restall [6]. An alternative is the American plan, recently defended by De and Omori [13]. On the American plan, negation is not interpreted using the star or a compatibility relation. Instead, one has a verification relation, $\Vdash^+$ and a falsification relation, $\Vdash^-$. One then has the conditions

- $a \Vdash^+ \neg B$ iff $a \Vdash^- B$, and
- $a \Vdash^- \neg B$ iff $a \Vdash^+ B$.

Restall [44], we think of b^* as the point that is maximally compatible with b, it is plausible that one may be unable to distinguish situations a and b while being able to distinguish a^* and b^*. While one's information does not permit one to distinguish a and b, the information that the agent is lacking may obscure important differences in the information of those two points. The maximal point compatible with a may then be distinguishable from the maximal point compatible with b, which is to say that $a \in [b]$ need not imply $a^* \in [b^*]$.

While it is plausible to suppose that the star need not respect equivalence relations, from a logical point of view it is worth considering the frames that result from the assumption that the star does respect equivalence classes. This would be the assumption that when two situations are indiscernible to an agent, the situations maximally compatible with them are also indiscernible. For this assumption to be plausible, agents would need more awareness of the information omitted from a situation, since that is salient to determining what maximally compatible situations. Intuitively, this would be an increased idealization of the agents' powers of negative introspection, but in order to make good on this idea, let us define the frames at issue.

Definition 12 (*Coordinated equivalence frames*) An equivalence frame $\langle K, N, R, \approx, ^* \rangle$ is a *coordinated equivalence frame* iff for all $a, b \in K$, $[a] = [b]$ implies $[a^*] = [b^*]$.

Let $\mathfrak{CEq}$ be the class of all coordinated equivalence frames.

We will write $\models_{\mathfrak{CEq}} A$ when A is valid over $\mathfrak{CEq}$.

Definition 13 (*The logic* CEq) Let CEq the set of all formulas valid over $\mathfrak{CEq}$, i.e. the set $\{A \in \mathscr{L} : \models_{\mathfrak{CEq}} A\}$.

Clearly, Eq $\subseteq$ CEq. In fact, this containment is proper, as the logic of coordinated equivalence frames is much stronger than that of equivalence frames.

Theorem 3 *The following are valid over coordinated equivalence frames*

(i) (B) $A \to \Box\neg\Box\neg A$
(ii) (5) $\neg\Box A \to \Box\neg\Box A$
(iii) (MC) $\Box(A \vee B) \to (\Box A \vee \neg\Box\neg B)$

Proof We will prove that (5) is valid, leaving the rest to the reader. Suppose that $a \Vdash \neg\Box A$ but $a \nVdash \Box\neg\Box A$. Then for some $b \in [a]$, $b \nVdash \neg\Box A$. Therefore, $b^* \Vdash \Box A$, so for all $c \in [b^*]$, $c \Vdash A$. As $a \Vdash \neg\Box A$, it follows that $a^* \nVdash \Box A$. Since $b \in [a]$, it follows that $[b^*] = [a^*]$. With $a^* \nVdash \Box A$, this implies that for some $c \in [b^*]$, $c \nVdash A$, which is a contradiction. Therefore, (5) is valid.

Corollary 2 Eq *is properly contained in* CEq.

If one adopts this approach, no analog of the logic Eq will arise, as (B) and (5) will be valid. This suggests that further consideration of both the Australian and American plans in the context of modal extensions of relevant logics would be worthwhile. I thank Hitoshi Omori for discussion of this point.

There is still some distance from the RS5-frames with which we started, but we can make a connection to another class of frames, relatives of some frames studied by Mares [33].

Definition 14 (*Mares-Meyer modal frame*) A Mares-Meyer modal frame is a quintuple $\langle K, N, R, S, ^* \rangle$, where $\langle K, N, R, ^* \rangle$ a ternary relational frame and $S \subseteq K^2$, obeying the following conditions.

(MM1) If Sbc and $a \leq b$, then Sac.
(MM2) If Sab, then $\exists x(x \leq b$, Sax, and $Sa^*x^*)$
(MM3) Saa.
(MM4) If Sab and Sbc, then Sac.
(MM4) If Sab, then Sb^*a^*.

We can define a new relation, T, on a frame as follows.

- Tab iff both Sab and Sa^*b^*.

Using this definition, (MM2) can be written as if Sab, then $\exists x(x \leq b$ and $Tax)$. These frames are related to another class of frames that will be of interest to us, Mares's mostly Meyer modal models.

Definition 15 (MMR-*frames*) An MMR-frame is a quintuple $\langle K, N, R, S, ^* \rangle$, where $\langle K, N, R, ^* \rangle$ an R-frame and $S \subseteq K^2$, obeying the following conditions.

(M1) If Sbc and $a \leq b$, then $\exists x(Sax$ and $x \leq c)$.
(M2) If Sab, then Sa^*b^*
(M3) Saa.
(M4) If Sab and Sbc, then Sac.
(M5) If Sab, then Sb^*a^*.

These are of interest for the following reason.

Theorem 4 MMR-*frames are coordinated equivalence frames.*

Proof First we show that S obeys symmetry. Suppose Sab. By (M2), Sa^*b^*. By (MM5), $Sb^{**}a^{**}$. As $c^{**} = c$, for all $c \in K$, Sba. Since S obeys reflexivity and transitivity, it follows that S is an equivalence relation.

Next, we show that if $[a] = [b]$, then $[a^*] = [b^*]$. Suppose $[a] = [b]$, so Sab. By (MM5), Sb^*a^*, so $[a^*] = [b^*]$.

Using the work of Mares [33], we have the following theorem.

Theorem 5 *Given a Mares-Meyer modal frame* $\langle K, N, R, S, ^* \rangle$, $\langle K, N, R, T, ^* \rangle$ *is a* MMR-*frame.*

Proof The proof follows that of section 7 of Mares [33]. The results extend to the frames obeying postulates (M3)–(M5), as noted by Mares.

Given a Mares-Meyer modal frame $\langle K, N, R, S, ^* \rangle$, one can construct an associated coordinated equivalence frame, $\langle K, N, R, T, ^* \rangle$ using the previous theorem. The addition of a valuation results in them agreeing pointwise on formulas.

Theorem 6 *Let $\langle K, N, R, S,^* \rangle$ be a Mares-Meyer modal frame and $\langle K, N, R, T,^* \rangle$ the constructed coordinated equivalence frame of Theorem 5. Let M be the model obtained by adding a valuation V to the Mares-Meyer modal frame and let N be the model obtained by adding V to the constructed coordinated equivalence frame. Then, for all $a \in K$ and all formulas A, in M, $a \Vdash A$ iff in N, $a \Vdash A$.*

Proof The proof is by induction on the construction of A. See Mares [33] lemma 7.3 for details.

Let HCEq be the logic R+{ (Agg), (Mono), (T), (4), (B), (5), (MC)}. The logic HCEq is sound with respect to the class of Mares-Meyer modal frames.

Theorem 7 *If A is a theorem of* HCEq, *then A is valid over the class of Mares-Meyer modal frame.*

Proof The proof is by induction on the construction of proof in HCEq. Condition (MM2) yields the validity of (MC). The rest are straightforward.

Corollary 3 *If A is a theorem of* HCEq, *then A is a theorem of* CEq.

Finally, we come to Completeness.

Theorem 8 *If A is valid over the class of Mares-Meyer modal frames, then A is a theorem of* HCEq.

Proof The proof is by a standard canonical model construction. See Mares [31, 32], Restall [45, Chap. 11], or Standefer [57] for examples.

Corollary 4 *If A is a theorem of* CEq, *then A is a theorem of* HCEq.

Proof Given a non-theorem A of HCEq, we can construct a counterexample Mares-Meyer modal model. Using Theorems 5 and 6, we can construct an associated coordinated equivalence frame that also serves as a counterexample.

Corollary 5 HCEq=CEq.

Given the axioms of HCEq, we can adapt the proof of variable-sharing due to Belnap to show that CEq has the variable-sharing property.[19]

Theorem 9 (Variable-sharing for CEq) CEq *enjoys the variable-sharing property.*

Proof We will use a matrix-based argument. The matrix M_0 is presented in Table 3.3.

In this matrix, the set of designated values are the ones marked with $+$. A valuation v assigns atoms values and the values of complex formulas are computed using the tables. For the modals, we will use the condition $v(\Box A) = v(A)$. The validity of a formula on this matrix is defined as having a designated value on all valuations. All

[19] See Anderson and Belnap [1, 252–254]. This proof was similarly adapted for use with a relevant modal logic by Standefer [61]. Robles and Méndez [47, 48] for a general characterization of matrices that can be used to demonstrate variable-sharing.

Table 3.3 Matrix M_0 for variable-sharing

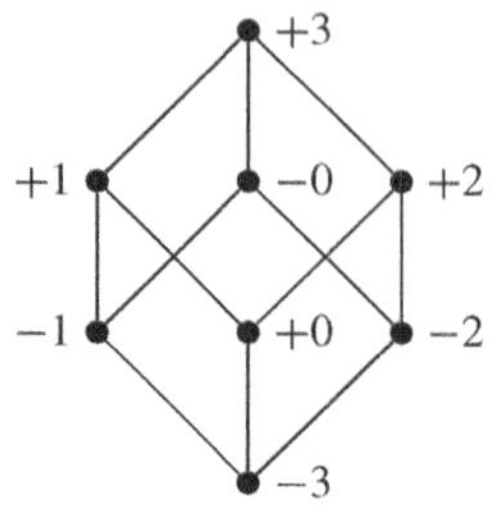

$\to$	−3	−2	−1	−0	+0	+1	+2	+3	$\neg$
−3	+3	+3	+3	+3	+3	+3	+3	+3	+3
−2	−3	+2	−3	+2	−3	−3	+2	+3	+2
−1	−3	−3	+1	+1	−3	+1	−3	+3	+1
−0	−3	−3	−3	+0	−3	−3	−3	+3	+0
+0	−3	−2	−1	−0	+0	+1	+2	+3	−0
+1	−3	−3	−1	−1	−3	+1	−3	+3	−1
+2	−3	−2	−3	−2	−3	−3	+2	+3	−2
+3	−3	−3	−3	−3	−3	−3	−3	+3	−3

the axioms of R are valid and the rules preserve validity. Further, the modal axioms of CEq are valid and the rule also preserves validity, therefore CEq is sound for this matrix.

The sets $\{+1, -1\}$ and $\{+2, -2\}$ are both closed under all the operations corresponding to the connectives. Therefore, by induction, if for all atoms p in a formula A, $v(p) \in \{+1, -1\}$, then $v(A) \in \{+1, -1\}$, and similarly for $\{+2, -2\}$. If $v(A) \in \{+1, -1\}$ and $v(B) \in \{+2, -2\}$, then $v(A \to B)$ is not designated. If A and B are assumed to have no atoms in common, then we can construct an assignment v such that $v(A) \in \{+1, -1\}$ and $v(B) \in \{+2, -2\}$, which then suffices for $v(A \to B)$ not being designated. Therefore, $A \to B$ is not a theorem of CEq.

As a corollary, this gives an alternative proof that Eq has the variable-sharing property.

Corollary 6 Eq *enjoys the variable-sharing property.*

Proof CEq enjoys the variable-sharing property. As Eq is a sublogic of CEq and the variable-sharing property is preserved to sublogics, Eq enjoys it as well.

The introduction of Mares-Meyer modal frames and MMR-frames is in part a technical device to permit an axiomatization and proof of Completeness for CEq. Nonetheless, we can make a few comments on the Mares-Meyer modal frames.

The accessibility relation of the Mares-Meyer modal frames gives a partial description of indistinguishability, in the sense that one can define an equivalence relation from it. The description is partial in the sense that Sab on its own does not say that the agent cannot distinguish a and b, but instead requires the addition of Sa^*b^* to conclude that the agent cannot distinguish a and b.

Next, we will comment on the relation between the class of Mares-Meyer modal frames and the class of RS5-frames. Both are subclasses of a more general class, namely the frames obeying the postulates.

- If Sbc and $a \leq b$, then Sac.
- Saa.
- If Sab and Sbc, then Sac.
- If Sab, then Sb^*a^*.

These are frames for the logic R.T4B, which includes (Mono) and (Agg). The RS5-frames add additional postulates to ensure the validity of (K) and (Nec), while the Mares-Meyer modal frame add a postulate for the validity of (MC). One could add (MC) to RS5, adding the corresponding postulate (MM2) to the RS5-frames to maintain Soundness and Completeness. Similarly, one could add postulates for (K) and (Nec) to the Mares-Meyer modal frames, which would transfer to the coordinated equivalence frames, by a result of Mares [33]. I have not added the extra conditions, because they would greatly restrict the equivalence relations available on the coordinated equivalence frames, which would undermine the intuitive interpretation of the equivalence relations as encoding indistinguishability.

Despite these connections, one should not think that Eq and CEq are sublogics of RS5. There is a formula that is a theorem of Eq that is not a theorem of RS5, $\Box(\Box A \vee B) \to (\Box A \vee \Box B)$.[20]

Theorem 10 *The formula* $\Box(\Box A \vee B) \to (\Box A \vee \Box B)$ *is a theorem of* Eq.

Proof Let M be an equivalence model. Suppose $a \Vdash \Box(\Box A \vee B)$. Suppose that $a \nVdash \Box A \vee \Box B$. It follows that $a \nVdash \Box A$ and $a \nVdash \Box B$. It follows that there are $b, c \in [a]$ such that $b \nVdash A$ and $c \nVdash B$. From the initial assumption, it follows that for all $d \in [a]$, $d \Vdash \Box A \vee B$, so $c \Vdash \Box A \vee B$. This implies $c \Vdash \Box A$, which in turn implies that for all $e \in [c]$, $e \Vdash A$. As $[c] = [a]$, it follows that $b \Vdash A$, which is a contradiction. Therefore, $a \Vdash \Box A \vee \Box B$, as desired.

There are instances of the scheme $\Box(\Box A \vee B) \to (\Box A \vee \Box B)$ that are not theorems of RS5.

Theorem 11 $\Box(\Box p \vee q) \to (\Box p \vee \Box q)$ *is not a theorem of* RS5.

Proof The matrix in Table 3.4 was found by MaGIC and provides a counterexample.[21] The set of designated elements is $\{2, 3, 4, 5\}$. A counterexample is any valuation that assigns a formula a value outside of the set of designated values.

To obtain the counterexample, we use a valuation v assigning $v(p) = 2$ and $v(q) = 1$. It follows that $v(\Box(\Box p \vee q)) = 3$ and $v(\Box p \vee \Box q) = 2$, which suffices for $v(\Box(\Box p \vee q) \to (\Box p \vee \Box q)) = 0$. As all the axioms of RS5 are designated on this matrix and the rules preserve designation, it follows that $\Box(\Box p \vee q) \to (\Box p \vee \Box q)$ is not a theorem.

There are theorems of RS5 that are not theorems of Eq, an example of which we can identify with an application of the rule (Nec).

[20] This formula is discussed by Ono [40] as axiom A_3 in the context of intuitionistic modal logics.

[21] John Slaney's program MaGIC can be found at http://users.cecs.anu.edu.au/~jks/magic.html.

Table 3.4 Six-element MaGIC counterexample

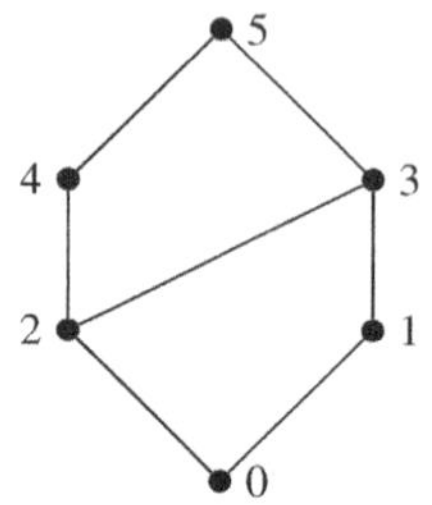

$\to$	0	1	2	3	4	5	$\neg$	$\Box$
0	5	5	5	5	5	5	5	0
1	0	2	0	4	0	5	4	0
2	0	1	2	3	4	5	3	2
3	0	0	0	2	0	5	2	3
4	0	0	0	1	2	5	1	2
5	0	0	0	0	0	5	0	5

Lemma 10 *The formula* $\Box(p \to p)$ *is not a theorem of* CEq.

Proof This is proved by noting that the frame used in the proof of Lemma 6 is, in fact, a coordinated equivalence frame.

Corollary 7 *The logic* RS5 *is incomparable with* Eq *and with* CEq.

While RS5 and the equivalence logics are incomparable, it is worth noting a logic that extends them all. Let RS5MC be the logic RS5 extended with (MC).

Theorem 12 *The logic* RS5MC *properly extends* CEq.

Proof All the axioms of HCEq are axioms of RS5MC, and they share the same rules. The containment is proper because the rule (Nec) permits the derivation of $\Box(p \to p)$, which is not a theorem of CEq by Lemma 10.

The relationships between four modal logics discussed in this paper are summarized in Table 3.5. Note that a complete axiomatization of Eq is not yet known, so the axioms and rules listed here should not be taken as exhaustive. Let us turn to some concluding discussion of these epistemic logics.

3.5 Discussion

So far, I have introduced three main epistemic relevant logics, RS5, Eq, and CEq, with a fourth logic, RS5MC, introduced as a logic extending all of them. Both equivalence frames and coordinated equivalence frames utilize the indistinguishability approach to knowledge. The coordinated equivalence frames validate some of the more controversial principles, (B) and (5). These are not valid over the class equivalence frames.

Table 3.5 Modal principles of the different logics

Name	Modal principle	Eq	CEq	RS5	RS5MC
(Agg)	$(\Box A \wedge \Box B) \rightarrow \Box(A \wedge B)$	✓	✓	✓	✓
(K)	$\Box(A \rightarrow B) \rightarrow (\Box A \rightarrow \Box B)$			✓	✓
(T)	$\Box A \rightarrow A$	✓	✓	✓	✓
(4)	$\Box A \rightarrow \Box\Box A$	✓	✓	✓	✓
(B)	$A \rightarrow \Box\neg\Box\neg A$		✓	✓	✓
(5)	$\neg\Box A \rightarrow \Box\neg\Box A$		✓	✓	✓
(MC)	$\Box(A \vee B) \rightarrow (\Box A \vee \neg\Box\neg B)$		✓		✓
	$\Box(\Box A \vee B) \rightarrow (\Box A \vee \Box B)$	✓	✓		✓
(Nec)	$A \Rightarrow \Box A$			✓	✓
(Mono)	$A \rightarrow B \Rightarrow \Box A \rightarrow \Box B$	✓	✓	✓	✓

From the point of view of the objections to S5 raised in Sect. 3.2, Eq appears to be the clear winner, avoiding (B) and (5), as well as (Nec). None of the logics canvassed avoid the Williamsonian objections to positive introspection, (4).[22] Indeed, it appears difficult to avoid (4) while using frames with equivalence relations. If one is less moved by the objections to negative introspection, perhaps the more S5-like logics, such as CEq or RS5MC would be more appealing.

While in some ways similar to S5, these epistemic logics offer distinctive responses to issues of logical omniscience. Logical omniscience comprises a cluster of issues regarding ways in which an agent's knowledge is closed under various logical operations. The issues do not all stand or fall together in the current setting.

Both the classes of equivalence frames and of coordinated equivalence frames invalidate (K) and (Nec). As a consequence, the sorts of agents they model do not automatically know all logical truths. Further, knowledge is not closed under implications in one sense, namely the sense of the (K) axiom $\Box(A \rightarrow B) \rightarrow (\Box A \rightarrow \Box B)$. There are related senses in which knowledge is closed, such as $(\Box(A \rightarrow B) \wedge \Box A) \rightarrow \Box B$, which is valid.[23] Since (Mono) is built in to this approach to epistemic relevant logics, knowledge will be closed under *logically valid* implications, although

[22] It is worth noting that one can, potentially, hold on to (Nec) while avoiding the Williamsonian objections. The reason is that in the present setting, the validity of $\Box A$ does not generally imply $A \rightarrow \Box A$. There are plausibly other reasons to avoid (Nec), discussed in this section, but it may be worth investigating how strong a logic with (Nec) can be while avoiding the Williamsonian objection. I would like to thank Lloyd Humberstone for suggesting this idea.

[23] Most of the features of the epistemic logics discussed in this paper do not depend on the choice of R as the base logic. In fact, the results carry over to all the other standard relevant logics. One exception is the validity of $(\Box(A \rightarrow B) \wedge \Box A) \rightarrow \Box B$, which does depend on the choice of R. If this seems like too much of a cost, it can be avoided by working with a weaker base logic.

not *merely assumed* implications. As a consequence of closure under (Mono), if an agent knows anything, they will know claims of arbitrarily high complexity, measured in terms of the number of logical connectives. Further, agents will know things involving arbitrarily many propositional variables. Thus, there is no sort of limitations in place on the basis of concepts or subject matters involved.[24]

Even though knowledge in these epistemic logics is closed under valid implications, this does not mean that agents know all logical truths. In fact, they can know some logical truths without knowing all of them. Further, they can know some classical tautologies without knowing all classical tautologies. Knowledge being closed under valid implications does not have as large negative effects as the knowledge of S5 being closed under valid implications. Even when knowledge is closed under valid relevant implications, there are still many logical truths that can remain unknown to the agents. In fact, agents can remain wholly ignorant of logical truths.

Adapting definitions from Williamson [70], one can understand a formula context C to be *hyperintensional* iff there are formulas A and B for which the implication $\Box(A \leftrightarrow B) \rightarrow \Box(C(A) \leftrightarrow C(B))$ is not valid, where $A \leftrightarrow B$ is defined as $(A \rightarrow B) \wedge (B \rightarrow A)$. Perhaps surprisingly, knowledge operators in all the epistemic relevant logics under consideration in the previous section exhibit a degree a hyperintensionality, which follows from the results of Standefer [60].[25], This means that equivalences that are assumed known will not always permit substitution into knowledge contexts. As a consequence, agents can know certain equivalences but this need not imply that their knowledge will be closed under substitution using those known equivalences. The logics are *congruential*, meaning they are closed under the rule (Cong), $A \leftrightarrow B \Rightarrow C(A) \leftrightarrow C(B)$. They are, therefore, not hyperintensional in the preferred sense of Odintsov and Wansing [39].

To summarize, there are some ways in which knowledge in relevant epistemic logics runs into issues of logical omniscience and there are some ways in which it avoids issues of logical omniscience. To be clear, knowledge and knowers in all the epistemic logics under consideration is highly idealized. That idealization will bring with it some amount of logical omniscience, but it seems less bad than in the classical setting, because logical implication and equivalence are harder to come by in the relevant logical setting than in the classical one. One can, of course, combine these epistemic logics with modal operators interpreted using other techniques, such as justifications or other modal operators.[26] These additional techniques for interpreting modal operators would provide further alternative responses to issues of logical omniscience.

---

[24] See Hawke et al. [23] for discussion of subject matter and knowledge, and see Ferguson [21] for an overview of conceptivist logics.

[25] See Berto and Nolan [5] and Berto and Jago [4] for more on hyperintensionality, and see Sedlár [53] for a general framework for hyperintensional logics.

[26] For relevant logics of justifications, see Savić and Studer [50] and Standefer [56, 58, 62]. For other modal operators interpreted using neighborhoods, see, for example, Standefer [56] or Ferenz and Tedder [20].

Importantly, since equivalence classes are used in the frames for Eq and CEq, one can make more or less direct contact with the extant work on common knowledge in classical partition frames, once one makes the straightforward extension to multi-agent systems where each agent is represented by their own equivalence relation which is used to interpret a distinct knowledge operator. There has been recent, pioneering work on common knowledge in relevant epistemic logics by Punčochář et al. [41], but that work does not use equivalence relations for its accessibility relation, so it arguably has a different conception of knowledge. Given the importance of common knowledge in formal epistemology and epistemic logic, further investigation of this topic seems promising.

To end, I will highlight a lingering open question.

Open Can the logic Eq be axiomatized in the vocabulary $\{\rightarrow, \neg, \wedge, \vee, \Box\}$? If so, what axioms need to be added? If not, how could one show that?

It appears that one can axiomatize the logic if one adds Boolean negation ($-$) to the language, with the verification clause

- $a \Vdash -B$ iff $a \nVdash B$,

but there are philosophical reasons not to do that. Among other things, the variable-sharing property is violated when Boolean negation is added. Further, the attractive features of the logic Eq fall away in the presence of Boolean negation, e.g. (B) becomes valid when it is formulated using $-$ rather than $\neg$, namely as $A \rightarrow \Box - \Box - A$ rather than $A \rightarrow \Box\neg\Box\neg A$. For these reasons, an axiomatization that does not rely on Boolean negation would be preferable.

Acknowledgements I would like to thank Greg Restall, Rohan French, Lloyd Humberstone, S.M. Love, Peter Hawke, Hitoshi Omori, NIkolaj Pedersen, Lavinia Picollo, Daniel Waxman, Hiroakira Ono, Katsuhiko Sano, Patrick Blackburn, Hanti Lin, anonymous referees, and the audiences at the Sixth Asian Workshop on Philosophical Logic and the Fifth Taiwan Philosophical Logic Colloquium for feedback on this work. This research was supported by the National Science and Technology Council of Taiwan grant 111-2410-H-002-006-MY3.

References

1. Anderson, A.R., Belnap, N.D.: Entailment: The Logic of Relevance and Necessity, Vol. I. Princeton University Press (1975)
2. Beall, J., Brady, R., Dunn, J.M., Hazen, A.P., Mares, E., Meyer, R.K., Priest, G., Restall, G., Ripley, D., Slaney, J., Sylvan, R.: On the ternary relation and conditionality. J. Philos. Log. **41**(3), 595–612 (2012)
3. Berto, F.: A modality called 'negation'. Mind **124**(495), 761–793 (2015)
4. Berto, F., Jago, M.: Impossible Worlds. Oxford University Press (2019)
5. Berto, F., Nolan, D.: Hyperintensionality. In: Zalta, E.N., (ed.), The Stanford Encyclopedia of Philosophy. Metaphysics Research Lab. Stanford University. Springer (2021)
6. Berto, F., Restall, G.: Negation on the Australian plan. J. Philos. Log. **48**(6), 1119–1144 (2019)
7. Bílková, M., Majer, O., Peliš, M.: Epistemic logics for sceptical agents. J. Log. Comput. **26**(6), 1815–1841 (2016)
8. Bílková, M., Majer, O., Peliš, M., Restall, G.: Relevant agents. In: Beklemishev, L., Goranko, V., Shehtman, V. (eds.) Advances in Modal Logic, vol. 8, pp. 22–38. College Publications (2010)
9. Bimbó, K.: Relevance logics. In: Jacquette, D. (ed.) Philosophy of Logic, volume 5 of Handbook of the Philosophy of Science, pp. 723–789. Elsevier (2007)
10. Brady, R., (ed.): Relevant Logics and Their Rivals, Volume II, A Continuation of the Work of Richard Sylvan, Robert Meyer, Val Plumwood and Ross Brady. Ashgate (2003)
11. Brady, R.T.: Rules in relevant logic - I: Semantic classification. J. Philos. Log. **23**(2), 111–137 (1994)
12. Brady, R.T.: Some concerns regarding ternary-relation semantics and truth-theoretic semantics in general. IfCoLog J. Logics Their Appl. **4**(3), 755–781 (2017)
13. De, M., Omori, H.: There is more to negation than modality. J. Philos. Log. **47**(2), 281–299 (2018)
14. Dunn, J.M.: Star and perp: two treatments of negation. Philos. Perspect. **7**, 331–357 (1993)
15. Dunn, J.M.: Positive modal logic. Stud. Logica. **55**(2), 301–317 (1995)
16. Dunn, J.M., Restall, G.: Relevance logic. In: Gabbay, D.M., Guenthner, F., (eds.), Handbook of Philosophical Logic, vol. 6, 2nd edn., pp. 1–136. Kluwer (2002)
17. Fagin, R., Halpern, J.Y.: Belief, awareness, and limited reasoning. Artif. Intell. **34**(1), 39–76 (1987)
18. Fan, J.: Axiomatizing Rumsfeld ignorance. J. Philos. Logic 1–19 (2023). Forthcoming
19. Ferenz, N.: First-order relevant reasoners in classical worlds. Rev. Symb. Logic 1–26 (2023). Forthcoming
20. Ferenz, N., Tedder, A.: Neighbourhood semantics for modal relevant logics. J. Philos, Logic (2022)
21. Ferguson, T.M.: Meaning and Proscription in Formal Logic: Variations on the Propositional Logic of William T. Parry. Springer, Cham (2017)
22. Fine, K.: Ignorance of ignorance. Synthese **195**(9), 4031–4045 (2018)
23. Hawke, P., Özgün, A., Berto, F.: The fundamental problem of logical omniscience. J. Philos. Log. **49**(4), 727–766 (2020)
24. Hintikka, J.: Knowledge and Belief: An Introduction to the Logic of the Two Notions. Cornell University Press, Ithaca (1962)
25. Humberstone, L.: Some epistemic capacities. Dialectica **42**(3), 183–200 (1988)
26. Humberstone, L.: Smiley's distinction between rules of inference and rules of proof. In: Smiley, T.J., Lear, J., Oliver, A. (eds.), The Force of Argument: Essays in Honor of Timothy Smiley, pp. 107–126. Routledge (2010)
27. Humberstone, L.: The Connectives. MIT Press (2011)
28. Humberstone, L.: Philosophical Applications of Modal Logic. College Publications, London (2016)
29. Levesque, H.J., Lakemeyer, G.: The Logic of Knowledge Bases. MIT Press (2001)

30. Logan, S.A.: Relevance logic. In: Elements in Philosophy and Logic. Cambridge Univeristy Press (2024)
31. Mares, E.D.: The semantic completeness of RK. Rep. Math. Logic 3–10 (1992)
32. Mares, E.D.: Classically complete modal relevant logics. Math. Log. Q. **39**(1), 165–177 (1993)
33. Mares, E.D.: Mostly Meyer modal models. Logique et Anal. (N.S.) **146**, 119–128 (1994)
34. Mares, E.D.: Relevant Logic: A Philosophical Interpretation. Cambridge University Press (2004)
35. Mares, E.D.: Relevance logic. In: Zalta, E.N., (ed.), The Stanford Encyclopedia of Philosophy. Metaphysics Research Lab, winter 2020 edition. Stanford University (2020)
36. Mares, E.D., Seligman, J., Restall, G.: Situations, constrains and channels. In: van Benthem, J., ter Meulen, A. (eds.), Handbook of Logic and Language, pp. 329–344. Elsevier (2010)
37. Meyer, J.C., van der Hoek, W.: Epistemic Logic for AI and Computer Science. Cambridge University Press (1995)
38. Meyer, R.K., Mares, E.D.: Semantics of entailment 0. In: Schroeder-Heister, P., Došen, K., (eds.), Substructural Logics, pp. 239–258. Oxford Science Publications (1993)
39. Odintsov, S., Wansing, H.: Routley star and hyperintensionality. J. Philos. Log. **50**(1), 33–56 (2021)
40. Ono, H.: On Some Intuitionistic Modal Logics. Publications of RIMS Kyoto University **13**, 687–722 (1977)
41. Punčochář, V., Sedlár, I., Tedder, A.: Relevant epistemic logic with public announcements and common knowledge. J. Log. Comput. **33**(2), 436–461 (2023)
42. Read, S.: Relevant Logic: A Philosophical Examination of Inference. Blackwell (1988)
43. Rendsvig, R., Symons, J., Wang, Y.: Epistemic logic. In: Zalta, E.N., Nodelman, U., (eds.), The Stanford Encyclopedia of Philosophy. Metaphysics Research Lab, Winter 2023 edition. Stanford University (2023)
44. Restall, G.: Negation in relevant logics (how I stopped worrying and learned to love the Routley star). In: Gabbay, D.M., Wansing, H. (eds.), What is Negation?, pp. 53–76. Kluwer Academic Publishers (1999)
45. Restall, G.: An Introduction to Substructural Logics. Routledge (2000)
46. Restall, G.: Logics, situations and channels. J. Cognit. Sci. **6**, 125–150 (2005)
47. Robles, G., Méndez, J.M.: A class of simpler logical matrices for the variable-sharing property. Logic Log. Philos. **20**(3), 241–249 (2011)
48. Robles, G., Méndez, J.M.: A general characterization of the variable-sharing property by means of logical matrices. Notre Dame J. Formal Logic **53**(2), 223–244 (2012)
49. Routley, R., Plumwood, V., Meyer, R.K., Brady, R.T.: Relevant Logics and Their Rivals, vol. 1. Ridgeview (1982)
50. Savíc, N., Studer, T.: Relevant justification logic. IfCoLog J. Logics Their Appl. 6(2) (2019). http://collegepublications.co.uk/ifcolog/?00031
51. Sedlár, I.: Substructural epistemic logics. J. Appl. Non-Classical Logics **25**(3), 256–285 (2015)
52. Sedlár, I.: Epistemic extensions of modal distributive substructural logics. J. Log. Comput. **26**(6), 1787–1813 (2016)
53. Sedlár, I.: Hyperintensional logics for everyone. Synthese **198**(2), 933–956 (2019)
54. Sedlár, I., Vigiani, P.: Relevant reasoning and implicit beliefs. In: Hansen, H.H., Scedrov, A., de Queiroz, R.J.G.B. (eds.) Logic, Language, Information, and Computation: 29th International Workshop, WoLLIC 2023, Halifax, NS, Canada, July 11–14, 2023, Proceedings, pp. 336–350. Springer Nature, Switzerland (2023)
55. Stalnaker, R.: On logics of knowledge and belief. Philos. Stud. **128**(1), 169–199 (2006)
56. Standefer, S.: Tracking reasons with extensions of relevant logics. Logic J. IGPL **27**(4), 543–569 (2019)
57. Standefer, S.: Actual issues for relevant logics. Ergo **7**(8), 241–276 (2020)
58. Standefer, S.: A substructural approach to explicit modal logic. J. Logic, Lang. Inf. **32**, 333–362 (2023)
59. Standefer, S.: What is a relevant connective? J. Philos. Log. **51**(4), 919–950 (2022)

60. Standefer, S.: Hyperintensionality in relevant logics. In: Alechina, N., Herzig, A., Liang, F. (eds.) Logic, Rationality, and Interaction: 9th International Workshop, LORI 2023, Jinan, China, October 26–29, 2023, Proceedings, pp. 238–250. Springer Nature, Switzerland (2023)
61. Standefer, S.: Varieties of relevant S5. Logic Logical Philos. **32**(1), 53–80 (2023)
62. Standefer, S.: Weak relevant justification logics. J. Log. Comput. **33**(7), 1665–1683 (2023)
63. Standefer, S.: Variable-sharing as relevance. In Tedder, A., Sedlár, I., Standefer, S., (eds.), New Directions in Relevant Logics, pp. 97–117. Springer (2025)
64. Standefer, S., Shear, T., French, R.: Getting some (non-classical) closure with justification logic. Asian J. Philos. **2**(2), 1–25 (2023)
65. Tedder, A.: Information flow in logics in the vicinity of BB. Australas. J. Log. **18**(1), 1–24 (2021)
66. Tedder, A. (2023). Situations, propositions, and information states. In Bimbó, K., (ed.), Relevance Logics and other Tools for Reasoning: Essays in Honor of J. Michael Dunn, pp. 410–426. College Publications
67. van Ditmarsch, H., Halpern, J.Y., van der Hoek, W., Kooi, B.: An introduction to logics of knowledge and belief. In: van Ditmarsch, H., Halpern, J.Y., van der Hoek, W., Kooi, B., (eds.), Handbook of Epistemic Logic, pp. 1–51. College Publications (2015)
68. van Ditmarsch, H., van der Hoek, W., Kooi, B.: Dynamic Epistemic Logic. Springer, Dordrecht, Netherland (2007)
69. Williamson, T.: Knowledge and its Limits. Oxford University Press, New York (2000)
70. Williamson, T.: Indicative versus subjunctive conditionals, congruential versus non-hyperintensional contexts. Philos. Issues **16**(1), 310–333 (2006)
71. Yap, A.: Idealization, epistemic logic, and epistemology. Synthese **191**(14), 3351–3366 (2014)

Chapter 4
Mereological Forcing

Santiago Jockwich Martinez

Abstract The main goal of this paper is to introduce a novel method that allows us to construct systematically models of mereology. We call the resulting method *mereological forcing*. Whereas set-theoretic forcing deals with the construction of models which witness the compatibility of ZFC with certain set-theoretic statements, mereological forcing, is about the construction of models that establish the compatibility of a given mereological theory with certain mereological statements. Consequently, we claim that such a method constitutes a powerful tool for establishing compatibility claims for the mereologist. More concretely, we showcase this by constructing models which display the compatibility of (atomic) classical mereology with the mereological statement that expresses that their exist inconsistent objects.

Keywords Mereology · Forcing · Inconsistent objects

4.1 Introduction

The application of Boolean-valued models—a more streamlined and intuitive presentation of set-theoretic forcing—to the study of philosophical notions has been attracting growing attention in recent years.[1] Among these notions we have arbitrary objects [3], logical connectives [6], mereological vagueness [16], the notion of possibility [11], iterative notions of sets [7], and model-theoretic concepts [15]. In this paper, we aim to further extend this research trend by outlining a systematic method for constructing models of mereology which is inspired by set-theoretic forcing.

[1] It is noticeable that a *ghost* wanders around the community of logicians and set theorists which consists in the belief that the poset approach to forcing is intrinsically superior to the algebra-valued models approach for carrying out "actual" forcing proofs. The only considerable advantage of the former approach stems from the fact that posets usually have better combinatorics than Boolean algebras, thus, slightly facilitating to work with them *initially*. However, at later stages of our construction, we can readily move back and forth between the two approaches.

S. J. Martinez (✉)
University of Oslo, IFIKK, Blindernveien 31, Oslo, Norway
e-mail: santijoxi@hotmail.com

K. Sano et al. (eds.), *Exploring Negation, Modality and Proof*,
Logic in Asia: Studia Logica Library,
https://doi.org/10.1007/978-981-96-8566-0_4

The original idea of applying Boolean-valued models to the construction of models of mereology is due to [16] and will serve as cornerstone for our own approach. More concretely, the authors of [16], introduced a particular kind of Boolean-valued models for the language of mereology, viz., atomic Boolean-valued models, with the purpose of modelling mereological indeterminacy while still retaining the laws of classical logic. Thus, providing an appealing alternative to mereological systems based on fuzzy logics (such as [12] or [9]).

We complement the approach of [16], essentially, in two aspects. Firstly, we spell out a philosophical narrative about the *rationale* that lies behind the systematic construction of models of mereology with very specific properties. This story is what we shall call *mereological forcing*. Secondly, we introduce atomic $(\mathbb{A}, D)$-valued models, where $(\mathbb{A}, D)$ denotes an ultra-designated Cobounded algebra (introduced in Sect. 4.2), rather than considering Boolean algebras. As a result we obtain a construction method that allows us also to produce non-classical models of mereology, i.e., models of mereology that employ a non-classical logic.

Now, before spelling out properly our narrative let us say something about (set-theoretic) forcing. Forcing was introduced by Paul Cohen in the the early 1960s and was conceived as a tool to prove the independence of the Continuum Hypothesis from Zermelo-Fraenkel set theory plus the axiom of choice (ZFC). The essential idea is that given a model $\mathcal{M}$ of ZFC (which we will call the ground model) we can construct in a very controlled way another model $\mathcal{M}'$ of ZFC that is properly contained in the ground model and which validates some set-theoretical sentence φ that we are interested in. It follows that there exists a model which validates ZFC $\cup \{\varphi\}$, or in other words, ZFC and φ are *compatible*. Thus, at its core, forcing aims to provide compatibility results.

Using the algebra-valued models approach, the essential steps of forcing can be stated as follows:

1. Fix a sentence φ in the language of set theory.
2. Choose a complete lattice $\mathbb{A} = \langle \mathbf{A}; \wedge, \vee, \rightarrow, ^*, \mathbf{0}, \mathbf{1} \rangle$, a filter D on $\mathbf{A}$ and construct an algebra-valued model $\mathbf{V}^{(\mathbb{A}, [\![\cdot]\!])}$ such that $\mathbf{V}^{(\mathbb{A}, [\![\cdot]\!])} \models_D$ ZFC and $\mathbf{V}^{(\mathbb{A}, [\![\cdot]\!])} \models_D \varphi$.
3. Conclude that ZFC and φ are compatible.

Notice that the difficulty of forcing resides precisely in Step 2, i.e., to choose an adequate algebraic structure that gives rise to a model which satisfies the desired sentence (specified in Step 1).[2]

Now, let us consider the case of mereological forcing. Just like its set-theoretic counterpart, mereological forcing is essentially concerned with generating compatibility results. Specifically, it is a method that effectively determines when a given mereological theory is compatible with a particular mereological statement. Unlike set theory, however, there is no consensus in the mereological community on which mereological theory should serve as the default (akin to ZFC in set theory) when

[2] For instance, for showing that ZFC and the negation of the Continuum hypothesis are compatible we may consider any Boolean algebra $\mathsf{RO}(2^{\omega \times \omega_\alpha})$ where $\alpha \geq 2$ and RO denotes the regular open sets. See [1, pp. 61-66], for further details.

exploring compatibility results. Nonetheless, the absence of a theory analogous to ZFC should not be seen as a weakness of our approach but rather as a strength, as it lends our method greater generality.

Thus, we have to consider an extra step when doing mereological forcing in order to get off the ground. In particular, we need to fix first a theory of mereology and this theory will then provide the particular language in which we can formulate our mereological statement of interest. Then, as in the case of set-theoretical forcing, we need to fix a mereological statement which we want to show to be compatible with our chosen mereological theory.

We believe that both of these choices should remain completely unconstrained to maintain the generality of our framework. In simple terms, different logicians may seek different compatibility results, and each of these pursuits is equally legitimate. Consequently, we should not rule out any of these possibilities as uninteresting a-priori.

For example, the main result of [15] consisted in showing that the theory of (atomic) classical mereology is compatible with vague identity statements, i.e., there exist identity statements that do not receive crisp truth values in our model (this is basically the content of Theorems 5.3 and 5.5 of [15]). Another interesting example is provided by [8], where the authors explored to what extent a mereological theory faithful to constructive reasoning is compatible with certain extensionality principles.

Moreover, the present author is particularly interested in showing that the theory of (atomic) classical mereology is, as well, compatible with inconsistent objects. The motivating background idea being that non-classical theories can be as mathematically expressive as purely classical theories. Thus, questioning the dividing line between classical and non-classical theories. On this conception of non-classicality, non-classical theories are valuable not because they significantly deviate from classical theories, but because they come as close as possible to classical theories without losing their characteristic non-classical features.[3]

For example, the mathematical ontology of Heyting-valued models (which validate intuitionistic logic as internal logic) resembles closely that of classical models of set theory (Cfr., [13, p. 852]) with exception of the subset relation. Consequently, many prominent theorems of classically analysis holds in these models. Additionally, the paraconistent models of set theory developed in [5] resemble very closely classical models. In fact, their mathematical ontology behaves entirely classically while still allowing for inconsistent sets, thus enabling us to demonstrate that full ZFC is compatible with inconsistent sets.

[3] This novel conception of non-classicality has been developed in detail within [4].

Before moving on, let us summarize mereological forcing as follows.

Mereological Forcing is the systematic study of models of mereology and the compatibility results that such models may give rise to.
1. Fix a mereological theory T.
2. Fix a sentence φ in the language of the chosen theory, i.e., $\mathcal{L}_{\mathsf{T}}$.
3. Choose a complete lattice $\mathbb{A} = \langle \mathbf{A}; \wedge, \vee, \rightarrow, ^*, \mathbf{0}, \mathbf{1} \rangle$, a filter D on $\mathbf{A}$ and construct an (atomic) algebra-valued model $\mathbf{V}^{(\mathbb{A}, [\![\cdot]\!])}$ (for the language $\mathcal{L}_{\mathsf{T}}$) such that $\mathbf{V}^{(\mathbb{A}, [\![\cdot]\!])} \models_D \mathsf{T}$ and $\mathbf{V}^{(\mathbb{A}, [\![\cdot]\!])} \models_D \varphi$.
4. Conclude that T and φ are compatible.

The paper is organized as follows: In Sect. 4.2, we provide an introduction to the fundamental mereological and algebraic definitions that underpin our work. Section 4.3 presents the central definition of the paper, specifically the concept of an atomic $\mathbb{A}$-valued model of mereology, and demonstrates the compatibility classical mereology and inconsistent objects. Furthermore, we show that none of these objects can be considered a (true) gluon. Thus, showing that our approach diverges from previously developed inconsistent mereologies such as the system of [10]. In the concluding section, we succinctly summarize our findings.

4.2 Technical Preliminaries

We begin by stating the basic mereological and algebraic notions that we shall use throughout the paper.

4.2.1 Mereological Notions

The language and axioms of mereology we shall employ closely adhere to the framework outlined in [16].

Definition 1 The language of mereology $\mathcal{L}_{\mathsf{M}}$ is the second-order language whose signature contains only a single binary relation $\leq$ which denotes parthood. We also use the following abbreviations:

(i) $v \lneq u := v \leq u \wedge \neg(u \leq v)$,
(ii) $\mathsf{E}(u) := \exists v\, \neg(u \leq v)$,
(iii) $u \circ v := \exists w(\mathsf{E}(w) \wedge w \leq u \wedge w \leq v)$,
(iv) $\mathsf{At}(u) := \mathsf{E}(u) \wedge \forall v(\mathsf{E}(v) \rightarrow \neg(v \lneq u))$,
(v) $\mathsf{FU}(v, X) := \forall w\big(X(w) \rightarrow w \leq v\big) \wedge \forall u\big(u \leq v \wedge \mathsf{E}(u) \rightarrow \exists s(X(s) \wedge u \circ s)\big)$.

The intuitive meaning of $v \lneq u$ is that v is a *proper* part of u, $\mathsf{E}(x)$ means that x exists (in the sense that x is non-zero or not a part of everything), $u \circ v$ means that u and v overlap, $\mathsf{At}(x)$ means that x is an atom, $\mathsf{FU}(v, X)$ means that v is the fusion consisting of all the X's.

Definition 2 The theory of classical mereology (CM) contains the following axioms:

$$\forall u \forall v \forall w \big(u \leq v \wedge v \leq w \rightarrow u \leq w\big) \qquad \text{(Transitivity)}$$
$$\forall u \forall v \big(v \nleq u \rightarrow \exists w(\mathsf{E}(w) \wedge w \leq v \wedge \neg(u \circ w))\big) \qquad \text{(Supplementation)}$$
$$\forall X \big(\exists v X(v) \rightarrow \exists w(\mathsf{FU}(w, X))\big) \qquad \text{(Fusion)}$$
$$\exists u \exists v \big(u \nleq v\big) \rightarrow \neg \exists w \neg \big(\mathsf{E}(w)\big) \qquad \text{(NoZero)}$$
$$\forall u \forall v \big(u \leq v \wedge v \leq u \rightarrow u = v\big) \qquad \text{(Anti-symmetry)}$$

From this theory we can obtain the theory of atomic classical mereology (ACM) if we add the following axiom to CM:

$$\forall u \big(\mathsf{E}(u) \rightarrow \exists v(\mathsf{At}(v) \wedge v \leq u)\big) \qquad \text{(Atomicity)}$$

4.2.2 Algebraic Notions

We presuppose basic algebraic notions such as complete bounded distributive lattices and filters. In particular, we will be working mainly with the following algebraic structures.

Definition 3 An algebraic structure $\mathbb{A} = \langle \mathbf{A}, \wedge, \vee, \Rightarrow, \mathbf{1}, \mathbf{0} \rangle$ is called a Cobounded-algebra if

(i) $\langle \mathbf{A}, \wedge, \vee, \mathbf{1}, \mathbf{0} \rangle$ is a complete distributive lattice,
(ii) if $\bigvee_{i \in I} a_i = \mathbf{1}$, then there exists $j \in I$ such that $a_j = \mathbf{1}$, where I is an index set and $a_i \in \mathbf{A}$ for all $i \in I$,
(iii) if $\bigwedge_{i \in I} a_i = \mathbf{0}$, then there exists $j \in I$ such that $a_j = \mathbf{0}$, where I is an index set and $a_i \in \mathbf{A}$ for all $i \in I$,
(iv) the binary operator $\Rightarrow$ is defined as follows: for $a, b \in \mathbf{A}$

$$a \Rightarrow b = \begin{cases} \mathbf{0}, & \text{if } a \neq \mathbf{0} \text{ and } b = \mathbf{0}; \\ \mathbf{1}, & \text{otherwise.} \end{cases}$$

However, we can extend the structure of these algebras by introducing an unary operator * with the help of a designated set.

Definition 4 Let $\mathbb{A} = \langle \mathbf{A}, \wedge, \vee, \Rightarrow, ^*, \mathbf{1}, \mathbf{0} \rangle$ be an algebraic structure and D be a designated set of $\mathbb{A}$. The pair $(\mathbb{A}, D)$ is called a designated Cobounded-algebra if

(i) $\langle \mathbf{A}, \wedge, \vee, \Rightarrow, \mathbf{1}, \mathbf{0} \rangle$ is a Cobounded-algebra,
(ii) the operator * is defined as follows: for $a, b \in A$

$$a^* = \begin{cases} \mathbf{0}, \text{ if } a = \mathbf{1}; \\ a, \text{ if } a \in D \setminus \{\mathbf{1}\}; \\ \mathbf{1}, \text{ if } a \notin D. \end{cases}$$

A designated Cobounded-algebra $(\mathbb{A}, D)$ is called an ultra-designated Cobounded-algebra, if the corresponding designated set D is an ultrafilter.

Let us now state an important example of a Cobounded-algebra, which is the three-element ultra-designated Cobounded-algebra $(\mathbb{A}_3, \{\mathbf{1}, \frac{1}{2}\})$. The operations of this algebra can be represented by the truth tables shown below. Further examples of Cobounded-algebras can be found in [5, Sect. 2.5].[4]

$\wedge$	$\mathbf{1}$	$\frac{1}{2}$	$\mathbf{0}$
$\mathbf{1}$	$\mathbf{1}$	$\frac{1}{2}$	$\mathbf{0}$
$\frac{1}{2}$	$\frac{1}{2}$	$\frac{1}{2}$	$\mathbf{0}$
$\mathbf{0}$	$\mathbf{0}$	$\mathbf{0}$	$\mathbf{0}$

$\vee$	$\mathbf{1}$	$\frac{1}{2}$	$\mathbf{0}$
$\mathbf{1}$	$\mathbf{1}$	$\mathbf{1}$	$\mathbf{1}$
$\frac{1}{2}$	$\mathbf{1}$	$\frac{1}{2}$	$\frac{1}{2}$
$\mathbf{0}$	$\mathbf{1}$	$\frac{1}{2}$	$\mathbf{0}$

$\Rightarrow$	$\mathbf{1}$	$\frac{1}{2}$	$\mathbf{0}$
$\mathbf{1}$	$\mathbf{1}$	$\mathbf{1}$	$\mathbf{0}$
$\frac{1}{2}$	$\mathbf{1}$	$\mathbf{1}$	$\mathbf{0}$
$\mathbf{0}$	$\mathbf{1}$	$\mathbf{1}$	$\mathbf{1}$

$*$	
$\mathbf{1}$	$\mathbf{0}$
$\frac{1}{2}$	$\frac{1}{2}$
$\mathbf{0}$	$\mathbf{1}$

Algebraic operations of $(\mathbb{A}_3, \{\mathbf{1}, \frac{1}{2}\})$.

Moreover, it was shown in [5] that any ultra-designated Cobounded algebra is sound and (weakly) complete with respect to the paraconsistent logic $\mathbb{L}$PS$_3$ which is *maximally consistent* with respect to classical logic.[5] Interestingly, for any $a \in \mathbb{A}_3$ we have $(a \vee a^*) \in D$, thus, the *law of excluded middle* holds. Further, we refer the interested reader to [14] where a detailed comparison of $\mathbb{L}$PS$_3$ to other prominent three-valued logics (such as LP) can be found.

Notice that the choice of ultra-designated Cobounded algebra for our models is, thus, motivated by the fact that the logic validated by these algebras is as close

[4] For an example of a Cobounded-algebra with an infinite domain simply fix a Boolean-algebra $\mathbb{B} = \langle \mathbf{A}; \wedge, \vee, \Rightarrow, ^*, \mathbf{0}, \mathbf{1} \rangle$ where the underlying universe $\mathbf{A}$ is infinite. If we now add a new bottom element $\mathbf{0}'$ and top element $\mathbf{1}'$ to $\mathbf{A}$ and extend the two operators $\wedge$ and $\vee$ on the resultant set $\mathbf{A}'$ as:

$$\mathbf{1}' \wedge a = a, \mathbf{0}' \wedge a = \mathbf{0}', \mathbf{1}' \vee a = \mathbf{1}', \text{ and } \mathbf{0}' \vee a = a.$$

Then, we can choose a set of designated values D on the stretched lattice $\mathbb{A} = \langle \mathbf{A}', \wedge, \vee, \mathbf{0}', \mathbf{1}' \rangle$ and extend it with a binary operation $\Rightarrow$ (as defined in Definition 3(iv)) and a unary operator * (as defined in Definition 4(ii)). It follows that $(\mathbb{A}, D)$ is a designated Cobounded-algebra with an infinite domain.

[5] We say that a logic $\mathbf{L}$ is maximally consistent with respect to classical logic if, upon adding to $\mathbf{L}$ any classical theorem that is not already a theorem of $\mathbf{L}$, the resulting system is classical logic.

as possible to classical logic. Thus, being an ideal candidate for developing non-classical models of mereology that converge with classical models of mereology with respect to their mathematical expressiveness.

4.3 $(\mathbb{A}, D)$-Valued Models of Mereology

We go on to show that both CM and ACM are compatible with the existence of inconsistent objects. We accomplish this by building atomic algebra-valued models based on ultra-designated Cobounded algebras for the language of mereology. These models are defined as follows.

Definition 5 Let $\mathbf{S}$ be a set of mereological atoms and $(\mathbb{A}, D)$ be an ultra-designated Cobounded algebra. The atomic $(\mathbb{A}, D)$-valued model on $\mathbf{S}$, $\mathbf{V}_{\mathbf{S}}^{(\mathbb{A},D)}$, is the atomic $(\mathbb{A}, D)$-valued model for $\mathcal{L}_{\mathsf{M}}$ with

(i) The domain:

$$\mathsf{N} = \{f : \mathbf{S} \to \mathbf{A} \mid \bigvee_{a\in\mathbf{S}} f(a) \neq \mathbf{0}\}.$$

(ii) For any $f_1, f_2 \in \mathsf{N}$,

$$[\![f_1 \leq f_2]\!] = \bigwedge_{a\in\mathbf{S}} \big(f_1(a) \Rightarrow f_2(a)\big).$$

(iii) For any $f_1, f_2 \in \mathsf{N}$,

$$\begin{aligned}[\![f_1 = f_2]\!] &= \mathbf{1}, \textit{ if } f_1 = f_2,\\ &= \bigwedge_{a\in\mathbf{S}}\{f_2(a) : f_1(a) > f_2(a)\} \wedge \bigwedge_{a\in\mathbf{S}}\{f_1(a) : f_2(a) > f_1(a)\}, \textit{ otherwise.}\end{aligned}$$

Moreover, given that the language of mereology is a second-order language we will also need to define the *second-order* domain of our atomic $(\mathbb{A}, D)$-valued models. However, for the purpose of checking the axioms of CM (in particular, for the case of Fusion) it is enough to consider relation symbols of arity 1. Given an $(\mathbb{A}, D)$-valued model, we define $\mathcal{D}^1_{\mathsf{N}}$ as the following set: $\mathcal{D}^1_{\mathsf{N}} = \{R : \mathsf{N} \to \mathbf{A}\}$. We call $\mathcal{D}^1_{\mathsf{N}}$ the 1-ary second-order domain of $\mathbf{V}_{\mathbf{S}}^{(\mathbb{A},D)}$.

Now, we are in a position to introduce a notion of validity. In particular, we say that a formula $\varphi \in \mathcal{L}_{\mathsf{M}}$ is *valid* in $\mathbf{V}_{\mathbf{S}}^{(\mathbb{A},D)}$ if $[\![\varphi]\!] \in D$. We denote this by $\mathbf{V}_{\mathbf{S}}^{(\mathbb{A},D)} \models \varphi$.

4.3.1 The Axioms of Classical Mereology in $\mathbf{V}_{\mathsf{S}}^{(\mathbb{A},D)}$

We begin by stating some useful properties of the parthood relation in atomic $(\mathbb{A}, D)$-valued models. In particular, we notice that the parthood relation, as well as the negation of the parthood relation, behave classically. Moreover, we will abbreviate the formula $\neg(f \leq g)$ as $f \nleq g$.

Lemma 1 *For any* $f_1, f_2 \in \mathsf{N}$ *we have*

(i) $[\![f_1 \leq f_2]\!] \in \{\mathbf{1}, \mathbf{0}\}$,
(ii) $[\![f_1 \leq f_2]\!] = \mathbf{0}$ *iff there exists some* $a \in \mathsf{S}$ *such that* $f_1(a) \neq \mathbf{0}$, $f_2(a) = \mathbf{0}$,
(iii) $[\![f_1 \leq f_1]\!] = \mathbf{1}$,
(iv) $[\![f_1 \leq f_2]\!] = \mathbf{1}$ *iff* $[\![f_1 \nleq f_2]\!] = \mathbf{0}$.

Proof (i) and (ii) are due to Definition 3 (iv). (iii) Follows from the fact that $(a \Rightarrow a) = \mathbf{1}$ for any $a \in \mathbf{A}$. (iv) Follows from item (i) and Definition 4(ii). □

We also notice that atomic $(\mathbb{A}, D)$-valued models differ from atomic Boolean-valued models with respect to the existence predicate. If we assume that the domain of atomic Boolean-valued models is defined analogously than in atomic $(\mathbb{A}, D)$-valued models then we allow for *vague existence*, i.e., the degree to which a function exists can range over all non-zero elements of the universe of a given Boolean algebra. On the other hand, in atomic $(\mathbb{A}, D)$-valued models we only have *sharp existence*, i.e., the degree to which a function exists is always $\mathbf{1}$.

Lemma 2 *For any* $f \in \mathsf{N}$, $[\![E(f)]\!] = \mathbf{1}$.

Proof Consider any $f \in \mathsf{N}$. We want to show that

$$[\![E(f)]\!] = \bigvee_{h \in \mathsf{N}} [\![(f \nleq h)]\!] = \mathbf{1}.$$

Since $f \in \mathsf{N}$ there must exist at least one $a_0 \in \mathsf{S}$ such that $f(a_0) \neq \mathbf{0}$. Fix h as $h(a_0) = \mathbf{0}$ and $h(a) = \mathbf{1}$ for all $a \neq a_0$. Then, $[\![f \nleq h]\!] = [\![E(f)]\!] = \mathbf{1}$. □

Additionally, we have the following lemma which we will be employed in further calculations.

Lemma 3 *For any* $f_1, f_2 \in \mathsf{N}$, $[\![f_1 \circ f_2]\!] \neq \mathbf{0}$ *iff* $\bigvee_{a \in \mathsf{S}} \big(f_1(a) \wedge f_2(a)\big) \neq \mathbf{0}$.

Proof Suppose we have $[\![f_1 \circ f_2]\!] \neq \mathbf{0}$ for some $f_1, f_2 \in \mathsf{N}$, i.e.,

$$\begin{aligned}[\![f_1 \circ f_2]\!] &= [\![\exists u\big(E(u) \wedge u \leq f_1 \wedge u \leq f_2\big)]\!] \\ &= \bigvee_{g \in \mathsf{N}} [\![E(g) \wedge g \leq f_1 \wedge g \leq f_2]\!] \neq \mathbf{0}.\end{aligned}$$

Hence, there exists $g \in \mathsf{N}$ such that $[\![g \leq f_1]\!] \neq \mathbf{0}$ and $[\![g \leq f_2]\!] \neq \mathbf{0}$. Moreover, since $g \in \mathsf{N}$ there exists some $a_0 \in \mathsf{S}$ such that $g(a_0) \neq \mathbf{0}$. Thus, $f_1(a_0) \neq \mathbf{0}$ and $f_2(a_0) \neq \mathbf{0}$. Therefore, $\bigvee_{a \in \mathsf{S}} \big(f_1(a) \wedge f_2(a)\big) \neq \mathbf{0}$.

Now, suppose

$$\bigvee_{a \in \mathsf{S}} \big(f_1(a) \wedge f_2(a)\big) \neq \mathbf{0}.$$

It follows that there exists some $a_0 \in \mathsf{S}$ such that $f_1(a_0) \neq \mathbf{0}$ and $f_2(a_0) \neq \mathbf{0}$. Then we can define the function j as $j(a_0) = \mathbf{1}$ and $j(a) = \mathbf{0}$ for all $a \neq a_0$. Clearly, $j \in \mathsf{N}$. By Lemma 2, we have $[\![E(j)]\!] = \mathbf{1}$ and we observe readily that $[\![j \leq f_1]\!] \neq \mathbf{0}$ and $[\![j \leq f_2]\!] \neq \mathbf{0}$. Thus, $[\![f_1 \circ f_2]\!] \neq \mathbf{0}$. □

Now we are in a position to show that the theory of classical mereology is valid in any atomic $(\mathbb{A}, D)$-valued model.

Theorem 1 *Let* $(\mathbb{A}, D)$ *be an ultra-designated* Cobounded*-algebra, then we have* $\mathbf{V}_{\mathsf{S}}^{(\mathbb{A},D)} \models \mathsf{CM}$. *In fact,* Transitivity, Supplementation, Fusion, NoZero *and* Anti-symmetry *receive the value* $\mathbf{1}$.

Proof Transitivity. Consider three arbitrary $f_1, f_2, f_3 \in \mathsf{N}$ such that

$$[\![f_1 \leq f_2]\!] \neq \mathbf{0} \text{ and } [\![f_2 \leq f_3]\!] \neq \mathbf{0}.$$

Take any $a_0 \in \mathsf{S}$ such that $f_1(a_0) \neq \mathbf{0}$. We are done if we can show that $f_3(a_0) \neq \mathbf{0}$. Since $[\![f_1 \leq f_2]\!] \neq \mathbf{0}$ we have $f_2(a) \neq \mathbf{0}$ and due to $[\![f_2 \leq f_3]\!] \neq \mathbf{0}$ it follows that $f_3(a_0) \neq \mathbf{0}$.

Supplementation. Consider two arbitrary $f_1, f_2 \in \mathsf{N}$ such that

$$[\![f_1 \not\leq f_2]\!] = \mathbf{1}.$$

We want to show that

$$\bigvee_{h \in \mathsf{N}} [\![E(h) \wedge h \leq f_1 \wedge \neg(h \circ f_2)]\!] \neq \mathbf{0}.$$

By Lemma 1(iv) it follows that $[\![f_1 \leq f_2]\!] = \mathbf{0}$ and thus by Lemma 1(ii) there exists some $a_0 \in \mathsf{S}$ such that $f_1(a_0) \neq \mathbf{0}$ and $f_2(a_0) = \mathbf{0}$. Then, fix h as $h(a_0) \neq \mathbf{0}$ and $h(a) = \mathbf{0}$ for $a \neq a_0$. Clearly, $h \in \mathsf{N}$, so by Lemma 2, $[\![E(h)]\!] = \mathbf{1}$. We readily observe that $[\![h \leq f_1]\!] \neq \mathbf{0}$. We are done if we can show that

$$[\![\neg(h \circ f_2)]\!] = \Big(\bigvee_{j \in \mathsf{N}} [\![(E(j) \wedge j \leq h \wedge j \leq f_2)]\!]\Big)^* \neq \mathbf{0}.$$

It is enough to show that for all $j \in \mathsf{N}$ at least one of the conjuncts of the formula above receives value $\mathbf{0}$. Now, notice that $f_2(a_0) = \mathbf{0}$ and $h(a_0) \neq \mathbf{0}$. Thus, if $j(a_0) \neq \mathbf{0}$ by Lemma 1(ii) we have $[\![j \leq f_2]\!] = \mathbf{0}$ and if $j(a_0) = \mathbf{0}$ there must exist some $a_1 \in \mathsf{S}$ such that $a_0 \neq a_1$ and $j(a_1) \neq \mathbf{0}$. However, $h(a_1) = \mathbf{0}$. So, by Lemma 1(ii) we have $[\![j \leq h]\!] = \mathbf{0}$. Thus, for any $j \in \mathsf{N}$ either $[\![j \leq f_2]\!] = \mathbf{0}$ or $[\![j \leq h]\!] = \mathbf{0}$. Hence, $[\![\neg(h \circ f_2)]\!] = \mathbf{1}$.

Fusion. Consider some $R \in \mathcal{D}^1_{\mathsf{N}}$ such that

$$\bigvee_{f \in \mathsf{N}} R(f) \neq \mathbf{0}.$$

We want to show that

$$\begin{aligned}
&[\![\exists v_2(\mathsf{FU}(v_2, R))]\!] \\
&= [\![\exists v_1(\forall v_2(R(v_2) \to v_2 \leq v_1) \wedge \forall v_3(v_3 \leq v_1 \wedge E(v_3) \to \exists v_4(R(v_4) \wedge v_3 \circ v_4)]\!] \\
&= \bigvee_{f \in \mathsf{N}} \Big(\bigwedge_{g \in \mathsf{N}} (R(g) \Rightarrow [\![g \leq f]\!]) \wedge \bigwedge_{h \in \mathsf{N}} ([\![h \leq f]\!] \wedge E(h) \Rightarrow \bigvee_{l \in \mathsf{N}} (R(l) \wedge [\![l \circ h]\!]))\Big) \neq \mathbf{0}.
\end{aligned}$$

Fix f^R as follows. For any $a \in \mathsf{S}$,

$$f^R(a) = \bigvee_{j \in \mathsf{N}} \big(R(j) \wedge j(a)\big).$$

By our initial assumption we know that there exists some f such that $R(f) \neq \mathbf{0}$ and since $f \in \mathsf{N}$ there exists some $a_0 \in \mathsf{S}$ such that $f(a_0) \neq \mathbf{0}$. Thus, $f^R \in \mathsf{N}$. We first show that

$$\bigwedge_{g \in \mathsf{N}} (R(g) \Rightarrow [\![g \leq f^R]\!]) = \bigwedge_{g \in \mathsf{N}} \Big(R(g) \Rightarrow \bigwedge_{a \in \mathsf{S}} (g(a) \Rightarrow f^R(a))\Big) \neq \mathbf{0}.$$

But this follows immediately since

$$\bigwedge_{g \in \mathsf{N}} \Big(R(g) \Rightarrow \bigwedge_{a \in \mathsf{S}} \Big(g(a) \Rightarrow \bigvee_{g \in \mathsf{N}} \big(R(g) \wedge g(a)\big)\Big)\Big) \neq \mathbf{0}.$$

It remains to be shown that

$$\bigwedge_{h \in \mathsf{N}} \Big([\![h \leq f^R]\!] \wedge E(h) \Rightarrow \bigvee_{l \in \mathsf{N}} (R(l) \wedge [\![l \circ h]\!])\Big) \neq \mathbf{0}.$$

Fix any $h_0 \in \mathsf{N}$ such that $[\![h_0 \leq f^R]\!] \neq \mathbf{0}$. Since $h_0 \in \mathsf{N}$ there exists a $a_0 \in \mathsf{S}$ such that $h_0(a_0) \neq \mathbf{0}$ and due to our assumption we have $f^R(a_0) \neq \mathbf{0}$. Thus,

$$\bigvee_{a\in \mathsf{S}} \big(f^R(a) \wedge h_0(a)\big) = \bigvee_{a\in \mathsf{S}} \big(\bigvee_{l\in \mathsf{N}} \big(R(l) \wedge l(a)\big) \wedge h_0(a)\big)$$
$$= \bigvee_{l\in \mathsf{N}} \bigvee_{a\in \mathsf{S}} \big(R(l) \wedge l(a) \wedge h_0(a)\big)$$
$$= \bigvee_{l\in \mathsf{N}} \big(R(l) \wedge \bigvee_{a\in \mathsf{S}} \big(l(a) \wedge h_0(a)\big)\big) \neq \mathbf{0}.$$

By Lemma 3, we conclude

$$\bigvee_{l\in \mathsf{N}} \big(R(l) \wedge [\![l \circ h_0]\!]\big) \neq \mathbf{0}.$$

NoZero. Suppose that we have $f_1, f_2 \in \mathsf{N}$ such that $[\![f_1 \nleq f_2]\!] = \mathbf{1}$. We go on to show that

$$\big(\bigvee_{g\in \mathsf{N}} \big(E(g)\big)^*\big)^* \neq \mathbf{0}.$$

By Lemma 2 we know that for any $g \in \mathsf{N}$ we have $[\![E(g)]\!] = \mathbf{1}$. This implies that

$$\bigvee_{g\in \mathsf{N}} \big(E(g)\big)^* = \mathbf{0}.$$

Therefore,

$$\big(\bigvee_{g\in \mathsf{N}} \big(E(g)\big)^*\big)^* = \mathbf{1}.$$

Anti-symmetry. We want to show that

$$[\![\forall v_1 \forall v_2 \big(v_1 \leq v_2 \wedge v_2 \leq v_1 \rightarrow v_1 = v_2\big)]\!] = \mathbf{1}.$$

Pick $f_1, f_2 \in \mathsf{N}$ such that $[\![f_1 \leq f_2]\!] = \mathbf{1}$ and $[\![f_2 \leq f_1]\!] = \mathbf{1}$. Consider any $a \in \mathsf{S}$ such that $f_1(a) \neq \mathbf{0}$. Then, since $[\![f_1 \leq f_2]\!] = \mathbf{1}$ we have $f_2(a) \neq \mathbf{0}$. Similarly, for any $a \in \mathsf{S}$ such that $f_2(a) \neq \mathbf{0}$. Then, by Definition 3(iii) we get immediately

$$\bigwedge_{a\in \mathsf{S}} \{f_2(a) : f_1(a) \neq \mathbf{0}\} \neq \mathbf{0} \text{ and } \bigwedge_{a\in \mathsf{S}} \{f_1(a) : f_2(a) \neq \mathbf{0}\} \neq \mathbf{0}.$$

Thus, by Definition 5(iii) it follows immediately that $[\![f_1 = f_2]\!] \neq \mathbf{0}$ and by Definition 3(iv) the desideratum follows immediately. □

It follows that we have shown that CM holds *consistently* in $(\mathbb{A}, D)$-valued model of mereology, i.e., all the mentioned axioms receive value $\mathbf{1}$. This implies, in particular, that the negation of these axioms will not hold in our models. Let us now have a look at additional axioms which give rise to stronger theories of mereology. For this purpose, we go on to show that in atomic $(\mathbb{A}, D)$-valued models the class of atoms (i.e., the class of functions $f \in \mathsf{N}$ such that $[\![\mathsf{At}(f)]\!] = \mathbf{1}$) collapses with the functions in N that assign a non-zero value only to a single atom. So, let us denote with $\mathrm{dom}^+(f) = \{a \in \mathsf{S} : f(a) \neq \mathbf{0}\}$ and further with $\mathsf{N}^{\mathsf{min}} = \{f \in \mathsf{N} : \mathrm{dom}^+(f) = \{a\}$ for some $a \in \mathsf{S}\}$.

Lemma 4 *Let $f, g \in \mathsf{N}$. Then, $[\![f \leq g]\!] = \mathbf{1}$ iff $\mathrm{dom}^+(f) \subseteq \mathrm{dom}^+(g)$.*

Proof Suppose we have $f, g \in \mathsf{N}$ such that $[\![f \leq g]\!] = \mathbf{1}$. This means that for every $a \in \mathsf{S}$ such that $f(a) \neq \mathbf{0}$ we have $g(a) \neq \mathbf{0}$. Thus, $\mathrm{dom}^+(f) \subseteq \mathrm{dom}^+(g)$. Now, suppose $\mathrm{dom}^+(f) \subseteq \mathrm{dom}^+(g)$ and towards contradiction $[\![f \leq g]\!] = \mathbf{0}$. By Lemma 1(ii) there must exist an $a_0 \in \mathsf{S}$ such that $f(a) \neq \mathbf{0}$ and $g(a) = \mathbf{0}$. But then $\mathrm{dom}^+(f) \nsubseteq \mathrm{dom}^+(g)$. Contradiction. □

Lemma 5 *$[\![\mathsf{At}(f)]\!] = \mathbf{1}$ iff $f \in \mathsf{N}^{\mathsf{min}}$.*

Proof Suppose that we have $f \in \mathsf{N}$ such that $[\![\mathsf{At}(f)]\!] = \mathbf{1}$. Thus for every $g \in \mathsf{N}$ we have $[\![g \lneq f]\!] = \mathbf{0}$. Now, suppose towards contradiction that $f \notin \mathsf{N}^{\mathsf{min}}$, so without loss of generalization suppose that $\mathrm{dom}^+(f) = \{a_0, a_1\}$. Then we can simply consider the function $j \in \mathsf{N}^{\mathsf{min}}$ where $\mathrm{dom}^+(j) = \{a_0\}$. Clearly $[\![j \leq f]\!] = \mathbf{1} = [\![f \nleq j]\!]$. Hence, $[\![j \lneq f]\!] = \mathbf{1}$. Contradiction. On the other hand, suppose that $f \in \mathsf{N}^{\mathsf{min}}$. We want to show that

$$[\![E(f) \wedge \forall v_1 \big(E(v_1) \rightarrow \neg(v_1 \lneq f)\big)]\!] \neq \mathbf{0}.$$

Given Lemma 2 it is enough to show that

$$\bigwedge_{g \in \mathsf{N}} [\![\neg(g \lneq f)]\!] = \mathbf{1}.$$

Fix any $g \in \mathsf{N}$. Then we have the following two cases: (i) $[\![g \leq f]\!] = \mathbf{0}$ or (ii) $[\![g \leq f]\!] = \mathbf{1}$. In the former case we get immediately $[\![\neg(g \lneq f)]\!] = \mathbf{1}$. In the latter case by Lemma 4 it follows that $\mathrm{dom}^+(g) = \mathrm{dom}^+(f) = \{a\}$ for some $a \in \mathsf{S}$. Therefore, $[\![f \leq g]\!] = \mathbf{1}$ and by Lemma 1(iv) we conclude $[\![f \nleq g]\!] = \mathbf{0}$. Hence, in both cases $[\![\neg(g \lneq f)]\!] = \mathbf{1}$. □

Finally, we are in a position to show that ACM holds as well in our atomic $(\mathbb{A}, D)$-valued models.

Theorem 2 *Let $(\mathbb{A}, D)$ be an ultra-designated* Cobounded*-algebra, then we have $\mathbf{V}_{\mathsf{S}}^{(\mathbb{A},D)} \models$ ACM. In fact,* Atomicity *receives value* $\mathbf{1}$.

Proof We want to show that

$$[\![\forall v_1 \big(E(v_1) \rightarrow \exists v_2(\mathsf{At}(v_2) \wedge v_2 \leq v_1)]\!] = \mathbf{1}.$$

Consider any $f \in \mathsf{N}$. It is enough to show that $[\![\exists v_2(\mathsf{At}(v_2) \wedge v_2 \leq f)]\!] = \mathbf{1}$. Since, $f \in \mathsf{N}$ we know that there exists at least one $a_0 \in \mathsf{S}$ such that $a_0 \in \mathrm{dom}^+(f)$. Now, simply consider $j \in \mathsf{N}^{\mathsf{min}}$ where $\mathrm{dom}^+(j) = \{a_0\}$. Clearly, we have that $[\![\mathsf{At}(j)]\!] = \mathbf{1} = [\![j \leq f]\!]$. Hence, we can conclude $[\![\exists v_2(\mathsf{At}(v_2) \wedge v_2 \leq f)]\!] = \mathbf{1}$. □

4.3.2 Inconsistent Objects in $\mathbf{V}_{\mathbf{S}}^{(\mathbb{A},D)}$

We have demonstrated that all the axioms of the theory of (atomic) classical mereology examined in [16] also hold in atomic $(\mathbb{A}, D)$-valued models. However, as the following theorem will reveal, atomic $(\mathbb{A}, D)$-valued models are fundamentally different from atomic Boolean-valued models, in that they permit the existence of objects with contradictory properties. In other words, our ontology contains inconsistent objects. Specifically, we will be utilizing the following definition of inconsistent objects.

$$\mathsf{Incon}(u) := \exists v(u = v \wedge u \neq v).$$

The following theorem shows that this definition is indeed non-trivial.

Theorem 3 *Let* $(\mathbb{A}, D)$ *be any ultra-designated* Cobounded *with more than two elements in its domain. Then,* $\mathbf{V}_{\mathbf{S}}^{(\mathbb{A},D)} \models \exists u\ \mathsf{Incon}(u)$.

Proof Consider $f_0, f_1 \in \mathsf{N}^{\mathsf{min}}$ where $\mathrm{dom}^+(f_0) = \mathrm{dom}^+(f_1) = \{a\}$ for some $a \in \mathsf{S}, f_0(a) = \mathbf{1}$ and $f_1(a) = c$ where $c \neq \mathbf{1}, \mathbf{0}$. Then, by Definition 5(iii) and Definition 4(ii) we have

$$[\![f_0 = f_1]\!] = c = c^* = [\![f_0 = f_1]\!]^* = [\![f_0 \neq f_1]\!].$$

□

Let us now compare our inconsistent mereology to previously developed approaches such as that of [10]. Specifically, we will demonstrate that gluons do not exist in our models, or at least not in the same sense as they were originally intended.

4.3.3 Gluons in $\mathbf{V}_{\mathbf{S}}^{(\mathbb{A},D)}$

One of the most notable features of the inconsistent mereology introduced in [10] is Priest's gluon theory. In essence, Priest defines the gluon of an object as follows.

> Given a partite object, x, a gluon for x is an object which is identical to all and only the parts of x. ([10], p. 20)

Further, gluons are philosophically appealing because they seem to offer a way out of *Bradley's Regress*[6] (Cfr., [10], p. 9). This follows from the fact, that on his

---

[6] Bradley's Regress can be phrased along the following lines: Suppose we have an object composed of finitely many parts $p_0, p_1, \ldots, p_n$. It is clear that the object in question is not merely a collection of these parts but forms a unity. To explain this unity, there must exist something (call it g_0) that makes these parts constitute a unified object. However, g_0 must also be considered a part of the initial object, meaning our object is now composed of the parts $p_0, p_1, \ldots, p_n, g_0$. But then, to explain the unity of these parts, there must exist something (call it g_1) that unifies $p_0, p_1, \ldots, p_n, g_0$. This reasoning can then be repeated indefinitely, leading to an infinite regress.

view, the unifiying "something" of an object (i.e., what Priest calls gluon) is identical to the object itself and to all the parts that compose the object. Thus, cutting off the infinite regress.

We aim to investigate whether it is possible to define gluons in our models and whether their corresponding axioms are upheld. For this purpose, we define the gluon g of an object u as follows:

$$\mathsf{Gluon}(u) := \exists g \forall w \big(w \leq u \leftrightarrow g = w \big).$$

The following theorem demonstrates that there are infinitely many gluons in each atomic $(\mathbb{A}, D)$-valued model. However, gluons exist only for atoms, and in fact, the gluon is the atom itself. Consequently, no composite object has a gluon in our models. This imposes a considerable obstacle for the intended interpretation of gluons within our models. According to the terminology introduced in [2], we can distinguish between *improper* gluons (gluons of an atom) and *proper* gluons (gluons of a composite object). It follows, therefore, that in our models we have only improper gluons, thus trivializing the notion of gluons to some extent since they were initially developed to make sense of composite objects.

Theorem 4 $[\![\mathsf{Gluon}(u)]\!] = \mathbf{1}$ *iff* $g \in \mathsf{N}^{\mathsf{min}}$ *and* $\mathrm{dom}^+(g) = \mathrm{dom}^+(u)$.

Proof Suppose $[\![\mathsf{Gluon}(f)]\!] = \mathbf{1}$ for some $f \in \mathsf{N}$ and that $g \notin \mathsf{N}^{\mathsf{min}}$ or $\mathrm{dom}^+(g) \neq \mathrm{dom}^+(f)$. In the former case, there exist at least two non-zero elements in the domain of g. Thus, without loss of generalization suppose that $\mathrm{dom}^+(g) = \{a_0, a_1\}$. But then consider $j \in \mathsf{N}^{\mathsf{min}}$ where $\mathrm{dom}^+(j) = \{a_0\}$. Clearly, $[\![j \leq g]\!] = \mathbf{1}$ and due to $[\![f \leq f]\!] = \mathbf{1}$ we have $[\![f = g]\!] \neq \mathbf{0}$. It follows that $[\![j \leq f]\!] = \mathbf{1}$. But then we get $[\![g = j]\!] = \mathbf{0}$ since $g(a_1) > j(a_1) = \mathbf{0}$. Thus, $[\![\mathsf{Gluon}(f)]\!] = \mathbf{0}$. Contradiction. In the latter case, it follows immediately that $[\![f \leq f]\!] = \mathbf{1}$ but $[\![g \neq f]\!] = \mathbf{0}$ since $\mathrm{dom}^+(g) \neq \mathrm{dom}^+(f)$. Thus, $[\![\mathsf{Gluon}(f)]\!] = \mathbf{0}$. Contradiction.

Now, suppose that $g \in \mathsf{N}^{\mathsf{min}}$ and $\mathrm{dom}^+(g) = \mathrm{dom}^+(f)$. So, $\mathrm{dom}^+(g) = \mathrm{dom}^+(f) = \{a\}$ for some $a \in \mathsf{S}$. We will show that $[\![\mathsf{Gluon}(f)]\!] = \mathbf{1}$. Consider any $j \in \mathsf{M}$ such that $[\![j \leq f]\!] = \mathbf{1}$, so $\mathrm{dom}^+(j) = \{a\}$. Thus, $[\![j = g]\!] \neq \mathbf{0}$ and therefore $[\![\forall w(w \leq f \rightarrow g = w)]\!] = \mathbf{1}$. On the other hand, consider any $j \in \mathsf{M}$ such that $[\![j = g]\!] \neq \mathbf{0}$ and suppose towards contradiction that $[\![j \leq f]\!] = \mathbf{0}$, i.e., there exists an $a_0 \in \mathsf{S}$ such that $j(a_0) \neq \mathbf{0}$ and $f(a_0) = \mathbf{0}$. But then $[\![j = g]\!] = \mathbf{0}$, contradiction. Therefore, $[\![j \leq f]\!] = \mathbf{1}$ and thus $[\![\forall w(g = w \rightarrow w \leq f)]\!] = \mathbf{1}$. We conclude, $[\![\mathsf{Gluon}(f)]\!] = \mathbf{1}$. □

There exists, however, another possible interpretation of gluon as Priest himself notes where we use the proper parthood relation (viz., $\lneq$) instead of the standard parthood relation (viz., $\leq$). Thus, we can define alternatively the gluon g of an object u as follows:

$$\mathsf{Gluon}^*(u) := \exists g \forall w \big(w \lneq u \leftrightarrow g = w \big).$$

We will show that this definition of gluon is unfeasible as well. In fact, we show that we have neither proper nor improper gluons in our models if we choose this definition.

Theorem 5 $[\![\mathsf{Gluon}^*(f)]\!] = \mathbf{0}$, *for any* $f \in \mathsf{N}$.

Proof We consider two cases: $f \notin \mathsf{N}^{\mathsf{min}}$ or $f \in \mathsf{N}^{\mathsf{min}}$. The former case follows immediately if we consider $h \in \mathsf{N}^{\mathsf{min}}$ such that $\mathrm{dom}^+(h) \subset \mathrm{dom}^+(f)$. In the latter case, suppose towards contradiction that $[\![\mathsf{Gluon}^*(f)]\!] = \mathbf{1}$ for some $f \in \mathsf{N}^{\mathsf{min}}$. Now, by Definition 5(iii) we have $[\![g = g]\!] = \mathbf{1}$. But since $\mathsf{At}(f) = \mathbf{1}$ it follows immediately that $[\![g \lneq f]\!] = \mathbf{0}$. Therefore, $[\![\forall w(g = w \rightarrow w \lneq f]\!] = \mathbf{0}$ and thus $[\![\mathsf{Gluon}^*(g, f)]\!] = \mathbf{0}$. Contradiction. □

4.4 Conclusion

Let us now come back to the opening theme of the introduction. What we have shown essentially is that the (atomic) theory of classical mereology (as introduced in Definition 2) and inconsistent objects (as defined at the beginning of Sect. 4.3.2) are *compatible*. This is the result of Theorems 1 and 2 which show that the axioms of classical and atomic classical mereology hold *consistently* in any atomic $(\mathbb{A}, D)$-valued model and Theorem 3 that there exist inconsistent objects in each $(\mathbb{A}, D)$-valued model where $(\mathbb{A}, D)$ has more than two elements.

Additionally, towards the end, we asked ourselves if these inconsistent objects (or any object of our ontology) could be interpreted as gluons (in the sense of [10]) and showed in Theorems 4 and 5 that this is not the case. Thus, showing that our inconsistent mereology is essentially different from that of [10]. It remains an open question whether this theory can be applied to model some phenomenon of interest.

Finally, we believe that the results of this paper are interesting mainly for the following reasons. First of all, we have delivered an intriguing compatibility result which questions the argument that the mathematical expressiveness of classical theories is intrinsically superior to that of non-classical theories. Consequently, providing examples that support the notion of logical pluralism that has been proposed in [4].[7]

References

1. Bell, J.L.: Set Theory: Boolean-Valued Models and Independence Proofs. Oxford University Press (2005). https://doi.org/10.1093/acprof:oso/9780198568520.001.0001
2. Carrara M., Mancini F., Smid J.: On the overlap between everything and nothing. Logic Log. Philos. 31(1), 143–160 (2021). https://doi.org/10.12775/LLP.2021.013
3. Horsten, L.: Boolean-Valued Sets as Arbitrary Objects. Mind **133**(529), 143–166 (2024). https://doi.org/10.1093/mind/fzad035
4. Jockwich, S.: Logical Pluralism by Mathematical Unification. Submitted
5. Jockwich S., Tarafder S., Venturi G.: ZF and its interpretations. Ann. Pure Appl. Logic **175**(6) (2024). https://doi.org/10.1016/j.apal.2024.103427

[7] The proposal's central idea is that the existence of non-classical theories, which match their classical counterparts in mathematical expressiveness, supports a notion of logical pluralism.

6. Jockwich, S., Venturi, G.: On negation for non-classical set theories. J. Philos. Logic **50**(3), 549–570 (2020). https://doi.org/10.1007/s10992-020-09576-3
7. Jockwich, S., Tarafder, S., Venturi, G.: Ideal objects for set theory. J. Philos. Logic **51**(3), 583–602 (2022). https://doi.org/10.1007/s10992-021-09642-4
8. Maffezioli, P., Varzi, A.C.: Intuitionistic mereology. Synthese **198**(Suppl. 18), 4277–4302 (2021). https://doi.org/10.1007/s11229-018-02035-2
9. Polkowski, L., Skowron, A.: Rough mereology: a new paradigm for approximate reasoning. Int. J. Approx. Reason. **15**(4), 333–365 (1996). https://doi.org/10.1016/S0888-613X(96)00072-2
10. Priest, G.: One: Being an Investigation Into the Unity of Reality and of its Parts. Oxford University Press, UK (2014). https://doi.org/10.1093/acprof:oso/9780199688258.001.0001
11. Roberts, S.: Sets as Possibilities, Submitted
12. Sadegh-Zadeh, K.: Fuzzy formal ontology. In: Seising, R., Sanz González, V. (eds.) Soft Computing in Humanities and Social Sciences, pp. 383–400. Springer, Heidelberg (2012). https://doi.org/10.1007/978-3-642-24672-2_20
13. Takeuti, G., Titani, S.: Intuitionistic fuzzy logic and intuitionistic fuzzy set theory. J. Symb. Logic **49**(3), 851–866 (1984). https://doi.org/10.2307/2274139
14. Tarafder, S., and Chakraborty, M.K.: A paraconsistent logic obtained from an algebra-valued model of set theory. In: Beziau J. Y., Chakraborty M., Dutta S. (eds.), New Directions in Paraconsistent Logic, pp. 165–183, Springer, New Deli (2015). https://doi.org/10.1007/978-81-322-2719-9_7
15. Wu, X.: Boolean-valued models, boolean valuations, and Löwenheim-Skolem theorems. J. Philos. Logic **53**(1), 293–330 (2023). https://doi.org/10.1007/s10992-023-09732-5
16. Wu, X.: Boolean mereology. J. Philos. Logic **52**(3), 731–766 (2023). https://doi.org/10.1007/s10992-022-09686-0

Chapter 5
Arbitrary Objects in a Bilateral Setting

Pedro Teixeira Yago

Abstract Arbitrary objects play an important role in both ordinary and mathematical reasoning. The reason for that is their distinct behavior: an arbitrary object presents those properties common to all individual objects in its range—what, following Kit Fine, has become known as the *Principle of Generic Attribution*, or simply *PGA*. Many philosophers have argued this precise quality to be contradictory, therefore rejecting arbitrary objects altogether—of which the most famous is George Berkeley's argument against Locke's *general ideas*. However, all the arguments that have been offered assume rejecting properties of an arbitrary object to be expressible in terms of asserting something of it—a unilateral setting. This paper has two main purposes: first, to show how, in the propositional bilateral system of Weak Rejectivist Logic, the PGA does not fall prey to the counter arguments which rise in tension with the nature of the logical connectives; second, to argue understanding sentences about arbitrary objects in terms of assertions which preserve (or not) generality better frames our reasoning about them, and is reflected by the Weak Rejectivist Logic. Then, following the relation between this propositional system and arbitrary objects, a further goal of this paper is to offer a characterization result for the Weak Rejectivist Logic with respect to first-order logic.

Keywords Arbitrary objects · Bilateralism · Modal logic · Negation

5.1 Introduction

An arbitrary object is an object which is representative of a certain class of individuals, insofar its properties reflect those shared in the class—which we call its *range*. An arbitrary natural number, for example, has a successor, is either even or odd, and is uniquely factored by primes, because each individual number is so, while it is not even, nor prime, because at least one individual number is not so. This distinguishing

P. Teixeira Yago (✉)
Scuola Normale Superiore, Pisa, PI, Italy
e-mail: pedro.tyago@outlook.com

K. Sano et al. (eds.), *Exploring Negation, Modality and Proof*,
Logic in Asia: Studia Logica Library,
https://doi.org/10.1007/978-981-96-8566-0_5

characteristic of arbitrary objects (henceforth, *A-objects*) is known as the *Principle of Generic Attribution* (henceforth, *PGA*)—a name which may be credited to Kit Fine [10]—, and is not only the most notable, but also the most defining.[1]

The PGA is so essential of A-objects that it not only has been present in some way in almost every conception of arbitrary entities—most famously, the concept of *general ideas* in John Locke [21], or of *prototype objects* in Alexius Meinong [23]—, but it may also allow us to historically identify the concept of A-objects in earlier philosophers such as Locke, Berkeley, Meinong, Frege, and Russell.[2] Contemporarily, the most famous and explicit elaboration of the PGA as a principle guiding the behavior of A-objects is due to Fine, first offered in [10]:

> Let $\varphi(x)$ be any condition with free variable x; let a be the name of an arbitrary object $\mathtt{a}$; and let i be a variable that ranges over the individuals in the range of $\mathtt{a}$. (I here follow a general convention whereby a names $\mathtt{a}$.) Then the required formulation of the principle is:
>
> $\varphi(a) \Leftrightarrow \forall i \varphi(i)$ (a φ's iff every individual [in its range] φ's)[3]

As simple and intuitive as the PGA is, problems quickly pile up when one takes it at face value. In this paper, I present the most common problems philosophers have appointed as stemming from the principle, and argue how those which involve logical connectives may be met by taking a framework for classical logic in which rejecting sentences about A-objects does not amount to asserting their negation.[4]

The paper is divided in two parts. The first part is strictly concerned with the PGA. In Sect. 5.2 I address the issues entailed by the PGA on which I shall focus, and in Sect. 5.3 I present the *Weak Rejectivist Logic* (henceforth, WRL), offered in [16], which as I argue, is the most adequate for reasoning about A-objects. The second part more openly relates WRL, A-objects, and first-order logic (henceforth, *FOL*). In Sect. 5.4, I present different semantics for the system, including one using first-order models which, to a certain extent, reflects how the PGA works; and in Sect. 5.5, I relate WRL to normal modal logics and FOL, offering a result for how WRL may be characterized within FOL. In Sect. 5.6, I give some prospectives for future work.

---

[1] With the only exception, to my knowledge, being Leon Horsten's view [15].

[2] For a brief survey of comments on arbitrary objects of earlier philosophers, and the different names by which they have gone, one may check [15] or [35].

[3] [10], p. 59. It is worth to note, in the same paper, Fine recognizes the inadequacy of this formulation, and settles for a different metatheoretical one. In later work, as we shall see, he offers an even more precise formulation.

[4] There are other problems raised for A-objects which do not follow from the PGA. For example, as exposed by Frege [14], the identity criterion for A-objects seem to be their range. However, for phrases such as *Let* $x, y \in \mathbb{R}$ *be such that* $x \neq y$ to make sense, that cannot be. Offering criteria of identity for A-objects which is compatible with mathematical practice and ordinary use is thus a recurring problem—though not unsolvable. For example, Fine argues there is a dependence relation between A-objects which individuates them, while Leon Horsten presents a vastly different view, in which they are individuated by their modal profile in a *sui generis* modality [15].

5.2 The Arguments Against the PGA

Any reader with a sharp eye might have already noticed the most clear inconsistency in the formulation of the PGA. It concerns presenting complex properties. Consider an arbitrary natural number: it cannot be even, since there is at least one number which is not even, and for a similar reason it cannot be odd. But certainly an arbitrary natural number must be either even or odd. How so? How can an arbitrary object validate a mutually exclusive and exhaustive disjunction, but not validate either of the disjuncts?

This counter argument—one of many against the conception of arbitrary entities, of which a more detailed exposition may be found in [10] and [15]—has become known as *Berkeley's argument*, for the seventeenth-century bishop was the first to elaborate it in [3], against Locke's theory of ideas. The worry was later echoed by Frege [14], and many others ever since.

A different inconsistency rises from *not* presenting a certain property. Consider an arbitrary planet. Every planet, but the Earth, presents the property of being different from the Earth. Therefore, following the PGA, an arbitrary planet should not present the property of being different from the Earth. But that means an arbitrary planet should then be identical to the Earth, which is certainly an absurd. Leśniewski exploits this precise tension in [20] to argue against A-objects.[5]

One last inconsistency follows from properties which are special in some sense. For example, every single pebble is a concrete object. It should follow that an arbitrary pebble is concrete. In the same way, an arbitrary keyboard is an abstract object, so it should be that any individual keyboard is equally abstract. That can hardly be so. Arbitrary pebbles are not usually seen being thrown by children into a lake any more than the keyboard I am using at this exact moment is immaterial. This sort of reasoning was exposed in [10, 19]. However, I do not address it in the current paper, for the problem, as we may see, lies in the nature of the properties involved, rather than on the interaction between the PGA and logical connectives. Therefore, from now on, whenever I mention problems or inconsistencies with the PGA, I shall be referring to those relating to arguments such as Berkeley's and Leśniewski's.

One way to maintain the PGA while avoiding inconsistencies is by arguing the background logic is not classical. As Berkeley's argument shows, to accept A-objects, and the PGA, one would have to accept that a disjunction may hold without either of the disjuncts holding, or even reject the Law of the Excluded Middle—as both a sentence and its negation may not hold for every object in the range of an A-object. Adopting a paracomplete, or even intuitionistic, logic would be a viable option. In [27], Marco Santambrogio explores that path.

Fine, in [9], offers a similar solution. He distinguishes between *generic* and *classical* occurrences of names of A-objects in sentences. An occurrence is generic when the name of the A-object plays a representative role of the objects in its range, and classical otherwise. For example, in the sentence *An arbitrary man has a father*, the name occurs representatively, whereas in *An arbitrary chair is an abstract entity*,

---

[5] A precise and thorough analysis of the argument may be seen in [25].

it occurs classically. Fine then makes a distinction between a relative (which we may call *classical*) and an absolute (*generic*) notion of truth. A sentence about an A-object is classically true, in a certain model, when it is validated by an assignment of A-objects to objects in their range, and it is generically true when it is validated by any such assignment. More specifically:

> Let φ be a formula whose [occurring names of A-objects denote the A-objects $a_1, ..., a_n$]. Then φ is true in the A-model $\mathcal{M}^*$ relative to [a valuation] v—in symbols, $\mathcal{M}^* \models_v \varphi$—if $a_1, ..., a_n$ [are in the domain of v] and $\mathcal{M} \models \varphi[v(a_1), ..., v(a_n)]$.[6] [...] On the other hand, φ is absolutely true in $\mathcal{M}^*$—in symbols, $\mathcal{M}^* \models \varphi$—if $\mathcal{M}^* \models_v \varphi$ for any [valuation encompassing $a_1, ..., a_n$]. In other words, a statement concerning A-objects is true just in case it is true for all of their values.[7]

He concludes, then, that sentences whose names of A-objects occur generically should be evaluated according to the absolute notion of truth, and analogously for classical occurrences and the relative notion. That, he says, is the PGA: for any sentence involving names of A-objects occurring generically, it is true iff it is generically true. This behavior of generic truth, in a way, mirrors validity of non-categorical theories—in that the truth of a sentence involving an A-object is for the truth of the corresponding sentences involving the objects in its scope as the validity of a formula is for the truth of that formula in each model. Thus, generic truth is defined by a global notion of validity, which means A-objects, in Fine's conception, behave according to a supervaluationist logic. Therefore, like Santambrogio, Fine similarly argues the logic of A-objects is not classical—despite validating all classical tautologies, as supervaluationist logics do [5]. However, abandoning classical logic is a road down which many philosophers might not go. Since A-objects are controversial as they stand, other options would be welcome.

Some other possible solutions that have been offered include keeping classical logic while disregarding the PGA, as Horsten does [15], and restricting the PGA to apply only to a fragment of the first-order language, but neither are unrestricting enough.[8]

A perhaps simpler solution than Fine's may be offered. To introduce it, we may consider what the PGA, in its most naive form, says. For simplicity, let us from now on assume, whenever the name of an A-object occurs, quantifiers occurring in the same context range only over its associated range. The first way of interpreting it is as applying to any formula whatsoever. In that way, for example, if a is the name of an A-object, then for any condition C we get (i) $C(a) \leftrightarrow \forall x C(x)$, and (ii) $\neg C(a) \leftrightarrow \forall x \neg C(x)$. However, from (i) we get (iii) $\neg C(a) \leftrightarrow \neg \forall x C(x)$, which presents a different condition for $\neg C(a)$ to hold than (ii).

A different example, from which Berkeley's argument stems, is that involving disjunctive conditions: when taking such a condition at face value, the PGA

[6] Here, $\mathcal{M}$ is the underlying classical first-order model, $\mathcal{M}^*$ the extend model of Fine's system of A-objects, and $\models$ is the classical relation of truth.

[7] [9], p. 42. The passage is largely simplified by me, for its notation and heavy terminology.

[8] In a formalization of Horsten's conception of A-objects, [33] show how a restricted form of PGA, where no disjunctive formulas and only restricted negations are allowed, follow directly from other features of A-objects.

gives us $(C(a) \vee \neg C(a)) \leftrightarrow \forall x(C(x) \vee \neg C(x))$. When analyzing each disjunct, however, we get $C(a) \leftrightarrow \forall x C(x)$ and $\neg C(a) \leftrightarrow \forall x \neg C(x)$. If the logic in question is classical logic (henceforth, CPL), that should mean $(\forall x C(x) \vee \forall x \neg C(x)) \leftrightarrow \forall x(C(x) \vee \neg C(x))$, which is blatantly wrong.

From each of these examples, we may learn something from how we conceive truth of sentences about A-objects, and consequently, how we assert or reject sentences about them. From the first, we get that asserting $\neg C(a)$—represented by (i)—and rejecting $C(a)$—represented by (iii)—do not amount to the same thing. From the second, that asserting a disjunction cannot be analyzed in terms of asserting the disjuncts. To follow up with this analysis in terms of assertion and rejection, we need to introduce a system capable of expressing that.

5.3 A-Objects and Bilateralism

For a long time following Frege [13], rejection was taken to be reducible to assertion. Prominently, Michael Dummett was of the view assertion was a primitive concept, which together with negation, gave rise to rejection—that is, rejecting A was nothing more than asserting *it is not the case that* A.[9] In [30], Timothy Smiley argued the two acts were on equal footing—even though both may be interdefinable.[10] Later, in [26], Ian Rumfitt offered formalizations of Smiley's view, where formulas are signed with either the "$+$" or the "$-$" symbol. In that system, the signs represent the assertoric forces of assertion and rejection, respectively. For example, $+\ A$ expresses that A is asserted, while $-\ A$, that A is rejected. Assertion and rejection are supposed to be interchangeable, in that $+\ \neg A$ is equivalent to $-\ A$.[11]

In [16], Incurvati and Schlöder argue for rejection to be understood weakly.[12] In that context, asserting and weakly rejecting are attitudes that reflect the *public commitment* of speakers with respect to sentences. Asserting A means being committed to it, while weakly rejecting A means *not* being committed to it—rather than being committed to *not* A. Thus, one may be committed to *neither* A nor $\neg A$—and thus *not* yield $+\ A$ nor $+\ \neg A$—, while still being committed to $A \vee \neg A$, yielding $+\ A \vee \neg A$.

Incurvati and Schlöder offer a Gentzen-style natural deduction system for their bilateral view, the WRL. Its language, which we may call $\mathcal{L}_S^0$, is obtained by adding

---

[9] [6], [7], and later in his response to Rumfitt, [8].

[10] "I treat assertion and rejection as distinct activities on all fours with one another. [...] asserting *not P* may be equivalent to rejecting P, but it is not the very same thing, any more than asserting *P and Q* is the same as the two assertions P and Q," ([30], p. 1.).

[11] To avoid confusion with the punctuation, I shall write signs and explicit signed formulas within quotation marks.

[12] In that work, weak rejection was marked by the minus sign "$-$". However, in [1, 17, 18] and latter work, the symbol was replaced by "$\ominus$", as opposed to a new marker of weak assertion, "$\oplus$". I follow the more recent use.

the two signs, "+" and "⊖", to the language of CPL, which we may call $\mathcal{L}^0$.[13] The well formed formulas of $\mathcal{L}^0_S$ are defined in a simple manner: for any formula A of $\mathcal{L}^0$, $+ A$ and $\ominus A$ are formulas of $\mathcal{L}^0_S$.

For a clearer presentation, from now on, I shall use the following conventions. I write capital letters of the beginning of the Latin alphabet, A, B, C, and so on, to represent propositional unsigned formulas, and capital letters with an accent, such as $\hat{A}$, $\hat{B}$, $\hat{C}$, for first-order unsigned formulas; lowercase letters of the Greek alphabet, φ, ψ, γ, for signed formulas; and capital letters of the Greek alphabet, $\varGamma$, $\varLambda$, $\varTheta$, for sets of formulas (either signed or unsigned, depending on the context). Furthermore, for any given language $\mathcal{L}$, I denote by $Form_{\mathcal{L}}$ the collection of its well formed formulas, and by $Var_{\mathcal{L}}$ the collection of its variables (either propositional or first-order, depending on the language being considered). I also follow the following convention:

$$\begin{array}{ccc} \varphi & & \varphi \\ \vdots & & +\vdots \\ \psi & & \psi \end{array}$$

I use the left for the assumption there is a derivation of ψ from φ, while the right for the further assumption the derivation has no undischarged rejected assumptions, and all premises are assertions.

To motivate the distinction between asserting and rejecting sentences of A-objects, I propose a different interpretation of the markers, rooted in mathematical practice. Take for example the elaboration of a proof of a theorem in arithmetic. It is routine to consider an arbitrary number, related to a collection of numbers of interest, and, *without loss of generality* (henceforth *w.l.g.*), make a serious of assertions about it. Thus, we assert sentences of A-objects whenever we do not lose generality (with respect to their range) when doing so. On the other hand, to falsify the conjecture that an arbitrary number of a collection has a certain property, one usually shows that, *in at least one case*—that is, *for at least one value of the arbitrary number*—, the property fails. In that way, we reject sentences of A-objects whenever the sentence is not true for some object in its range—that is, when in asserting it we lose generality. Therefore, the markers "+" and "⊖" can represent the two modes—so that $+ A$ is read as "without loss of generality, A" while $\ominus A$, as "losing generality, A." Alternatively, "+" and "⊖" may indicate preservation and loss of generality, respectively.

While the new interpretation for the "+" marker seems entirely natural, that of "⊖", although valid, is less so. Thus, to aid our intuition, I shall also make an analogy between each marker and an appropriate quantifier, and between each unsigned formula and a sentence about A-objects, following the PGA. For the purpose of this section, let us use the same letter, A and $\hat{A}$, to indicate that, despite one formula being propositional and the other of first-order, they nevertheless express the same content. Then, more specifically, $+ A$ is analogous to $\forall x \hat{A}$—as to conclude A of an A-object, w.l.g., A must obtain for every object in its range—, and accordingly, we may see $\ominus A$ as $\exists x \neg \hat{A}$—since, if we lose generality by having A of an A-object, then

[13] $\mathcal{L}^0_S$ is therefore the language composed by a countable collection of propositional variables; a propositional constant $\bot$; the logical constants $\neg$, $\vee$, $\wedge$, and $\rightarrow$; and the signs $+$ and $\ominus$.

there must be some object in its range for which $\neg A$ holds. We shall see how this helps us in motivating the inference rules. I start with those governing the relation between the two markers, and which are known as *coordination principles*:

$$\frac{+A \qquad \ominus A}{\bot}\,Rej \qquad \frac{\begin{array}{c}[+A]\\ \vdots\\ \bot\end{array}}{\ominus A}\,SR_1 \qquad \frac{\begin{array}{c}[\ominus A]\\ \vdots\\ \bot\end{array}}{+A}\,SR_2 \qquad \frac{\begin{array}{c}[+\neg A]\\ +\vdots\\ +\neg B\end{array} \qquad \ominus A}{\ominus B}\,SI$$

The rule Rej reflects the Law of Non-contradiction: a sentence about an A-object cannot be simultaneously asserted with *and* without loss of generality—it cannot be the case that both $\forall x \hat{A}$ and $\exists x \neg \hat{A}$. The rules SR_1 and SR_2 are forms of *reductio*. If assuming something of an A-object leads to a contradiction, then the contrary must be the case. SI is a weak for of *modus tollens*. If a series of assertions about A-objects allows us to assert, w.l.g., the negation of a sentence from asserting, w.l.g., the negation of another, then the fact we lose generality by asserting the first sentence means we lose generality by asserting the second. Alternatively, if one derives $\forall x \neg \hat{B}$ from $\forall x \neg \hat{A}$, then what one in fact gets is $\forall x(\neg \hat{A} \rightarrow \neg \hat{B})$, so with $\exists x \neg \hat{A}$, we get $\exists x \neg \hat{B}$.

Next, we have the assertive rules for conjunction.

$$\frac{+A \qquad +B}{+A \wedge B}\,\wedge\mathrm{I}^{+} \qquad \frac{+A \wedge B}{+A}\,\wedge\mathrm{E}_1^{+} \qquad \frac{+A \wedge B}{+B}\,\wedge\mathrm{E}_2^{+}$$

These rules are uncontroversial enough. Asserting w.l.g. that an arbitrary dog has four legs, and then that it has a tail, is equivalent to asserting w.l.g. an arbitrary dog has four legs *and* a tail—$\forall x \hat{A}$ and $\forall x \hat{B}$ iff $\forall x(\hat{A} \wedge \hat{B})$.

Now, the rules for implication.

$$\frac{\begin{array}{c}[+A]\\ +\vdots\\ +B\end{array}}{+A \rightarrow B}\rightarrow \mathrm{I}^{+} \qquad \frac{+A \rightarrow B \qquad +A}{+B}\rightarrow \mathrm{E}^{+}$$

If one asserts A w.l.g. of an A-object, and one derives an assertion of B w.l.g., then, w.l.g., A implies B . Analogously, as in the case of rule SI , if one derives $\forall x \hat{B}$ from $\forall x \hat{A}$, what we in fact have is $\forall x(\hat{A} \rightarrow \hat{B})$. A different justification is the following. If one asserts A of an A-objects, then A holds for every object in its range. If from further assertions, one derives an assertion of B of the same A-object—meaning B also holds of every object in the range—, then it means asserting A of an object in the range implies asserting B of that same object. But that just means $A \rightarrow B$ holds of every object in the range, and thus, by the PGA, we may assert it of the A-object. That licenses $\rightarrow I^{+}$. The adequacy of rule $\rightarrow \mathrm{E}^{+}$, on the other hand, is easy to grasp, as it is just an instance of *modus ponens*, which is preserved in the described manner by assertions w.l.g.—for the unconvinced reader, I believe it suffices to apply the first-order analogy to see so.

For disjunction, we have the rules below.

$$\frac{+\,A}{+\,A \vee B}\ \vee\mathrm{I}_1^+ \qquad \frac{+\,B}{+\,A \vee B}\ \vee\mathrm{I}_2^+ \qquad \frac{+\,A \vee B \qquad \begin{matrix}[+\,A]\\ +\vdots\\ \varphi\end{matrix} \qquad \begin{matrix}[+\,B]\\ +\vdots\\ \varphi\end{matrix}}{\varphi}\ \vee\mathrm{E}^+$$

For the rules $\vee\mathrm{I}^+$, consider the following. If we assert A w.l.g., there is no loss in generality in asserting $A \vee B$. For example, if an arbitrary chair is made of wood, then every individual chair is made of wood. That means every individual char is either made of wood or steel, and so must be the arbitrary chair—if $\forall x \hat{A}$, then $\forall x(\hat{A} \vee \hat{B})$. The case of rule $\vee\mathrm{E}^+$ is a bit less clear, so we might try to understand the restriction on the subderivations. For one thing, with the unrestricted version, together with the rule *Rej* , it is possible to derive a contradiction.[14] The rule *Rej* certainly cannot be discarded, so the disjunction elimination rule for assertion must be changed. Furthermore, as we have seen, we may assert $A \vee B$ w.l.g., while losing generality when asserting either A or B—as in the case of an arbitrary natural number's being either even or odd. That means it must *not* be the case that from $+\,A \vee B$, $\ominus\,A$, and $\ominus\,B$, one may derive $\bot$. Restricting the subderivations in that way allows that.

The last rules are those for negation.

$$\frac{\begin{matrix}+\vdots\\ \ominus\,A\end{matrix}}{+\,\neg A}\ \neg\mathrm{I}^+ \qquad \frac{+\,\neg A}{\ominus\,A}\ \neg\mathrm{E}^+$$

Rule $\neg\mathrm{I}^+$ may be the less straightforward rule of the system. Its justification lies in understanding the restriction on the derivation in that specific instance. According to it, if one derives that we lose generality in asserting A, and in a way all undischarged assumptions are assertions preserving generality, then one may assert w.l.g. $\neg A$. But that is the case because the only way for such a derivation of $\ominus\,A$ to happen, according to the other rules, is if $\neg A$ has already been asserted w.l.g.—for only then may one derive $\ominus\,A$, in the so restricted derivation. In other words, if we have arrived to the fact at least one object in the range of an A-object does not satisfy A, and if we do that solely by asserting sentences about the A-object itself, then we *must* have already asserted (or at least committed to) the fact the A-object itself does not satisfy A. Still, as seen in $\neg\mathrm{E}^+$, if we assert $\neg A$ w.l.g. of A-object, then we lose generality by asserting A. For then, that sentence does not hold for at least

[14] This is a derivation of a contradiction by the unrestricted rule of disjunction elimination, that is, *without* the restriction that premises and assumptions must be asserted:

$$\frac{+\,A \vee B \qquad \dfrac{\ominus\,A \qquad [+\,A]}{\bot}\,\mathrm{Rej} \qquad \dfrac{\ominus\,B \qquad [+\,B]}{\bot}\,\mathrm{Rej}}{\bot}\ \text{unrestricted } \vee\mathrm{E}^+$$

one object in its range, and therefore we have grounds to reject it—if $\forall x \neg \hat{A}$, then $\neg \forall x \hat{A}$.[15]

Let us denote by $\vdash^{WRL}$ the derivability relation of WRL. With the rules just presented, one may derive the rules for rejection (or the marker for loss of generality).[16] An important point to make in the present case, and which Incurvati and Schlöder note, is that the principle of classical negation introduction for natural deduction fails in WRL:

$$\frac{\begin{array}{c}[+\, A]\\ \vdots\\ \bot\end{array}}{+\, \neg A}\ \text{CNI}$$

In the present case, that may be seen as a desideratum. The fact that asserting A w.l.g. gets us a contradiction does not mean there is no loss of generality in asserting $\neg A$. For, consider a mammal, and suppose I assert it is a bat. That means every single mammal is a bat, so that a particular cat must be a bat. However, it is not possible for a concrete animal to be both a cat and a bat—that is a taxonomic contradiction. Nevertheless, from that,—if the PGA is to hold—, one cannot assert that it is *not* a bat, for that would mean every mammal is not a bat, which is false.

We may now see how this framework avoids the problems we have pointed out in Sect. 5.2. Let O be a sentence stating that an arbitrary natural number is odd, and E , that it is even. Then, it may be that $+\, O \vee E$, but neither $+\, O$ nor $+\, E$, since $+\, O \vee E, \ominus\, O, \ominus\, E \nvdash^{WRL} \bot$. More generally, let $A_1, \ldots, A_n$ be sentences which are mutually exclusive and exhaustive of all the objects in the range of an arbitrary object, and $A'_1, \ldots, A'_n$ the analogous sentences about the A-object. Then, $+\, \bigvee_{i \leq n} A'_i, \ominus\, A'_1 . \ominus\, A'_n \nvdash^{WRL} \bot$.[17] This accounts for Berkeley's problem.

Suppose now D is a sentence stating of an arbitrary planet that it is identical to the Earth, so that $\neg D$ states that it is different from the Earth. Since not every planet is different from the Earth, we lose generality in asserting $\neg D$, but that just means $\ominus\, \neg D$—instead of $+\, \neg D$, as the usual analysis goes. Furthermore, not every planet is identical to the Earth, so that D cannot be asserted w.l.g. either, which means $\ominus\, D$. But we have, in WRL, that $\ominus\, D, \ominus\, \neg D \nvdash^{WRL} \bot$, so there is no contradiction in that position. This dissolves Leśniewski's argument.

5.4 Semantics for WRL

I have provided an interpretation of signed formulas in terms of assertions of A-objects which either preserve or lose generality. We may now see how that is reflected

[15] Regarding the analogy with FOL, the argument relies on the non-emptiness of the domain of quantification, which is implicit when one talks about an A-object, since if there is such object, then it must be an arbitrary *something*.

[16] For the rules, one may see [16].

[17] The multiple disjunction is not defined in the language of WRL, but given a sequence of formulas, it may always be inductively defined by $\bigvee_{i \leq 1} A_i = A_i$ and $\bigvee_{i \leq n+1} A_i = (\bigvee_{i \leq n} A_i) \vee A_{n+1}$.

in the semantics. To do that, I first introduce the semantics originally provided for WRL.

Definition 5.1 (*ω-pointed model, (*[16], p. 755*)*) Let $\mathbb{C}_{\text{CPL}}$ be the class of models of CPL. An ω-pointed model is a mapping $\mathcal{V} : \omega \to \mathbb{C}_{\text{CPL}}$.

In that way, the definition of local truth is inherited from the definition of truth in models of CPL—that is, a formula $A \in Form_{\mathcal{L}^0}$ is locally true in a ω-pointed model $\mathcal{V}$ iff $\mathcal{V}(n) \models A$ for some $n \in \omega$. Global truth, for signed formulas, in a ω-pointed model is defined as:

$$\begin{array}{l} \mathcal{V} \not\models \bot \\ \mathcal{V} \models +A \text{ iff } \forall n \in \omega(\mathcal{V}(n) \models A) \\ \mathcal{V} \models \ominus A \text{ iff } \exists n \in \omega(\mathcal{V}(n) \not\models A) \end{array}$$

We may call this the *standard semantics for* WRL. As usual, given a set of formulas Γ , I write $\mathcal{V} \models \Gamma$ to mean that for all $\psi \in \Gamma$, $\mathcal{V} \models \psi$.

Let us write $\mathbb{C}_\omega$ for the class of all ω-pointed models, and $\models^\omega$ for the entailment relation of the above semantics defined as local preservation of truth: $\Gamma \models^\omega \varphi$ if for all $\mathcal{V} \in \mathbb{C}_\omega$, whenever $\mathcal{V} \models \Gamma$, we have $\mathcal{V} \models \varphi$.[18] The following result is then provided.

Proposition 5.1 (Completeness, ([16], p. 756)) *Let $\Gamma \subseteq Form_{\mathcal{L}^0_S}$ and $\varphi \in Form_{\mathcal{L}^0_S}$. Then,*

$$\Gamma \models^\omega \varphi \textit{ iff } \Gamma \vdash^{\text{WRL}} \varphi.$$

To further drive my argument, I now offer a different semantics by using first-order models which reflect precisely the previous interpretation of the force markers. Let us consider the fragment of the first-order language, $\mathcal{L}^1$, with a single variable x , no function symbols, and only unary predicates. Let us denote such fragment by $\mathcal{L}^1_-$. I start with the following definitions.

Definition 5.2 A first-order model (for $\mathcal{L}^1_-$) is a tuple $\mathcal{N} = \langle N, I \rangle$ such that $N \neq \varnothing$ and for each n-ary predicate $P \in \mathcal{L}^1$, $I(P) \subseteq N^n$. An assignment on $\mathcal{N}$ is a function $\cdot^\mu : \{x\} \to N$.

[18] If the set Γ is a singleton $\{\psi\}$, I shall drop the brackets and write simply, for example, $\psi \models^\omega \varphi$. Furthermore, instead of writing, for example, $\Gamma \cup \{\psi\} \models^\omega \varphi$, I shall simply write $\Gamma, \psi \models^\omega \varphi$

For formulas of $\mathcal{L}^1_-$, truth is defined as usual, relative to an assignment μ:

$$
\begin{array}{ll}
\mathcal{N} \models_\mu Px & \text{iff } x^\mu \in I(P) \\
\mathcal{N} \models_\mu \neg\hat{A} & \text{iff } \mathcal{N} \not\models_\mu \hat{A} \\
\mathcal{N} \models_\mu \hat{A} \wedge \hat{B} & \text{iff } \mathcal{N} \models_\mu \hat{A} \text{ and } \mathcal{N} \models_\mu \hat{B} \\
\mathcal{N} \models_\mu \forall x \hat{A} & \text{iff for any assignment } \mu' \text{ , } \mathcal{N} \models_{\mu'} \hat{A}
\end{array}
$$

$\hat{A}$ is defined to be true in the model $\mathcal{N}$—that is, $\mathcal{N} \models \hat{A}$—when for any assignment μ , $\mathcal{N} \models_\mu \hat{A}$.

Notice the truth condition for the universal closure of a formula is the same as that for the truth of the formula in the whole model.[19] I denote by $\mathbb{C}_{\text{FOL}}$ the class of all first-order models, and by $\models^{\text{FOL}}$ its associated entailment relation.[20]

Let now $p_1, p_2, ..., p_n, ...$ and $P_1, P_2, ..., P_n, ...$ be some given enumerations of the propositional variables of $\mathcal{L}^0$ and the unary predicates of $\mathcal{L}^1_-$, respectively. Local and global truth, for formulas of $\mathcal{L}^0$ and $\mathcal{L}^0_S$, respectively, are defined in a first-order model as:

$$
\begin{array}{l}
\mathcal{N} \models_\mu p_i \ \text{ iff } x^\mu \in I(P_i) \\
\mathcal{N} \not\models \bot \\
\mathcal{N} \models + A \ \text{ iff for all } \mu \text{ , } \mathcal{N} \models_\mu A \\
\mathcal{N} \models \ominus A \ \text{ iff for some } \mu \text{ , } \mathcal{N} \not\models_\mu A
\end{array}
$$

We may call this the *first-order semantics for* WRL.

Consider now $\langle N, I \rangle \in \mathbb{C}_{\text{FOL}}$, and a set of individuals $X \subseteq N$. Naturally, an arbitrary X is an A-object whose range is X . In a first-order semantics, the true sentences about an arbitrary X may then be characterized by what sentences are true when one restricts $\mathcal{N}$ to the individuals in X—that is, for $\hat{A} \in Form_{\mathcal{L}^1_-}$, $\hat{A}$ holds of an arbitrary X iff $\mathcal{N} \upharpoonright_X \models \hat{A}$. Now, notice that restricting any first-order model to a subset of its domain is equivalent to simply building a submodel which embeds into the original model. Likewise, any first-order model may be extended with a larger domain, of which its original domain and interpretation are subsets. Therefore, given a set of individuals X and a interpretation I of the predicates of $\mathcal{L}^1_-$ with respect to X , to gather what sentences are true of every single individual in X , there is no difference, on one hand, in taking X as being the domain of a given model with interpretation I and seeing what sentences are true in that model, and, on the other

[19] As we know, in a first-order language with multiple variables, assignments in a model are defined for every variable, and truth for universally quantified formulas is usually defined as $\mathcal{N} \models_\mu \forall x \hat{A}$ iff for any $o \in N$, $\mathcal{N} \models_{\mu[o/x]} \hat{A}$, where $\mu[o/x]$ is an assignment which agrees with μ on every variable except for x , to which it assigns the value o . However, since $\mathcal{L}^1_-$ has a single variable, any two assignments whatsoever vacuously satisfy the condition of agreeing on every variable except for x . Hence the collapse of truth in the model and truth of universally quantified formulas.

[20] Notice I shall use $\models^{\text{FOL}}$ for the entailment relation related to the defined semantics independently of the language.

hand, taking X to be the subset of the domain of a larger model $\mathcal{N}$, of which I is the restriction of the interpretation of $\mathcal{N}$ to X, and seeing what sentences are true in $\mathcal{N} \upharpoonright_X$. I conclude, in the most general case, the logical behavior of A-objects may be analyzed by what sentences (of, in the present case, $\mathcal{L}^1_{-}$) are true in first-order models. We may see the above semantics for formulas of $\mathcal{L}^0_S$, the language of WRL, reflects that. According to it, $\mathcal{N} \models +A$ precisely when each individual in N satisfies A, and therefore when we may assert of an arbitrary N that A. Analogously, we get $\mathcal{N} \models \ominus A$ whenever A does not hold for at least one individual N, so that we may reject A of an arbitrary N—as desired.

To show the adequacy of the alternative semantics with respect to WRL, we have the following results.

Theorem 5.1 *Let* $\varphi \in Form_{\mathcal{L}^0_S}$. *Then:*

(1) *for any* $\mathcal{N} = \langle N, I \rangle \in \mathbb{C}_{FOL}$ *there is* $\mathcal{V}_{\mathcal{N}} \in \mathbb{C}_\omega$ *such that* $\mathcal{N} \models \varphi$ *iff* $\mathcal{V}_{\mathcal{N}} \models \varphi$;
(2) *for any* $\mathcal{V} \in \mathbb{C}_\omega$ *there is* $\mathcal{N}_{\mathcal{V}} \in \mathbb{C}_{FOL}$ *such that* $\mathcal{N}_{\mathcal{V}} \models \varphi$ *iff* $\mathcal{V} \models \varphi$.

Proof Sketch For (1), if $\mathcal{N}$ is countable, enumerate the members of N. If it is not, by the Löwenheim-Skolem theorem, there is a countable model equivalent to $\mathcal{N}$, so work with that instead. Then, build an ω-pointed model $\mathcal{V}_{\mathcal{N}}$ such that, for each $o_i \in N$ and $p_j \in \mathcal{V}ar_{\mathcal{L}^0}$, $\mathcal{V}_{\mathcal{N}}(i) \models p_j$ iff $o_i \in I(P_j)$, and if N is finite, then for $k > |N|$, let $\mathcal{V}_{\mathcal{N}}(k) = \mathcal{V}_{\mathcal{N}}(1)$. That way, for any signed formula φ, $\mathcal{N} \models \varphi$ iff $\mathcal{V}_{\mathcal{N}} \models \varphi$. For (2), notice for each ω-model a countable first-order model may be built using the same definition. □

From Proposition 5.1 and the above theorem, we immediately get the following.

Corollary 5.1 *Let* $\Gamma \subseteq Form_{\mathcal{L}^0_S}$ *and* $\varphi \in Form_{\mathcal{L}^0_S}$. *Then,*

$$\Gamma \models^{\text{FOL}} \varphi \textit{ iff } \Gamma \vdash^{\text{WRL}} \varphi.$$

With the dissolution of the problems in Sect. 5.3, and with Theorem 5.1, the following result provided in [16] is of especial interest. Let $\vdash^{\text{CPL}}$ be the derivability relation of CPL.

Proposition 5.2 (Classicality, ([16], p. 754)) *For* $\Gamma \subseteq Form_{\mathcal{L}^0}$, *let* $+(\Gamma) = \{+A \mid A \in \Gamma\}$. *Then,*

$$\Gamma \vdash^{\text{CPL}} B \textit{ iff } +(\Gamma) \vdash^{\text{WRL}} +B.$$

Notice the above result is stronger than simply stating $+A$ is a theorem of WRL iff A is a tautology. Rather, as the authors point out, "we in fact obtain the classical derivability relation" ([16], p. 754).

Now, recall a *metainference* is an inference whose antecedents and consequents are themselves inferences. In fact, a hierarchy of inferential levels may be recursively defined for any logic by taking a successor level of metainferences to contain—either cumulatively or not—inferences whose antecedents and consequents are themselves metainferences of the preceding level. More precisely:

Definition 5.3 (*Inferential hierarchy*) For $n \in \omega$, the *inferential level* n *of* CPL *(*L_n^{CPL}*) and* WRL *(*L_n^{WRL}*)* is inductively defined in the following way:

- $L_0^{\text{CPL}} = Form_{\mathcal{L}^0}$;
- $L_{n+1}^{\text{CPL}} = \{\Theta \Rightarrow_n \gamma \mid \Theta \subseteq L_n^{\text{CPL}} \;\&\; \gamma \in L_n^{\text{CPL}}\}$;

and likewise,

- $L_0^{\text{WRL}} = Form_{\mathcal{L}_S^0}$;
- $L_{n+1}^{\text{WRL}} = \{\Theta \Rightarrow_n \gamma \mid \Theta \subseteq L_n^{\text{WRL}} \;\&\; \gamma \in L_n^{\text{WRL}}\}$.[21]

Let $\mathbb{C}$ be a class of structures of a given semantics for CPL (or WRL) and $n > 0$.[22] Validity of metainferences of level n of CPL (respectively, WRL) may be defined as:

- $\mathbb{C} \models \Theta \Rightarrow_n \gamma$ iff $\forall \mathcal{M} \in \mathbb{C}(\mathcal{M} \models \Theta \Rightarrow \mathcal{M} \models \gamma)$ (and analogously for WRL).[23]

If $\mathbb{C}$ is the class of all structures of a semantics with respect to which CPL is sound and complete, we may simply say $\Theta \Rightarrow_n \gamma$ *is valid according to* CPL, and write $\models^{\text{CPL}} \Theta \Rightarrow_n \gamma$ (and analogously for WRL).

Consider now the fact that, for any $n \in \omega$, there is a logic that agrees on the validity of all metainferences up to level n with classical logic, but disagrees with it at the level $n + 1$.[24] The matter of identifying a given logic with classical logic is therefore not uncontroversial. Since a more detailed discussion of this subject is outside the scope of this paper, I shall be brief on the topic.

For $\Theta \Rightarrow_n \gamma \in L_n^{\text{CPL}}$, we may define a translation $+ : L_n^{\text{CPL}} \to L_n^{\text{WRL}}$ between the two inferential hierarchies by setting the base case $\Gamma \subseteq L_0^{\text{CPL}} = Form_{\mathcal{L}^0}$ as in Proposition 5.2, and $+(\Theta \Rightarrow_n \gamma) = +(\Theta) \Rightarrow_n +(\gamma)$. Given the above considerations, it is not hard to see the coincidence between the asserted fragment of WRL and CPL goes beyond that stated in Proposition 5.2. In fact:

Proposition 5.3 *Let* $n \in \omega$ *and* $\Theta \Rightarrow_n \gamma \in L_n^{\text{CPL}}$. *Then,*

$$\models^{\text{CPL}} \Theta \Rightarrow_n \gamma \textit{ iff } \models^{\text{WRL}} +(\Theta \Rightarrow_n \gamma).$$ [25]

---

[21] This definition is the one used in [2], and is non-cumulative. To define a cumulative hierarchy (such as the one in [28]), it suffices to change the definition of the inductive step—so that, for example, $L_{n+1}^{\text{CPL}} = \{\Theta \Rightarrow_n \Phi \mid \Theta \subseteq \bigcup_{i \leq n} L_i^{\text{CPL}} \;\&\; \gamma \in L_n^{\text{CPL}}\}$.

[22] Metainferences are only generated at the first level of the hierarchy, since the 0th inferential level is composed by sentences of the object language. Hence, the definition does not encompass the case in which $n = 0$.

[23] Notice the recursive definition makes sense since the base case, where $n = 1$, accords to the usual notion of first-level metainference: $\Theta \Rightarrow_1^{\text{CPL}} A$ is valid iff $\Theta \models A$ iff $\Theta \vdash^{\text{CPL}} A$ is valid (preserves truth).

[24] Those are the so called *strict-tolerant logics*. For results relating the characterizability of classical logic and its coincidence with other logics up to many levels of metainference, one may see [2, 11, 22, 28].

[25] It is not hard to see how this may be proven by induction on the inferential levels, given that Proposition 5.2 settles the base case.

Given our previous considerations, I shall not claim the above result implies the asserted fragment of WRL corresponds to classical logic. However, it does mean the logic of asserted formulas of WRL is robustly classical.

For a quick comparison with a unilateral perspective of the PGA, we may contrast the above results with Fine's solution for the problems we have discussed so far. As pointed out in Sect. 5.2, Fine's solution for the tension between the PGA and classical logic is to drop the latter, and claim A-objects behave according to supervaluationist logics. Supervaluationist logics coincide with CPL with respect to validity of 0th level inferences. However, they fail to verify classically valid patterns of inference like:[26]

- *Conditional proof*: if $\Gamma, A \vdash B$, then $\Gamma \vdash A \to B$;
- *Reductio ad absurdum*: if $\Gamma, A \vdash \bot$, then $\Gamma \vdash \neg A$;
- *Contraposition*: if $\Gamma, A \vdash B$, then $\Gamma, \neg B \vdash \neg A$;
- *Argument by cases*: $\Gamma, A \vdash C$ and $\Gamma, B \vdash C$, then $\Gamma, A \vee B \vdash C$.

However, as obtained from Proposition 5.3, all of the counterparts of the above hold in WRL:

- if $+(\Gamma), +A \vdash^{\mathrm{WRL}} +B$, then $+(\Gamma) \vdash^{\mathrm{WRL}} +A \to B$;
- if $+(\Gamma), +A \vdash^{\mathrm{WRL}} \bot$, then $+(\Gamma) \vdash^{\mathrm{WRL}} +\neg A$;
- if $+(\Gamma), +A \vdash^{\mathrm{WRL}} +B$, then $+(\Gamma), +\neg B \vdash^{\mathrm{WRL}} +\neg A$;
- if $+(\Gamma), +A \vdash^{\mathrm{WRL}} +C$ and $+(\Gamma), +B \vdash^{\mathrm{WRL}} +C$, then $+(\Gamma), +A \vee B \vdash^{\mathrm{WRL}} +C$.

I conclude that, if one adopts a bilateral perspective, then one can accept the PGA while retaining classical logic—or, at least, staying as close to it as possible. One does not have to abandon classical logic to maintain the possibility of A-objects as they are usually conceived.

5.5 Characterizing WRL Within FOL

A question which may be raised by the adequacy of the first-order semantics is the relation between WRL and the fragment of FOL expressed by $\mathcal{L}^1_-$. To provide an answer, I now turn my attention to that matter.

To characterize WRL within FOL, the ultimate goal would be to mirror the van Benthem characterization theorem for modal logics [32], according to which modal logic corresponds to the fragment of $\mathcal{L}^1$ restricted to finite variables, unary predicates and a binary relation, and invariant under bisimulations.[27] If the language of WRL

[26] For a brief overview of counterexamples of classically valid arguments in supervaluationist logics, one may check [34] and [5].

[27] More precisely, a first-order formula (of the fragment of $\mathcal{L}^1$ restricted to finite variables, unary predicates and a binary relation) is equivalent to a modal formula iff it is invariant under bisimulations.

corresponds to a fragment of the normal modal language at the level of models, then the van Benthem theorem could be adapted to offer the desired characterization.[28] I therefore start by investigating the relation between normal modal logics and WRL, to see if such strategy is available.

As one may see, ω-pointed models are essentially modal structures. Each model is composed by many models of CPL. As such, they are maximally consistent valuations of propositional formulas, and therefore equivalent to states in a modal structure. In fact, by using a translation from $\mathcal{L}^0_S$ into the modal language $\mathcal{L}^0_\Box$, it is shown in [16] that WRL is sound and complete with respect to the class of frames for **KD45** . However, the extent to which WRL relates to the minimal normal modal logic **K** is unclear. For example, we do not know if the class of frames for **KD45** is the largest class of frames in which both logics (and languages) may be compared, or even *if* they are comparable in the class of all frames. I thus make a slight detour into that matter.

Definition 5.4 A modal frame is a tuple $\langle W, R\rangle$ such that $W \neq \varnothing$ is a set of states, and $R \subseteq W \times W$. A modal model is a tuple $\langle W, R, V\rangle$, where $\langle W, R\rangle$ is a modal frame and $V : Var_{\mathcal{L}^0} \to \mathcal{P}(W)$ is a valuation of the propositional variables.[29]

For formulas of $\mathcal{L}^0_\Box$, truth in a state $w \in W$ is defined by:

$$\begin{array}{ll} \mathcal{M}, w \models p & \text{iff } w \in V(p) \\ \mathcal{M}, w \models \neg A & \text{iff } \mathcal{M}, w \not\models A \\ \mathcal{M}, w \models A \wedge B & \text{iff } \mathcal{M}, w \models A \text{ and } \mathcal{M}, w \models B \\ \mathcal{M}, w \models \Box A & \text{iff } \forall u \in W(wRu \Rightarrow \mathcal{M}, u \models A) \end{array}$$

And in a model, and frame, as:

$$\begin{array}{l} \mathcal{M} \models A \text{ iff } \forall w \in W(\mathcal{M}, w \models A) \\ F \models A \text{ iff for all valuation } V \text{ on } F\text{ , } \langle F, V\rangle \models A \end{array}$$

Here, I need to make a distinction on two notions of validity. One says that a formula is valid if it is true in every frame. The other may be defined as truth in every model. Let us use $\mathbb{C}^F_{\mathbf{K}}$ and $\mathbb{C}^{\mathcal{M}}_{\mathbf{K}}$ to denote the class of all modal frames and models, respectively. More precisely, two entailment relations rise from the two notions:

[28] I follow the convention of using *minimal normal modal logic* to refer to the system **K**—composed by CPL extended with modal axiom K and the *Necessitation* rule, and sound and complete with respect to the class of all relational frames.

[29] These structures are usually called *Kripke* or *relational frames and models*. However, to emphasize the sort of semantics I am defining, I call them differently.

$$\Gamma \models^{\mathbf{K}}_{F} A \text{ iff } \forall F \in \mathbb{C}^{F}_{\mathbf{K}}(F \models \Gamma \Rightarrow F \models A)\,;$$
$$\Gamma \models^{\mathbf{K}}_{\mathcal{M}} A \text{ iff } \forall \mathcal{M} \in \mathbb{C}^{\mathcal{M}}_{\mathbf{K}}(\mathcal{M} \models \Gamma \Rightarrow \mathcal{M} \models A)\,;$$
$$\text{iff } \forall V \forall F \in \mathbb{C}^{F}_{\mathbf{K}}(\langle F, V\rangle \models \Gamma \Rightarrow \langle F, V\rangle \models A)\,.$$

Given the universal[30] character of the assertion sign—and the existential character of the rejection sign—, the first definition of truth of formulas of $\mathcal{L}^0_S$ in modal structures I might entertain is a local one[31]:

$$\mathcal{M}, w \not\models \bot$$
$$\mathcal{M}, w \models +A \text{ iff } \mathcal{M}, w \models \Box A$$
$$\mathcal{M}, w \models \ominus A \text{ iff } \mathcal{M}, w \models \neg\Box A$$

However, that does not amount to an adequate interpretation of the force markers in either of the modal entailment relations. For example, it is vacuously the case that $+\, p \wedge \neg p \models^{\omega} \ominus\, q$, since no ω-pointed model may verify $+\, p \wedge \neg p$. Consider however a frame $F = \langle W, R\rangle$ such that $R = \varnothing$. Then, $F \models \Box(p \wedge \neg p)$, which means $F \models +\, p \wedge \neg p$, but $F \not\models \neg\Box q$, so that $F \not\models \ominus\, w$. One may also check, for any valuation V , that likewise $\langle F, V\rangle \models +\, p \wedge \neg p$, but $\langle F, V\rangle \not\models \ominus\, q$. Therefore, $+\, p \wedge \neg p \not\models^{\mathbf{K}}_{F} \ominus\, q$ and $+\, p \wedge \neg p \not\models^{\mathbf{K}}_{\mathcal{M}} \ominus\, q$, so by Proposition 5.1, we may conclude the above semantics is not adequate for WRL—at least when considering the class of all modal structures. If one considers, alternatively, the class of those which are serial, the frame and model considered as counterexamples above are excluded. However, as stated earlier, our interest is in the class of all modal structures, so we need to change the definition of truth of signed formulas.

Let us retain the definition of truth in a state of non-modal formulas—that is, formulas of $\mathcal{L}^0$. For formulas of $\mathcal{L}^0_S$, define:

$$\mathcal{M} \not\models \bot$$
$$\mathcal{M} \models +A \text{ iff } \forall w \in W(\mathcal{M}, w \models A)$$
$$\mathcal{M} \models \ominus A \text{ iff } \exists w \in W(\mathcal{M}, w \not\models A)$$
$$F \models \varphi \quad \text{ iff for all valuation } V \text{ on } F\,,\ \langle F, V\rangle \models \varphi$$

Notice, in this semantics, the accessibility relation of frames and models is irrelevant to how formulas of WRL are evaluated. A direct consequence of this definition is the following.

Corollary 5.2 *Let $A \in Form_{\mathcal{L}^0}$, $F \in \mathbb{C}^{F}_{\mathbf{K}}$, and $\mathcal{M} \in \mathbb{C}^{\mathcal{M}}_{\mathbf{K}}$. Then:*

---

[30] That is, just as defined for then entailment relation of the ω-pointed models semantics, for a class of structures (models or frames) $\mathbb{C}$, these entailment relations are defined by $\Gamma \models \varphi$ iff for all $F \in \mathbb{C}$, if $F \models \Gamma$, then $F \models \varphi$.

[31] Notice, by the definition of truth in a model as truth in every state, at the level of models, this definition is equivalent to setting $\mathcal{M} \models +A$ iff $\mathcal{M} \models \Box A$.

$$\mathcal{M} \models +A \textit{ iff } \mathcal{M} \models A \text{, and } F \models +A \textit{ iff } F \models A$$

Given the entailment relations, validity of a formula φ may be expressed by simply $\models^{\mathbf{K}}_{F} \varphi$ (respectively, $\models^{\mathbf{K}}_{\mathcal{M}} \varphi$).[32] The first thing we may observe is that both notions of entailment amount to equivalent notions of validity. By the definitions, we have the following sequence of equivalences: $\models^{\mathbf{K}}_{F} \varphi$ iff $\forall F \in \mathbb{C}^{F}_{\mathbf{K}}(F \models \varphi)$ iff $\forall F \in \mathbb{C}^{F}_{\mathbf{K}} \forall V$ on $F(\langle F, V\rangle \models \varphi)$ iff, since every frame, with every possible valuation on it, amount to every model, $\mathcal{M} \in \mathbb{C}^{\mathcal{M}}_{\mathbf{K}}(\mathcal{M} \models \varphi)$ iff $\models^{\mathbf{K}}_{\mathcal{M}} \varphi$. However, the entailment relations differ elsewhere, as $\models^{\mathbf{K}}_{F}$ is defined by stronger conditions than $\models^{\mathbf{K}}_{\mathcal{M}}$. Consider that, given $p \in Var_{\mathcal{L}^0}$, for any frame F there is a valuation making p false at some state. Therefore, no frame satisfies $+ p$, which means $+ p \models^{\mathbf{K}}_{F} \bot$ is vacuously true. Since it is trivial to build a model verifying p at every state, we have $+ p \not\models^{\mathbf{K}}_{\mathcal{M}} \bot$. Thus, at least one of the two is not an adequate entailment relation for WRL.

In fact, despite the entailment relation defined by truth in frames being the standard definition for normal modal logic, we find it once again to be lacking with respect to WRL: $+ p \not\models^{\omega} \bot$, so by the above argument and Proposition 5.1, $\models^{\mathbf{K}}_{F}$ is not adequate with respect to the relation of derivability between sentences of WRL. Nevertheless, as one would expect (by its resemblance to ω-pointed models), the modal semantics for $\mathcal{L}^0_S$ is adequate when we consider $\models^{\mathbf{K}}_{\mathcal{M}}$.

Theorem 5.2 *Let* $\varphi \in Form_{\mathcal{L}^0_S}$. *Then:*

(1) *for any* $\mathcal{M} \in \mathbb{C}^{\mathcal{M}}_{\mathbf{K}}$ *there is* $\mathcal{V}_{\mathcal{M}} \in \mathbb{C}_{\omega}$ *such that* $\mathcal{M} \models \varphi$ *iff* $\mathcal{V}_{\mathcal{M}} \models \varphi$ *;*
(2) *for any* $\mathcal{V} \in \mathbb{C}_{\omega}$ *there is* $\mathcal{M}_{\mathcal{V}} \in \mathbb{C}^{\mathcal{M}}_{\mathbf{K}}$ *such that* $\mathcal{M}_{\mathcal{V}} \models \varphi$ *iff* $\mathcal{V} \models \varphi$ *.*

Proof Sketch Analogous to that of Theorem 5.1: let $\mathcal{M} = \langle W, R, V\rangle \in \mathbb{C}^{\mathcal{M}}_{\mathbf{K}}$, and $w_1, w_2, \ldots$ be an enumeration of the members of W (since the Löwenheim-Skolem theorem applies to modal logics); then, build $\mathcal{V}_{\mathcal{M}}$ such that, for each $w_i \in W$ and $p \in Var_{\mathcal{L}^0}$, $\mathcal{V}_{\mathcal{M}}(i) \models p$ iff $w_i \in V(p)$, and if W is finite, for $n > |W|$, we let $\mathcal{V}_{\mathcal{M}}(n) = \mathcal{V}_{\mathcal{M}}(1)$. Once again, this yields $\mathcal{M} \models \varphi$ iff $\mathcal{V}_{\mathcal{M}} \models \varphi$, and goes for both directions. □

Corollary 5.3 *Let* $\Gamma \subseteq Form_{\mathcal{L}^0_S}$ *and* $\varphi \in Form_{\mathcal{L}^0_S}$ *. Then:*

$$\Gamma \models^{\mathbf{K}}_{\mathcal{M}} \varphi \textit{ iff } \Gamma \models^{\omega} \varphi \textit{ (iff } \Gamma \vdash^{\mathrm{WRL}} \varphi).$$

Therefore—while in hindsight, this is a trivial result—, WRL is sound and complete with respect to the entailment relation defined by preservation of truth in the class of all modal models. With Corollary 5.2, we then get the following.

Corollary 5.4 *Let* $\Gamma \subseteq Form_{\mathcal{L}^0}$ *and* $A \in Form_{\mathcal{L}^0}$ *. Then:*

$$\Gamma \models^{\mathbf{K}}_{\mathcal{M}} A \textit{ iff } +(\Gamma) \models^{\omega} +A .$$

[32] That is, as entailment from an empty set of formulas.

With the above result, we might wonder if WRL corresponds to a proper fragment—more specifically, to a modal-free fragment—of the modal language, in the defined semantics. Upon further inspection, we may see that cannot be the case.

Definition 5.5 Let $\mathbb{C}$ be a class of structures, $A \in Form_{\mathcal{L}}$ and $B \in Form_{\mathcal{L}'}$. We say A and B are *equivalent in* $\mathbb{C}$ when it holds that for all $S \in \mathbb{C}$, $S \models A$ iff $S \models B$.

Theorem 5.3 *There is no $A \in Form_{\mathcal{L}^0_\Box}$ such that A is equivalent to $\ominus\, p$ in $\mathbb{C}^{\mathcal{M}}_{\mathbf{K}}$. Likewise, there is no $\varphi \in Form_{\mathcal{L}^0_S}$ such that φ is equivalent to $\Box(p \wedge \neg p)$ in $\mathbb{C}^{\mathcal{M}}_{\mathbf{K}}$.*

Proof We start with the former claim. Suppose there is $A \in Form_{\mathcal{L}^0_\Box}$ equivalent to $\ominus\, p$ in $\mathbb{C}^{\mathcal{M}}_{\mathbf{K}}$. Let $\mathcal{M} \in \mathbb{C}^{\mathcal{M}}_{\mathbf{K}}$. Now, clearly, $\ominus\, p$ is falsifiable, so that there is $\mathcal{M} = \langle W, R, V\rangle \in \mathbb{C}^{\mathcal{M}}_{\mathbf{K}}$ such that $\mathcal{M} \not\models \ominus\, p$. Since A is equivalent to $\ominus\, p$, $\mathcal{M} \not\models A$, and therefore there is $w \in W$ such that $\mathcal{M}, w \not\models A$. Consider then $\mathcal{M}' = \langle W \cup \{u\}, R, V\rangle$ such that $u \notin W$. It is a well known result that, in such case, for any modal formula B and $v \in W$, $\mathcal{M}, v \models B$ iff $\mathcal{M}', v \models B$.[33] Therefore, $\mathcal{M}', w \not\models A$, and so $\mathcal{M}' \not\models A$. However, since $u \notin W$, $u \notin V(p)$, and thus $\mathcal{M}', u \not\models p$. But that means $\mathcal{M}' \models \ominus\, p$, contradicting our supposition. Now, for the latter claim, once again suppose otherwise. If $\varphi = +\, A$ for some $A \in Form_{\mathcal{L}^0}$, since $\Box(p \wedge \neg p)$ is not valid, neither is $+\, A$, which means there is $\mathcal{M} = \langle W, R, V\rangle \in \mathbb{C}^{\mathcal{M}}_{\mathbf{K}}$ such that $\mathcal{M} \not\models +\, A$, and therefore $\mathcal{M} \models \ominus\, A$. That means, for some $w \in W$, $\mathcal{M}, w \not\models A$. Since $\Box(p \wedge \neg p)$ is equivalent to $+\, A$, we have $\mathcal{M} \not\models \Box(p \wedge \neg p)$, which means, for some $u \in W$, $\mathcal{M}, u \not\models \Box(p \wedge \neg p)$. But that is the case iff $\mathcal{M}, u \models \neg\Box\neg(p \vee \neg p)$. One may check that implies $R \neq \varnothing$. Consider $\mathcal{M}' = \langle W, \varnothing, V\rangle$. Then, $\mathcal{M}' \models \Box(p \wedge \neg p)$. However, since $A \in Form_{\mathcal{L}^0}$, A is non-modal, and therefore, $\mathcal{M}', w \not\models A$, so that $\mathcal{M}' \models \ominus\, A$, a contradiction.[34] Similarly, if $\varphi = \ominus\, A$, since $\Box(p \wedge \neg p)$ is not valid and equivalent to $\ominus\, A$, there must be $\mathcal{M} = \langle W, R, V\rangle \in \mathbb{C}^{\mathcal{M}}_{\mathbf{K}}$ such that $\mathcal{M} \not\models \ominus\, A$, and thus $\mathcal{M} \models +\, A$ and $\mathcal{M} \not\models \Box(p \wedge \neg p)$. Once again, consider the model $\mathcal{M}'$ above. There, we have $\mathcal{M}' \models \Box(p \wedge \neg p)$, but since A is non-modal, we still have $\mathcal{M}' \models \ominus\, A$, a contradiction. □

In other words, when considering validity in the class of all models, each language has a formula which is not expressible by the other. However, when considering the expressivity in the class of all frames, we can indeed compare both languages.

Theorem 5.4 *For all $\varphi \in Form_{\mathcal{L}^0_S}$ there is $A \in Form_{\mathcal{L}^0_\Box}$ such that φ is equivalent to A in $\mathbb{C}^F_{\mathbf{K}}$. However, there is no $\varphi \in Form_{\mathcal{L}^0_S}$ such that φ is equivalent to $\Box p \vee \Box\neg p$ in $\mathbb{C}^F_{\mathbf{K}}$.*

Proof For the former claim, we start with two observations: first, that for any $A \in Form_{\mathcal{L}^0}$, frame $F = \langle W, R\rangle$, and $w \in W$, if A is classically satisfiable, then it can be made true in w under some valuation; second, that for any $A \in Form_{\mathcal{L}^0}$,

[33] For brevity, I shall just mention this is not hard to obtain by using bisimulation, for which one may check [4] Chapter 2.

[34] To see that, notice the semantics for non-modal formulas in modal models is defined solely by their valuations of the propositional variables.

the truth of A at one state of W is independent from its truth at a different state—that is, non-modal formulas are evaluated independently at each state of a model, and depend only on the valuation V . Let $F \in \mathbb{C}^F_{\mathbf{K}}$ and $\mathcal{M} = \langle F, V \rangle$, and suppose $F \models \ominus A$. That means for any V , $\mathcal{M} \models \ominus A$. But that implies A is not classically satisfiable, otherwise, by our observations, we could make A true at every state of F with some valuation. Since A is not classically satisfiable, it must be false at every state and under any valuation, which means $F \models + \neg A$. Since $+ \neg A \models^{\mathbf{K}}_F \ominus A$, by the arbitrariness of A , we conclude $+ \neg A$ is equivalent to $\ominus A$ in $\mathbb{C}^F_{\mathbf{K}}$. Let now $\varphi \in Form_{\mathcal{L}^0_S}$ and $F \in \mathbb{C}^F_{\mathbf{K}}$. If $\varphi = + A$, by Corollary 5.2 we have $F \models + A$ iff $F \models A$. If now $\varphi = \ominus A$, we get $F \models \ominus A$ iff $F \models + \neg A$ iff, by Corollary 5.2, $F \models \neg A$. For the latter, it is easy to check a modal frame validates $\Box p \vee \Box \neg p$ iff its accessibility is functional—that is, each state accesses at most one other state. If there is φ equivalent to that formula, then it should likewise be the case φ is validated on a frame iff the frame is functional. However, as we have expressed, formulas of $\mathcal{L}^0_S$ are insensitive to the accessibility relation of modal structures (in the semantics we have specified). So let F be a functional frame, and F' a non-functional extension of it possessing the same set of states—that is, only their accessibility relations differ. Then, $F' \models \varphi$, while $F' \not\models \Box p \vee \Box \neg p$, contradicting our supposition. □

What we may conclude is that despite the entailment relation related to the class of all modal frames being inadequate for WRL, the normal modal language is more expressive than the language of WRL in that class with respect to the semantics provided. On the other hand, WRL is sound and complete with respect to the entailment relation related to the class of all modal models, the expressivity of the two languages is incomparable in that class, and therefore neither language encompass the other in the semantics provided. As such, the strategy laid out at the beginning of this section is not available. Nevertheless, we may try to mirror it. Were we to follow that strategy, we would need to translate the formulas of WRL into $\mathcal{L}^0_\Box$, and then use the standard translation of $\mathcal{L}^0_\Box$ into a given fragment of $\mathcal{L}^1$ and the van Benthem theorem [4] to obtain our result. Therefore, I shall skip the intermediate steps, and define a different translation to offer a similar theorem.

For modal logics, the translation language for the standard translation is constructed by restricting the first-order language, $\mathcal{L}^1$, to its fragment with only two variables, and without function symbols or predicates of arity greater than 1 . Similarly, in the present case, we use $\mathcal{L}^1_-$ as defined in Sect. 5.4.

Define now the following translation $t : Form_{\mathcal{L}^0} \to Form_{\mathcal{L}^1_-}$:

$$
\begin{aligned}
t(p_i) &= P_i x \\
t(\neg B) &= \neg t(B) \\
t(A * B) &= t(A) * t(B) , \text{ for any binary operator } *
\end{aligned}
$$

That is, given the enumerations presented in Sect. 5.4, t maps the i-th propositional variable to the first-order formula which is the application of the i-th unary predicate to the sole variable of $\mathcal{L}^1_-$.

Theorem 5.5 *Let $A \in Form_{\mathcal{L}^0}$ and $\mathcal{N} \in \mathbb{C}_{FOL}$. Then:*
$\mathcal{N} \models + A$ iff $\mathcal{N} \models \forall x t(A)$ (equivalently, $\mathcal{N} \models \ominus A$ iff $\mathcal{N} \models \exists x \neg t(A)$)

Proof By induction on A we show that $\mathcal{N} \models_{\mu} A$ iff $\mathcal{N} \models_{\mu} t(A)$. Suppose $\mathcal{N} \models_{\mu} p_i$. That is the case iff $x^{\mu} \in I(P_i)$, iff $\mathcal{N} \models_{\mu} P_i x$. The cases for negation and conjunction are straightforward by induction hypothesis. Let now $\mathcal{N} \models + A$. That is the case iff for all μ , $\mathcal{N} \models_{\mu} A$, iff, by the induction above, for all μ , $\mathcal{N} \models_{\mu} t(A)$, iff $\mathcal{N} \models \forall x t(A)$. □

Corollary 5.5 *Every formula of $\mathcal{L}^0_S$ is equivalent to a formula of $\mathcal{L}^1_-$ in $\mathbb{C}_{FOL}$.*

Therefore, t faithfully embeds formulas of $\mathcal{L}^0_S$ into formulas of $\mathcal{L}^1_-$. As with the van Benthem theorem, we need now to find out what fragment of $\mathcal{L}^1_-$ is the image of $\mathcal{L}^0_S$ under t .

Lemma 5.1 *Let $\hat{A} \in Form_{\mathcal{L}^1_-}$ be quantifier-free. Then, for some $B \in Form_{\mathcal{L}^0}$, $\hat{A} = t(B)$.*

Proof By induction on $\hat{A}$. If $\hat{A} = P_i x$, then clearly $\hat{A} = t(p_i)$. If $\hat{A} = \neg\hat{C}$, then by induction hypothesis $\hat{C} = t(C')$ for some $C' \in Form_{\mathcal{L}^0}$, and so $\hat{A} = \neg t(C') = t(\neg C')$. Likewise, if $\hat{A} = \hat{C} * \hat{D}$ for a binary operator $*$, then by induction hypothesis $\hat{C} = t(C')$ and $\hat{D} = t(D')$ for some $C', D' \in Form_{\mathcal{L}^0}$, and thus $\hat{A} = t(C') * t(D') = t(C' * D')$. □

Theorem 5.6 *Let $\hat{A} \in Form_{\mathcal{L}^1}$. Then, the following are equivalent:*

(1) *$\hat{A}$ is equivalent to a formula of $\mathcal{L}^0_S$ in $\mathbb{C}_{FOL}$;*
(2) *$\hat{A}$ is equivalent to a formula of $\mathcal{L}^1_-$ with prenex normal form having exactly one quantifier.*[35]

Proof ((1) $\Rightarrow$ (2)) Let $\hat{A}$ be equivalent to $\varphi \in Form_{\mathcal{L}^0_S}$. If $\varphi = + B$ for $B \in Form_{\mathcal{L}^0}$, by Theorem 5.5, $+ B$ is equivalent to $\forall x t(B)$. Likewise, if $\varphi = \ominus B$, by Theorem 5.5, $\ominus B$ is equivalent to $\exists x \neg t(B)$. Since the image of $\mathcal{L}^0$ under t is quantifier-free, we are done. ((2) $\Rightarrow$ (1)) Let $\mathcal{N} \in \mathbb{C}_{\text{FOL}}$, and $\hat{A}$ be equivalent to a formula with prenex normal form $\forall x \hat{B} \in Form_{\mathcal{L}^1_-}$. Since $\hat{B}$ is quantifier-free, by Lemma 5.1, $\hat{B} = t(B')$ for some $B' \in Form_{\mathcal{L}^0}$, so $\hat{A}$ is equivalent to $\forall x t(B')$, which by Theorem 5.5, means $\hat{A}$ is equivalent to $+ B'$. Likewise, if $\hat{A}$ is equivalent to a formula with prenex normal form $\exists x \hat{B} \in Form_{\mathcal{L}^1_-}$, then $\hat{A}$ is equivalent to $\exists x t(B')$, which is equivalent to $\exists x \neg\neg t(B')$. Since $\exists x \neg\neg t(B') = \exists x \neg t(\neg B')$, by Theorem 5.5, that means $\hat{A}$ is equivalent to $\ominus \neg B'$. □

[35] Notice this is different from saying $\hat{A}$ has a prenex normal form having exactly one quantifier. For example, the prenex normal form of $\forall x P_1 x \wedge \forall x P_2 x$ is $\forall x \forall y (P_1 x \wedge P_2 y)$. Nevertheless, $\forall x P_1 x \wedge \forall x P_2 x$ is equivalent to $\forall x (P_1 x \wedge x P_2 x)$, which is in prenex normal form and has exactly one quantifier.

We find thus the desired result: WRL corresponds to the fragment of FOL characterized by its formulas being equivalent to some other with prenex normal form having exactly one quantifier, a single variable, no predicates of arity 1 , and without identity or function symbols—that is, in $\mathcal{L}^1_-$.

5.6 Conclusion

I hope to have offered a way to see how a bilateral system—specifically, WRL—better reflects our reasoning with A-objects, and how that dissolves some issues that lie in the core of their conception—those of the interaction between the PGA and logical connectives—while maintaining classical logic. Furthermore, although obvious, I also hope to have made explicit the relation between WRL, FOL, and normal modal logic—and, more specifically, to have elucidated how WRL is a fragment of FOL.

In the present work, I have relied on a specific interpretation of signed formulas in WRL, at a propositional level, and on a limited first-order semantics which reflects the PGA. It may be thus suggested, for future work, to understand to what extent extensions of WRL to first-order may accommodate a naive reading of the PGA—such as those provided in [1, 18, 29]—, and how the PGA would fair in such extensions as a proper rule of inference. To provide an example of how that may be done, let $\hat{C}$ be a certain condition delimiting a definite collection, and let us name an arbitrary $\hat{C}$ by $\mathrm{a}^{\hat{C}}$. Then, if the first-order language is extended with countably many constants $a_i^{\hat{C}}$, for each condition $\hat{C}$, and take the $a_i^{\hat{C}}$ as new logical constants, the PGA might take the following schematic elaboration:

$$\frac{+\ \forall x_n(\hat{C}(x_n) \to A(x_n))}{+\ A(a_n^{\hat{C}})}\ \mathrm{PGA}_1^+ \qquad \frac{+\ A(a_n^{\hat{C}})}{+\ \forall x_n(\hat{C}(x_n) \to A(x_n))}\ \mathrm{PGA}_2^+$$

Here, in PGA_2^+ , x_n is the n-th variable of some fixed enumeration. Equivalently, the rules for rejection may be derived in WRL.[36]

$$\frac{+\ \neg\forall x_n(\gamma(x_n) \to A(x_n))}{\ominus\ A(a_n^{\gamma})}\ \mathrm{PGA}_1^{\ominus} \qquad \frac{\ominus\ A(a_n^{\gamma})}{+\ \neg\forall x_n(\gamma(x_n) \to A(x_n))}\ \mathrm{PGA}_2^{\ominus}$$

Although, to our knowledge, there are no specific criteria for *harmony* for rules for terms in bilateral systems, there are reasons to believe these rules, understood as those governing $a_i^{\hat{C}}$, to be harmonious. The criterion for bilateral harmony in

---

36 I offer here the derivation of $\mathrm{PGA}_1^{\ominus}$:

$$\dfrac{\dfrac{+\ \neg\forall x_n(\hat{C}(x_n) \to A(x_n))}{\ominus\ \forall x_n(\hat{C}(x_n) \to A(x_n))}\ \neg E^+ \qquad \dfrac{[+\ A(a_n^{\hat{C}})]^i}{+\ \forall x_n(\hat{C}(x_n) \to A(x_n))}\ \mathrm{PGA}_2^+}{i\ \dfrac{\bot}{\ominus\ A(a_n^{\hat{C}})}\ SR_1}\ Rej$$

[26] is arguably satisfied, just as the criterion for unilateral harmony, elaborated in [12].[37] However, a more rigorous criterion of bilateral harmony, presented in [31], requires not only that the introduction and elimination rules for the relevant logical constant, with respect to both force markers, be unilaterally harmonious, but they must also preserve all coordination principles.[38] Although the former is satisfied by the schematic rules, it remains to be shown if the latter is also. Nevertheless, if it is not, that concerns the PGA only in the above elaboration, whereas a different one may be possibly offered which successfully satisfies the criterion.

Notice the above rules reflect the ones for the universal quantifier in regular natural deduction systems for FOL—where a name of an A-object takes the place of a parameter. Fine, in [9], offers a thorough application of his theory of A-objects to natural deduction systems, such as Gentzen's, Quine's, and Copi's. However, as Fine interprets them, the rules of those systems involving names of A-objects are not harmonious, and thus not suitable for a bilateral system.[39]

The conclusion we may take is there seems to be much to investigate on the relation between A-objects and bilateral systems.

Notes and Comments. I would like to thank Bas Kortenbach, Giorgio Venturi, and an anonymous reviewer for their helpful feedback on this paper. I am also grateful for the reception and comments from the audience of the 6th Asian Workshop on Philosophical Logic, where this paper was presented.

References

1. Aloni, M., Incurvati, L., Schlöder, J.J.: Epistemic modals in hypothetical reasoning. Erkenntnis **88**(8), 3551–3581 (2023). https://doi.org/10.1007/s10670-022-00517-x
2. Barrio, E.A., Pailos, F., Szmuc, D.: A hierarchy of classical and paraconsistent logics. J. Philos. Log. **49**(1), 93–120 (2020). https://doi.org/10.1007/s10992-019-09513-z
3. Berkeley, G.: A Treatise Concerning the Principles of Human Knowledge. Hacking Publishing Company, Indianapolis (1982)
4. Blackburn, P., Rijke, M.D., Venema, Y.: Modal Logic. Cambridge University Press (2001)
5. Cobreros, P.: Supervaluationism and classical logic. In: Nouwen, R., van Rooij, R., Sauerland, U., Schmitz, H.C. (eds.) Vagueness in Communication, pp. 51–63. Springer, Berlin (2011)
6. Dummett, M.: What is a theory of meaning? (II). In: Evans, Gareth, McDowell, John (eds.) What is a Theory of Meaning?, pp. 67–137. Clarendon Press, Oxford (1976)
7. Dummett, M.: Frege: Philosophy of Language, 2nd edn. Duckworth, London (1981)
8. Dummett, M.: Yes', 'no' and 'can't say. Mind **111**(442), 289–95 (2002)
9. Fine, K.: Reasoning with Arbitrary Objects. Blackwell, Oxford (1985)
10. Fine, K., Tennant, N.: A defence of arbitrary objects. Aristotel. Soc. Supplem. **57**(1), 55–89 (1983). https://doi.org/10.1093/aristoteliansupp/57.1.55
11. Fjellstad, A.: Metainferential reasoning on strong kleene models. J. Philos. Log. **51**(6), 1327–1344 (2021). https://doi.org/10.1007/s10992-021-09627-3

[37] The criterion is based on Dag Prawitz's remark that elimination rules for a logical constant should be in a sense "implicit" in its introduction rules [24]

[38] A coordination principle is preserved if, when restricted to atomic sentences, all its instances for sentences containing the logical constant as their main operator are derivable.

[39] It is worth to note Gentzen's system, for example, is not harmonious in itself.

12. Francez, N., Dyckhoff, R.: A note on harmony. J. Philos. Log. **41**(3), 613–628 (2012)
13. Frege, G.: Negation. In: Geach, P., Black, M. (eds.), Translations from the Philosophical Writings of Gottlob Frege, 2nd edn., pp. 117–135. Blackwell, Oxford (1952)
14. Frege, G.: What is a function? In: Geach, P., Black, M. (eds.), Translations from the Philosophical Writings of Gottlob Frege, 3rd edn., pp. 107–116. Blackwell, Oxford (1960)
15. Horsten, L.: The Metaphysics and Mathematics of Arbitrary Objects. Cambridge University Press, Cambridge (2019)
16. Incurvati, L., Schlöder, J.J.: Weak rejection. Australas. J. Philos. **95**(4), 741–760 (2017). https://doi.org/10.1080/00048402.2016.1277771
17. Incurvati, L., Schlöder, J.J.: Weak assertion. Philos. Quart. **69**(277), 741–770 (2019). https://doi.org/10.1093/pq/pqz016
18. Incurvati, L., Schlöder, J.J.: Meta-inferences and supervaluationism. J. Philos. Log. **51**(6), 1549–1582 (2021). https://doi.org/10.1007/s10992-021-09618-4
19. Küng, G.: Ontology and the Logistic Analysis of Language: An Enquiry Into the Contemporary Views on Universals. Springer, Dordrecht (2013)
20. Leśniewski, S.: The critique of the logical principle of the excluded middle. In: Leśniewski, S. Collected Works, vol. 1, pp. 47–85. Polish Scientific Publishers-Kluwers, Dordrecht (1992). translated by S.J. Surma, J. Wójcik
21. Locke, J.: An essay concerning human understading. In: Nidditch, P.H. (ed.) The Clarendon Edition of the Works of John Locke. Oxford University Press, Oxford (1975)
22. McAllister, I.: Classical logic is not uniquely characterizable. J. Philos. Log. **51**(6), 1345–1365 (2022). https://doi.org/10.1007/s10992-021-09633-5
23. Meinong, A.: A theory of objects. In: Chisholm, R.M. (ed.), Realism and the Background of Phenomenology, pp. 76–117. Glencoe, Ill., Free Press (1960), translated by Levi, I., Terrell, D. B. and Chisholm, R. M
24. Prawitz, D.: Proofs and the meaning and completeness of logical constants. In: Hintikka, J., Niiniluoto, I., Saarinen, E. (eds.) Essays in Mathematical and Philosophical Logic, pp. 25–40. Reidel, Dordrecht (1978)
25. Richard, S.: A meinongian way out of the polish proofs against general objects. J. Philos. Log. **47**(6), 1061–1084 (2018). https://doi.org/10.1007/s10992-018-9462-5
26. Rumfitt, I.: Yes and no. Mind **109**(436), 781–823 (2000). https://doi.org/10.1093/mind/109.436.781
27. Santambrogio, M.: Generic and intensional objects. Synthese **73**, 637–663 (1987). https://doi.org/10.1007/BF00484996
28. Scambler, C.: Classical logic and the strict tolerant hierarchy. J. Philos. Logic **49**(2), 351–370 (2020). https://doi.org/10.1007/s10992-019-09520-0
29. Schlöder, J.J.: Identity and harmony and modality. J. Philos. Log. **52**(5), 1269–1294 (2023). https://doi.org/10.1007/s10992-023-09705-8
30. Smiley, T.: Rejection. Analysis **56**(1), 1–9 (1996). https://doi.org/10.1111/j.0003-2638.1996.00001.x
31. del Valle-Inclan, P., Schlöder, J.J.: Coordination and harmony in bilateral logic. Mind **132**(525), 192–207 (2023). https://doi.org/10.1093/mind/fzac012
32. Van Benthem, J.: Correspondence Theory, pp. 167–247. Springer Netherlands, Dordrecht (1984). https://doi.org/10.1007/978-94-009-6259-0_4
33. Venturi, G., Yago, P.: Modelling afthairetic modality. J. Philos. Log. **53**(4), 1027–1065 (2024). https://doi.org/10.1007/s10992-024-09757-4
34. Williamson, T.: Vagueness. Routledge, New York (1994)
35. Yago, P.: Essence and arbitrariness. Master's thesis, University of Campinas (UNICAMP), Campinas, Brazil (2020)

Chapter 6
Measurement-Theoretic Foundations of Logic of Epistemic Modals

Satoru Suzuki

Abstract Yalcin (2010) shows that Kratzer's model (1991) does not validate some intuitively valid inferences and validates some intuitively invalid ones. In order to handle this problem, he adopts a model based directly on a probability measure. However, as Kratzer (2012) says, "Our semantic knowledge alone does not give us the precise quantitative notions of probability and desirability that mathematicians and scientists work with", Yalcin's model seems to be unnatural as a model for comparative epistemic modals. We (2013) proposed a new version of logic—modal-qualitative-probability logic (MQPL)—that has the following four merits: (i) The model of MQPL reflects Kratzer's intuition above in the sense that it is not based directly on probability measures, but based on qualitative-probability orderings that represent probability measures. (ii) MQPL does not cause Yalcin's problem. (iii) The model has no limitation of the size of the domain. (iv) The model, which is based on qualitative-probability orderings that represent nonstandard-real-valued probability measures, can deal with the cases that Kolmogorov probability theory cannot. However, we (2013) did not prove the completeness of MQPL. Segerberg (1971)'s proof based on a probability measure model has been the standard proof of the completeness of logic of qualitative probability. This proof uses Scott (1964)'s representation theorem of measurement theory. The aim of this paper is (i) to give the reason why we adopt not a probability measure model but a qualitative-probability model as a model for a comparative epistemic modal, which essentially contributes to simplification of the proof of the completeness of MQPL, in terms of a representation theorem and arbitrary reference, (ii) to show that the proof system of MQPL is simpler than that of Segerberg's PK, and (iii) to prove the completeness of MQPL. Because the construction of finite Boolean algebras to make it possible to apply Scott's representation theorem covers a considerable part of Segerberg's proof, whereas our proof does not require this construction, our proof has the merit that it is by far simpler than Segerberg's proof.

Keywords Arbitrary object · Arbitrary reference · Epistemic modal · Instantial reasoning · Measurement theory · Modal logic · Natural language ·

S. Suzuki (✉)
Faculty of Arts and Sciences, Komazawa University, Komazawa, Setagaya-ku, Tokyo, Japan
e-mail: bxs05253@nifty.com

K. Sano et al. (eds.), *Exploring Negation, Modality and Proof*,
Logic in Asia: Studia Logica Library,
https://doi.org/10.1007/978-981-96-8566-0_6

Nonstandard-real-valued probability · Philosophy of language · Qualitative probability · Representation theorem

6.1 Motivation

Kratzer [24] provides comparative epistemic modals such as

at least as likely as

with their models in terms of a qualitative ordering on propositions derived from a qualitative ordering on possible worlds. Yalcin [43] shows that Kratzer's model does not validate some intuitively valid inference schemata and validates some intuitively invalid ones. He adopts a model based directly on a probability measure for comparative epistemic modals. His model does not cause this problem. Lassiter [26] also gives a detailed analysis of epistemic modals based directly on a probability measure. However, as Kratzer [25, p. 25] says, "Our semantic knowledge alone does not give us the precise quantitative notions of probability and desirability that mathematicians and scientists work with", Yalcin's model seems to be unnatural as a model of comparative epistemic modals. Segerberg [36] proposes a logic PK of qualitative probability. Gärdenfors [15] simplifies the model of PK. Holliday and Icard [19] prove that not only a probability measure model but also a qualitatively additive measure model and a revised version of Kratzer's model do not cause Yalcin's problem. Marushak [29] points out that Holliday and Icard's revised version of Kratzer's model faces some problems in case of infinite domains. PK has a binary sentential operator as a qualitative-probability operator, whereas Delgrande and Renne [5] propose a logic of qualitative probability with (m, n)-ary sentential operator. PK is based on a single probability measure, whereas both Harrison-Trainer et al. [16]'s and Ding et al. [6]'s logics of qualitative probability are based on a set of probability measures. Delgrande and Renne [5]'s, Harrison-Trainer et al. [16]'s and Ding et al. [6]'s models are all based on finite probability measures. We [40] proposed a new version of logic—modal-qualitative-probability logic (MQPL)—that has the following four merits:

1. The model of MQPL reflects Kratzer's intuition above in the sense that it is not based directly on probability measures, but based on qualitative-probability orderings that represent probability measures.
2. MQPL does not cause Yalcin's problem.
3. The model has no limitation of the size of the domain.
4. The model, which is based on qualitative-probability orderings that represent nonstandard-real-valued probability measures, can deal with the cases that Kolmogorov probability theory cannot.

However, we [40] did not prove the completeness of MQPL. Segerberg [36]'s proof based on a probability measure model has been the standard proof of the completeness of logic of qualitative probability. This proof uses Scott [35]'s representation theorem of measurement theory. As the title of this paper shows, its aim is not to provide a version of logic that gives new valid or invalid patterns of reasoning but to explore measurement-theoretic foundations of logic of epistemic modals. In fact, MQPL gives much the same valid or invalid patterns of reasoning as Segerberg [36]'s PK and Holliday and Icard [19]'s $\mathbf{FP}_{\infty}$. Then what are the aims of this paper? The aims of this paper are

1. to show that because MQPL has not a probability measure model but a qualitative-probability model, MQPL can avoid the two problems that probability measure models can invite: (i) the problem of unnaturalness of preciseness of probability-value assignment and (ii) the problem of arbitrary objects and reference,
2. to show that the proof system of MQPL is simpler than that of Segerberg's PK, and
3. to show that the qualitative-probability model brings about a metalogical merit in the proof of its completeness: Because the construction of finite Boolean algebras to make it possible to apply Scott [35]'s representation theorem covers a considerable part of Segerberg [36]'s and Gärdenfors [15]' proof, whereas our proof does not require this construction, our proof has the merit that it is by far simpler than Segerberg's and Gärdenfors' proofs. Moreover, our proof is also simpler than Delgrande and Renne [5]'s, Harrison-Trainer et al. [16]'s, and Ding [6]'s proofs of completeness.

The structure of this paper is as follows: In Sect. 6.2, we give a brief introduction of measurement theory. In Sect. 6.3, we argue on some relations between a qualitative-probability ordering and a probability measure, and state a representation theorem of a qualitative-probability ordering by a nonstandard probability measure. In Sect. 6.4, we define the language $\mathscr{L}_{\mathsf{MQPL}}$ of MQPL. In Sect. 6.5, we consider the semantics of MQPL. In Sect. 6.5.1, we make clear the semantic status of the model of MQPL by comparing it with the models of several logics. In Sect. 6.5.2, we give the reason why we adopt a qualitative-probability model in terms of representation theorem and arbitrary reference. In Sect. 6.5.3, we define a model $\mathfrak{M}$ of MQPL, provide MQPL with a truth definition and a validity definition, show that MQPL justifies the (in)validity of Yalcin's formulae. In Sect. 6.6, we consider the proof theory of MQPL. In Sect. 6.6.1, we provide MQPL with its proof system. In Sect. 6.6.2, we compare the axiomatic counterpart of **Finite Cancellation (S)** with that of **Finite Cancellation (D)**, In Sect. 6.7, we prove the completeness of MQPL. In Sect. 6.8, we finish with brief concluding remarks.[1]

---

[1] The reason why we arrange statement in this paper in the above order is as follows: Customarily, it may be better to arrange in such a way as the definitions, model, proof system, metalogical theorems, and, lastly, technical or philosophical discussions. However, in order to persuade, *in the early stages of reading*, readers of this paper to understand our *motivation* for adoption of a qualitative-probability model, which is what we would like to convey most in this paper especially in the sense that this model makes it possible to simplify the proof of the completeness of MQPL, instead of probability measure models that all other authors of the papers about logic of qualitative probability adopt, we would like to dare to arrange in the above order, though at first glance it might seem rather complicated and lengthy.

6.2 Brief Introduction to Measurement Theory

Measurement plays an essential role not only in science but also in our daily lives. Measurement theory investigates not concrete methods for measurement but conditions for *measurability*. The mathematical foundation of measurement had not been studied before Hölder [18] developed his axiomatization for the measurement of mass. Krantz et al. [23], Suppes et al. [39], and Luce et al. [28] are seen as milestones in the history of measurement theory. We would like to give brief explanations of the following important concepts of measurement theory:

Homomorphism Suppose a relational structure $\mathfrak{A} := (A, R_1, R_2, \ldots, R_p, \circ_1, \circ_2, \ldots, \circ_q)$ and another $\mathfrak{B} := (B, R'_1, R'_2, \ldots, R'_p, \circ'_1, \circ'_2, \ldots, \circ'_q)$, where A and B are sets, $R_1, R_2, \ldots, R_p$ are relations on A, $R'_1, R'_2, \ldots, R'_p$ are relations on B, $\circ_1, \circ_2, \ldots, \circ_q$ are operations on A, and $\circ'_1, \circ'_2, \ldots, \circ'_q$ are operations on B. f is called a *homomorphism* from $\mathfrak{A}$ into $\mathfrak{B}$ if, for any $a_1, a_2, \ldots, a_{r_i} \in A$, $R_i(a_1, a_2, \ldots, a_{r_i})$ iff $R'_i(f(a_1), f(a_2), \ldots, f(a_{r_i}))$, $i = 1, 2, \ldots, p$, and for any $a, b \in A$, $f(a \circ_i b) = f(a) \circ'_i f(b)$, $i = 1, 2, \ldots, q$.

Scale A *scale* is a triple $(\mathfrak{A}, \mathfrak{B}, f)$ where $\mathfrak{A}$ is an observed relational structure that is qualitative, $\mathfrak{B}$ is a numerical relational structure that is quantitative, and f is a homomorphism from $\mathfrak{A}$ into $\mathfrak{B}$. .

Representation and Uniqueness Theorems There are two main problems in *measurement theory*[2]:

1. the *representation problem*: Given a *numerical* relational structure $\mathfrak{B}$, find conditions on an *observed* relational structure $\mathfrak{A}$ (necessary and) sufficient for the *existence* of a homomorphism f from $\mathfrak{A}$ to $\mathfrak{B}$ that preserves all the relations and operations in $\mathfrak{A}$.

2. the *uniqueness problem*: Find the type of transformation of the homomorphism f under which all the relations and operations in $\mathfrak{A}$ are preserved.

The theorem stating conditions on $\mathfrak{A}$ are (necessary and) sufficient for the existence of f is called a *representation theorem* that can furnish a solution to the representation problem. Holman [20] classifies representations of, for example, $\mathfrak{A} := (A, \succsim)$ into the following two types:

1. *Strong representation* is defined as the construction of such a nonconstant function f on A that, for any $a, b \in A$,

$$a \succsim b \quad \text{iff} \quad f(a) \geq f(b).$$

2. *Weak representation*[3] is defined as the construction of such a nonconstant function f on A that, for any $a, b \in A$,

---

[2] Roberts [34] gives a comprehensive survey of measurement theory.

[3] About merits of weak representation, refer to Mundy [30].

$$\text{if} \quad a \succsim b, \quad \text{then} \quad f(a) \geq f(b).$$

On the other hand, the theorem stating the type of transformation up to which f above is unique is called a *uniqueness theorem*, which can furnish a solution to the uniqueness problem.

6.3 Qualitative Probability and Nonstandard Probability

Suppose that Ω is a non-empty set of elementary events, $\mathscr{F}$ a Boolean algebra of subsets of Ω, and $\succsim$ a qualitative-probability ordering on $\mathscr{F}$, and that, for any $A, B \in \mathscr{F}$,

1. $A \succ B := A \succsim B$ and not $(B \succsim A)$, and
2. $A \sim B := A \succsim B$ and $B \succsim A$.

De Finetti [13] specifies the following four necessary conditions on $(\Omega, \mathscr{F}, \succsim)$ for the existence of such a finitely additive probability measure satisfying strong representation:

1. **Nontriviality**: $\Omega \succ \emptyset$.
2. **Nonnegativity**: $A \succsim \emptyset$ for any $A \in \mathscr{F}$.
3. **Connectedness**: $A \succsim B$ or $B \succsim A$ for any $A, B \in \mathscr{F}$.
4. **Additivity**: For $A, B, C \in \mathscr{F}$, if $A \cap B = B \cap C = \emptyset$, then $A \succ B$ iff $A \cup C \succ B \cup C$.

Kraft et al. [22] show that de Finetti's conditions on $(\Omega, \mathscr{F}, \succsim)$ are not sufficient, and *strengthen*[4] **Additivity** in order to give necessary and sufficient conditions on $(\Omega, \mathscr{F}, \succsim)$ for the existence of a finitely additive probability measure satisfying *strong representation* in the case that Ω is *finite*. Scott [35] presents much the same condition in a clearer format:

Definition 1 (*Finite Cancellation (S)*) For any $n \geq 1$ and any $A_1, \ldots, A_n, B_1, \ldots, B_n \in \mathscr{F}$, if, for any $i < n$, $A_i \succsim B_i$, then $B_n \succsim A_n$, given that

$$\sum_{i=1}^{n} \chi_{A_i} = \sum_{i=1}^{n} \chi_{B_i}$$

holds, where χ_A is a *characteristic function* of $A \in \mathscr{F}$. χ_A is a function from Ω to $\{0, 1\}$ such that

$$\chi_A(\omega) := \begin{cases} 1 & \text{if} \quad \omega \in A, \\ 0 & \text{if} \quad \omega \notin A. \end{cases}$$

---

[4] In fact, Fishburn [14, p. 337] calls Finite Cancellation below *Strong Additivity*.

Remark 1 (*Intuitive Meaning of Cancellation (S)*) Intuitively, **Finite Cancellation (S)** says that whenever, for any $\omega \in \Omega$, ω is in *exactly as many* A_i's *as* B_i's, if, for any $i < n$, $A_i \succsim B_i$, then $B_n \succsim A_n$.

Scott [35, pp. 246–247] proves *strong representation* in terms of **Finite Cancellation (S)**: When Ω is finite, then there exists a finitely additive real-valued probability measure $P : \mathscr{F} \to \mathbb{R}$ satisfying *strong representation*, that is,

$$A \succsim B \text{ iff } P(A) \geq P(B)$$

iff $(\Omega, \mathscr{F}, \succsim)$ satisfies **Nontriviality**, **Nonnegativity**, **Connectedness**, and **Finite Cancellation (S)**. Domotor [7, pp. 25–35] proves that Scott's representation theorem follows from Mazur-Orlicz theorem. According to Takahashi [41, pp. 168–171], Mazur-Orlicz theorem follows from Markov-Kakutani fixed-point theorem and Fan theorem. So Scott's representation theorem follows from Markov-Kakutani fixed-point theorem and Fan theorem. Narens [31, pp. 414–415] shows that the same conditions as Scott are *sufficient* for the existence of a *nonstandard* probability measure satisfying *weak representation without the limitation of the size* of Ω. However, **Finite Cancellation (S)** is unpleasant because it is stated in terms of *characteristic functions* rather than in terms of union or other primitive notions. Domotor [8] shows the following lemma:

Lemma 1 (Equivalence)

$$\sum_{i=1}^{n} \chi_{A_i} = \sum_{i=1}^{n} \chi_{B_i} \quad \textit{iff} \quad \bigcup_{1 \leq i_1 < \cdots < i_k \leq n} (A_{i_1} \cap \cdots \cap A_{i_k}) = \bigcup_{1 \leq i_1 < \cdots < i_k \leq n} (B_{i_1} \cap \cdots \cap B_{i_k})$$

holds for any k *with* $1 \leq k \leq n$.

Proof Refer to Domotor [8, pp. 204–205]. ∎

On the basis of Lemma 1, Domotor states finite cancellation in terms of intersection, union, and $\succsim$ as follows:

Definition 2 (*Finite Cancellation (D)*)
For any $n \geq 1$ and any $A_1, \ldots, A_n, B_1, \ldots, B_n \in \mathscr{F}$, if for any $i < n$, $A_i \succsim B_i$, then $B_n \succsim A_n$, given that

$$\bigcup_{1 \leq i_1 < \cdots < i_k \leq n} (A_{i_1} \cap \cdots \cap A_{i_k}) = \bigcup_{1 \leq i_1 < \cdots < i_k \leq n} (B_{i_1} \cap \cdots \cap B_{i_k})$$

holds for any k with $1 \leq k \leq n$.

When ${}^*\mathbb{R}$ denotes a *nonstandard field* defined by the usual ultrapower construction on reals[5], we can state *Naren's theorem* by using *Domotor's notation* as follows:

---

[5] About the details of ultrapower construction on reals and (qualitative) nonstandard-real-valued probability theory, refer to Narens [33].

Fact 1 (Weak Representation for Qualitative-Probability Ordering) *There exists a finitely additive* nonstandard-real-valued *probability measure* $P : \mathscr{F} \to^* \mathbb{R}$ *satisfying* weak representation, *that is,*

$$\text{if } A \succsim B, \text{ then } P(A) \geq P(B)$$

if $(\Omega, \mathscr{F}, \succsim)$ *satisfies **Nontriviality**, **Nonnegativity**, **Connectedness**, and **Finite Cancellation (D)**.*[6]

Proof Refer to Narens [31, pp. 414–415]. ■

In Fact 1, it may seem that the introduction of nonstandard reals is too heavy. However, many philosophers (e.g., Skyrms [37, Appendix 4], Benci et al. [1], and Wenmackers and Horsten [42]) argue for nonstandard probability theory. For example, in order to support this theory, Wenmackers and Horsten [42, p. 37] appeal to the de Finetti lottery:

Example 1 (*De Finetti Lottery*) Although a fair lottery over the natural numbers may not objectively exist, we have strong intuitions about it. We may even give quantitative answers to questions such as: "What is the probability that the winning number is odd?" We may answer "Fifty per cent", by considering the probability of even (odd) numbers in finite lotteries.

According to Kolmogorov probability theory, the answer is forced to be $\frac{0}{0}$ (*undefined*). On the other hand, according to nonstandard probability theory, the answer can be $\frac{1}{2}$, which corresponds with our intuitions above. By the way, when there is no limitation of the size of Ω, is it possible to give (necessary and) sufficient conditions for the existence of a probability measure satisfying *strong representation*? Chateauneuf and Jaffray [3, p. 198][7] gives a positive answer to this question: There exists a finitely additive real-valued probability measure $P : \mathscr{F} \to \mathbb{R}$ satisfying *strong representation* iff $(\Omega, \mathscr{F}, \succsim)$ satisfies **Nontriviality**, **Nonnegativity**, **Connectedness**, and **Archimedeanity (CJ)**[8]: For any $m \geq 2$ and any $A_1, \ldots, A_m, B_1, \ldots, B_m \in \mathscr{F}$, if $A_1 \succsim B_1$, then there is $N \in \mathbb{N}^+$ such that if

$$k\chi_{\emptyset} + n\chi_{A_1} + \sum_{i=2}^{m} \chi_{A_i} = k\chi_{\Omega} + n\chi_{B_1} + \sum_{i=2}^{m} \chi_{B_i} \text{ with } k \geq 0 \text{ and } n \geq 1$$

[6] In this paper, "Fact" means a theorem already proven by others.

[7] Suppes and Zanotti [38] also try to prove strong representation without limitation of the size of Ω. According to Cohen [4, p. 244], their conditions depend on the existence and uniqueness of an extension of the original ordering. However, these conditions are not definable in terms of the original ordering.

[8] Although not all forms of Archimedeanity imply Finite Cancellation, **Archimedeanity (CJ)** implies **Finite Cancellation (S)**.

and

$$\text{for any } j \geq 2, A_j \precsim B_j,$$

then $\frac{k}{n} > \frac{1}{N}$. We would like to provide MQPL with its model to which neither Scott's nor Chateauneuf and Jaffray's representation theorem but Fact 1 can be applied. The reason why we do not apply Scott's representation theorem is as follows: The domain Ω of the model (Definition 4 below) of MQPL has no limitation of size. However, the domain of this theorem is finite. So in order to make it possible to apply Scott's representation theorem, we must construct *finite* Boolean algebras. This construction is a lengthy process. The reason why we do not apply Chateauneuf and Jaffray's representation theorem is as follows: We would like to prove the completeness of MQPL. However, the axiomatic counterpart of **Archimedeanity (CJ)** is not expressible in the language $\mathscr{L}_{\mathsf{MQPL}}$ (cf. Definition 3 below) of MQPL. Because **Archimedeanity (CJ)** has an existential quantification on $N \in \mathbb{N}^+$, it is not expressible as an infinite schema in propositional logic, whereas **Finite Cancellation (D)** is expressible as an infinite schema in propositional logic. Since MQPL is a version of propositional logic, **Archimedeanity (CJ)** is not expressible in $\mathscr{L}_{\mathsf{MQPL}}$. So the model $\mathfrak{M}$ (cf. Definition 4 below) of MQPL cannot presuppose **Archimedeanity (CJ)**. We would like to close this section with a uniqueness theorem for probability measures. Roberts [34, pp. 391–392] and Narens [32, pp. 291–293] independently prove the uniqueness theorem. In order to prove the uniqueness theorem, both of them require some conditions for $\mathscr{F}$. The reason why we do not apply the uniqueness theorem is the same as the reason why we do not apply Chateauneuf and Jaffray's representation theorem. For the axiomatic counterparts of these conditions are not expressible in $\mathscr{L}_{\mathsf{MQPL}}$.

6.4 Language of Modal-Qualitative-Probability Logic (MQPL)

We define the language $\mathscr{L}_{\mathsf{MQPL}}$ of MQPL, which is the *same language* as Gärdenfors [15], as follows:

Definition 3 (*Language*)

- (i) Let $\mathscr{S}$ denote a set of sentential variables and $\succcurlyeq$ a binary sentential operator.
- (ii) The language $\mathscr{L}_{\mathsf{MQPL}}$ of MQPL is given by the following BNF grammar:

$$\varphi ::= s \mid \top \mid \neg\varphi \mid (\varphi \wedge \varphi) \mid (\varphi \succcurlyeq \varphi)$$

  such that $s \in \mathscr{S}$.
- (iii) $\bot$, $\vee$, $\rightarrow$ and $\leftrightarrow$ are introduced by the standard definitions.
- (iv) $\varphi \succcurlyeq \psi$ is interpreted to mean that φ is *at least as likely as* ψ.
- (v) $\varphi \succ \psi := (\varphi \succcurlyeq \psi) \wedge \neg(\psi \succcurlyeq \varphi)$.

- (vi) $\varphi > \psi$ is interpreted to mean that φ is *more likely than* ψ.
- (vii) $\varphi \approx \psi := (\varphi \geqslant \psi) \wedge (\psi \geqslant \varphi)$.
- (viii) $\varphi \approx \psi$ is interpreted to mean that φ is *as likely as* ψ.
- (ix) $\Delta\varphi := \varphi > \neg\varphi$.
- (x) $\Delta\varphi$ is interpreted to mean that *probably* φ.
- (xi) $\Box\varphi := \varphi \approx \top$.
- (xii) $\Box\varphi$ is interpreted to mean that *necessarily* φ.
- (xiii) $\Diamond\varphi := \neg\Box\neg\varphi$.
- (xiv) $\Diamond\varphi$ is interpreted to mean that *possibly* φ.
- (xv) The set of all well-formed formulae of $\mathscr{L}_{\mathsf{MQPL}}$ is denoted by $\Phi_{\mathscr{L}_{\mathsf{MQPL}}}$.

6.5 Semantics of MQPL

6.5.1 *Semantic Status of Model of* MQPL

In this subsection, we would like to make clear the semantic status of the model of MQPL by comparing it with the models of several logics of Holliday and Icard [19]. We have adopted the same language $\mathscr{L}_{\mathsf{MQPL}}$ as $\mathscr{L}$ of [19]. Holliday and Icard consider *three* kinds of models of logics of epistemic modals:

1. measure model,
   (a) *finitely additive probability measure model,*
   (b) *qualitatively additive measure model,*
2. *event-ordering model,*
3. world-ordering model
   (a) Kratzer's world-ordering model,
   (b) *revised version of Kratzer's world-ordering model.*

They show that a complete logic $\mathbf{FP}_\infty$ based on a *finitely additive probability measure model*, a complete logic **FA** based on a *qualitatively additive measure model*, and a logic $\mathbf{WP}_\infty\mathbf{R}$ based on a *revised version of Kratzer's world-ordering model* each validate all of (V1)–(V12) below and validate neither (I1) nor (I2) below.[9] On the other hand, in this paper, we would like to propose a complete logic MQPL based on a kind of *event-ordering* model—*qualitative-probability* model. Whereas this model does not have as a component a *quantitative* probability space assignment that any finitely additive probability measure model has, Fact 1 says that if a qualitative-probability ordering $\succsim_w$ satisfies all of the conditions of Fact 1, though

[9] Marushak [29] points out that Holliday and Icard's revised version of Kratzer's model faces some problems in case of infinite domains.

actual qualitative-probability orderings may not satisfy all of them, it can *represent* a probability measure. Fact 1 is an *existence* theorem. So the essential difference between a finitely additive probability measure model and a qualitative-probability model is that whereas the former directly gives an *arbitrary*[10] probability measure, the latter provides not an arbitrary probability measure but the *existence* of a probability measure in terms of *representation*.

6.5.2 Motivation to Adopt Qualitative-Probability Model: Representation Theorem and Arbitrary Objects or Reference

As Kratzer [25, p. 25] says, "Our semantic knowledge alone does not give us the precise quantitative notions of probability and desirability that mathematicians and scientists work with", the first problem in adopting a probability measure model as a model for epistemic modals arises form the unnaturalness of the preciseness of probability-value assignment. For, in fact, speakers of epistemic modals do not always assign such precise probability-values as probability measure models require. On the other hand, let us consider a qualitative-probability model. The representation theorem (Fact 1) is a *conditional* sentence. It states that if $\succsim_\omega$ in the qualitative-probability model $\mathfrak{M}$ (Definition 4) satisfies such and such conditions, then there exists a probability measure. So $\succsim_\omega$ itself does not require the direct (absolute) preciseness of probability-value assignment. In this way $\mathfrak{M}$ can avoid the first problem. The second problem in adopting a probability measure model as a model for epistemic modals arises from instantial reasoning via a representation theorem implicitly presupposed by this model that may invite the semantic problems of arbitrary objects or reference. For example, Segerberg [36] and Gärdenfors [15] adopt a probability measure model for logic of qualitative probability. These models are based on Scott [35]'s representation theorem. They implicitly presuppose the following reasoning:

> By Scott's representation theorem, *there exists* a probability measure P satisfying $A \succsim B$ iff $P(A) \geq P(B)$ iff the conditions for representation are satisfied. So let P_1 be such an *arbitrary* probability measure. This measure P_1 constitutes the probability measure model.

Of course, if the proof of Scott's representation theorem were a *constructive proof*, then the above reasoning would have no problem. But in fact this proof [35, pp. 246–247]is no constructive proof. Fine calls "instantial reasoning" reasoning based on such rules as Universal Generalization (UG) and Existential Instantiation (EI). The above reasoning presupposed in probability measure models is an example of EI. According to Breckenridge and Magidor [2, p. 378], instantial reasoning implicitly presupposes such a stipulation as "Let P_1 be such an arbitrary probability measure". But the *semantics* underlying such a stipulation is far from clear. Breckenridge and Magidor [2, p. 383] classify existing accounts of instantial reasoning into three kinds:

---

[10] About arbitrary reference, refer to Sect. 6.5.2.

1. the instrumentalist account: According to it, instantial terms and thus the lines in which they occur are simply meaningless. The reason why we nevertheless include such lines in our arguments is that they are instrumentally valuable.
2. quantificational accounts (e.g., King [21]): According to them, we should think of instantial terms as variables, ones that are implicitly bound by quantifiers.
3. Fine [10–12]'s referential account that is the most prominent view: According to Fine, instantial terms refer to special kinds of objects—"arbitrary objects"—which are distinct from any particular object.

Breckenridge and Magidor criticize [2, pp. 383–397] each of these accounts and give their own account called the arbitrary reference (AR) account: It is possible to fix the reference of an expression arbitrarily. When we do so, the expression receives its ordinary kind of semantic value, though we do not and cannot know which value in particular it receives. But Haze [17] gives a counterexample to the AR account. In this way, instantial reasoning is very polemic from a semantic point of view. On the other hand, because the qualitative-probability model $\mathfrak{M}$ has directly as a constituent a qualitative-probability ordering $\succsim_\omega$, it does not presuppose instantial reasoning. So $\mathfrak{M}$ can avoid the second problem. So we would like to adopt a qualitative-probability ordering model as a model for a comparative epistemic modal "at least as likely as", which avoids both the first and second problems.

6.5.3 Model of MQPL

We define a structured model $\mathfrak{M}$ of MQPL as follows:

Definition 4 (*Model*) $\mathfrak{M}$ is a triple (Ω, ρ, V) in which

1. Ω is a non-empty set of elementary events,
2. ρ is a qualitative-probability space assignment that assigns to each $\omega \in \Omega$ a qualitative-probability space $(\Omega, \mathscr{F}, \succsim_\omega)$ in which $\succsim_\omega$ is a qualitative-probability ordering on a Boolean algebra $\mathscr{F}$ of subsets of Ω that satisfies all of **Nontriviality**, **Nonnegativity**, **Connectedness**, and **Finite Cancellation (D)** of Fact 1, and
3. V is a truth assignment to each $s \in \mathscr{S}$ for each $\omega \in \Omega$.

We provide MQPL with the following truth definition at $\omega \in \Omega$ in $\mathfrak{M}$, define the truth in $\mathfrak{M}$, and then define validity as follows:

Definition 5 (*Truth and Validity*) The notion of $\varphi \in \Phi_{\mathscr{L}_{\mathsf{MQPL}}}$ being true at $\omega \in \Omega$ in $\mathfrak{M}$, in symbols $(\mathfrak{M}, \omega) \models_{\mathsf{MQPL}} \varphi$, is inductively defined as follows:

1. (a) $(\mathfrak{M}, \omega) \models_{\mathsf{MQPL}} s$ iff $V(\omega)(s) = \mathit{true}$.
   (b) $(\mathfrak{M}, \omega) \models_{\mathsf{MQPL}} \top$.
   (c) $(\mathfrak{M}, \omega) \models_{\mathsf{MQPL}} \neg\varphi$ iff $(\mathfrak{M}, \omega) \not\models_{\mathsf{MQPL}} \varphi$.
   (d) $(\mathfrak{M}, \omega) \models_{\mathsf{MQPL}} \varphi \land \psi$ iff $(\mathfrak{M}, \omega) \models_{\mathsf{MQPL}} \varphi$ and $(\mathfrak{M}, \omega) \models_{\mathsf{MQPL}} \psi$.

(e) $(\mathfrak{M}, \omega) \models_{\mathsf{MQPL}} \varphi \geqslant \psi$ iff $[\![\varphi]\!]^{\mathfrak{M}} \succsim_{\omega} [\![\psi]\!]^{\mathfrak{M}}$, where $[\![\varphi]\!]^{\mathfrak{M}} := \{\omega' \in \Omega : (\mathfrak{M}, \omega') \models_{\mathsf{MQPL}} \varphi\}$.

2. If $(\mathfrak{M}, \omega) \models_{\mathsf{MQPL}} \varphi$ for all $\omega \in \Omega$, we write $\mathfrak{M} \models_{\mathsf{MQPL}} \varphi$ and say that φ is true in $\mathfrak{M}$.

3. If φ is true in all models of MQPL, we write $\models_{\mathsf{MQPL}} \varphi$ and say that φ is valid.

Remark 2 (*Independence*) Like Lewis [27], for any $\omega_1, \omega_2 \in \Omega$, $\succsim_{\omega_1}$ is given independently of $\succsim_{\omega_2}$.

Remark 3 (*Qualitative-Probability Ordering and Truth of* φ) In the way of the truth clause of $\varphi \geqslant \psi$, the qualitative-probability ordering $\succsim_{\omega}$ bears some relation to the truth of φ defined by the truth-value assignment V to each $s \in \mathscr{S}$ for each $\omega \in \Omega$.

For example, by Definition 5, we have the following truth condition of $(\varphi \geqslant \psi_1) \geqslant (\varphi \geqslant \psi_2)$. This truth condition could help readers to understand the semantics of MQPL:

Example 2 (*Truth Condition of* $(\varphi \geqslant \psi_1) \geqslant (\varphi \geqslant \psi_2)$)

$$\begin{aligned}
&(\mathfrak{M}, \omega) \models_{\mathsf{MQPL}} (\varphi \geqslant \psi_1) \geqslant (\varphi \geqslant \psi_2)\\
&\text{iff } [\![\varphi \geqslant \psi_1]\!]^{\mathfrak{M}} \succsim_{\omega} [\![\varphi \geqslant \psi_2]\!]^{\mathfrak{M}}\\
&\text{iff } \{\omega_1 \in \Omega : (\mathfrak{M}, \omega_1) \models_{\mathsf{MQPL}} \varphi \geqslant \psi_1\} \succsim_{\omega} \{\omega_2 \in \Omega : (\mathfrak{M}, \omega_2) \models_{\mathsf{MQPL}} \varphi \geqslant \psi_2\}\\
&\text{iff } \{\omega_1 \in \Omega : [\![\varphi]\!]^{\mathfrak{M}} \succsim_{\omega_1} [\![\psi_1]\!]^{\mathfrak{M}}\} \succsim_{\omega} \{\omega_2 \in \Omega : [\![\varphi]\!]^{\mathfrak{M}} \succsim_{\omega_2} [\![\psi_2]\!]^{\mathfrak{M}}\}.
\end{aligned}$$

Remark 4 (*Outermost* $\geqslant$) Pay attention to the fact that in the truth condition above of $(\varphi \geqslant \psi_1) \geqslant (\varphi \geqslant \psi_2)$ an evaluation world (elementary event) is referred not in all the semantic counterparts of three $\geqslant$'s but in only that of the outermost $\geqslant$.

The next corollary follows from Definitions 3 and 5:

Corollary 1 *(Truth Condition of* $\Box\varphi$*, That of* $\Diamond\varphi$*, and That of* $\triangle\varphi$*)*

1. $(\mathfrak{M}, \omega) \models_{\mathsf{MQPL}} \Box\varphi$ *iff* $[\![\varphi]\!]^{\mathfrak{M}} \sim_{\omega} \Omega$.

2. $(\mathfrak{M}, \omega) \models_{\mathsf{MQPL}} \Diamond\varphi$ *iff* $[\![\varphi]\!]^{\mathfrak{M}} \not\sim_{\omega} \emptyset$.

3. $(\mathfrak{M}, \omega) \models_{\mathsf{MQPL}} \triangle\varphi$ *iff* $[\![\varphi]\!]^{\mathfrak{M}} \succ_{\omega} \overline{[\![\varphi]\!]^{\mathfrak{M}}}$.

The representation theorem (Fact 1) plays an essential role in the next corollary. It follows from Fact 1 and Definition 5:

Corollary 2 *(Truth Condition of* $\varphi \geqslant \psi$ *by Probability Measure) There exists a finitely additive* nonstandard-real-valued *probability measure* $P_{\omega} : \mathscr{F} \rightarrow^{*} \mathbb{R}$ *satisfying*

$$(\mathfrak{M}, \omega) \models_{\mathsf{MQPL}} \varphi \geqslant \psi \;\; \textit{iff} \;\; [\![\varphi]\!]^{\mathfrak{M}} \succsim_{\omega} [\![\psi]\!]^{\mathfrak{M}} \;\; \textit{only if} \;\; P_{\omega}([\![\varphi]\!]^{\mathfrak{M}}) \geq P_{\omega}([\![\psi]\!]^{\mathfrak{M}})$$

if ***Nontriviality, Nonnegativity, Connectedness,*** *and* ***Finite Cancellation (D)*** *are satisfied.*

Yalcin [43] presents the following list of intuitively valid formulae (V1)–(V11) and intuitively invalid formulae (I1) and (I2):

$$
\begin{aligned}
&(V1)\ \Delta\varphi \to \neg\Delta\neg\varphi,\ (V2)\ \Delta(\varphi \wedge \psi) \to (\Delta\varphi \wedge \Delta\psi),\\
&(V3)\ \Delta\varphi \to \Delta(\varphi \vee \psi),\\
&(V4)\ \varphi \geqslant \bot,\\
&(V5)\ \top \geqslant \varphi,\\
&(V6)\ \Box\varphi \to \Delta\varphi,\\
&(V7)\ \Delta\varphi \to \Diamond\varphi,\\
&(V8)\ (\varphi \to \psi) \to (\Delta\varphi \to \Delta\psi),\\
&(V9)\ (\varphi \to \psi) \to (\neg\Delta\psi \to \neg\Delta\varphi),\\
&(V10)\ (\varphi \to \psi) \to (\psi \geqslant \varphi),\\
&(V11)\ (\psi \geqslant \varphi) \to (\Delta\varphi \to \Delta\psi),\\
&(V12)\ (\psi \geqslant \varphi) \to ((\varphi \geqslant \neg\varphi) \to (\psi \geqslant \neg\psi)),\\
&(I1)\ ((\varphi \geqslant \psi) \wedge (\varphi \geqslant \chi)) \to (\varphi \geqslant (\psi \vee \chi)),\ \text{and}\\
&(I2)\ ((\varphi \geqslant \neg\varphi) \wedge (\neg\varphi \geqslant \varphi)) \to (\varphi \geqslant \psi).
\end{aligned}
$$

MQPL justifies the (in)validity of Yalcin's formulae as follows:

Proposition 1 (Justification of Yalcin's Formulae) MQPL validates all of (V1)–(V12) and validate neither (I1) nor (I2).

Proof First, we would like to take up (V12) in order to demonstrate that **Finite Cancellation (D)** of Definition 2 plays an essential role in deducing valid formulae, though it seems to be only a structural axiom: In order to demonstrate (V12), we first demonstrate $\models_{\mathsf{MQPL}} ((\varphi \geqslant \psi) \wedge (\psi \geqslant \chi)) \to (\varphi \geqslant \chi)$ (Transitivity of $\geqslant$). In **Finite Cancellation (D)** of Definition 2, let us put $n = 3$ and $A_1 = [\![\varphi]\!]^{\mathfrak{M}}$, $A_2 = [\![\psi]\!]^{\mathfrak{M}}$, $A_3 = [\![\chi]\!]^{\mathfrak{M}}$, $B_1 = [\![\psi]\!]^{\mathfrak{M}}$, $B_2 = [\![\chi]\!]^{\mathfrak{M}}$, $B_3 = [\![\varphi]\!]^{\mathfrak{M}}$.

(i) $k = 1$

$[\![\varphi]\!]^{\mathfrak{M}} \cup [\![\psi]\!]^{\mathfrak{M}} \cup [\![\chi]\!]^{\mathfrak{M}} = [\![\psi]\!]^{\mathfrak{M}} \cup [\![\chi]\!]^{\mathfrak{M}} \cup [\![\varphi]\!]^{\mathfrak{M}}$.

(ii) $k = 2$

$([\![\varphi]\!]^{\mathfrak{M}} \cap [\![\psi]\!]^{\mathfrak{M}}) \cup ([\![\varphi]\!]^{\mathfrak{M}} \cap [\![\chi]\!]^{\mathfrak{M}}) \cup ([\![\psi]\!]^{\mathfrak{M}} \cap [\![\chi]\!]^{\mathfrak{M}}) = ([\![\psi]\!]^{\mathfrak{M}} \cap [\![\chi]\!]^{\mathfrak{M}}) \cup ([\![\psi]\!]^{\mathfrak{M}} \cap [\![\varphi]\!]^{\mathfrak{M}}) \cup ([\![\chi]\!]^{\mathfrak{M}} \cap [\![\varphi]\!]^{\mathfrak{M}})$.

(iii) $k = 3$.

$[\![\varphi]\!]^{\mathfrak{M}} \cap [\![\psi]\!]^{\mathfrak{M}} \cap [\![\chi]\!]^{\mathfrak{M}} = [\![\psi]\!]^{\mathfrak{M}} \cap [\![\chi]\!]^{\mathfrak{M}} \cap [\![\varphi]\!]^{\mathfrak{M}}$.

Then since the equation of **Finite Cancellation (D)** holds, we get Transitivity of $\geqslant$. In order to demonstrate (V12), we second demonstrate $\models_{\mathsf{MQPL}} (\varphi \geqslant \psi) \to (\neg\psi \geqslant \neg\varphi)$ (Contraposition of $\varphi \geqslant \psi$). In **Finite Cancellation (D)**, let us put $n = 2$ and $A_1 = [\![\varphi]\!]^{\mathfrak{M}}$, $A_2 = \overline{[\![\varphi]\!]^{\mathfrak{M}}}$, $B_1 = [\![\psi]\!]^{\mathfrak{M}}$, $B_2 = \overline{[\![\psi]\!]^{\mathfrak{M}}}$.

(i) $k = 1$

$[\![\varphi]\!]^{\mathfrak{M}} \cup \overline{[\![\varphi]\!]^{\mathfrak{M}}} = [\![\psi]\!]^{\mathfrak{M}} \cup \overline{[\![\psi]\!]^{\mathfrak{M}}} = \Omega$.

(ii) $k = 2$

$[\![\varphi]\!]^{\mathfrak{M}} \cap \overline{[\![\varphi]\!]^{\mathfrak{M}}} = [\![\psi]\!]^{\mathfrak{M}} \cap \overline{[\![\psi]\!]^{\mathfrak{M}}} = \emptyset$.

Then since the equation of **Finite Cancellation (D)** holds, we get Contraposition of $\varphi \geqslant \psi$. Suppose that $\psi \geqslant \varphi$ and $\varphi \geqslant \neg\varphi$. Then from Contraposition of $\psi \geqslant \varphi$, $\psi \geqslant \varphi \geqslant \neg\varphi \geqslant \neg\psi$ follows. From this and Transitivity of $\geqslant$, we get (V12).

Second, we would like to give a countermodel $\mathfrak{U}$ to (I1) and (I2). $\mathfrak{U}$ is a triple (Ω, ρ, V) based on Definition 4 where:

1. Ω is a set of elementary events ω_i of getting a number $i = 1, 2, \ldots, 6$ when one rolls a dice.
2. $\{\omega_1\} \sim_{\omega_i} \{\omega_2\} \sim_{\omega_i} \cdots \sim_{\omega_i} \{\omega_6\}$, for $i = 1, 2, \ldots, 6$.
3. In (I1):
   φ: One gets a number more than 2.
   ψ: One gets an odd number.
   χ: One gets an even number.
4. In (I2):
   φ: One gets a number more than 3.
   ψ: One gets a number more than 2.

Then we have the following:

$$\mathfrak{U} \not\models_{\mathsf{MQPL}} ((\varphi \geqslant \psi) \wedge (\varphi \geqslant \chi)) \rightarrow (\varphi \geqslant (\psi \vee \chi)), \text{ and}$$
$$\mathfrak{U} \not\models_{\mathsf{MQPL}} ((\varphi \geqslant \neg\varphi) \wedge (\neg\varphi \geqslant \varphi)) \rightarrow (\varphi \geqslant \psi).$$

■

6.6 Axiomatization of MQPL

6.6.1 Proof System of MQPL

The proof system of MQPL consists of the following:

Definition 6 (*Proof System*)

1. All tautologies of classical propositional logic,
2. $\Box(\varphi \rightarrow \psi) \rightarrow (\Box\varphi \rightarrow \Box\psi)$ (K),
3. $(\Box(\varphi_1 \leftrightarrow \varphi_2) \wedge \Box(\psi_1 \leftrightarrow \psi_2)) \rightarrow ((\varphi_1 \geqslant \psi_1) \leftrightarrow (\varphi_2 \geqslant \psi_2))$ (Replacement of Necessary Equivalents on $\geqslant$),
4. $\top > \bot$ (Axiomatic Counterpart of **Nontriviality**),
5. $\varphi \geqslant \bot$ (Axiomatic Counterpart of **Nonnegativity**),
6. $(\varphi \geqslant \psi) \vee (\psi \geqslant \varphi)$ (Axiomatic Counterpart of **Connectedness**),
7.
$$\left(\bigvee_{1 \leq i_1 < \cdots < i_k \leq n} (\varphi_{i_1} \wedge \cdots \wedge \varphi_{i_k}) \leftrightarrow \bigvee_{1 \leq i_1 < \cdots < i_k \leq n} (\psi_{i_1} \wedge \cdots \wedge \psi_{i_k})\right)$$
$$\rightarrow \left(\bigwedge_{i=1}^{n-1} (\varphi_i \geqslant \psi_i) \rightarrow (\psi_n \geqslant \varphi_n)\right), \text{ for any } n \geq 1 \text{ and any } k \text{ with } 1 \leq k \leq n,$$

(Axiomatic Counterpart of **Finite Cancellation (D)**),
8. Modus Ponens, and
9. Necessitation.
10. A proof of $\varphi \in \Phi_{\mathscr{L}_{\mathsf{MQPL}}}$ is a finite sequence of $\mathscr{L}_{\mathsf{MQPL}}$-formulae having φ as the last formula such that either each formula is an instance of an axiom or it can be obtained from formulae that appear earlier in the sequence by applying an inference rule.
11. If there is a proof of φ, we write $\vdash_{\mathsf{MQPL}} \varphi$.

Remark 5 (*Infinite Schema*) The axiomatic counterpart of **Finite Cancellation (D)** is an *infinite schema* of axioms for any $n \geq 1$ and any k with $1 \leq k \leq n$.

6.6.2 Axiomatic Counterpart of Finite Cancellation (S) and That of Finite Cancellation (D)

Segerberg [36] introduces the following abbreviation schema:

Definition 7 (*Abbreviation Schema*)

$$\varphi_1 \ldots \varphi_n \mathbb{E} \psi_1 \ldots \psi_n := \Box(C_0 \vee \ldots \vee C_n),$$

where $n \geq 1$ and, for $0 \leq i \leq n$, the schema C_i is the disjunction of all the following conjunctions:

$$(\bigstar) \quad \delta_1\varphi_1 \wedge \ldots \wedge \delta_n\varphi_n \wedge \epsilon_1\psi \wedge \ldots \wedge \epsilon_n\psi_n$$

such that exactly i of the δ's and i of the ϵ's are the empty symbol string, the rest of them being the negation signs.

Remark 6 (*Characteristic Function*) δ's and ϵ's play the role of the characteristic function χ in Definition 1.

In terms of Definition 7, Segerberg gives the following axiomatic counterpart of **Finite Cancellation (S)** with the antecedent accompanied by $\Box$: For any $n \geq 1$,

$$\varphi_1 \ldots \varphi_n \mathbb{E} \psi_1 \ldots \psi_n \rightarrow \left(\bigwedge_{i=1}^{n-1} (\varphi_i \gtrsim \psi_i) \rightarrow (\psi_n \gtrsim \varphi_n) \right).$$

We have the following proposition:

Proposition 2 (Equivalence) For any $n \geq 1$, When

$$\varphi_1 \ldots \varphi_n \mathbb{E}^{-\Box} \psi_1 \ldots \psi_n := C_0 \vee \ldots \vee C_n,$$

$$\varphi_1 \dots \varphi_n \mathbb{E}^{-\Box} \psi_1 \dots \psi_n \text{ iff } \bigvee_{1 \le i_1 < \dots < i_k \le n} (\varphi_{i_1} \wedge \dots \wedge \varphi_{i_k}) \leftrightarrow \bigvee_{1 \le i_1 < \dots < i_k \le n} (\psi_{i_1} \wedge \dots \wedge \psi_{i_k}), \text{ for any } k \text{ with} 1 \le k \le n.$$

Remark 7 (*Axiomatic Counterpart of Lemma* 1) Proposition 2 s a axiomatic counterpart of Lemma 1.

Proposition 2 is a axiomatic counterpart of Lemma 1. Because $\varphi_1 \dots \varphi_n \mathbb{E}^{-\Box} \psi_1 \dots \psi_n$ is the *disjunction* of the *disjunctions* of the *conjunctions* (★), by Proposition 2, we can say that the axiomatic counterpart of **Finite Cancellation (D)** is an *equivalent* and *simple* version of the axiomatic counterpart of **Finite Cancellation (S)**.

By Definition 7, $\varphi_1 \dots \varphi_n \mathbb{E} \psi_1 \dots \psi_n$ is the *disjunction* of the *disjunctions* of the *conjunctions* (★). So by Proposition 2, we can say that the axiomatic counterpart of **Finite Cancellation (D)** is an *equivalent* and *simple* version of the axiomatic counterpart of **Finite Cancellation (S)**.

6.7 Metalogic of MQPL

6.7.1 Soundness

We prove the soundness of MQPL:

Theorem 1 (Soundness of MQPL) *For any* $\psi \in \Phi_{\mathscr{L}_{\mathsf{MQPL}}}$, *if* $\vdash_{\mathsf{MQPL}} \psi$, *then* $\models_{\mathsf{MQPL}} \psi$.

Proof It is trivial to verify the conditions except the axiomatic counterpart of **Finite Cancellation (D)**. By Proposition 2, this verification becomes much the same as Segerberg [36, pp. 344–346]'s. ■

6.7.2 Completeness of MQPL

We now turn to the proof of the completeness of MQPL. Both Segerberg [36]'s proof of the completeness of PK and Gärdenfors [15]' proof of the completeness of QP use Scott [35]'s representation theorem with *finite* domain, whereas our proof the completeness of MQPL does not *directly* use any representation theorem including Fact 1, though Fact 1, that is, a representation theorem with *no limitation of size* can be applied in the model of MQPL. So because the construction of *finite* Boolean algebras to make it possible to apply Scott's representation theorem covers a considerable part of both Segerberg's proof and Gärdenfors' proof, whereas our proof does not require this construction, our proof has the merit that it is by far simpler than Segerberg's proof and Gärdenfors' proof. Moreover, our proof is also simpler than Delgrande and Renne [5] 's, Harrison-Trainer et al. [16]'s, and Ding [6]'s proofs of completeness.

We would like to prove the completeness of MQPL as follows. Let $\varphi \in \Phi_{\mathscr{L}_{\mathsf{MQPL}}}$ any such formula that $\nvdash_{\mathsf{MQPL}} \varphi$. We would like to construct a structural model $\mathfrak{M}^{\varphi} := (\Omega^{\varphi}, \rho^{\varphi}, V^{\varphi})$ where $\rho^{\varphi} := (\Omega^{\varphi}, \mathscr{F}^{\varphi}. \precsim_{\omega})$ such that $\mathfrak{M}^{\varphi} \not\models_{\mathsf{MQPL}} \varphi$. First, we define equivalence class $[\varphi]$:

Definition 8 (*Equivalence Class*) For any $\varphi \in \Phi_{\mathscr{L}_{\mathsf{MQPL}}}$, the *equivalence class* $[\varphi] := \{\psi \in \Phi_{\mathscr{L}_{\mathsf{MQPL}}} : \vdash_{\mathsf{MQPL}} \varphi \leftrightarrow \psi\}$.

The following lemma follows directly from Definition 8:

Lemma 2 (Provability and Equivalence Class) $\vdash_{\mathsf{MQPL}} \varphi$ *iff* $[\varphi] = [\top]$.

We define Boolean operators $\overline{\cdots}$ and $\cap$ on the set of equivalence classes:

Definition 9 ($\overline{\cdots}$ *and* $\cap$) $\overline{\cdots}$ and $\cap$ on the set of equivalence classes are defined in such a way that $\overline{[\varphi]} := [\neg\varphi]$ and $[\varphi] \cap [\psi] := [\varphi \wedge \psi]$ for any $\varphi, \psi \in \Phi_{\mathscr{L}_{\mathsf{MQPL}}}$.

By Definitions 8 and 9, we define ρ^{φ}:

Definition 10 (ρ^{φ}) For any $[s]$ where $s \in \mathscr{S}$ we can choose s^* as its representative. $\rho^{\varphi} := (\Omega^{\varphi}, \mathscr{F}^{\varphi}. \precsim_{\omega})$ such that $\mathfrak{M}^{\varphi} \not\models_{\mathsf{MQPL}} \varphi$, where Ω^{φ} is the set of all s^*'s, and $\mathscr{F}^{\varphi}$ is the infinite Boolean algebra of s^*'s with $[\top]$ and $[\bot]$ as unit and zero elements respectively.

Remark 8 (*Representative as Elementary Event*) Generally speaking, in probability theory, the status of an elementary event ω is not specified. This non-specificity guarantees that we can choose as an elementary event ω a representative s^* of the equivalence class of sentential variables that is a syntactic object.

Then we define $[\psi] \succsim_{\omega} [\chi]$ in terms of $[\psi \geqslant \chi]$:

Definition 11 ($\succsim_{\omega}$ *on Set of Equivalence Classes*) For any $\omega \in \Omega^{\varphi}$, for any $\psi, \chi \in \Phi_{\mathscr{L}_{\mathsf{MQPL}}}$, $[\psi] \succsim_{\omega} [\chi]$ iff $\omega \in [\psi \geqslant \chi]$.

Then we prove the next lemma:

Lemma 3 (Satisfaction of Conditions for Representation) *For any* $\omega \in \Omega^{\varphi}$, $\succsim_{\omega}$ *of Definition 11 satisfies* ***Nontriviality**, **Nonnegativity**, **Connectedness**, and **Finite Cancellation (D)** of Fact 1.*

Proof It is trivial to verify that **Nontriviality**, **Nonnegativity**, **Connectedness** are satisfied given their axiomatic counterparts. For the verification of **Finite Cancellation (D)**, we make the following three assumptions:

1. Assume that $[\psi_1], \ldots, [\psi_n], [\chi_1], \ldots, [\chi_n] \in \mathscr{F}^{\varphi}$.

2. Assume that the equation part of **Finite Cancellation (D)** holds for these elements:

$$\bigcup_{1 \leq i_1 < \cdots < i_k \leq n} ([\psi_{i_1}] \cap \cdots \cap [\psi_{i_k}]) = \bigcup_{1 \leq i_1 < \cdots < i_k \leq n} ([\chi_{i_1}] \cap \cdots \cap [\chi_{i_k}]),$$

   for any $n \geq 1$ and any k with $1 \leq k \leq n$, which is the same as assuming that any $\omega \in \Omega^{\varphi}$ belongs to exactly as many $[\psi_i]$'s as $[\chi_i]$'s (cf. Remark 1).

3. Assume that $[\psi_1] \succsim_\omega [\chi_1], \ldots, [\psi_{n-1}] \succsim_\omega [\chi_{n-1}]$.

Then by Assumption 2 and Definition 9,

$$\omega \in \left[\bigvee_{1 \le i_1 < \cdots < i_k \le n} (\psi_{i_1} \wedge \cdots \wedge \psi_{i_k}) \leftrightarrow \bigvee_{1 \le i_1 < \cdots < i_k \le n} (\chi_{i_1} \wedge \cdots \wedge \chi_{i_k}) \right],$$

for any $n \geq 1$ and any k with $1 \leq k \leq n$. This holds for any $\omega \in \Omega^\varphi$. So

$$\left[\bigvee_{1 \le i_1 < \cdots < i_k \le n} (\psi_{i_1} \wedge \cdots \wedge \psi_{i_k}) \leftrightarrow \bigvee_{1 \le i_1 < \cdots < i_k \le n} (\chi_{i_1} \wedge \cdots \wedge \chi_{i_k}) \right] = [\top],$$

for any $n \geq 1$ and any k with $1 \leq k \leq n$. So from the axiomatic counterpart of **Finite Cancellation (D)**, Lemma 2 and Modus Ponens

$$\left[\bigwedge_{i=1}^{n-1} (\psi_i \geqslant \chi_i) \rightarrow (\chi_n \geqslant \psi_n) \right] = [\top],$$

for any $n \geq 1$. So from Assumption 3, Definition 8, Lemma 2 and Modus Ponens, we have $\omega \in [\chi_n \geqslant \psi_n]$. So from Definition 11, $[\chi_n] \succsim_\omega [\psi_n]$. ■

Remark 9 (*Weak Representation*) By Fact 1, the qualitative-probability ordering $\succsim_\omega$ of Lemma 3 *represents* a finitely additive nonstandard-real-valued probability measure P_ω. However, we do not *directly* use any probability measure in the proof of the completeness of MQPL. In other words, we do not directly use Fact 1. So Fact 1 does not have to possess strong representation.

Then we prove the following lemma from Lemma 3 as follows:

Lemma 4 (Truth Lemma) *For any* $\psi \in \Phi_{\mathscr{L}_{\mathsf{MQPL}}}$, $[\psi] = [\![\psi]\!]^{\mathfrak{M}^\varphi}$.

Proof Let V^φ be *any* truth assignment such that, for any sentential variable $s \in \mathscr{S}$, $[s] = \{\omega \in \Omega^\varphi : V^\varphi(\omega)(s) = \mathit{true}\} = [\![s]\!]^{\mathfrak{M}^\varphi}$.[11] This equivalence class belongs to $\mathscr{F}^\varphi$. We show by induction on the length of ψ that $[\psi] = [\![\psi]\!]^{\mathfrak{M}^\varphi}$ as follows:

1. When ψ is a sentential variable, it holds by definition.
2. When ψ is a $\top$, trivially $[\top] = [\![\top]\!]^{\mathfrak{M}^\varphi}$.
3. When ψ is $\neg\chi$, $[\neg\chi] = \overline{[\chi]}$ by Definition 9. $\overline{[\chi]} = \overline{[\![\chi]\!]^{\mathfrak{M}^\varphi}}$ by the induction hypotheses. $\overline{[\![\chi]\!]^{\mathfrak{M}^\varphi}} = [\![\neg\chi]\!]^{\mathfrak{M}^\varphi}$ by Definition 5.
4. When ψ is $\chi \wedge \tau$, $[\chi \wedge \tau] = [\chi] \cap [\tau]$ by Definition 9. $[\chi] \cap [\tau] = [\![\chi]\!]^{\mathfrak{M}^\varphi} \cap [\![\tau]\!]^{\mathfrak{M}^\varphi}$ by the induction hypotheses. $[\![\chi]\!]^{\mathfrak{M}^\varphi} \cap [\![\tau]\!]^{\mathfrak{M}^\varphi} = [\![\chi \wedge \tau]\!]^{\mathfrak{M}^\varphi}$ by Definition 5.

[11] In Definition 10, we choose a representative s^* of $[s]$ as an elementary event ω.

5. When ψ is $\chi \geqslant \tau$, $[\chi \geqslant \tau] = \{\omega \in \Omega^{\varphi} : \omega \in [\chi \geqslant \tau]\}$ by the property of equivalence classes. $\{\omega \in \Omega^{\varphi} : \omega \in [\chi \geqslant \tau]\} = \{\omega \in \Omega^{\varphi} : [\chi] \succsim_{\omega} [\tau]\}$ by Definition 11. $\{\omega \in \Omega^{\varphi} : [\chi] \succsim_{\omega} [\tau]\} = \{\omega \in \Omega^{\varphi} : [\![\chi]\!]^{\mathfrak{M}^{\varphi}} \succsim_{\omega} [\![\tau]\!]^{\mathfrak{M}^{\varphi}}\}$ by the induction hypotheses. By Lemma 3, we can apply Definition 5 to $\succsim_{\omega}$. So $\{\omega \in \Omega^{\varphi} : [\![\chi]\!]^{\mathfrak{M}^{\varphi}} \succsim_{\omega} [\![\tau]\!]^{\mathfrak{M}^{\varphi}}\} = [\![\chi \geqslant \tau]\!]^{\mathfrak{M}^{\varphi}}$. ■

Finally, the completeness of MQPL follows from Lemmas 2 and 4

Theorem 2 (Completeness) *For any* $\psi \in \Phi_{\mathscr{L}_{\mathsf{MQPL}}}$, *if* $\models_{\mathsf{MQPL}} \psi$, *then* $\vdash_{\mathsf{MQPL}} \psi$.

Proof Suppose that $\nvdash_{\mathsf{MQPL}} \psi$. Then, if $\nvdash_{\mathsf{MQPL}} \psi$, then $[\psi] \neq [\top]$ by Lemma 2. So $[\![\psi]\!]^{\mathfrak{M}^{\varphi}} \neq \Omega^{\varphi}$ by Lemma 4. So $\mathfrak{M}^{\varphi} \not\models_{\mathsf{MQPL}} \psi$. Therefore, there is a model $\mathfrak{M}$ such that $\mathfrak{M} \not\models_{\mathsf{MQPL}} \psi$. ■

6.8 Concluding Remarks

In this paper,

1. we have given the reason why we adopt not a probability measure model but a qualitative-probability model as a model for a comparative epistemic modal, which contributes essentially to simplification of the proof of the completeness theorem of MQPL, in terms of a representation theorem and arbitrary reference,
2. we have shown that the proof system of MQPL is simpler than that of Segerberg's PK, and
3. we have proven the completeness of MQPL. Because the construction of finite Boolean algebras to make it possible to apply Scott's representation theorem covers a considerable part of Segerberg's proof, whereas our proof does not require this construction, our proof has the merit that it is by far simpler than Segerberg's and Gärdenfors' proofs. Moreover, our proof is also simpler than Delgrande and Renne's, Harrison-Trainer et al.'s, and Ding's proofs of completeness.

In the near future, we would like to provide logic of epistemic modal of "$\varphi \gg \psi$" that is interpreted to mean that

φ is *definitely* more likely than ψ.

We can provide this logic with a model based on an additively-semiordered-qualitative-probability structure in [7, pp. 37–59] and [9] in the following way. When Ω is *finite*, there exist a finitely additive probability measure $P_w : \mathscr{F} \to \mathbb{R}$ relative to $\omega \in \Omega$ and $\delta \in \mathbb{R}$ satisfying

$$A \succ_w B \text{ iff } P_w(A) \geq P_w(B) + \delta, \text{ where } 0 < \delta \leq 1$$

iff $(\Omega, \mathscr{F}, \succ_{\omega})$ is an additively-semiordered-qualitative-probability structure. δ is interpreted to mean a *just noticeable difference* (JND). δ can play the semantic role of "definitely". We are now studying whether we can resort to a finitely additive nonstandard-real-valued probability measure when Ω is countably infinite. To examine logics of various epistemic modals including $\varphi \gg \psi$ from a logical point of view would require another paper.

Acknowledgements The author would like to thank the three reviewers of AWPL 2024 for their very helpful comments.

References

1. Benci, V., et al.: Infinitesimal probabilities. British J. Philos. Sci. **69**, 509–552 (2018)
2. Breckenridge, W., Magidor, O.: Arbitrary reference. Philos. Stud. **158**, 377–400 (2012)
3. Chateauneuf, A., Jaffray, J.V.: Archimedean qualitative probabilities. J. Math. Psychol. **28**, 191–204 (1984)
4. Cohen, M.A.: Necessary and sufficient conditions for existence and uniqueness of weak qualitative probability structures. J. Math. Psychol. **35**, 242–259 (1991)
5. Delgrande, J.P., Renne, B.: The logic of qualitative probability. In: Yang, Q., Wooldridge, M. (eds.) Proceedings of the 24th International Joint Conference on Artificial Intelligence (IJCAI-15), pp. 2904–2910. AAAI Press, Buenos Aires (2015)
6. Ding, Y., et al.: Logics of impricise comparative probability. Int. J. Approx. Reason. **132**, 154–180 (2021)
7. Domotor, Z.: Probabilistic Relational Structures and Their Applications, technical Report No. 144, Institute for Mathematical Studies in the Social Sciences, Stanford University (1969)
8. Domotor, Z.: Qualitative probabilities revisited. In: Humphreys, P. (ed.), Patrick Suppes: Scientific Philosopher, Vol.1. Probability and Probabilistic Causality, pp. 197–238. Kluwer, Dordrecht (1994)
9. Domotor, Z., Stelzer, J.: Representation of finitely additive semiordered qualitative probability structures. J. Math. Psychol. **8**, 145–158 (1971)
10. Fine, K.: A defence of arbitrary objects. Proc. Aristotel. Soc., Supplement. **57**, 55–77 (1983)
11. Fine, K.: Natural deduction and arbitrary objects. J. Philos. Log. **14**, 57–107 (1985)
12. Fine, K.: Reasoning with Arbitrary Objects. Basil Blackwell, Oxford (1985)
13. de Finetti, B.: La prévision: Ses lois logiques, ses sources subjectives. Annales de l'Institut Henri Poincaré **7**, 1–68 (1937)
14. Fishburn, P.C.: The axioms of subjective probability. Stat. Sci. **1**, 335–358 (1986)
15. Gärdenfors, P.: Qualitative probability as an intensional logic. J. Philos. Log. **4**, 171–185 (1975)
16. Harrison-Trainor, M., et al.: Preferential structures for comparative probabilistic reasoning. In: Proceedings of the Thirty-First AAAI Conference on Artificial Intelligence, pp. 1135–1141 (2017)
17. Haze, T.: A counterexample to the breckenridge-magidor account of instantial reasoning. J. Philos. Res. **41**, 539–541 (2016)
18. Hölder, O.: Die Axiome der Quantität und die Lehre vom Mass. Berichte über die Verhandlungen der Königlich Sächsischen Gesellschaft der Wissenschaften zu Leipzig. Mathematisch-Physikalische Klasse **53**, 1–64 (1901)
19. Holliday, W.H., Icard, III, T.F.: Measure semantics and qualitative semantics for epistemic modals. In: Snider, T. (ed.), Proceedings of SALT 23. pp. 514–534. Linguistic Society of America (2013)
20. Holman, E.W.: Strong and weak extensive measurement. J. Math. Psychol. **6**, 286–293 (1969)
21. King, J.: Instantial terms, anaphora and arbitrary objects. Philos. Stud. **61**, 239–265 (1991)
22. Kraft, C.H., et al.: Intuitive probability on finite sets. Ann. Math. Stat. **30**, 408–419 (1959)

23. Krantz, D.H., et al.: Foundations of Measurement, vol. 1. Academic Press, New York (1971)
24. Kratzer, A.: Modality. In: von Stechow, A., Wunderlich, D. (eds.) Semantics: An International Handbook of Contemporary Research, pp. 639–50. de Gruyter, Berlin (1991)
25. Kratzer, A.: Modals and Conditionals. Oxford University Press, Oxford (2012)
26. Lassiter, D.: Graded Modality. Oxford University Press, Oxford (2017)
27. Lewis, D.: Conutefactuals. Blackwell, Oxford (1973)
28. Luce, R.D., et al.: Foundations of Measurement, vol. 3. Academic Press, San Diego (1990)
29. Marushak, A.: Probability modals and infinite domains. J. Philos. Log. **49**, 1041–1055 (2020)
30. Mundy, B.: Faithful representation, physical extensive measurement theory and archimedean axioms. Synthese **70**, 373–400 (1987)
31. Narens, L.: Minimal conditions for additive conjoint measurement and qualitative probability. J. Math. Psychol. **11**, 404–430 (1974)
32. Narens, L.: Abstract Measurement Theory. The MIT Press, Cambridge (1985)
33. Narens, L.: Theories of Probability. World Scientific, Singapore (2007)
34. Roberts, F.S.: Measurement Theory. Addison-Wesley, Reading (1979)
35. Scott, D.: Measurement structures and linear inequalities. J. Math. Psychol. **1**, 233–247 (1964)
36. Segerberg, K.: Qualitative probability in a modal setting. In: Fenstad, J.E. (ed.), Proceedings of the Second Scandinavian Logic Symposium, pp. 341–352. North-Holland, Amsterdam (1971)
37. Skyrms, B.: Causal Necessity. Yale University Press, New Haven, Conn (1980)
38. Suppes, P., Zanotti, M.: Necessary and sufficient conditions for existence of a unique measure strictly agreeing with a qualitative probability ordering. J. Philos. Log. **5**, 431–438 (1976)
39. Suppes, P., et al.: Foundations of Measurement, vol. 2. Academic Press, San Diego (1989)
40. Suzuki, S.: Epistemic modals, qualitative probability, and nonstandard probability. In: Aloni, M., et al. (eds.), Proceedings of the 19th Amsterdam Colloquium (AC 2013), pp. 211–218 (2013)
41. Takahashi, W.: Nonlinear Functional Analysis. Kindaikagakusha, Tokyo (1988). in Japanese
42. Wenmackers, S., Horsten, L.: Fair infinite lotteries. Synthese **190**, 37–61 (2013)
43. Yalcin, S.: Probability operators. Philos Compass **5**, 916–937 (2010)

Chapter 7
Perspective Shifts: Formalizing Epistemic *Might* in Multi-agent Models

Jialiang Yan, Maria Aloni, and Fenrong Liu

Abstract Epistemic *might* is argued to be sensitive to agents' perspective. A same *might* claim could be evaluated differently by different agents. We propose a framework to formalize *might* to capture this feature. The resulted model is a multi-agent model which combines a team-based semantics and a two-dimensional semantics. Furthermore we investigate properties of epistemic *might* in the model and its interaction with knowledge and belief in multi-agent contexts.

Keywords Epistemic *might* · Perspective shifts · Multi-agent model

7.1 Introduction

Consider the following sentence:

(1) # It is not raining and it might be raining.

Scholars in philosophy and semantics have noted infelicity of the sentence in (1), specifically its inconsistency in asserting not-p and *might*-p at the same time. In the example *might* has epistemic reading. Sentences that follow this pattern are argued to lead to so-called **epistemic contradictions**, as highlighted by [6, 9, 17–19] and others. However, if we place this sentence in a dialogue context, the contradiction seems to disappear. For example,

J. Yan (✉) · F. Liu
Tsinghua University, Beijing, Haidian, China
e-mail: jialiang_yan@mail.tsinghua.edu.cn

F. Liu
e-mail: fenrong@tsinghua.edu.cn

M. Aloni · F. Liu
ILLC, University of Amsterdam, Amsterdam, Netherlands
e-mail: M.D.Aloni@uva.nl

K. Sano et al. (eds.), *Exploring Negation, Modality and Proof*,
Logic in Asia: Studia Logica Library,
https://doi.org/10.1007/978-981-96-8566-0_7

(2) a. Baoyu: "It is not raining."
b. Daiyu: "It might be raining."

In the dialogue, Baoyu states, "it is not raining," and Daiyu counters with "it might be raining," without contradictions. This is because *might* in (2-b) reflects only Daiyu's belief, not a fact, allowing Baoyu to assert a fact while Daiyu shares her viewpoint. Conversely, in (1), using *might* implies the speaker's knowledge, creating inconsistency when an agent affirms "not raining" and the possibility of "raining" simultaneously

These examples highlight the same *might*-claims uttered by different individuals can influence the context variably. Moreover, they suggest that interpreting epistemic might needs information about the perspective from which assertions are made.

In the literature, this phenomenon is recognized as perspective-sensitivity where certain elements, termed Perspective-Sensitive Items (PSIs), induce dependency on perspective (e.g., [7, 10]). A typical PSI, the predicate of taste, ties the interpretation of expressions like "This girl is beautiful" or "The cake is tasty" to the speaker's individual preferences or opinions, making the truth or falsity of such assertions uncontestable by others due to their inherent subjectivity. Epistemic might, with its inherent dependency on the speaker's epistemic state, is also suggested to be a Perspective-Sensitive Item (PSI), as discussed in various works (e.g., [2, 8, 14, 15]).

In [5] the author emphasizes two key aspects of PSIs: (i) default speaker-orientation and (ii) shiftability.[1] Both of these features are present in epistemic might. In a default cases, as seen in (1) and (2), *might* aligns with the speaker's perspective. Yet, in attitude reports like "Baoyu believes it might be raining," the perspective shifts to Baoyu. So the first step for a suitable semantic framework for *might* may need to presuppose a knower(s) of the *might*-claims. Despite wide discussions on PSIs, identifying the knower, as MacFarlane highlights ([8]), remains a challenge and is one of this paper's focus.

Moreover, we also need to integrate a mechanism for shiftability in the semantic framework to capture the cases of embedding *might* in attitude reports. In addition, [20] has observed that the use of *might* under *know* and *believe* highlights a clear difference between the philosophical concepts of knowledge and belief. This observation suggests that the epistemic *might* may exhibit different behaviors of shiftability when embedded under different modalities. Consider the following the sentences:

(3) a. Baoyu believes that -5 might be less than -6.
b. # Baoyu knows that -5 might be less than -6.

In (3-a), the *might*-claim is tied to Baoyu's epistemic state, expressing his belief and the speakers attitude is not taken into account. However, due to the factivity of knowledge, the *might*-claim embedded under *know* needs to be presupposed as truth, which means it requires agreement with the knowledge content by anyone asserting

[1] It means that, the perspective associated with *might* will undergo a transition, shifting from the speaker to the other agent.

it. Thus, in (3-b), the knower of the *might*-claim shifts to include not only Baoyu but also the speaker. To formalize this, we introduce a multi-agent model that merges state-based semantics for epistemic might [20] with the two-dimensional semantics of Epistemic Friendship Logic (EFL, [11, 12]), which facilitates transformations between agents to capture perspective shifts. In this framework, we also explore the interaction between epistemic might, knowledge, and belief in multi-agent settings.

This paper is organized as follows: Sect. 7.2 delves into the perspective-sensitivity of epistemic might. Section 7.3 introduces the state-based semantics for might and the two-dimensional semantics of Epistemic Friendship Logic, followed by an attempt to merge the models to capture the perspective-sensitivity in multi-agent situations. Finally, Sect. 7.4 extends the model to include belief and knowledge.

7.2 Perspective-Sensitivity of Epistemic *might*

As highlighted in Sect. 7.1, the ability to shift perspective is a key characteristic of perspective-sensitive items, particularly evident in attitude reports.

When analyzing a perspective-sensitive item (PSI), it's essential to account for the perspective of at least one specific agent, which informs the interpretation's domain. Typically this agent is the speaker, who possess knowledge, or holds a belief, or has some other relevant attitude. However, in attitude reports, the perspective shifts from the speaker to the attitude holder. For example:

(4) a. The eggplant is tasty.
b. Liulaolao believes the eggplant is tasty.

In (4-a), the adjective "tasty," as a predicate of personal taste, is a perspective-sensitive item, which reflects the speaker's personal preference. When switched into the context of an attitude report as in (4-b) involving belief, it articulates Liulaolao's personal taste about the eggplant, suggesting that the interpretation of "tasty" is subjective to Liulaolao's perspective, rather than the speaker's. So in this context of embedding under belief, the epistemic agent of the PSI shifts to the attitude holder from the speaker.

Similarly, consider these expressions concerning epistemic possibilities:

(5) a. It might be an eggplant.
b. Liulaolao believes it might be an eggplant.

In (5-a), the interpretation of *might* is linked to the speaker's knowledge, which expresses the object is possible being an eggplant for the speaker. While in (5-b), the phrase captures Liulaolao's belief about the possibility of it being an eggplant, framed by her perspective. This assertion holds true as long as Liulaolao does not dismiss the possibility of its being an eggplant.

The complexity increases as statements involve multiple nested beliefs:

(6) a. Daiyu believes that Liulaolao believes the eggplant is tasty.
b. Daiyu believes that Liulaolao believes it might be an eggplant.

In these sentences, the first statement in (6-a) only specifies that Liulaolao finds the eggplant tasty, without implying Daiyu's attitude to the taste of the eggplant. The second indicates that Daiyu recognizes Liulaolao's belief in the possible identity of the eggplant but does not reveal Daiyu's own beliefs regarding if it is an eggplant.

These examples illustrate how perspective shifts are handled in complex sentences with nested clauses. Each layer of belief adjusts the perspective from which the item is viewed. So in attitude reports, the rule is that the interpretation of perspective-sensitive items is always linked to the immediate subject, who is also the immediate holder of the attitude.

In conclusion, understanding the shift in perspective in attitude reports is crucial for interpreting PSIs accurately. This understanding allows us to recognize whose perspective is being presented, especially in the complex sentences.

7.3 A State-Based Multi-agent Model

This section introduces the state-based semantics from Bilateral State-based Epistemic Logic (BSEL, [20]), and the perspective-shifting capabilities of Epistemic Friendship Logic (EFL, [11]). Then we propose multi-agent models that integrate both to formalize perspective-sensitivity.

7.3.1 *Bilateral State-Based Epistemic Logic for Might*

BSEL, an epistemic variant of Bilateral State-based Modal Logic ([1]), is designed to explore epistemic modalities and provides a non-truth conditional analysis for epistemic *might*. Inspired by [16], the fundamental concept of epistemic *might* is that a statement *might*-ϕ is acceptable if and only if ϕ is consistent with the relevant agent's information state, viewed as a set of possibilities ([17]). Here, we introduce a BSEL fragment BSEL$^{\blacklozenge}$ containing only the epistemic *might* operator.

Definition 7.1 (*Language $\mathcal{L}_B$ of BSEL$^{\blacklozenge}$*)

$$\phi ::= p \mid \neg\phi \mid \phi \wedge \phi \mid \phi \vee \phi \mid \blacklozenge\phi$$

The operator $\blacklozenge$ stands for the epistemic *might*. The disjunction is not the classical one, and it is from team semantics which is also called *tensor* disjunction (see [21]).

A model for $\mathcal{L}_B$ is a triple $\mathcal{M}_B = \langle W, R, V\rangle$, where W is a set of possible worlds. R is an accessibility relation on W. V is a world-dependent valuation function:

$Prop \times W \to \{0, 1\}$, where *Prop* is a set of propositions. Unlike classical modal logic, formulas in BSEL$^{\blacklozenge}$ are evaluated w.r.t an information state $s \subseteq W$, which is a set of possible worlds. Moreover, BSEL$^{\blacklozenge}$ is bilateral in the sense that the assertability and rejectability of a sentence are modeled by *support* ($\vDash$) and *anti-support* ($\Dashv$) conditions respectively rather than truth. The semantic clauses are defined as follows ([1, 20]).

Definition 7.2 (*Semantic clause*)

$$
\begin{array}{ll}
\mathcal{M}_B, s \vDash p & \text{iff } \forall w \in s : V(w, p) = 1 \\
\mathcal{M}_B, s \Dashv p & \text{iff } \forall w \in s : V(w, p) = 0 \\
\mathcal{M}_B, s \vDash \neg\phi & \text{iff } \mathcal{M}_B, s \Dashv \phi \\
\mathcal{M}_B, s \Dashv \neg\phi & \text{iff } \mathcal{M}_B, s \vDash \phi \\
\mathcal{M}_B, s \vDash \phi \vee \psi & \text{iff } \exists t, t' \colon t \cup t' = s \text{ and } \mathcal{M}_B, t \vDash \phi \text{ and } \mathcal{M}_B, t' \vDash \psi \\
\mathcal{M}_B, s \Dashv \phi \vee \psi & \text{iff } \mathcal{M}_B, s \Dashv \phi \text{ and } \mathcal{M}_B, s \Dashv \psi \\
\mathcal{M}_B, s \vDash \phi \wedge \psi & \text{iff } \mathcal{M}_B, s \vDash \phi \text{ and } \mathcal{M}_B, s \vDash \psi \\
\mathcal{M}_B, s \Dashv \phi \wedge \psi & \text{iff } \exists t, t' \colon t \cup t' = s \text{ and } \mathcal{M}_B, t \Dashv \phi \text{ and } \mathcal{M}_B, t' \Dashv \psi \\
\mathcal{M}_B, s \vDash \blacklozenge\phi & \text{iff } \exists s' \subseteq s \text{ and } s' \neq \emptyset \text{ s.t. } \mathcal{M}_B, s' \vDash \phi \\
\mathcal{M}_B, s \Dashv \blacklozenge\phi & \text{iff } \mathcal{M}_B, s \Dashv \phi
\end{array}
$$

A proposition p is supported by the information state s iff p is true in every world in s; and p is anti-supported by s iff it is false in every world in s. Specially, a disjunction is supported by s iff s is a union of two substates each supporting one of the disjuncts. And $\phi \vee \psi$ is anti-supported iff both ϕ and ψ are anti-supported by s. The semantics of disjunction is from team logic, termed as *split disjunction* (see [21]).

The formula $\blacklozenge\phi$ is supported by an information state s iff there is a non-empty substate of s supporting ϕ. And $\blacklozenge\phi$ is anti-supported by s iff ϕ is anti-supported by s, which means there is no possibility of ϕ in s.

In BSEL, epistemic contradictions can be derived.

- **Epistemic contradictions**
  $\blacklozenge\phi \wedge \neg\phi \vDash \bot$

The proof of derivability and further details of the logic of BSEL can be found in [20]. The graph in Fig. 7.1 illustrates a toy model of BSEL$^{\blacklozenge}$.

In the model, the information state s consists of two possible worlds, $\{w_{pq}, w_q\}$. The proposition p only holds in the world w_{pq}, while q holds everywhere in s. So in this model, both $\blacklozenge p$ and $\blacklozenge q$ are supported by s since there are non-empty subsets supporting p or q. In addition, in this model, q is supported, while $\blacklozenge\neg q$ is anti-supported. So $q \wedge \blacklozenge\neg q$ derives a contradiction, which shows the epistemic contradiction pertaining to q can be derived.

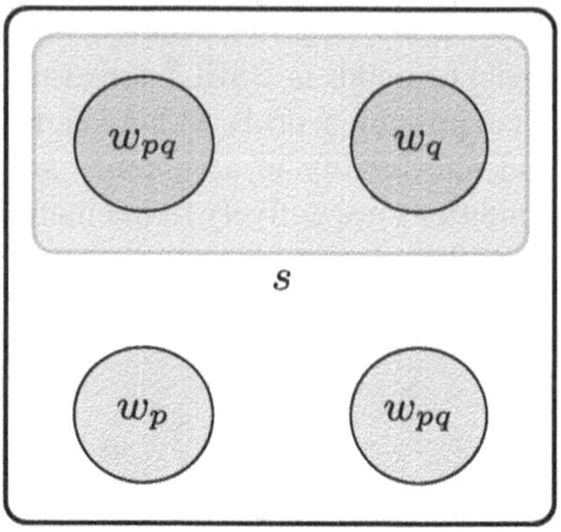

Fig. 7.1 A toy model of BSEL$^{\blacklozenge}$

7.3.2 Epistemic Friendship Logic

In [11, 12], a two-dimensional modal logic was proposed for analyzing and reasoning about knowledge and social relationships in social networks. In this logic, agents are placed in the model, and formulas are evaluated with respect to both possible worlds and agents. So it allows us to express aspects of agency. In this chapter, we only focus on the static fragment of EFL.

The language $\mathcal{L}_E$ is based on atoms of two types: propositional variables $p \in Prop$ and (a finite set of) agent nominals $n \in ANom$. The propositional variables can represent the indexical propositions like "I am hungry," and the nominals name the agents.

Definition 7.3 (*Language* $\mathcal{L}_E$ [12])

$$\varphi ::= p \mid n \mid \neg\varphi \mid \varphi \wedge \varphi \mid K\varphi \mid F\varphi \mid \mathbf{A}\varphi$$

The operator K stands for knowledge. We read the formula $F\phi$ as "all my friends have the property ϕ" and the formula $\mathbf{A}\phi$ as "every agent has the property ϕ."

Models for $\mathcal{L}_E$ are Kripke-style models of the form $\mathcal{M}_E = \langle W, A, R, f, V \rangle$.

- W is a set of possible worlds, and A is a set of agents.
- R is a family of accessibility relations, which is reflexive, symmetric, and transitive. R_a is the epistemic indistinguishability relation for each agent $a \in A$.
- f is a family of symmetric and irreflexive relations. f_w for each $w \in W$ represents the friendship relation in state w.
- V is a valuation function: mapping propositional variables $p \in Prop$ to subsets of $W \times A$, namely $(w, a) \in V(p)$ which means that the proposition p holds true for agent a in world w.

We define g to be an assignment function on $\mathcal{M}_E$ mapping each agent nominal $n \in ANom$ to the agent $g(n) \in A$ named by n.

We interpret EFL on models two-dimensionally. That is, formulas will be evaluated at pairs (w, a). This differs from classical modal logic, where formulas use only one evaluation index.

Definition 7.4 (*Semantics* ([12])) The EFL models are used to interpret $\mathcal{L}_E$ in a double-indexical way:

$$
\begin{array}{ll}
\mathcal{M}_E, w, a \models p & \text{iff } (w, a) \in V(p), \text{for } p \in Prop \\
\mathcal{M}_E, w, a \models n & \text{iff } g(n) = a \text{ for } n \in ANom \\
\mathcal{M}_E, w, a \models \neg\phi & \text{iff } \mathcal{M}_E, w, a \not\models \phi \\
\mathcal{M}_E, w, a \models \phi \wedge \psi & \text{iff } \mathcal{M}_E, w, a \models \phi \text{ and } \mathcal{M}_E, w, a \models \psi \\
\mathcal{M}_E, w, a \models K\phi & \text{iff } \mathcal{M}_E, v, a \models \phi \text{ for every } v \in W \text{ s.t. } R_a(w, v) \\
\mathcal{M}_E, w, a \models F\phi & \text{iff } \mathcal{M}_E, w, b \models \phi \text{ for every } b \in A \text{ s.t. } f_w(a, b) \\
\mathcal{M}_E, w, a \models A\phi & \text{iff } \mathcal{M}_E, w, b \models \phi \text{ for every } b \in A
\end{array}
$$

A proportional letter p is true with respect to indexes w, a if and only if p holds at w for agent a. A nominal n is true with respect to w, a if and only if n is the name of the agent a.

To illustrate how this semantics works, consider the following toy example.

In the model depicted by Fig. 7.2, $W = \{w, w'\}$ and $A = \{a, b\}$. There four points in the model $\{(w, a), (w', a), (w, b), (w', b)\}$, and $V(p) = \{(w, a), (w', a), (w, b)\}$. The line represents the equivalence relation R. As the model shows, $R_{a,b} = \{w, w'\}$. Suppose the agent a has a name n, viz $g(n) = a$. In this model, we can get the following results:

- $\mathcal{M}_E, w, a \models p$
- $\mathcal{M}_E, w, a \models n$
- $\mathcal{M}_E, w, a \models Kp$
- $\mathcal{M}_E, w, b \models \neg Kp$

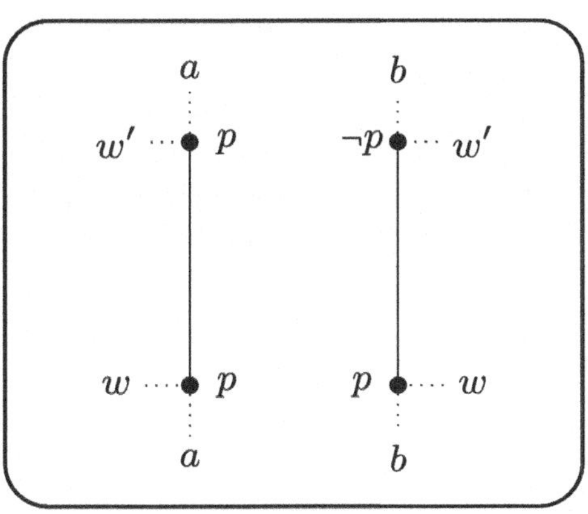

Fig. 7.2 EFL model

Agent nominals play an important role in EFL. Assuming the rigid reading, every nominal n specifies an agent. This is inspired by hybrid logic (see [4, 13], etc., and [3] provides a good overview) which uses nominals to name possible worlds. In addition, another hybrid-style operator @ can be defined as a derived one: $@_n\phi = \langle\mathbf{A}\rangle(n \wedge \phi)$, which means "For n, ϕ holds."

$$\mathcal{M}_E, w, a \models @_n\phi \text{ iff } \mathcal{M}_E, w, g(n) \models \phi$$

The formula $@_n\phi$ is true relative to indexes w and a if and only if ϕ is true relative to the reference of n. Notice that the point a in fact does not matter in the truth-condition for $@_n\phi$, since the operator $@_n$ moves the current index of agent to the reference of n whatever a is.

$\mathcal{M}_E$ is a named agent model, if every agent in $\mathcal{M}_E$ has a name, viz, for each $a \in A$, there is a nominal $n \in ANom$ such that $g(n) = a$. By assuming a named agent model,[2] another hybrid-style operator can be defined in EFL, $\downarrow n$, which provides a way of indexically referring to the current agent.

$$\downarrow n\phi = \bigvee_{m \in ANom} (m \wedge \phi^n_m)$$

The semantics for $\downarrow n$ is defined as follows. where $\mathcal{M}^n_a$ is the result of changing

$$\mathcal{M}_E, w, a \models \downarrow n\, \phi \text{ iff } \mathcal{M}^n_a, w, a \models \phi$$

$\mathcal{M}_E$ by forcing n to name a. More precisely, the assignment function g defined on $\mathcal{M}_E$ now is changed to g^n_a, and g^n_a(m)=a if $m = n$ and $g(m)$ otherwise. The operator $\downarrow n$ behaves like the $\downarrow$ binder in hybrid logic which binds a nominal to the point of evaluation. In EFL, $\downarrow n\ \phi$ is true relative to w and a if and only if ϕ is true relative to w and a when n refers to a.

This semantics allows us to express indexicals under knowledge in EFL. For example, let p be the proposition "I'm hungry" and nominal m be Daiyu.

(7) $\downarrow n @_m K @_n p$: Daiyu knows that I am hungry.

The operator $\downarrow n$ binds the nominal n that occurs in the bound formula, which force n refer back to the current agent in the pair of evaluation. In this way, "I" represented by n refers to the speaker rather than the subject of the sentence.

[2] Otherwise, the operator can be introduced as a primitive.

7.3.3 The Multi-agent Model for Epistemic Might

In this section, we extend BSEL using the semantics of EFL introduced in the previous section. We start with a simple language $\mathcal{L}$ which is a combination of fragments of EFL and BSEL.

Definition 7.5 (*Language* $\mathcal{L}$) The language is based on three types of atoms: propositional variables $p \in Prop$, nominols $n \in ANom$.

$$\varphi ::= p \mid n \mid \neg\varphi \mid \varphi \wedge \varphi \mid \varphi \vee \varphi \mid \blacklozenge\varphi \mid @_n\varphi \mid \downarrow n\varphi$$

A model of $\mathcal{L}$ is a tuple $\mathcal{M} = \langle W, A, R, g, V\rangle$. Such models are like models for EFL without the friendship relations. W is the set of possible worlds and A is the set of agents. R is a family of equivalence relations, and R_a represents the relation for agent $a \in A$. g is an assignment function: $ANom \rightarrow A$. And V is an evaluation function from Prop to the subsets of $W \times A$, with $(w, a) \in V(p)$.

It is a state-based system, viz, formulas of $\mathcal{L}$ are interpreted with respect to an information state s. To achieve this, it is necessary to lift the single evaluation pair (w, a) to a set. One approach to lift is: an information state s is defined as a set of indices i. An index i is a pair $i = \langle w_i, a_i\rangle$. So an information state, in fact, is a set of pairs of world w and agent a. For the sake of simplicity, we assume the agents in an information state s are the same one. Hence we have the constraint that for all $i, j \in s$, $a_i = a_j$. Because of the identity of agents in s, we label the agent as a subscript of s: s_a, highlighting the epistemic agent (knower) of the information to be a. Now we can define the semantic clauses. We skip the semantics of connectives, as they will be defined in a similar way to BSEL.

Definition 7.6 (*Semantics*) Formulas of $\mathcal{L}$ are interpreted in models $\mathcal{M}$ with respect to a state $s \subseteq W \times A$. Both support $\vDash$, and anti-support $\dashv$ conditions are specified.

$$\begin{array}{ll}
\mathcal{M}, s_a \vDash p & \text{iff } \forall i \in s_a : (w_i, a_i) \in V(p) \\
\mathcal{M}, s_a \dashv p & \text{iff } \forall i \in s_a : (w_i, a_i) \notin V(p) \\
\mathcal{M}, s_a \vDash n & \text{iff } \forall i \in s_a, g(n) = a \\
\mathcal{M}, s_a \dashv n & \text{iff } \forall i \in s_a, g(n) \neq a \\
\mathcal{M}, s_a \vDash @_n\phi & \text{iff } \mathcal{M}, s_a[g(n)] \vDash \phi \\
\mathcal{M}, s_a \dashv @_n\phi & \text{iff } \mathcal{M}, s_a[g(n)] \dashv \phi \\
\mathcal{M}, s_a \vDash \downarrow n\ \phi & \text{iff } \mathcal{M}^n_a, s_a \vDash \phi \\
\mathcal{M}, s_a \dashv \downarrow n\ \phi & \text{iff } \mathcal{M}^n_a, s_a \dashv \phi \\
\mathcal{M}, s_a \vDash \blacklozenge\phi & \text{iff there is a non-empty subset } s'_a \text{ of } s_a\text{, such that } s'_a \vDash \phi \\
\mathcal{M}, s_a \dashv \blacklozenge\phi & \text{iff } \mathcal{M}, s_a \dashv \phi
\end{array}$$

- $i[g(n)] = \langle w_i, g[n]\rangle$
- $s_a[g(n)] = \{i[g(n)] \mid i \in s\}$
- $\mathcal{M}^n_a = \langle W, A, R, g^n_a, V\rangle$, where g^n_a is a n-variant of g.

As shown in Definition 7.6, it is a two-dimensional state-based semantics. A propositional variable p is supported in information state s_a if and only if for every index i in s_a, p holds for agent a_i at world w_i. A nominal n is supported by s_a if and only if n is the name of the agent a who is the knower of s_a.

The formula $@_n\phi$ is supported in s_a if and only if ϕ is supported in $s_a[g(n)]$ which is the information state after switching the agent a of s_a to the agent named n.[3] The conditions of $\downarrow n$ can be reduced to the conditions of ϕ interpreted in the model $\mathcal{M}^n_a$ which is the resulting model by substituting g with g^n_a. The formula $\blacklozenge\phi$ is supported in s_a if and only if there is at least one pair of (w, a) supporting ϕ.

In this framework, perspectives are represented by the information states that incorporate agency. The agent to whom the *might*-claim is linked can be specified by the $@_n$ operator. The interpretation of $@_n\phi$ requires a move to the agent named n from the current agent. Perspective shifting involves transitioning from one agent to another against a consistent information background. This means altering only the agent parameter while maintaining constant possible worlds, reflecting varied judgments even with same information.

Reconsider example (2), repeated below.

(8) a. Baoyu: "It is not raining."
b. Daiyu: "It might be raining."

In the multi-agent model, (8-a) is formalized as $@_b\neg p$ and (8-b) as $@_d\blacklozenge p$, where b and d represent Baoyu and Daiyu, respectively, and p denotes "it is raining." The conjunction $@_b\neg p \wedge @_d\blacklozenge p$ is consistent since $\blacklozenge p$ and $\neg p$ are evaluated with respect to s_d and s_b, respectively. However, $@_b(\neg p \wedge \blacklozenge p)$ is inconsistent as both $\neg p$ and $\blacklozenge p$ are interpreted in the same state s_b, resulting in epistemic contradictions.

In addition, the model validates the principles of hybrid logic and EFL.

$$@_i @_j \phi \leftrightarrow @_j \phi$$

[3] It could assume that agent a is maximally informed while agent $g(n)$ is maximally ignorant. Normally this can be captured by assigning a singleton to a's information state and the set of all possible worlds to $g(n)$'s information state. However, we have to handle the situation differently in our model since all states have the same cardinality. The underlying idea behind this approach is that different agents have access to the same information base, but they can consider different possibilities and make different assertions. The model assumes that the information state refers to the current information possessed by the agents, and it is important to note that this does not encompass the entirety of their knowledge background. In fact, it is a technical issue. In EFL, the @ operator is only used to switch between agents, and we preserve this setting in our model. But EFL is a world-based logic, so it does not have the same problems that arise in a state-based account such like dealing with different information bases of varying cardinalities. If we set the shifting to occur between different states of different cardinalities, i.e., @ operator switches between information states that have different worlds and agents, then we may encounter various technical problems that need to be resolved, such as defining the difference between information states, determining whether the agent or cardinality is different, or whether both are different. These questions are left for future research.

This accounts for situations where no shifting occurs, such as when another agent assesses sentence (8-b), maintaining the link of *might* to Daiyu. For a formula $@_d\blacklozenge p$, regardless of the number of added @ operators, $\blacklozenge p$ will ultimately be evaluated with reference to $s_a[g(d)]$.

7.3.4 Epistemic Contradictions in Multi-Agent Settings

To explain how our model functions, we revisit the epistemic contradictions, and examine how the contradiction resolves in our model.

The epistemic contradictions mentioned in Sect. 7.1 are reiterated here, along with their formalization in our proposed model.

(9) a. Epistemic contradiction: # It is raining and it might not be raining.
b. $\phi \wedge \blacklozenge\neg\phi \models \bot$

(10) a. Conversational epistemic contradiction: Baoyu said "It is raining." and Daiyu said "It might not be raining."
b. $@_n\phi \wedge @_m\blacklozenge\neg\phi \not\models \bot$, where $g(n) \neq g(m)$.

As mentioned in Sect. 7.1, we consider it consistent to have conversational epistemic contradictions. This is formalized using the @ operator, as shown in (10-b). Specifically, for agent a named n, ϕ holds, while for agent b named m, $\blacklozenge\neg\phi$ holds.

We illustrate these examples with the graph in Fig. 7.3, which depicts a toy model of multi-agent model for *might*. In this model, $W = w_1, w_2$ and $A = a, b$, resulting in two information states s_a and $s_b \subset W \times A$, where $s_a = (w_1, a), (w_2, a)$ and $s_b = (w_1, b), (w_2, b)$. The valuation $V(p) = (w_1, a), (w_2, a), (w_1, b)$, and (w_2, b) supports $\neg p$.

Now suppose agents a and b have names n and m, respectively, such that $g(n) = a$ and $g(m) = b$. We can derive that epistemic contradictions are true contradictions for a single agent (a or b), but not for two agents when one asserts ϕ and the other disagrees by asserting might-not-ϕ. The following formulas can be derived using the model in Fig.7.3:

- $\mathcal{M}_{\mathfrak{M}}, s_a \models p$
- $\mathcal{M}_{\mathfrak{M}}, s_a \dashv p \wedge \blacklozenge\neg p$
- $\mathcal{M}_{\mathfrak{M}}, s_b \dashv p \wedge \blacklozenge\neg p$
- $\mathcal{M}_{\mathfrak{M}}, s_a \models p \wedge @_m\blacklozenge\neg p$

In the model, since $s_a \subseteq V(p)$, agent a can assert p in the current situation, but agent b cannot. In an information state, epistemic contradictions are rejected for a single agent. If p holds in the information state, then $\neg p$ cannot hold in that same state, since there is no non-empty substate to support $\blacklozenge\neg p$. This is verified by the semantics of $\blacklozenge$.

Fig. 7.3 A toy model of the multi-agent model

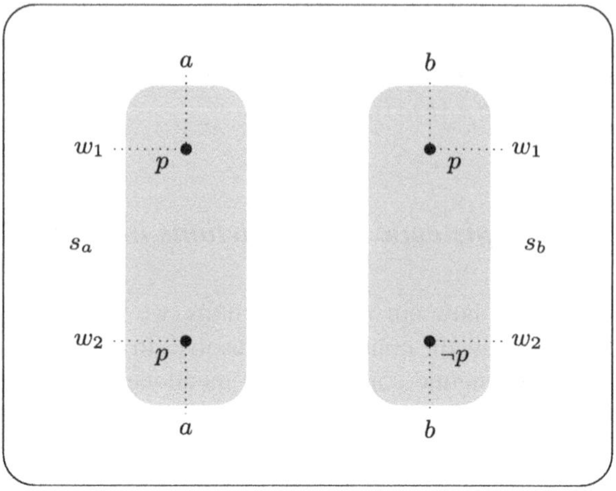

However, on the information state s_a, $@_m \blacklozenge \neg p$ is supported. This is because eliminating the $@_m$ operator shifts the evaluation parameter to s_b, where $\blacklozenge \neg p$ is assertable. In this way, the conversational epistemic contradictions are supported.

This distinction highlights the difference between normal epistemic contradictions and conversational epistemic contradictions.

7.4 Whose Knowledge Is Being Expressed?

7.4.1 *Epistemic Might Under Belief*

As previously discussed in Sect. 7.2, the knower of *might*-claims in the doxastic context can be recognized.

- In a believe-*might* sentence, the knower to whom the embedded *might* clause is linked is the holder of the attitude.
- If a believe-*might* sentence embedded under another attitude verbs, the knower is the immediate subject of the *might*-claim.

However in some cases, the speaker plays a role in interpretation of believe-might statement. For example, we distinguish the following two sentences in (11-a) and (11-b).

(11) Daiyu, as the speaker, asserts the following two sentences:

a. Baoyu believes that it might be raining.

b. I believe that Baoyu believes that it might be raining.

The assertion in (11-a) reflects Baoyu's belief, while (11-b) conveys Daiyu's belief about Baoyu's doxastic state. In (11-a), the speaker's perspective doesn't influence sentence evaluation, merely describing Baoyu's belief. Conversely, in (11-b), while the *might*-claim is linked only to Baoyu's perspective, Daiyu's perspective is implicated since her belief attributes to Baoyu's belief.

To formalize cases of belief reports, we extend the language $\mathcal{L}$ to include informational and doxastic modalities **I** and **B**. The expressions $\mathbf{I}\phi$ and $\mathbf{B}\phi$ are respectively interpreted as "the agent has information of ϕ" and "the agent believes that ϕ." The duals are defined in the standard manner, such that $\langle \mathbf{I} \rangle = \neg\mathbf{I}\neg$ and $\langle \mathbf{B} \rangle = \neg\mathbf{B}\neg$.

The modality **I** is the standard **S5** modality,[4] which is defined in terms of the equivalence relation that is already included in the multi-agent model $\mathcal{M}$. The operator **B** is represented by means of plausibility relations, hence it is necessary to incorporate a plausibility ordering $\leq_{a,w}$ into the model.

Now the multi-agent model is defined as: $\mathcal{M} = \langle W, A, R, \leq_{a,w}, g, V \rangle$, where $\leq_{a,w}$ is the plausibility ordering among possible worlds at w for agent $a \in A$. Intuitively, for every world v and v', $v \leq_{a,w} v'$ says that at the world w, v is at least as plausible as v' for the agent a. Then the semantic clauses for **I** and **B** can be defined. where the abbreviations are defined as follows:

$$\mathcal{M}, s_a \vDash \mathbf{I}\phi \text{ iff } \forall i \in s_a : \mathcal{M}, R_{a_i}(w_i)[a_i] \vDash \phi$$
$$\mathcal{M}, s_a \mathrel{=\!\!|} \mathbf{I}\phi \text{ iff } \forall i \in s_a\text{: there is a non-empty subset } t \subseteq R_{a_i}(w_i)[a_i],\ t \mathrel{=\!\!|} \phi$$
$$\mathcal{M}, s_a \vDash \mathbf{B}\phi \text{ iff } \forall i \in s_a\text{: } Dox_{a_i}(w_i)[a_i] \vDash \phi$$
$$\mathcal{M}, s_a \mathrel{=\!\!|} \mathbf{B}\phi \text{ iff } \forall i \in s_a\text{: there is a non-empty subset } t \subseteq Dox_{a_i}(w_i)[a_i],\ t \mathrel{=\!\!|} \phi$$

- For a set Γ, $\Gamma[a_i] = \{< v, a_i > \mid v \in \Gamma\}$
- $R_a(w) = \{v \mid wR_a v\}$
- $R_{a_i}(w_i)[a_i] = \{< v, a_i > \mid v \in R_{a_i}(w_i)\}$
- $Min_{\leq_{a,w}}(R_a(w)) := \{v \in R_a(w) \mid \forall w' \in R_a(w) : w' \leq_{a,w} v \Rightarrow v \leq_{a,w} w'\}$
- $Dox_a(w) = Min_{\leq_{a,w}}(R_a(w))$
- $Dox_{a_i}(w_i)[a_i] = \{< v, a_i > \mid v \in Dox_{a_i}(w_i)\}$

The informational formula $\mathbf{I}\phi$ is supported in s_a if and only if for every $i \in s_a$, ϕ is supported in every equivalence class of w_i for agent a which is generated in term of relation R_a. A doxistic formula $\mathbf{B}\phi$ is supported in s_a if and only if ϕ is supported in every doxistic set of w_i for agent a.

Associated with the operator $@_n$, we can capture the difference shown in (11). Let variable p be the proposition "it is raining," and nominal n and m indicate "Daiyu" and "Baoyu," respectively.

(12) a. "Baoyu believes that it might be raining" uttered by Daiyu.
b. Formalization: $@_n @_m \mathbf{B}\blacklozenge p$

[4] In BSEL, for some technical aims, it is not sufficient to use S5 modality to represent knowledge. Knowledge is defined as information and presupposition there, see [20] for more arguments.

(13) a. "I believe that Baoyu believes that it might be raining" uttered by Daiyu.
b. Formalization: $\downarrow n\mathbf{B}@_m\mathbf{B}\blacklozenge p$

In (13-b), $\mathbf{B}\blacklozenge p$ is evaluated by state Dox_m through eliminating B in s_n. This means Baoyu's information state for interpreting $\blacklozenge p$ originates from the speaker's belief state.

7.4.2 *Knowledge*

In this section, we proceed to formalize the concept of knowledge in multi-agent models. To begin, we consider the epistemic agents who possess knowledge of know-*might* sentences.

Let us revisit example (3), which is repeated below.

(14) a. Baoyu believes that -5 might be less than -6.
b. # Baoyu knows that -5 might be less than -6.

As argued, the interpretation of sentence (14-b) needs to consider both the speaker's and Baoyu's perspectives, due to the factivity of knowledge. In BSEL, the factivity of knowledge is treated as a presupposition, interpreting "knowing p" as the agent having information of p and presupposing p to be true. Employing this approach, we propose a knowledge definition in the multi-agent models, initially introducing the formula ϕ_ψ to represent "ϕ presupposes ψ."

$$\mathcal{M}, s_a \vDash \phi_\psi \text{ iff } \mathcal{M}, s_a \vDash \phi \text{ and } \mathcal{M}, s_a \vDash \psi$$
$$\mathcal{M}, s_a \mathrel{=\!\!|} \phi_\psi \text{ iff } \mathcal{M}, s_a \mathrel{=\!\!|} \phi \text{ and } \mathcal{M}, s_a \vDash \psi$$

ϕ_ψ is supported in s_a if and only if ϕ is supported and also its presupposition is supported; and ϕ_ψ is anti-supported if and only if ϕ is anti-supported but its presupposition is still supported.[5]

In BSEL, the knowledge operator is defined as $(\mathbf{I}\phi)_\phi$, but this poses a challenge in multi-agent models. Consider the sentence "Daiyu (m) knows that it might be raining," formalized as $@_m K\blacklozenge p$. Given our semantics, the distribution of operator $@_n$ applies to presuppositions, making $@n(\phi_\psi) \vDash (@_n\phi)@_n\psi$. Thus, $@_m K\blacklozenge p$ = $@_m((\mathbf{I}\blacklozenge p)_{\blacklozenge p})$, implying $(@_m\mathbf{I}\blacklozenge p)_{@_m\blacklozenge p}$. Both $\mathbf{I}\blacklozenge p$ and $\blacklozenge p$ are evaluated with respect to $s_{g(d)}$, considering only Daiyu's perspective and excluding the speaker's, represented by s_a.

Another limitation of this definition is that it only enables the derivation of factivity of knowledge in restricted situations. So we propose the following definition of knowledge modality in the multi-agent model.

---

[5] This is because negation of an expression does not change its presuppositions.

$$\mathbf{K}_n\phi := (@_n\mathbf{I}\phi)_\phi$$

We read the formula $K_n\phi$ as "the agent named n knows ϕ."

$$\mathcal{M}, s_a \vDash \mathbf{K}_n\phi \text{ iff } \mathcal{M}, s_a \vDash @_n\mathbf{I}\phi \text{ and } \mathcal{M}, s_a \vDash \phi$$
$$\mathcal{M}, s_a \dashv \mathbf{K}_n\phi \text{ iff } \mathcal{M}, s_a \dashv @_n\mathbf{I}\phi \text{ and } \mathcal{M}, s_a \vDash \phi$$

Based on this semantics, it can be inferred that

$$K_n\phi \models \phi$$

Intuitively, "n knows ϕ" means n has information of ϕ and ϕ is presupposed by the speaker.

Furthermore, the placement of a binder $\downarrow n$ in front of $K_n\phi$ expresses the indexical proposition "I (the agent with the index n) know that ϕ."

Based on this definition, the factivity of knowledge now indicates that a given agent knows ϕ if and only if ϕ is assertable for that agent. The following inferences of factivity can be derived in our model.

Fact 7.1 *$K_a\phi \vDash \phi$, in particular $K_a\blacklozenge\phi \vDash \blacklozenge\phi$.*

Fact 7.1 can be shown directly by the definition of knowledge.

Fact 7.2

1. $K_nK_m\phi \vDash \phi$.
2. $K_nK_m\phi \vDash K_n\phi$
3. $K_nK_m\phi \vDash K_m\phi$

Proof 1. Suppose $\mathcal{M}, s_a \vDash K_nK_m\phi$. This means that $\mathcal{M}, s_a \vDash (@_n\mathrm{IK}_\mathrm{m}\phi)_{\mathrm{K}_\mathrm{m}\phi}$, which implies that $\mathcal{M}, s_a \vDash K_m\phi$ and $\mathcal{M}, s_a \vDash @_n\mathrm{IK}_\mathrm{m}\phi$. By the first condition, we can conclude that $\mathcal{M}, s_a \vDash @_mI\phi$ and $\mathcal{M}, s_a \vDash \phi$. Therefore it derives that ϕ is supported by s_a in $\mathcal{M}$.

2. Suppose $\mathcal{M}, s_a \vDash K_nK_m\phi$. It follows that $\mathcal{M}, s_a \vDash @_n\mathrm{IK}_\mathrm{m}\phi$, which implies that $s_a[g(n)] \vDash \mathrm{IK}_\mathrm{m}\phi$, namely $s_n \vDash \mathrm{IK}_\mathrm{m}\phi$. Thus it implies that $\forall i \in s_n$: $R_n(w_i)[a_i] \vDash K_m\phi$, which means $\forall i \in s_n$: $R_n(w_i)[a_i] \vDash \phi$ and $R_n(w_i)[a_i] \vDash @_m\mathrm{I}\phi$. By the first condition, we can conclude that $s_n \vDash \mathrm{I}\phi$, so $s_a \vDash @_n\mathrm{I}\phi$. By the former conclusion $s_a \vDash \phi$, so we have $s_a \vDash @_n\mathrm{I}\phi$ and ϕ, therefore $s_a \vDash K_n\phi$ by the definition of knowledge.
3. By Fact 1, the last item can be derived directly.

□

The inferences in Fact 7.2 demonstrate factivity of knowledge, which is similar to that in classical multi-agent epistemic logic.

In our framework, we also observe that the following inferences are blocked. Specifically, knowledge claims in our framework presuppose truth only from the

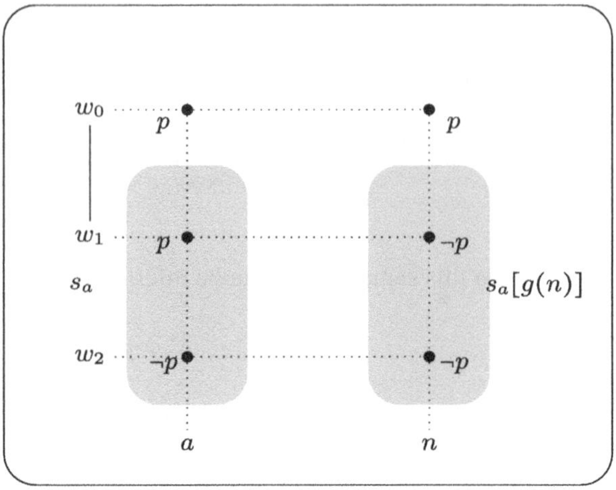

Fig. 7.4 A counterexample

speaker's perspective. This means that only the speaker presupposes the truth of the knowledge component, and this assumption does not extend to other agents. For instance, when a speaker articulates another agent's knowledge, they cannot guarantee the truth of the content for that agent. This highlights the limitations in assuming a uniform perspective across different agents.

$$K_n\phi \not\models @_n\phi, \text{ if } s_a \neq s_a[g(n)].$$

The model illustrated in the following figure serves as a counterexample to this inference (Fig. 7.4).

In this model, there are two information states $s_a = \{\langle w_1, a\rangle, \langle w_2, a\rangle\}$ and $s_a[g(n)] = \{\langle w_1, n\rangle, \langle w_2, n\rangle\}$. The line connecting w_1 to w_0 denotes the accessibility relation R between these worlds. The model validates that $s_a \models \mathrm{K}_n\blacklozenge p$, indicating that $s_a \models @_n\mathrm{I}\blacklozenge p$ and $s_a \models \blacklozenge p$. Consequently, $s_a[g(n)] \models \mathrm{I}\blacklozenge p$ holds, meaning that for all i in $s_a[g(n)]$, $R_n(w_i)[n] \models \blacklozenge p$. This is confirmed as $\langle w_0, n\rangle \in s_a[g(n)]$ and p is true there. However, no i in $s_a[g(n)]$ supports p. Thus, $s_a[g(n)] \not\models \blacklozenge p$, leading to the conclusion $s_a \not\models @_n\blacklozenge p$.

7.5 Conclusion

In this paper, we proposed a multi-agent framework for epistemic *might* and its extension with knowledge and belief. It integrates BSEL models and EFL models to provide a comprehensive framework for analyzing aspects of agency of *might*-claims, and its interaction with belief and knowledge in multi-agent situations.

In the resulted model, perspective shifting is achieved using the hybrid operator @. Different information states represent different agents' perspectives, and @ allows for the interpretation of shifts from one agent's perspective to another when evaluating epistemic claims. This approach distinguishes between epistemic contradictions and conversational epistemic contradictions, with the latter not being considered contradictions in this context.

Moreover, when considering knowledge and belief, we highlighted a key distinction between knowledge and belief: interpreting epistemic *might* under knowledge always requires both the speaker's and the immediate subject's perspectives, whereas belief only requires the subject's perspective. Moreover, our model captures how perspective plays a role in reasoning about knowledge.

For future research, we plan to further investigate the notion of knowledge in this semantics by exploring concepts such as common knowledge and distributed knowledge.

References

1. Aloni, M.: Logic and conversation: the case of free choice. Semant. Pragmat. **15**, 5 (2022)
2. Anand, P., Hacquard, V.: Epistemics and attitudes. Semant. Pragmat. **6** (2013)
3. Areces, C., ten Cate, B.: Hybrid logics. In: Blackburn, P., Van Benthem, J., Wolter, F. (eds.), Handbook of Modal Logic, Studies in Logic and Practical Reasoning, vol. 3, pp. 821–868. Elsevier (2007)
4. Blackburn, P., Seligman, J.: Hybrid languages. J. Logic Lang. Inform. **4**, 251–272 (1995)
5. Bylinina, L., Sudo, Y., McCready, E.: Notes on Perspective-Sensitivity. LRC Publishing (2015)
6. DeRose, K.: Epistemic possibilities. Philos. Rev. **100**(4), 581–605 (1991)
7. Heim, I., Kratzer, A.: Semantics in Generative Grammar. Blackwell, Malden (1998)
8. MacFarlane, J.: Epistemic modals are assessment-sensitive. In: Egan, A., Weatherson, B. (eds.), Epistemic Modality. Oxford University Press (2011)
9. Mandelkern, M.: Bounded modality. Philos. Rev. **128**(1), 1–61 (2019). https://doi.org/10.1215/00318108-7213001
10. Partee, B.H.: Binding implicit variables in quantified contexts. In: CLS. vol. 25, pp. 342–365 (1989). https://doi.org/10.1002/9780470751305.ch13
11. Seligman, J., Liu, F., Girard, P.: Logic in the community. In: Banerjee, M., Seth, A. (eds.) Logic and Its Applications, pp. 178–188. Springer, Berlin (2011)
12. Seligman, J., Liu, F., Girard, P.: Facebook and the epistemic logic of friendship. In: Poster presentation at TARK 2013 (2013). (arXiv preprint arXiv:1310.6440)
13. Seligman, J.: The logic of correct description. In: Advances in Intensional Logic, pp. 107–135 (1997)
14. Speas, P., Tenny, C.: Configurational properties of point of view roles. In: Asymmetry in grammar, pp. 315–344 (2003)
15. Stephenson, T.: Judge dependence, epistemic modals, and predicates of personal taste. Linguist. Philos. **30**(4), 487–525 (2007). https://doi.org/10.1007/s10988-008-9023-4
16. Veltman, F.: Defaults in update semantics. J. Philos. Log. **25**(3), 221–261 (1996)
17. Veltman, F., Groenendijk, J., Stokhof, M.: This might be it. In: Westerståhl, D., Seligman, J. (eds.), Language, Logic, and Computation: the 1994 Moraga Proceedings, pp. 255–70. CSLI (1996)
18. Wittgenstein, L.: Philosophical Investigations. Blackwell Publishing (1953)
19. Yalcin, S.: Epistemic modals. Mind **116**(464), 983–1026 (2007). https://doi.org/10.1093/mind/fzm983
20. Yan, J., Aloni, M.: Knowing and believing an epistemic possibility. Manuscript (2022)
21. Yang, F., Väänänen, J.: Propositional team logics. Ann. Pure Appl. Logic **168**(7), 1406–1441 (2017)

Chapter 8
This Time As Grandfather

Patrick Blackburn, Torben Braüner, and Julie Lundbak Kofod

Abstract Arthur Prior used modal logics with propositional quantifiers throughout his career. As we now know, propositional quantifiers easily give rise to highly complex logics. But propositional quantifiers are syntactically simple and were philosophically attractive to Prior. We discuss how Prior put such quantifiers to work, both in his early work, and in his creation of hybrid tense logic and egocentric logic. We also discuss what we call Prior's Ideal Language (PIL), a hybrid modal logic enriched with propositional quantifiers.

Keywords Arthur prior · Propositional quantifiers · General models · Tense logic · Hybrid logic · Egocentric logic · Description logic · Protothetic

8.1 Introduction

Arthur Prior is often called the father of tense logic: its basic ideas were created by him, and he was the first to explore many of its most challenging problems, for example, in the logic of branching time. He is also deservedly called the father of hybrid logic and, as we shall note later, we can view him as a forgotten father of description logic too. But another form of logic is used throughout Prior's work: various forms of modal logic with propositional quantifiers, logics in which you can write things like

$$\exists p \Box p,$$

("there is at least one necessary proposition"), and

$$F \forall p (p \to Gp),$$

P. Blackburn · J. L. Kofod
Department of Philosophy and Science Studies, Roskilde University, Roskilde, Denmark

T. Braüner (✉)
Department of People and Technology, Roskilde University, Roskilde, Denmark
e-mail: torben@ruc.dk

K. Sano et al. (eds.), *Exploring Negation, Modality and Proof*,
Logic in Asia: Studia Logica Library,
https://doi.org/10.1007/978-981-96-8566-0_8

("at some future time, all true proposition will continue to be true at all times thereafter"), and

$$\forall p(B_{arthur}p \rightarrow p)$$

("every proposition that Arthur believes is true").

Such logics were used by Arthur Prior throughout his career, and played an increasingly prominent role in his work. Prior is not the "father" of such logics—he did not invent them, or discuss them systematically—but his persistent use of them helped establish them as part of the toolkit of working philosophical logicians, and two of his PhD students, Robert Bull and Kit Fine, did important technical work on them in the late 1960s. Hence "grandfather".

Modal logics with propositional quantifiers are interesting for several reasons. For a start, they can lead to unexpected philosophical territory. Prior's adoption of them in the 1950s was rather unusual. Willard Quine, then a leading analytic philosopher, argued that modal logic was conceived in sin (the sin of confusing use and mention), and that propositions did not exist (as they could not be individuated).[1] Prior's unabashed use of modal logics that allowed quantification over ontologically unfashionable entities was thus doubly eccentric. But Prior's use of these logics was not some contrarian impulse: it reflected his deep interest in both propositions and modality, a combination that towards the end of his career led him (somewhat unwillingly) to open a novel door on idealism.

Modal logics with propositional quantifiers are also *logically* surprising. On the face of it, they seem straightforward: simply start with propositional modal logic, and add object-language quantification over the propositional symbols. All very harmless, seemingly a baby extension. But—as Kit Fine was to show—appearances are deceptive: these logics are typically extremely strong fragments of higher-order logic.

In this paper we sketch the role such logics played in Prior's work. As we shall see, modal logic with propositional quantifiers provided Prior with a useful tool for thinking, one he used in making the transition from tense logic, via hybrid tense logic, to description logic. We begin in Sect. 8.2 with some logical background: What are modal logics with propositional quantifiers? What do we know about them? And how do they relate to modal logics *without* propositional quantifiers? We then turn in Sect. 8.3 to the role of propositional quantifiers in Prior's early work. Where did Prior learn about them? Why did he find them interesting? How did he use them? In Sect. 8.4 we turn to the final years of his career, focussing on key papers from *Papers on Time and Tense* [43, 46] that chart the path from tense logic to hybrid logic and beyond. This leads us, in Sect. 8.5 to what we call "Prior's Ideal Logic". Section 8.6 concludes.

[1] Chaper 1 of Quine's [49] and some of his essays in [48] present these themes clearly.

8.2 Propositional Quantifiers

What happens when we add propositional quantifiers to (classical) propositional logic? That is, what happens if we let ourselves write expressions such as:

$$\forall p \exists q (p \leftrightarrow \neg q)?$$

Not so much actually. For a start, the new quantifiers are eliminable. It is easy to see that:

$$\forall p \varphi \text{ is truth functionally equivalent to } \varphi[\top/p] \wedge \varphi[\bot/p]$$

and

$$\exists p \varphi \text{ is truth functionally equivalent to } \varphi[\top/p] \vee \varphi[\bot/p].$$

Consider, for example, $\forall p \exists q (p \leftrightarrow \neg q)$. This says that every proposition p is equivalent to a negated proposition, which is true in classical logic. The propositional quantifiers let us state this concisely, but we can make a similar claim without their help: simply eliminate them using the two equivalences:

$$\begin{gathered}
\forall p \exists q (p \leftrightarrow \neg q) \\
\exists q (\top \leftrightarrow \neg q) \wedge \exists q (\bot \leftrightarrow \neg q) \\
((\top \leftrightarrow \neg \top) \vee (\top \leftrightarrow \neg \bot)) \wedge \exists q (\bot \leftrightarrow \neg q) \\
((\top \leftrightarrow \neg \top) \vee (\top \leftrightarrow \neg \bot)) \wedge ((\bot \leftrightarrow \neg \top) \vee (\bot \leftrightarrow \neg \bot)) \\
(\top \leftrightarrow \neg \bot) \wedge (\bot \leftrightarrow \neg \top).
\end{gathered}$$

While $\forall p \exists q (p \leftrightarrow \neg q)$ and $(\top \leftrightarrow \neg \bot) \wedge (\bot \leftrightarrow \neg \top)$ don't say exactly same thing, they both truthfully reflect the same point: there are only two classical truth values, and classical negation toggles between them. A computer scientist might point out that the brevity offered by propositional quantifiers comes at a computational price (assuming $P \neq NP$) as it takes us from an NP-complete satisfiability problem (without the quantifiers) to a PSPACE-complete satisfiability problem with the quantifiers). Otherwise, there is little drama in the addition of propositional quantifiers in this simple setting.

Matters change spectacularly when we add propositional quantifiers in the setting of modal logic, that is, if we allow ourselves write expressions like:

$$\forall p \Box \exists q (p \leftrightarrow \forall p \Diamond q).$$

It turns out that we have entered a realm of extremely complex—typically highly undecidable—logics, as Kit Fine showed in his groundbreaking 1970 paper [19]. For example, Fine showed that if the underlying modal logic was **K**, **T**, **S4**, **S4.2** or **B**, adding (standardly interpreted) propositional quantifiers resulted in undecidable—indeed unaxiomatisable—logics. Yet he also showed (using a quantifier elimination arguement) that if the underlying logic was **S5**, the resulting logic remained decidable. Fine's pioneering work on modal logics with propositional quantifiers has since been expanded in multiple directions; see Peter Fritz's [25] for a broad overview, and his [23] for a detailed treatment of axiomatisability results.

But why this complexity explosion? Matters become clear if we think in terms of Kripke semantics, where modalities such as $\Box$ and $\Diamond$ (or the temporal modalities G, F, H and P of Arthur Prior's tense logic) are interpreted with respect to a set of worlds W (or a set of times T) and a binary accessibility relation R (or a temporal ordering $<$). At the very heart of this interpretation lies the following idea: *propositions should be thought of as sets of worlds (or times) namely those worlds (or times) where they are true*. This means that propositional variables should range over subsets of W (or T), and that the following *standard interpretation* of the propositional quantifiers is the most obvious choice:

$\forall p\varphi$ is true at some world $w \in W$, if φ is true no matter which *subset of* W (that is: no matter which proposition) interprets p, and

$\exists p\varphi$ is true at some world $w \in W$, if there is at least one *subset of* W (that is: at least one proposition) that makes φ true when it interprets p.

The standard interpretation is certainly natural. But note: the quantification over subsets of W that it rests on is a form of monadic second-order quantification. Fine's results show that (despite the rather simple syntax) adding propositional quantifiers to modal logic packs a powerful second-order punch.

But with this noted, it is time to turn our attention back to modal logic *without* propositional quantifiers. This is because the second-order phenomena just noted are already present in ordinary modal logic, albeit in disguised form: we encounter them when we examine such topics as *frame definability* and *frame incompleteness*.[2] Our brief return to ordinary modal logic will show just how powerful even a restricted form of propositional quantification (universal quantification over a single variable) can be; it will also lead to a second interpretation of propositional quantifiers, namely a *general* (or *Henkin-style*) interpretation.

Recall that a *Kripke frame* is a pair (W, R), a set of worlds together with a binary accessibility relation R over them. Given a Kripke frame (W, R), a valuation V on that frame is a mapping from the propositional variables to a subset of W; that is, V assigns every propositional variable a proposition. A structure (W, R, V) is called a *Kripke model*. Similarly, a temporal frame is a pair $(T, <)$ and a temporal Kripke model is a pair $(T, <, V)$. We say that a formula φ is *globally true* in a model (modal or temporal) iff it true at all worlds (or times) in that model. And now for the key definition, namely frame validity:

> A modal formula φ is *valid on a frame* (W, R) iff φ is globally true on (W, R, V). A temporal formula φ is valid on a temporal frame $(T, <)$ iff for all valuations V on $(T, <)$, φ is globally true on $(T, <, V)$.

For example, $\Diamond\Diamond p \to \Diamond p$ is valid on every transitive frame: it is simple to check that given any frame (W, R) where R is transitive, then no matter which V we choose,

[2] The relationship between modal logics with and without propositional quantifiers, and in particular, the relationship between incompleteness and non-conservativity, are explored by Peter Fritz in [24].

$\Diamond\Diamond p \to \Diamond p$ will be globally true. Indeed, we also have that given any non-transitive frame (W, R), it is always possible to find a valuation in which $\Diamond\Diamond p \to \Diamond p$ is *not* globally true. So $\Diamond\Diamond p \to \Diamond p$ is valid on precisely the transitive frames, and thus modal logicians say it *defines* transitivity.

So far no surprises. Most philosophical logicians (even those who do not work on modal logic) are probably aware that $\Diamond\Diamond p \to \Diamond p$ is the **4** axiom and has something to do with transitivity, that $\Box p \to p$ is the **T** axiom and has some something to do with reflexivity, and so on. For, after the pioneering work of Kripke in the late 1950s and early 1960s [31, 32], most of the familiar modal logics suddenly had characterisations in terms of relatively simple first-order properties of frames, such as transitivity, reflexivity, symmetry, and so on.

But the definition of frame validity given above should make us suspicious. Note the "iff for all valuations V on (W, R)". This quantifies over valuations in the mathematical metalanguage, and as valuations are functions assigning propositions to the propositional variables, this is metalinguistic quantification over propositions. If we had propositional quantifiers in the object language, we could restate the definition of frame validity as follows:

> A modal formula φ is *valid on a frame* (W, R) iff, $\forall p_1 \cdots \forall p_n \varphi$ is globally true on (W, R), where $p_1, \ldots p_n$, are all the propositional variables in φ.

The validity definition for temporal formulas is analogous. Note that as all variables in φ have been bound, we have no need of the valuation V. Also note the restricted syntactic form: all the quantifiers are universal, and all occur in an unbroken block preceding φ (so: universal prenex form). Thus, the frame validity of the ordinary modal formula

$$\Diamond\Diamond p \to \Diamond p,$$

corresponds to the global truth of

$$\forall p(\Diamond\Diamond p \to \Diamond p),$$

and the frame validity of the temporal formula

$$(Fp \vee Fq) \to F(p \wedge Fq) \vee F(p \wedge q) \vee F(q \wedge Fp),$$

(which says that time-does-not-branch-towards-the-future), corresponds to the global truth of

$$\forall p((Fp \vee Fq) \to F(p \wedge Fq) \vee F(p \wedge q) \vee F(q \wedge Fp)).$$

Both transitivity and time-does-not-branch-towards-the-future are first-order conditions,[3] so the quantifier $\forall p$ in these two examples is not doing any genuine second-order work. But some very simple formulas do some very heavy lifting. An elegant

[3] No future branching is $\forall x \forall y \forall z((x < y \wedge x < z) \to (y < z \vee y = z \vee z < y))$.

example is the Löb axiom, here written using Arthur Prior's "at all past times" modality H, as

$$H(Hp \to p) \to Hp.$$

This is valid on precisely those temporal frames $(T, <)$ in which $<$ is transitive and there are no infinitely descending chains (for example, the natural numbers under their usual ordering); this is *not* a first-order definable property (by the Compactness Theorem for first-order logic).[4] Thus

$$\forall p H(Hp \to p) \to Hp$$

is an syntactically simple example (universal prenex form with just one variable) of a modal formula with propositional quantifiers that specifies a non-elementary class of frames. The second-order nature of frame definability is usually explained using the Standard Translation into the classical correspondence language.[5] That is, the usual explanation is that an an ordinary modal formulas φ is valid on some frame iff the classical second-order formula

$$\forall x \forall P_1 \cdots \forall P_n \mathrm{ST}_x(\varphi),$$

is true on that frame; here $\mathrm{ST}_x(\varphi)$ is the standard translation of φ. The explanation in terms of propositional quantifiers is somewhat more direct.

But let us move on to frame incomplete logics. The Löb formula (presented in temporal notation) was one of the axioms used in the first frame incomplete logic, which appeared in 1972 in a groundbreaking paper by S. K. Thomason. The other key axioms were the time-does-not-branch-to-the-future formula which we saw above, the no-last-point-of-time axiom $Gp \to Fp$, and the McKinsey formula

$$GFp \to FGp.$$

Like the Löb formula, the McKinsey formula defines a genuinely second-order property—but it is quite hard to state what this property is and to prove that it is non-elementary.[6] Thomason showed that no frame could validate all four formulas. So it seemed that adding all four as axioms to basic tense logic should yield an inconsistent logic (or to put it semantically: that this logic should be the logic of the empty class of frames). Surprisingly, Thomason proved that the logic was *consistent*. Thus Thomason's tense logic was the first example of a modal logic that

---

[4] See Example 9, pages 48–49 of [57], the published version of Johan van Benthem's 1976 PhD thesis. Example 3.9 in [10] also presents full proof details.

[5] That is, the classical language with a binary relation symbol R (or $<$ in the temporal case) and a predicate symbol P for each propositional symbol p. See [58] or [12] for introductory overviews of correspondence theory.

[6] See Example 3.11 in [10] for a proof that the McKinsey formula defines a property that violates the downward Löwenheim-Skolem theorem for first-order logic.

was not semantically characterised by *any* class of frames—that is, the first example of a *frame incomplete* modal logic. In 1974, both S. K. Thomason [54] and Kit Fine [20] published examples of frame incomplete modal logics in the simpler $\Box$ and $\Diamond$ language.

How did Thomason show that his logic was consistent? His key insight was that although no class of frames validated the four crucial axioms, there were more fine-grained semantic structures, called *general frames*, that did. In fact, Thomason proved that every normal modal logic was characterised by a class of general frames—there is no such thing as a "general frame incomplete" modal logic. Now, a lot more could be said about general frames, but what is important for our purposes is that they are a example of a Henkin-style semantics for "taming" higher order logic.[7] In particular, they give us a different way of thinking about the interaction of propositions and modality.

Suppose we are working with the $\Box$ and $\Diamond$ language. Let (W, R) be a frame. Then a general frame is a triple (W, R, Π), where Π is a non-empty collection of subsets of W (the admissible subsets) closed under the following operations:

(1) if $X \in \Pi$, then $W - X \in \Pi$

(2) if $X, Y \in \Pi$, then $X \cap Y \in \Pi$

(3) if $X \in \Pi$, then $\{w \in W : \forall v(wRv \rightarrow v \in X)\} \in \Pi$

First, note that these conditions ensure that Π behaves like a set of (modal) propositions: Clause 1 says any proposition can be negated, Clause 2 says that any two propositions can be conjoined, and Clause 3 says that any proposition can be necessitated. Thus admissible sets are modally-closed sets of *propositions*. We make a *general model* out of a general frame by adding a valuation V, we insist that valuations assign elements of Π to all propositional variables. As the atomic formulas are all admissible, the three closure conditions ensure that all complex formulas will be admissible too.

Second, note that every ordinary frame (W, R) gives rise to multiple general frames: it is straightforward to see that choosing $\Pi = \emptyset$ or $\Pi = \mathscr{P}(W)$—the powerset of W—satisfies the three conditions, and typically there are many other choices that work too. So if we define *general validity* of a formula φ to mean its global truth on all *general* frames, it is reasonable to expect that there may be *fewer* general validities, as there are more general frames than frames. Furthermore, note that every general frame of the form $(W, R, \mathscr{P}(W))$ is essentially the familiar Kripke frame (W, R). Hence any general validity φ will be standardly valid, but the converse is not true.

So S. K. Thomason's use of general frames is elegant example of the Leon Henkin's method of "taming" high-order logic: instead of working with propositions dictated by set theory (where ordinary frames *implicitly* restrict us to one fixed

---

[7] The most important omission is that S. K. Thomason also linked general frames and algebraic semantics for modal logic; see [10] for historical and technical discussion. Interestingly, in Robert Bull's 1969 paper which proves completeness for a tableaux system for **S5** with propositional quantifiers [17], Bull states that he is more interested in algebraic models than Kripke models (page 258), citing earlier work by E. J. Lemmon [33]. Bull proves completeness by building what he calls *secondary models* representing algebras, and notes the analogy with Henkin's work (pp. 258–259).

collection, namely $\mathscr{P}(W)$), we work with an *explicitly* pre-structured collections of admissible propositions instead. Leon Henkin [27] used his method to prove a general completeness result for classical type theory; S. K. Thomason used it to prove a general completeness result for normal modal logics. But, as we shall see in Sect. 8.5, general frames are also useful when it comes to hybrid modal languages with propositional quantifiers.

To sum up: the language of modal logic with propositional quantifiers is simple syntactically: just add propositional quantifiers to some language modal logic. However, if you give the quantifiers the standard interpretation—they range over *all* subsets of worlds—this typically results in highly complex logics. This is because modal languages with propositional quantifiers are genuinely second order. Indeed some simple modal formulas, such as the Löb axiom or the McKinsey formula, which contain only a single propositional variable, define non-elementary frame classes when prefixed by a universal propositional quantifier. General frames extend Kripke semantics in a semantically natural way. S. K. Thomason used them to "tame" modal frame-incompleteness results, and they will be useful for thinking about Prior's later work too.

8.3 Arthur Prior, Early Adopter

Prior's interest in propositional quantification reflected his belief in the importance of both propositions and modality. His interest in these topics is abundantly clear in his first big logical project, which began life in 1949 as a "Dictionary of Formal Logic". At the suggestion of Oxford University Press (OUP), Prior converted the dictionary into a textbook with the working title *The Craft of Formal Logic*. He completed this (roughly 220,000 word) text on 6th December 1951, but it was never published in full; in 1976, however, selected chapters (lightly edited by Peter Geach and Anthony Kenny) appeared under the title *The Doctrine of Propositions and Terms*.[8] This gives us some insight into what Prior was thinking about propositions, terms and modality just before he fathered tense logic—Chap. 1, especially Sect. 4, "Proposition types and proposition instances" with its concluding reference to J.D Findlay is noteworthy here. Nonetheless, the text is largely historical (predominantly Aristotelian and medieval logic) and contains virtually no logical symbolism. Neither "quantifier" nor "quantification" appears in the index.

All this was soon to change. Encouraged by OUP to shorten *The Craft*, Prior produced an essentially new text, called *Formal Logic* [38]; this was to be his first published book on logic.[9] Syllogistic logic was not neglected in the new textbook, but the emphasis is on teaching students the value of symbolism and giving them a

---

[8] For more background, see the introduction by Geach and Kenny to the volume; we have taken 6th December 1951 as the finishing date for *The Craft of Formal Logic* from this source.

[9] Despite its title, Prior's first published book, *Logic and the Basis of Ethics* [37], is very much a book on metaethics, not logic.

glimpse of the wide range of logics under discussion in the mid-20th century, including modal logic, intuitionistic logic, and the type theory of Russell and Whitehead's *Principia Mathematica*. The most obvious contrast with *The Craft* is the heavy use of sophisticated logical symbolism, mostly drawn from the Polish tradition,[10] and the many deductive proofs it contains. *Formal Logic* is an impressive achievement, and Prior produced it fast: OUP published it in 1955, but the preface is dated May 1953. But for our purposes, what is most important about *Formal Logic* is that we have here perhaps the earliest example of Prior talking about propositional quantifiers and using a formal result about them to discuss an issue in modal logic.

In a section called *Quantification* ***within*** *the Propositional Calculus* (our emphasis),[11] Prior mentions the "standard modern approach" to logic. By this he means by the approach to teaching logic we are familiar with today: building up from a base of (classical) propositional calculus towards richer systems, such first-order logic and modal logic. He then remarks that the standard approach "has already been varied a little" in his own book, for although he does introduce ordinary propositional calculus in *Formal Logic*, he also introduces Leśniewski's *protothetic*, a syntactically more elaborate version of propositional logic with variables over propositional operators, typically symbolised by δ. Now, the syntax of protothetic need not concern us here; more relevant is Prior's next comment (again, our emphasis):

> And we might easily have departed from the standard practice still more radically, by introducing [...] quantifier symbols **while still in the realm of propositions unanalysed into predicates and names**. This was in fact the procedure for the original 'protothetic' of Leśniewski. [38, page 91]

That is: adding propositional quantifiers is a somewhat "radical" step, but it has a history: Leśniewski already took this step in the protothetic. Prior seems to have viewed propositional quantification as a Polish invention, and he found the idea appealing. He liked the representation of falsum it offered: why not think of $\bot$ as $\forall pp$, that is, "For all p, p". Or as Prior puts it:

> It is doubtful whether the precise sense of this proposition can be expressed without using variables, but it is near enough to 'Everything is true', which is quite an interesting candidate for the position of Standard False Proposition. [38, page 91]

Citing work by Meredith and Łukasiewicz, Prior devotes the remainder of the short section discussing completeness, and notes that eliminability of propositional quantifiers holds for Leśniewski's protothetic; in essence, this is the same eliminability result that we noted earlier for propositional logic, but eliminability is much trickier to prove for protothetic logic.

However, as yet we have no modalities. Prior adds them in the opening of Part III ("Modal, Three Values and Extensional Systems")[12]:

---

[10] Łukasciewicz style Polish notation predominates in *Formal Logic*, but Russell and Whitehead style notation makes an appearance in Part 3. Prior remarks in the Preface that: "It does students no harm to learn to use two different notations, and to employ the one that is better for whatever they may have in hand at the time".

[11] Part 1, Chap. 2, Sect. 3 (pages 90–94) of *Formal Logic*.

[12] We have written Prior's $KMpMNp$ as $\Diamond p \vee \Diamond\neg p$, and $CQpQNp$ as $Qp \to Q\neg p$.

> [O]ne small but fruitful discovery which Aristotle made about 'contingency' is that if any proposition is contingent, then so is its contradictory. Using Qp for 'It is contingent that p' ($\Diamond p \vee \Diamond\neg p$), Aristotle's thesis may be written $Qp \to Q\neg p$. The point is simple enough; since 'It is contingent that p' amounts to 'Both p and not-p are possible', and 'It is contingent that not-p' amounts to 'Both not-p and not-not-p are possible', and 'Both p and not-p are possible implies this by double negation and the convertibility of conjunction.
> [38, page 190–191]

Prior then discusses an issue Łukasiewicz raised about Aristotle's law and the contingency modality: if you assume Aristotle's thesis $Qp \to Q\neg p$, and you also assume that "modal functions are truth functions" then, using general properties of the protothetic, you can prove:

$$C\Sigma pQp\Pi pQp,$$

that is, in contemporary notation:

$$\exists pQp \to \forall pQp,$$

or in words:

If some proposition is contingent, then every proposition is contingent.

Łukasiewicz concluded that no modal system can consistently contain an operator such as Q. Prior remarks (on page 192) that "This seems hardly reasonable", and draws the opposite conclusion: it shows that modal functions are clearly *not* truth functions of propositions (it is worth remembering that this was written well before Kripke's revolution in modal semantics).

So Prior appreciated propositional quantification from the start of his logical career—but before we turn to his use of them in the final phase of his career, let us briefly note some middle-period discussions of propositions and propositional quantification. There are two handy sources here. One is the posthumous book *Objects of Thought* [44], which appeared in 1971 and the other is the posthumous collection *Papers in Logic and Ethics* [45] from 1976. Both were edited by Peter Geach and Anthony Kenny.

Objects of Thought has eleven chapters; the editors state that Chaps. 1–6 were for (the most part) complete, and finished by 1965; the remaining chapters were in various draft stages. It gives a fascinating snapshot of Prior's thinking about propositions in the mid 1960s. Prior is keen to engage with contemporaries (such as Quine) and propositional quantification is a recurring motif. For example, in Chap. 6 he discusses the logic of falsehood, error and insincerity. He introduces a belief operator B and a says operator S, and then defines $S(i)xp$,[13] 'x says insincerely that p', by

[13] The i in $S(i)$ is not a nominal; it is simply Prior's shorthand for "insincerely".

$$Sxp \wedge \neg Bxp.$$

Then Ix, 'x is insincere', just requires a propositional quantifier:

$$\exists pS(i)xp.$$

Again, we have converted Prior's formulas from Polish to contemporary notation; from now on we will do so without comment. Perhaps what is most striking about propositional quantification in *Objects of Thought* is how completely Prior takes it for granted. Just before this example he says "We shall presuppose **the usual** Łukasiewicz symbols for truth-functions and quantifiers" (our emphasis again), and individual variables x and propositional variables p are both listed as quantifiable. Indeed Prior's discussion in *Objects of Thought* takes him beyond propositional quantification to other forms of higher-order quantification—in Chap. 3, "Platonism and Quantification", he considers using λ-operators as the only variable binders, as this would let 'abstracts' of various types (such as predicate abstracts) be formed in a uniform way: on the third page of the chapter he give as examples

$$\forall(\lambda x(\varphi x \rightarrow \psi x))$$

("Everything that φ's, ψ's")

$$\exists(\lambda p \neg p)$$

("Something is not the case"). In the first example, a non-binding $\forall$ combines with a predicate abstract, and in the second a non-binding $\exists$ combines with a propositional abstract.

But to conclude our discussion of propositional quantifiers in Prior's middle period work, we note three papers from *Papers on Logic and Ethics* where they play an important role.[14] For example, in "Definitions, Rules, and Axioms", which originally appeared shortly after *Formal Logic* was published,[15] Prior takes as his starting point an (at first glance, rather strange) looking definition of conjunction due to Sobociński,[16] namely: Prior reformulates the definition in protothetic with propo-

> I. Every truth function of q and p has the same truth-value as that truth-function of p and the statement 'Every statement has the same truth-value as itself'. [45, page 39]

[14] The editors P.T. Geach and A.J.P Kenny selected the 22 papers in this posthumous collection, which includes two of Prior's best known papers, "Thank Goodness That's Over" and "The Runabout Inference-ticket".

[15] In *Proceedings of the Aristotelean Society*, n.s. volume 56 (1955–6).

[16] Prior refers to a 1949 paper with the same English title, published in Brussels, but gives no further information. It seems to be [52], which first appeared in Polish in 1939. For some online demonstrations of protothetic defintions being unpacked, see https://demonstrations.wolfram.com/Protothetic/.

sitional quantification and then the technical discussion turns (in typical Prior style) into a philosophical discussion of Russelian versus Leśniewski style definition.

Two other noteworthy papers from the collection are "Epimenides the Cretan" (originally published in [40]) and "The Cogito of Descartes and the Concept of Self-Confirmation" (originally published in [41]). Both give formal proofs involving propositional quantifiers in protothetic. The first derives the well-known Epimenides paradox concerning the truth value of an assertion made by a Cretan that nothing true is ever asserted by a Cretan.[17] The second critically examines A. J. Ayers' reconstruction of Descartes *Cogito ergo sum*. Here Prior focusses on the following sentence: There is an interesting parallelism to Prior's argumentation in these two

> I. If I think that there is something that I think, then there *is* something that I think.

papers. In the Epimenides paper, Prior points out that paradox has little to do with *assertion*, but rather leads to the conclusion that there are affirmations which cannot be *feared, appreciated, loathed*, ..., or indeed *asserted* in isolation. In the *Cogito* paper, Prior points out that Ayers' argument has little to do with the logic of *I think that*; it also works with *I fear that*, and indeed with any logic of psychological attitudes at all.[18] The parallelism between the papers is no accident. In both cases Prior's formalisations and proofs are in *protothetic* with propositional quantifiers. Prior in essence observes that the protothetic functor δ (which ranges over all monadic proposition-forming functors of propositions) appearing in these proofs is open to a wide range of interpretations, both extensional and intentional. Glossing this in contemporary terminology: *Prior's observes that his formal proofs go through for a wide range of modalities.* Prior's sensitivity to the wide range of interpretations of modal logic (which meshed well with his use of Polish protothetic) would soon lead him to "egocentric logic" and the realisation that modal logic could be used to talk out absolutely anything.

8.4 From Tense Logic to Egocentric Logic

Prior's use of propositional quantifiers reflected his deep interest in propositions and modality, and nowhere is this more evident than in his work on the philosophy and logic of time. Prior was not the first to emphasise the importance of A-series (dynamic, present-centred) language over B-series (static, de-centred) language; for

[17] Prior calls it a paradox, but as he emphasizes [45, page 74] there is "nothing actually contradictory" about the propositionally quantified formula that expresses it. Prior cites Alonzo Church's earlier observation that the oddity of the Epimenides paradox is that it seems to show "that it is possible to settle an empirical question on logical grounds" [45, page 70].

[18] In addition, Prior links his discussion of the *Cogito* back to his analysis of the Epimenides paradox. He does so by using the Socrates versus Protagoras debate from Plato's *Theaetetus*.

example, Moritz Schlick had already done so 25 years earlier.[19] Nor was Prior the first to devise a temporal logic; for example, Jerzy Łoś had earlier devised a B-series temporal logic, which Prior used to good effect in his earliest work on tense logic.[20] But Prior seems to have been the first logician of the symbolic era to insist that that propositions *changed truth values over time*, and that the logic of time was a *modal* logic. His (at that time, unorthodox) view of propositions he acquired from his study of medieval logic; his view that the tenses of natural language effectively amounted to a modal logic seems to have been inspired by his mentor J. D. Findlay. Prior wove these two convictions together to create tense logic, a novel A-theoretic logic of time. And, right from the start, he viewed tense logic as *metaphysically* important.

We will not discuss the origins and development of tense logic; Øhrstrøm and Hasle's [34] monograph and Copeland's [18] SEP article are good points of entry. Here we merely note that Prior's first talk on tense logic was given on 27th August 1954 in New Zealand (the talk was published in 1957 as "The Syntax of Time Distinctions" [36]); that his first book on the tense logic was *Time and Modality* [39], which also appeared in 1957, but was based on his 1956 John Locke lectures at Oxford; and that the book which established tense logic as a research field was his *Past, Present and Future* [42], which appeared in 1967. However, we will focus on two slightly later papers, both from his 1968 collection *Papers on Time and Tense* [42], the last of Prior's books to appear in his lifetime. For the story we will tell here is how Prior turned his tense logic into a logic of *everything*.

The two papers that witness this transition are "Tense Logic and the Logic of Earlier and Later" and "Quasi-Propositions and Quasi-Individuals". The first is Prior's most detailed account of the transition from tense logic to what is nowadays called *hybrid* tense logic; the second then "converts" hybrid tense logic into what Prior called *egocentric logic*—an (extremely!) early example of what we would nowadays call *description logic* [3]. We will discuss why Prior found the transition from tense logic to hybrid logic important, and how it led him—somewhat ambivalently—to egocentric logic. These two papers contain some of Prior's most interesting logico-philosophical reflections on propositions and modality, and propositional quantification—indeed, quantification over two different sorts of proposition—features heavily in both. [21]

So: why did Prior invent *hybrid* tense logic? In essence, to solve a problem that he had already clearly formulated in "The Syntax of Time Distinctions" [36]. As Prior says there, his tense logic—his PF-calculus as he sometimes called it—neatly expressed A-series properties of time, "the movement of events from the future through the present to the past" [36, page 116]. But there was an alternative to tense logic, and Prior showed how to interpret tense logic in a B-theoretic language called

[19] As Florian Fischer showed in his "Prior to Prior" [21].

[20] See [55] and [35] for some spirited discussion.

[21] For more detailed treatments of the transition from hybrid tense logic to egocentric logic, see [5] and Chap. 6 of Julie Kofod's PhD thesis [29].

the I-calculus. [22] For example, Prior interpreted Fp as $\exists x(lxz \wedge px)$; here z is a 'date variable' representing the time of utterance, and lxz means 'x is later than z'. Note that his is essentially the familiar clause for F from the contemporary standard translation.

Nonetheless, the I-calculus, the UT-calculus, and the contemporary correspondence language are all classical languages that talk about time "from the outside", rather than "from the inside" modal style. Prior concedes that "the interpretation of the PF-calculus within the I-calculus is clearly a device of considerable metalogical utility" [36, page 115]. But does it have metaphysical significance? Absolutely not! In Prior's words (and with our emphasis):

> As an 'interpretation' in the metaphysical sense of the 'now' which is understood in all the 'propositions' with which the PF-calculus is designed to deal ...the 'z' we have used in the I-calculus is certainly **a complete sham**. [36, page 116]

And this leads to the the problem formulation:

> If there is to be to be any 'interpretation' of our calculi in the metaphysical sense, it will probably need to go the other way round; that is, the I-calculus should be exhibited as a logical construction out of the PF-calculus rather than *vice-versa*. How this could be achieved in detail has yet to be investigated[...] [36, page 116]

In short: the PF-calculus is the metaphysically significant one, so it should interpret the I-calculus. But this is not a trivial task: as Prior was well aware, his PF-calculus is *weaker* than the -calculus. For example, while the simple axiom $FFp \to Fp$ captures the idea of the temporal flow being transitive, there is no PF-axiom that captures its irreflexivity. The I-calculus, of course, has no problem with irreflexivity: $\forall z \neg Izz$.

Prior eventually solved his problem—and he did so by quantifying over propositions.[23] But not just any propositions: he introduced a second sort of proposition, nowadays called nominals, and which Prior called "world propositions" or "world-state propositions" or "world variables". Here (using contemporary notation and terminology) is Prior's recipe for building a strong hybrid tense logic capable of interpreting the the I-calculus.

Step 1: First, *sort* the atomic symbols of tense logic. Ordinary propositional variable are usually written p, q, r, and so on. Let us write the second sort as i, j, k and so on. Call these nominals. Both sorts of symbol can be freely combined to form more complex formulas in the usual way. For example,

$$F(i \wedge p) \wedge F(i \wedge q) \to F(p \wedge q)$$

[22] Note: in [36, page 116] Prior sometimes calls this system I-calculus and sometimes l-calculus; we will write I-calculus. The I-calculus, the UT-calculus, and Rescher and Urquhart's topological logics [50] are all versions of Jerzy Łoś temporal logic. We can also think of them as variants of what we nowadays would call the temporal *correspondence language*; see [56] for (a lot) more on temporal correspondence theory.

[23] Prior had solved his problem by 1967; all the necessary details are presented in Chap. 5 and Appendix B3 of *Past, Present and Future* [42].

is a well formed formula. Insist that each nominal be true at *exactly* one time. A nominal "names" a time by being true there and nowhere else.[24]

Step 2: Add formulas of the form $@_i\varphi$. These assert that φ is satisfied at the point named by the nominal i; symbols of the form $@_i$ are called satisfaction operators. Actually, Prior's step 2 was a little different: he instead added the *universal modality*: its box form $A\varphi$ means that φ is true at *all* times, and the diamond form $E\varphi$ means that φ is true at *some* time. So Prior could define $@_i\varphi$ either as $A(i \rightarrow \varphi)$ or as $E(i \wedge \varphi)$.[25]

Step 3: Add quantifiers over nominals. Actually, in a way it almost misleading to say that Prior "added" them. Prior took propositional quantifiers pretty much for granted. So *of course* you can quantify over nominals too—they are just propositions.[26] But however you put it, Prior allowed expressions like

$$F\exists i\exists j(F(i \wedge \neg j \wedge G\bot) \wedge F(j \wedge \neg i \wedge G\bot)).$$

This says: there is a future time with two distinct future times, both of which have no successors (so: somewhere in the future time forks into two dead-ends).

Clearly Prior's hybrid language is strong. Here we use it to interpret the first-order correspondence language. Assume we have 1-1 pairings between first-order variables x and nominals i, and between predicate symbol P and propositional symbols p. Then: Note that the real work takes place in the first three clauses, where the atomic

$$\begin{aligned}
HT(x < y) &= @_i\mathsf{F}j \\
HT(\mathsf{P}x) &= @_i p \\
HT(x = y) &= @_i j \\
HT(\neg\varphi) &= \neg HT(\varphi) \\
HT(\varphi \wedge \psi) &= HT(\varphi) \wedge HT(\psi) \\
HT(\exists x\varphi) &= \exists i HT(\varphi) \\
HT(\forall x\varphi) &= \forall i HT(\varphi).
\end{aligned}$$

formulas are translated. The use of @ in these clauses is essential. If we don't have @ in the language (or something like the universal modality) the B-language is stronger and no translation is possible (see Theorem 4.1 in [11]). To put it another way, step 2 of Prior's recipe is important. Just adding quantification over nominals (step 3) is not sufficient.

---

[24] Thus the displayed formula is valid: if $F(i \wedge p)$ is true, then p is true at the same future time as i, and if $F(i \wedge q)$ is true, then q is true at the same future time as i. But there is only one time where i is true, so p and q are true at the same future time, which is what $F(p \wedge q)$ says.

[25] Prior typically presented the universal modality as an extra **S5** modality pair $\Box$ and $\Diamond$. For further discussion, see Chap. 6 of Julie Kofod's PhD thesis [29]. For a systematic overview of contemporary hybrid logic, see [16].

[26] Prior puts it like this:. "Certainly this revision of the concept of an instant turns quantification over instants into quantification over a certain sort of propositional variable …but, while I will not argue the pros and cons here, it seems to me, for reasons that have nothing specially to do with tense logic, that we have to admit such quantifications anyhow." [46, pages 124–125]

So Prior successfully solved his problem, and the paper "Tense Logic and the Logic of Earlier and Later" [26, page 134] is Prior's most detailed account of the stages by which (as Prior poetically puts it) "tense-logic so swells as to encompass the earlier-later logic". Mimicking Quine's "Three Grades of Modal Involvement" [48, pages 158-176] the paper stepwise deconstructs the B-series language into strong hybrid tense logic, carefully noting several variations and alternatives along the way.

One of these alternatives concerns nominals. In some earlier work, Prior seems to have taken what we call a "telescope" view of nominals, viewing them as the (infinite) conjunction of all the information at some time. But by the time this paper was written, he seems to have settled on what we call the "tag" interpretation, in which nominals are more like "propositional labels", which is the way nominals are typically viewed in contemporary hybrid logic. This becomes clear when Prior introduces the Q operator,[27] which he defines using propositional quantifiers as follows:

$$Qp := Ep \wedge \forall q(A(p \rightarrow q) \vee A(p \rightarrow \neg q)).$$

This says: p is a proposition which is true somewhere, and for every proposition q, the truth p of "fixes" the truth of q. To use Prior's terminology: p acts as an index of an instant. To use ours: p is a tag nominal. Prior remarks that with the help of the Q operator he can "desort" his language, and just work with ordinary proposition symbols (see [46, pages 129–130]).

And now comes a dramatic transition: in "Quasi-Propositions and Quasi-Individuals" (the paper that immediately follows "Tense Logic and the Logic of Earlier and Later" in the first edition of *Papers on Time and Tense*) Prior shows that he can use his strong hybrid tense to talk about ...*absolutely anything*. It is a superb paper: short, wide-ranging, and lively. Prior invents what he calls "ego-centric logic", giving as his example a logic of people and their relative heights. He talks of "person-propositions", and with the help of the "quasi-modalities" $\langle shorter \rangle$, $[shorter]$, $\langle taller \rangle$, and $[taller]$ he formalizes this logic. But is not so much that he has "converted" hybrid tense logic into egocentric logic, rather he looking at the same formal system differently: if you think of people as propositions, and you use these unusual "modalities" to reason about relative height, egocentric logic emerges. Prior is explicit that you can capture the full effect of *any* first-order theory whatsoever by working with propositions, suitable quasi-modalities, and adding the universal modality and quantification over nominals ([46, pages 215–216]).

There is a lot to discuss here. First the easy part. Yes, egocentric logic really is description logic. Description logic was developed in computer science for reasoning about objects their relationships, and has been widely applied; see [3]. In 1991 Klaus Schild [51] showed that the basic description logic (called $\mathcal{ALC}$) was a notational variant of the standard multimodal logic **K**, and Carlos Areces [1] later showed that important parts of description logic can be thought of as multimodal logics with nominals and the universal modality. Prior's short-tall egocentric logic is a bimodal example of $\mathcal{ALCI}$; here the $\mathcal{I}$, indicates that it uses "inverse roles", namely

[27] This is not the Q operator we briefly met in Sect. 8.3 when discussing *Formal Logic*.

short and tall. Nowadays, thanks to the influence of Johan van Benthem and the Amsterdam school, many modal logicians would happily call ⟨*shorter*⟩, [*shorter*], ⟨*taller*⟩, and [*taller*] modalities. Prior called them quasi-modalities, but he saw their universal utility well before the emergence of either Amsterdam-style modal logic or description logic.

Prior the philosopher was somewhat ambivalent about his discovery: the interpretation of B-series logic in A-series logic was meant to hold metaphysical significance. Prior asks (emphasis ours):

> Does not the possibility of this reversal mark off tense logic from other calculi that may be presented as by products of first-order theories? Or can this reversal equally be performed in these other theories? **I am afraid the answer is that it can.** [46, page 216]

That is, the possibility of interpretational reversals in other settings muddies the philosophical waters. In the setting of tense logic, it was taken to affirm the primacy of A-series notions. But when carried out in the setting of egocentric logic, Prior sees links with ideas from Leibnizian monadology [46, pages 219]). Prior returns to the Leibnizian theme in his paper "Egocentric Logic" [46, pages 223–240]. There he doubles down on his primary claim that tense logic is "*metaphysically fundamental*" (Prior's italics) but rejects similar significance in the case of egocentric logic. But he makes a concession (emphasis ours):

> Nevertheless the fact that we can have a consistent and comprehensive egocentric logic as well as a logic of tenses does suggest that **some sort of idealism or relativism is a more defensible position than it once looked**. [46, page 232]

Prior the logician, on the other hand, seems to have been fascinated: he wrote several papers exploring egocentric and related topics, and started writing a new book to be called *Worlds, Times and Selves*. Unfortunately, he only finished the first chapter before his death on 6 October 1969. Kit Fine took this chapter, added four late period papers (plus unpublished supplements to them) and wrote a lengthy Postscript as Chap. 8; the resulting book appeared in 1977 [47]. In 2003, the same four papers appeared in the new expanded version of *Papers on Time and Tense*, though the supplements did not. Kit Fine's Chap. 8 remains the most important response to this aspect of Prior's work.

We have covered a lot of ground; a summary may be helpful. Prior's tense logic is built around propositions that change truth values and modalities that modify them. Prior was happy to quantify over propositions, and showed that his A-series tense logic could interpret B-series earlier-later logic; he did this by quantifying over special propositions (nominals) and using the universal modality. But this strong hybrid tense logic, celebrated in "Tense Logic and the Logic of Earlier and Later", could be used to talk about *anything*. Prior demonstrated this in "Quasi-propositions and Quasi-individuals", the paper in which hybrid tense logic became egocentric logic, an early (the first?) example of description logic. Prior was somewhat ambivalent about where these developments took him philosophically, but in "Egocentric logic" and other later papers he explored "quasi-modalities" , "quasi-propositions", and "person-propositions" and other novel ideas with evident enthusiasm.

8.5 Prior's Ideal Logic

From Prior's late-period papers, something like the following picture emerges: some kind of hybrid logic (with tag nominals), the universal modality, and quantification over both ordinary propositional variables and nominals may be a suitable—or at least an *interesting*—language for talking about worlds, times, and selves, and perhaps other things as well. In this section we present our first attempt to construct what we call "Priors Ideal Logic", or PIL. Actually, since we here discuss only the bare essentials, it would probably be better to call this PIL version 1.0, but we will stick with PIL for the sake of brevity. In what follows we outline the syntax and semantics of PIL and draw attention to some of its feature that seem particularly Priorean in spirit.

First some formalities. Fix a set of propositional variables $PVAR = \{p, q, r, \dots\}$ and a set of nominals $NOM = \{i, j, k \dots\}$. We assume that both sets are non-empty and countable, that they are disjoint, and write $ATOM$ for $PVAR \cup NOM$. We will call any nominal or propositional variable an atomic symbol. We then define the wffs of PIL over $ATOM$ as follows:

$$\varphi ::= \ p \mid i \mid \neg\varphi \mid \varphi \wedge \varphi \mid \Box\varphi \mid @_i\varphi \mid \forall p\varphi \mid A\varphi \mid \ \downarrow i\varphi,$$

where $p \in PVAR$ and $i \in NOM$. Note that nominals can occur in two syntactic positions: as a standalone atomic formula i (formula position), or as a subscript to @ (operator position). We define other booleans in the usual way, and $\Diamond$ and E are defined by $\neg\Box\neg \equiv \Diamond$ and $\neg A\neg \equiv E$. Similarly, $\exists p$ is defined by $\neg\forall p\neg \equiv \exists p$. In short, PIL is basic hybrid logic augmented by the propositional quantifiers, the universal modality A, and the yet-to-be-explained symbol $\downarrow$, the *downarrow* binder. Note that the universal quantifier $\forall$ is only used to bind propositional variables (not nominals) and the downarrow binder $\downarrow$ is only used with nominals (not propositional variables). As its names suggests, the downarrow binder binds all nominals in its scope, but it does not bind ordinary propositional variables. So it seems that quantification over nominals is not allowed—but as we shall see later, that is illusory.

Now for the semantics. Because of their simplicity, and the fact that they give rise to completeness results for all normal modal logics (recall our discussion of S. K. Thomason's work in Sect. 8.2) we here base our semantics on general frames (W, R, Π), and define a general model $\mathcal{M}$ to be a tuple (W, R, Π, N, V) where $N : NOM \to W$ and $V : PROP \to \Pi$.

With one important exception, we have already discussed the ideas underlying our general models: the exception is the nomination N. Note that as we have defined them here, *nominations directly tag worlds with nominals*. Moreover, if a world w is tagged by a nominal, we do *not* demand that $\{w\} \in \Pi$.[28] Recall that admissible propositions Π are used by V to interpret ordinary propositional variables, and although some

[28] We remark that if $\{w\} \in \Pi$ for any $w \in W$, that is, all singleton sets of worlds are admissible, then the general frame in question is called *discrete*. Of course, any standard frame is discrete, but a discrete frame need not be standard. See the discussion in [8].

general frames Π will contain "singleton world propositions", not all do. Nominals are propositions—but they are special. Our atomic tag nominals *ignore* Π; they link directly to the underlying worlds, unmediated by Π. One way of thinking about this is that nominals are "logical propositions", a bit like $\top$ and $\bot$. Perhaps a better way of viewing them is as 0-place modalities, "deictic modalities" that both point to and are true at unique worlds (or times or selves). But however you think about them, the different ways nominations and valuations assign values in our general models, is the most fundamental design decision underlying PIL.

Now for the Henkin-style satisfaction definition. Let $\mathcal{M} = (\mathcal{G}, N, V)$ be a general model based on a general frame $\mathcal{G} = (W, R, \Pi)$. Some notation: for any nominations N and N', $N' \sim_i N$ means they are identical save possibly for the value they assign to i, and for any valuations V and V', $V' \sim_p V$ means they are identical save possibly for the value they assign to p. Then $\mathcal{M}$ satisfies a formula φ at a world w is inductively defined by:

$$
\begin{array}{ll}
\mathcal{M}, w \models i & \text{iff } w = N(i), \text{ for any } i \in NOM \\
\mathcal{M}, w \models c & \text{iff } w \in V(c), \text{ for any } c \in PVAR \\
\mathcal{M}, w \models \neg\varphi & \text{iff } \mathcal{M}, w \not\models \varphi \\
\mathcal{M}, w \models \varphi \wedge \psi & \text{iff } \mathcal{M}, w \models \varphi \text{ and } \mathcal{M}, w \models \psi \\
\mathcal{M}, w \models \Box\varphi & \text{iff for all } v \in W, \text{ if } wRv \text{ then } \mathcal{M}, v \models \varphi \\
\mathcal{M}, w \models @_i\varphi & \text{iff } \mathcal{M}, N(i) \models \varphi \\
\mathcal{M}, w \models \forall p\varphi & \text{iff for all } \mathcal{M}' = (\mathcal{G}, N, V') \\
 & \quad \text{where } V' \sim_p V, \text{ we have } \mathcal{M}', w \models \varphi \\
\mathcal{M}, w \models A\varphi & \text{iff for all } v \in W, \mathcal{M}, v \models \varphi \\
\mathcal{M}, w \models \downarrow i\varphi & \text{iff } \mathcal{M}', w \models \varphi \\
 & \quad \text{where } \mathcal{M}' = (\mathcal{G}, N', V) \text{ and } N' \sim_i N \text{ such that } N'(i) = w.
\end{array}
$$

Note that nominals are linked directly to worlds, that propositional quantifiers range over the admissible propositions in Π, and that A is the universal modality; we shall discuss the clause for $\downarrow$ later. We say that a PIL formula is globally true in a general model iff it true at all worlds in that model, that it is valid on a general frame (W, R, Π) iff for all valuations V on (W, R, Π), φ is globally true on (W, R, V), and that it is valid iff it is valid on all general frames.

In what follows, we shall need the following terminology. A *soft-QF* formula has no nominals in formula position (that's the soft part) and no propositional quantifiers (that's the QF part). Note that a *soft-QF* formula can contain both A and the $\downarrow$ binder.

Basic PIL

Let us take a closer look at the logic of PIL—and let us do so in two stages. First we restrict our attention to the fragment without A and $\downarrow$; we call this basic PIL. Once we have discussed this important fragment, we will see what A and $\downarrow$ add to the mix.

Basic PIL is just basic hybrid logic augmented with propositional quantifiers, and its axiomatisation is a simple extension of the axiomatization of basic hybrid logic. Thus, to axiomatise basic PIL, take a (sound and purely complete) axiomatisation of basic hybrid logic and add the following three axiom schemes:

In addition, add the proof rule Gen$_\forall$: If $\vdash \varphi$ then $\vdash \forall p\varphi$.

$$\begin{aligned} &Q1\colon \forall p(\varphi \to \psi) \to (\varphi \to \forall p\psi) \\ &\qquad \textit{where } \varphi \textit{ contains no free occurrences of } p, \\ &Q2\text{-}sqf\colon \forall p\varphi \to \varphi[\psi/p] \\ &\qquad \textit{where } \varphi[\psi/p] \textit{ is a soft-QF substitution}, \\ &Barcan_{@}\colon \forall p @_i\varphi \leftrightarrow @_i\forall p\varphi. \end{aligned}$$

The enriched system is sound and complete with respect to all general models. The soundness of *Q1*, $Barcan_{@}$, and $\text{Gen}_{\forall}$ are straighforward, but note that the restriction to soft-QF formulas is essential for proving the soundness of *Q2-sqf*. For although this axiom without the restriction to soft-QF formulas is sound with respect to *standard* models, in our general model semantics we cannot allow soft substitutions (as not all choices of Π include "single world propositions") and we cannot allow substitutions of formulas containing quantifiers either (as general frames have no closure condition for propositional quantifiers). Despite this, the proof system is a complete—indeed a *purely* complete—axiom system for basic PIL; the proof details can be found in [6].[29] The proof reuses the machinery used for the pure completeness proof for basic hybrid logic in Sect. 7.3 of [10], and adapts it to define a suitable Π.

But what does *purely* complete mean? This is where our decision to "hardwire" nominals to worlds pays off: it means that the completeness result not only shows that the proof system generates all valid fomulas (that is: generates the minimal logic), it shows that it is also complete for all *pure extensions* of the minimal logic as well.

In hybrid logic, a pure formula is a formula whose only atoms are nominals, for example: $\Diamond\Diamond i \to \Diamond i$ or $@_i\neg\Diamond i$. It is easy to check that $\Diamond\Diamond i \to \Diamond i$ is valid on all (and only on) transitive frames, and that $@_i\neg\Diamond i$ is valid on all (and only on) irreflexive frames. Thus, taken together, these two formulas define the class of strictly partially ordered frames. In hybrid logic, such definability results automatically give rise to completeness results: if we extend the basic logic of PIL with these two pure formulas as extra axioms, we have a complete axiomatisation of the PIL logic of strictly partially ordered frames. Indeed, *any* pure axiomatic extension of PIL is complete with respect to the class of frames that that the extra pure axioms define.

One final point before we turn to strong PIL: basic PIL not only contains the tag nominals that we have so carefully hardwired into it, it also contains the telescope nominals defined in the previous section. Why? Because we are working with *general* models. Recall from the previous section that Prior defined Qp by

$$Ep \wedge \forall q(A(p \to q) \vee A(p \to \neg q)).$$

As we noted, under the *standard* interpretation of propositional quantifiers, this definition guarantees that Qp is a tag nominal, that is, that p is true at a singleton

[29] We remark that there is a lot of choice here: For example, the axiom system of [6] includes standard hybrid-logical rules like NAME and PASTE, but not uniform nominal substitution, however, due to completeness of the system, the latter is admissible.

set $\{w\}$, where w is some world.[30] Under our general interpretation, however, this definition of Qp says that p is an *atom* that is, a non-empty minimal element of Π. This may well be a singleton set $\{w\}$, but it may also be a minimal set of worlds that make the same admissible propositions true, which seems like a reasonable candidate for Prior's telescope nominals.

Strong PIL

But it is time to turn to full PIL. As we have seen, Prior worked with a strong form of hybrid logic which allowed direct quantification over nominals. That is, Prior was happy with expressions like

$$\forall i\, @_i \neg\Diamond i.$$

Note that this is the universal closure of the pure axiom for irreflexivity. Moreover, note that if we place the universal modality A in front of the we get

$$A\forall i\, @_i \neg\Diamond i$$

which is a formula true only in models where every point is irreflexive. So the concept of pure frame definability is elegantly expressible in the Prior's object language. Moreover, as we discussed in the previous section, Prior's version of strong hybrid logic is very strong indeed: he wanted an A-language that was as strong as the B-language, and the universal modality in combination with quantification over nominals gave him what he wanted.

But in PIL, direct quantification over nominals is *not* permitted. Instead we must use $\downarrow$ to bind nominals. What is going on? If we look at the semantic clause for A, we see that A is indeed the "at all worlds" modality and its dual E is the "at some world" modality. But what does the $\downarrow$ clause give us? If we unpack the definition, we see that when we evaluate a formula of the form $\downarrow i\varphi$ at a world w, we effectively assign nominal i as a name for world w, and then evaluate φ with i interpreted in this way. Here is the simplest possible example:

$$\downarrow ii.$$

This is a valid formula: it names the present world i, which (obviously!) makes i true there. As we shall now show, using A and $\downarrow$ together results in a very powerful system: it gives us *precisely* the same expressive power as Prior's direct quantification over nominals, and arguably does so in a very Priorean fashion.

Let's first check that Prior's original language is strong enough to define $\downarrow$. This is pretty straightforward:

$$\downarrow i\varphi := \exists i(i \wedge \varphi).$$

[30] Thus, under the standard interpretation of propositional quantifiers, Qp is equivalent to Up, called the uniqueness operator, defined as $E(p \wedge \neg Dp)$, where D is the usual difference modality.

It is unimportant whether or not φ contains the nominal i. Either way, $\exists i$ binds all occurrences of i—including the the first conjunct—which guarantees that i must be true at the world of evaluation.

Next we must show that $\downarrow$—working together with E—is strong enough to capture Priors direct quantification over nominals. This definition is from [11]:

$$\exists i\varphi := \downarrow jE\downarrow iE(j \wedge \varphi), \textit{ where } j \textit{ is a nominal not occurring in } \varphi.$$

Suppose we are evaluating this formula at a world w. The first thing we do is give this world a brand new name: $\downarrow j$ does this as j does not occur in φ. Then, using the "at some world modality" E followed by $\downarrow i$, we move to some world w' and bind i to that world; in effect the $E\downarrow i$ combination corresponds to the action of the classsical quantifier $\exists i$ which binds the nominal i to *some* world. With this done, we use E again to take us safely back to the world of evaluation w (conveniently labelled j) and evaluate φ there — but with all occurrences i now safely bound to w'.

So there is a no drama here: at the end of the day it makes no formal difference whether we work with $A+\downarrow$ (our choice) or $A+\forall$ (Prior's choice). Still, we like $\downarrow$. For a start, its locality—bind i to the *actual* world, or the *present* time, or the *current* speaker—is Priorean in spirit. We also like the way that expressing $\exists i\varphi$ as $\downarrow jE\downarrow iE(j \wedge \varphi)$ breaks down the classical quantifier $\exists$ in a suggestive way: it decomposes it into global search for truth (using E) coupled with local binding (using $\downarrow$). Moreover, although Prior himself did not isolate downarrow, he came close to doing so. In particular, he was fond of the strong validity

$$\exists ii,$$

"some proposition is true right here and now", and uses it as an axiom on several occasions. Indeed, at the climax of the triumphal paper "Tense Logic and the Logic of Earlier and Later" discussed in the previous section, right at the moment when "tense-logic so swells as to encompass the earlier-later logic", $\exists aa$ (clearly a close syntactic relative of $\downarrow ii$) appears as axiom T9 [26, page 135].

It is straightforward to extend the axiomatisation of basic PIL the full PIL language. In fact there are several ways of doing so (for a selection, see [9]) as both E and $\downarrow$ are proof theoretically well-behaved. Here we will simply note two key axioms that can be used. For the universal modality the axiom

$$Ei$$

is simple and powerful. It says that i is true at *some* world—or better: no matter which world i names, the E modality can see it—which pins down the required universality. As for $\downarrow$, the following axiom is revealing:

$$@_i(\downarrow j\varphi \leftrightarrow \varphi[i/j]).$$

Here $\varphi[i/j]$ is the result of substituting nominal i for all free occurrences of nominal j in φ, thus the axioms neatly captures the interplay between @ and $\downarrow$. A completeness proof for PIL which makes use of both these axioms is presented in [7]. Once again, this is a *pure* completeness proof, so given that the expressive of A and $\downarrow$ is equal to that of Prior's system, we have completeness for any first-order definable property of Kripke frames (W, R).

We close our discussion by noting an elegant result due to Balder ten Cate. Every formula of modal logic with propositional quantifiers (interpreted over standard models) expresses a second-order property; as we discussed in Sect. 8.2 some of these properties don't have elementary equivalents, but some do. Balder ten Cate showed that every such elementary property is equivalent to a pure sentence of hybrid logic containing $\downarrow$ and @ (but without A). Here 'sentence' means that all nominals are bound by $\downarrow$; see (see Theorem 12.0.3, page 179, in [53]). Thus our completeness result for PIL covers all elementary frame classes definable in modal logic with propositional quantifiers under the *standard* interpretation.

8.6 Curious Hybrids

We have discussed the role of propositional quantifiers in Arthur Prior's work, and said a little about our own Prior-inspired work on the topic. But where might this all lead?

Multimodal versions of PIL may be useful in applied logic. We have seen that the *pure* completeness results for PIL give us automatic completeness results for various classes of frame; in multimodal settings, they can make it easy to control interactions between modalities. Here is a simple example: in a system containing two primitive modalities, we can force them to interact like Priorean tense operators (or shorter-taller quasi modalities) by adding the pure axioms $@_i\Box_1\Diamond_2 i$ and $@_i\Box_2\Diamond_1 i$. Multimodal systems with propositional quantifiers have been proposed by Belardinelli, van der Hoek, and Kuijer [4] for epistemic reasoning; multimodal PIL might be useful here too.

Sorted versions of PIL, in which reasoning about abstract object (like worlds and times) and are mixed together with reasoning about ordinary individuals, may be interesting too. In effect: why not add Prior's person-propositions (or penguin propositions, or indeed any-ordinary-object propositions that we find interesting) into the mix? Since the 20th century, the most common option for combining reasoning about abstract objects (like worlds and times) with everyday objects (like people) has been to combine modal logic (for reasoning about the abstract entities) with first-order logic (for the everyday stuff)—that is, the default has been to use some version of first-order modal logic [22]. In the 1980s, during the heyday of logic in AI and Natural Language Processing, it was fairly common to eliminate the modal component and do everything with first-order quantification over both abstract and everyday entities—in short, to use sorted first-order logic for everything (see [28] for classic examples). But the third option, Prior's option—*do everything with quasi-modalities*—remains underexplored, though there are some examples of combining tense logic with description logic (see for example [2]). Kit Fine briefly discusses the philosophical implications of sorting from the perspective of modal actualism, remarking in his Chap. 8 postscript to *Worlds, Times and Selves*, remarking that

> "[T]he new language is not above suspicion..... Possible world quantifiers and modal operators are usually regarded as two competing means of expressing modal truths. The new language is a curious hybrid in which both both means of expression are combined" [47, page 142]

Analytic proof-theory (tableau, sequent, and natural deduction style proof-systems) for basic hybrid logic work very well compared to ordinary modal logic, so an obvious line of work is analytic proof-theory for PIL. Tableau systems have already been developed, see [8, 30]. It would be obvious also to develop natural deduction systems for PIL, see the brief note [14] for some initial considerations. Another related line of work is the following: Hybrid logic as presented here has a classical logical basis, that is, the principle *tertium non datur* holds, concretely, it can be proved in our axiom system for PIL. If this principle is removed from basic hybrid logic (in a natural deduction presentation this can be done simply by removing a specific proof-rule), an *intuitionistic* version of hybrid logic is obtained. Semantically, this step amounts to interpreting the first-order correspondence language intuitionistically rather than classically. See [15] and Chap. 8 of [13]. It could be interesting to investigate what happens if we add propositional quantifiers to this picture. This seems straightforward to do syntactically, at least in a natural deduction system for PIL (punch out a specific proof-rule), but what happens semantically?

All in all, we think that "curious hybrids" such as PIL may be a fruitful way of developing Arthur Prior's philosophical ideas. Here is our final example. Throughout his career, Prior was interested in propositions that referred to non-permanent entities (like people). In 1957, Prior developed a solution he found satisfactory in modal settings, namely "System Q".[31] In "Tense Logic for Non-Permament Existents" (which first appeared in the 1968 edition of *Papers on Time and Tense*) he explores tense logical versions of system Q, and speculates on the usefulness of adding nominals. In "Modal Logic and the Logic of Applicability"[32] he developed his ideas on System Q further, and in "Supplement to Modal Logic and the Logic of Applicability"[33] he speculates on using the Q-operator to explore System Q. Can such hybrids flourish? We think they deserve further attention.

Acknowledgements We would like to thank Katsuhiko Sano for his support (and patience) during the writing of this paper.

[31] This system motivated and introduced in Chaps. 4 and 5 of [39]. The Q here does not refer to the nominal-creating Q operator discussed earlier.

[32] Published as Chap. 6 of *Worlds, Times and Selves* [47], and reprinted in the new expanded edition of *Papers on Time and Tense* [46].

[33] Chapter 7 of of *Worlds, Times and Selves* [47], but not included in [46].

References

1. Areces, C.E.: Logic engineering: the case of description and hybrid logics. Ph.D. thesis, Institute for Logic, Language and Computation (2000)
2. Artale, A., Franconi, E.: A temporal description logic for reasoning about actions and plans. J. Artif. Intell. Res. 463–507 (1998)
3. Baader, F., Horrocks, I., Sattler, U.: Description logics. In: Handbook of Knowledge Representation, pp. 135–179. Elsevier (2008)
4. Belardinelli, F., Van Der Hoek, W., Kuijer, L.B.: Second-order propositional modal logic: expressiveness and completeness results. Artif. Intell. **263**, 3–45 (2018)
5. Blackburn, P.: Arthur Prior and hybrid logic. Synthese **150**(3), 329–372 (2006)
6. Blackburn, P., Braüner, T., Kofod, J.L.: An axiom system for basic hybrid logic with propositional quantifiers. In: International Workshop on Logic, Language, Information, and Computation, pp. 118–134. Springer (2023)
7. Blackburn, P., Braüner, T., Kofod, J.L.: Prior's ideal language. Mathematical Structures in Computer Science. Math. Struct. Comput. Sci. **35**, e7. Cambridge University Press (2025)
8. Blackburn, P., Braüner, T., Kofod, J.L.: Remarks on Hybrid Modal Logic with Propositional Quantifiers. Logic and Philosophy of Time. Aalborg Universitetsforlag (2020)
9. Blackburn, P., ten Cate, B.: Pure extensions, proof rules, and hybrid axiomatics. Stud. Logical **84**, 277–322 (2006)
10. Blackburn, P., de Rijke, M., Venema, Y.: Modal Logic. Cambridge University Press (2001)
11. Blackburn, P., Seligman, J.: Hybrid languages. J. Logic Lang. Inform. **4**, 251–272 (1995)
12. Blackburn, P., Van Benthem, J.: Modal logic: a semantic perspective. In: Handbook of Modal Logic, pp. 1–84. Elsevier (2007)
13. Braüner, T.: Hybrid Logic and its Proof-Theory, Applied Logic Series, vol. 37. Springer (2011)
14. Braüner, T., Blackburn, P., Kofod, J.: Hybrid logic with propositional quantiers: Natural deduction style (work in progress). In: Non-Classical Modal and Predicate Logic : Book of Abstracts, pp. 29–32. Ruhr University Bochum (2021)
15. Braüner, T., de Paiva, V.: Intuitionistic hybrid logic. J. Appl. Log. **4**, 231–255 (2006)
16. Braüner, T.: Hybrid logic. In: Zalta, E. (ed.), The Stanford Encyclopedia of Philosophy. Stanford University (2021), on-line encyclopedia article available at http://plato.stanford.edu/entries/logic-hybrid
17. Bull, R.: On modal logic with propositional quantifiers. J. Symb. Logic **34**(2), 257–263 (1969)
18. Copeland, B.J.: Arthur prior. In: Zalta, E.N., Nodelman, U. (eds.), The Stanford Encyclopedia of Philosophy. Metaphysics Research Lab. Stanford University, Winter (2022)
19. Fine, K.: Propositional quantifiers in modal logic. Theoria **36**(3), 336–346 (1970)
20. Fine, K.: An incomplete logic containing S4. Theoria **40**(1), 23–29 (1974)
21. Fischer, F.: Prior to Prior. In: Hasle, P., Jakobsen, D., Øhrstrøm, P. (eds.), The Metaphysics of Time, Logic and Philosophy of Time, vol. 4, pp. 165–178. Aalborg University Press (2020)
22. Fitting, M., Mendelsohn, R.L.: First-Order Modal Logic. Kluwer Academic Publishers (1998)
23. Fritz, P.: Axiomatizability of propositionally quantified modal logics on relational frames. J. Symb. Logic **89**(2), 758–793 (2024)
24. Fritz, P.: Nonconservative extensions by propositional quantifiers and modal incompleteness. J. Logic Comput. (2024)
25. Fritz, P.: Propositional Quantifiers. Cambridge University Press (2024)
26. Hasle, P., Øhrstrøm, P., Braüner, T., Copeland, J. (eds.), Revised and Expanded Edition of Arthur N. Prior: Papers on Time and Tense. Oxford University Press (2003)
27. Henkin, L.: Completeness in the theory of types. J. Symb. Logic **15**(2), 81–91 (1950)
28. Hobbs, J.R., Moore, R.C.: Formal theories of the commonsense world. Ablex (1985)
29. Kofod, J.L.: What on earth was Arthur Prior thinking?: About physics and hybrid logic. Ph.D. thesis, IKH, Roskilde University (2022)
30. Kofod, J.L., Blackburn, P., Braüner, T.: A complete tableau system for basic hybrid logic with propositional quantication. Logic J. IGPL. (2025), accepted for publication

31. Kripke, S.A.: A completeness theorem in modal logic. J. Symb. Logic **24**(1), 1–14 (1959)
32. Kripke, S.A.: Semantical analysis of modal logic I: normal modal propositional calculi. Math. Log. Q. **9**(5–6), 67–96 (1963)
33. Lemmon, E.J.: Algebraic semantics for modal logics I. J. Symb. Logic **31**(1), 46–65 (1966)
34. Øhrstrøm, P., Hasle, P.: Temporal Logic: From Ancient Ideas to Artificial Intelligence, vol. 57. Springer (2007)
35. Øhrstrøm, P., Hasle, P.: The significance of the contributions of A.N. Prior and Jerzy Łoś in the early history of modern temporal logic. In: Blackburn, P., Hasle, P., Øhrstrøm, P. (eds.), Further Themes from Prior, Logic and Philosophy of Time, vol. 2. Aalborg Universitetsforlag (2019)
36. Prior, A.N.: The syntax of time-distinctions. Franciscan Stud. **18**(2), 105–120 (1958). http://www.jstor.org/stable/41974665
37. Prior, A.N.: Logic and the Basis of Ethics. Oxford University Press (1949)
38. Prior, A.N.: Formal Logic. Oxford University Press (1955)
39. Prior, A.N.: Time and Modality. Oxford University Press (1957)
40. Prior, A.N.: Epimenides the Cretan. J. Symb. Logic **23**(3), 261–266 (1958)
41. Prior, A.N.: The cogito of descartes and the concept of self-confirmation. In: The Foundations of Statements and Decisions, pp. 47–53. Warsaw (1965)
42. Prior, A.N.: Past, Present and Future. Oxford University Press (1967)
43. Prior, A.N.: Papers on Time and Tense. Oxford University Press (1968)
44. Prior, A.N.: Objects of Thought, ed. by P.T. Geach, A.J.P. Kenny. Oxford University Press (1971)
45. Prior, A.N.: Papers in Logic and Ethics, ed. by P.T. Geach, A.J.P. Kenny. University of Massachusetts Press (1976)
46. Prior, A.N.: Papers on Time and Tense, new expanded edn. ed, by P. Hasle, P. Øhrstrøm, T. Braüner, J. Copeland. Oxford University Press (2003)
47. Prior, A.N., Fine, K.: Worlds, Times, and Selves. University of Massachusetts Press (1977)
48. Quine, W.V.O.: The Ways of Paradox. Harvard University Press (1966)
49. Quine, W.V.O.: Philosophy of Logic. Prentice-Hall (1970)
50. Rescher, N., Urquhart, A.: Temporal Logic. Springer Science (1971)
51. Schild, K.: A correspondence theory for terminological logics: preliminary report. In: IJCAI'91: Proceedings of the 12th International Joint Conference on Artificial Intelligence, pp. 466–471 (1991)
52. Sobociński, B.: An investigation of protothetic. Leśniewski's Systems Protothetic, pp. 69–83 (1998)
53. Ten Cate, B.: Model Theory for Extended Modal Languages. Ph.D. thesis, Institute for Logic, Language and Computation (2005)
54. Thomason, S.K.: An incompleteness theorem in modal logic. Theoria **40**(1), 30–34 (1974)
55. Tkaczyk, M., Jarmużek, T.: Jerzy łoś positional calculus and the origin of temporal logic. Logic Log. Philos. **28**(2), 259–276 (2019)
56. Van Benthem, J.: The Logic of Time: A Model-theoretic Investigation into the Varieties of Temporal Ontology and Temporal Discourse. Springer (1983)
57. Van Benthem, J.: Modal Logic and Classical Logic. Bibliopolis, Napoli (1983)
58. Van Benthem, J.: Correspondence theory. In: Handbook of Philosophical Logic, pp. 167–247. Springer (1984)

Chapter 9
Complete and Terminating Tableau Calculus for Undirected Graph

Yuki Nishimura and Tsubasa Takagi

Abstract Hybrid logic is a modal logic with additional operators specifying nominals and is highly expressive. For example, there is no formula corresponding to the irreflexivity of Kripke frames in basic modal logic, but there is in hybrid logic. Irreflexivity is significant in that irreflexive and symmetric Kripke frames can be regarded as undirected graphs reviewed from a graph-theoretic point of view. Thus, the study of the hybrid logic with axioms corresponding to irreflexivity and symmetry can help to elucidate the logical properties of undirected graphs. In this paper, we formulate the tableau method of the hybrid logic for undirected graphs. Our main result is to show the completeness theorem and the termination property of the tableau method, which leads us to prove the decidability.

Keywords Hybrid logic · Tableau calculus · Completeness · Termination property · Undirected graph · Orthoframe

9.1 Introduction

9.1.1 Background

Hybrid logic is an extension of basic modal logic with additional propositional symbols called nominals and an operator called a satisfaction operator @. A nominal is a formula that is true only in a single possible world in a Kripke model. Nominals are regarded as syntactic names of possible worlds. Hybrid logic is highly expressive because nominals allow arbitrary possible worlds to be designated. Hybrid logic was

Y. Nishimura (✉)
Institute of Science Tokyo, Tokyo, Japan
e-mail: nishimura.y.as@m.titech.ac.jp
URL: https://storage.googleapis.com/nishimura-yuki-website/index-en.html

T. Takagi
Japan Advanced Institute of Science and Technology, Ishikawa, Japan
e-mail: tsubasa@jaist.ac.jp

K. Sano et al. (eds.), *Exploring Negation, Modality and Proof*,
Logic in Asia: Studia Logica Library,
https://doi.org/10.1007/978-981-96-8566-0_9

first invented by Prior [17, 18] and has developed in various ways since then. (See [1, 6, 10, 16, 23].)

One of the advantages of hybrid logic compared to basic modal logic is that we can treat the irreflexivity of Kripke frames. In basic modal logic, no axiom corresponds to the irreflexivity of the model [13]. In hybrid logic, however, we have a simple axiom, $@_i \neg \Diamond i$, where i is a nominal [7].

This advantage makes hybrid logic well-suited for dealing with undirected graphs, namely relational structures with irreflexivity and symmetry. There are some logics whose semantics are given by undirected graphs; for example, orthoframes used in (the minimal) quantum logic [14] and social networks used in Facebook logic [20, 21] can be regarded as undirected graphs.

9.1.2 Motivation

The motivation of this paper is to find a procedure that can automatically determine the validity of formulas for undirected graphs. In general, tableau calculus gives such a procedure if its completeness and the termination property are shown. That is, the decidability of the validity of formulas follows from the existence of a complete and terminating tableau calculus.

There are some previous research on tableau calculi for hybrid logic. Tableau calculi for hybrid logic was first proposed by Tzakova [24] as a prefixed tableau calculus, which was based on the method of Fitting [11]. Later, Bolander and Blackburn [8] proposed tableau calculi for hybrid logic in another way, as a labeled sequent calculus. It worked better with hybrid logic than the previous method because we can *internalize* labels owing to the satisfaction operator, and we no longer need labels to construct a proof system. A subsequent paper [9] of [8] proposed a tableau calculus for hybrid logic based on **K4**, and claimed that the loop-checking approach can be extended in it to the hybrid logics based on **S4** and **S5**.

Our study, however, focuses on a combination of axioms corresponding to irreflexivity and symmetry, namely (irr) and (sym) in [9]. This combination is worth considering from the graph-theoretic point of view because it characterizes undirected graphs as Kripke frames. On the other hand, Bolander and Blackburn [9] have not dealt with the combination of (irr) and (sym). The reason is that they considered it challenging: in [9], they wrote, "But then we are faced with the task of combining such conditions as (irr), (sym), (asym), (antisym), (intrans), (uniq) and (tree) with (trans), and here matters are likely to be much trickier."

9.1.3 Results

With the above motivation, we propose a complete and terminating tableau calculus with respect to the class of irreflexive and symmetric frames: its completeness is shown in Theorem 9.4, and its termination property is shown in Theorem 9.1. For this purpose, we show key lemmas called the model existence lemma (Lemma 9.10) and the bulldozing lemma (Lemma 9.14).

To address a hybrid logic with the axioms (irr) and (sym), we employ the bulldozing method, which was originally proposed in Segerberg [19]. (See also [15].) Blackburn [4] used the bulldozing method in proving the completeness of hybrid logic with the axiom (irr). Our work shows that the bulldozing method also works in dealing with the tableau calculus for hybrid logic with the rule of irreflexivity [*Irr*] in Sect. 9.3.

Our work can be regarded as an extension of Blackburn [4] in the following sense. The paper [4] showed that none of the reflexive points in the model created by a maximal consistent set has a nominal and checked that there is no matter to apply bulldozing. In our work, however, all the models created by an open saturated branch are named—each world makes at least one nominal true. We show that this situation causes no problem since all the nominals true in reflexive worlds are not an element that determines the validity of a formula but just "labels" to point to worlds.

9.1.4 Related Work

Examples of previous study on graph theory using hybrid logic can be found in [2, 3, 12]. They studied hybrid graph logic, a hybrid logic with a special operator $\Diamond^+$, as a tool for analyzing some properties of graph theory. In particular, for undirected graphs, Takeuti and Sano [22] discussed how to express the planarity of undirected graphs. (Note that [22] also showed the decidability of hybrid logic for undirected graphs.)

All of these studies adopted Hilbert-style axiomatization as their proof system. On the other hand, we adopt tableau calculus as a proof system. As is well known, tableau calculi have a clear advantage compared to Hilbert-style axiomatizations in that the proof construction can be automated.

9.1.5 Organization of the Paper

The rest of this paper is organized as follows. Section 9.2.1 introduces the semantics of hybrid logic, and Sect. 9.2.2 shows the tableau system of basic hybrid logic. After that, we introduce the tableau calculus corresponding to undirected graphs in Sect. 9.3.1, and then we show the termination property and the completeness theorem in Sects. 9.3.2 and 9.3.3, respectively. Finally, in Sect. 9.4, we summarize the main result of the paper and propose future work.

9.2 Preliminaries

9.2.1 *Kripke Semantics of Hybrid Logic*

Here, we review a hybrid logic with an operator @. (See [10] for more details.)

Definition 9.1 (*Language*) We have a countably infinite set **Prop** of *propositional variables* and another countably infinite set **Nom** of *nominals*, which is disjoint from **Prop**. A *formula* φ of hybrid logic is defined inductively as follows:

$$\varphi ::= p \mid i \mid \neg\varphi \mid \varphi \wedge \varphi \mid \Diamond\varphi \mid @_i\varphi,$$

where $p \in$ **Prop** and $i \in$ **Nom**.

We use the following abbreviations:

$$\varphi \vee \psi := \neg(\neg\varphi \wedge \neg\psi),\ \varphi \to \psi := \neg(\varphi \wedge \neg\psi),\ \Box\varphi := \neg\Diamond\neg\varphi.$$

Definition 9.2 A *Kripke model* (we call it *model*, in short) $\mathcal{M} = (W, R, V)$ is defined as follows:

- W is a non-empty set,
- R is a binary relation on W,
- V is a function $V : \mathbf{Prop} \cup \mathbf{Nom} \to \mathcal{P}(W)$ such that $V(i) = \{w\}$ for some $w \in W$ for each $i \in$ **Nom**, where $\mathcal{P}(W)$ denotes the powerset of W.

Furthermore, we call $\mathcal{F} = (W, R)$ that satisfies the conditions above a *Kripke frame* (or shortly *frame*).

This definition reflects the property of nominals: one nominal is true in only one world. In this paper, we write wRv to mean $(w, v) \in R$. Moreover, We abbreviate $w \in W$ such that $V(i) = \{w\}$ by i^V.

Definition 9.3 Given a model $\mathcal{M}$, a possible world w in $\mathcal{M}$, and a formula φ, the *satisfaction relation* $\mathcal{M}, w \models \varphi$ is defined inductively as follows:

$$\begin{aligned}
\mathcal{M}, w \models p &\iff w \in V(p), \text{ where } p \in \mathbf{Prop};\\
\mathcal{M}, w \models i &\iff w = i^V, \text{ where } i \in \mathbf{Nom};\\
\mathcal{M}, w \models \neg\varphi &\iff \text{not } \mathcal{M}, w \models \varphi;\\
\mathcal{M}, w \models \varphi \wedge \psi &\iff \mathcal{M}, w \models \varphi \text{ and } \mathcal{M}, w \models \psi;\\
\mathcal{M}, w \models \Diamond\varphi &\iff \text{there exists } v \text{ such that } wRv \text{ and } \mathcal{M}, v \models \varphi;\\
\mathcal{M}, w \models @_i\varphi &\iff \mathcal{M}, i^V \models \varphi.
\end{aligned}$$

Definition 9.4 A formula φ is said to be *valid* (denoted as $\models \varphi$) if $\mathcal{M}, w \models \varphi$ holds for all models $\mathcal{M}$ and all of its worlds w.

9.2.2 *Tableau Calculus for* K(@)

K(@) is an axiomatization of hybrid logic with an operator @, based on the minimal normal modal logic **K**. (See [6], for example.) We write **TAB** to indicate the tableau calculus for hybrid logic **K**(@). This section is based on [8].

Definition 9.5 A *tableau* is a well-founded tree constructed in the following way:

- Start with a formula of the form $@_i\varphi$ (called the *root formula*), where φ is a formula of hybrid logic and $i \in \mathbf{Nom}$ does not occur in φ.
- For each branch, extend it by applying rules (see Definition 9.7) to all nodes as often as possible. However, we can no longer add any formula in a branch if at least one of the following conditions is satisfied:

(i) Every new formula generated by applying any rule already exists in the branch.
(ii) The branch is closed. (See Definition 9.6)

Here, a *branch* means a maximal path of a tableau. If a formula φ occurs in a branch Θ, we write $\varphi \in \Theta$.

Definition 9.6 A branch of a tableau Θ is *closed* if there exists a formula φ and a nominal i such that $@_i\varphi, @_i\neg\varphi \in \Theta$. We say that Θ is *open* if it is not closed. A tableau is called *closed* if all branches in the tableau are closed.

Definition 9.7 We provide the rules of **TAB** in Fig. 9.1.

Definition 9.8 (*Probability*) Given a formula φ, we say that φ is *provable* in **TAB** if there is a closed tableau whose root formula is $@_i\neg\varphi$, where $i \in \mathbf{Nom}$ does not occur in φ.

$$\frac{@_i\neg j}{@_j j}\,[\neg] \qquad \frac{@_i\neg\neg\varphi}{@_i\varphi}\,[\neg\neg] \qquad \frac{@_i(\varphi\wedge\psi)}{\begin{array}{c}@_i\varphi\\ @_i\psi\end{array}}\,[\wedge] \qquad \frac{@_i\neg(\varphi\wedge\psi)}{@_i\neg\varphi \mid @_i\neg\psi}\,[\neg\wedge]$$

$$\frac{@_i\Diamond\varphi}{\begin{array}{c}@_i\Diamond j\\ @_j\varphi\end{array}}\,[\Diamond]^{*1,*2,*3} \qquad \frac{\begin{array}{c}@_i\neg\Diamond\varphi\\ @_i\Diamond j\end{array}}{@_j\neg\varphi}\,[\neg\Diamond] \qquad \frac{@_i@_j\varphi}{@_j\varphi}\,[@] \qquad \frac{@_i\neg@_j\varphi}{@_j\neg\varphi}\,[\neg@] \qquad \frac{\begin{array}{c}@_i\varphi\\ @_i j\end{array}}{@_j\varphi}\,[Id]^{*3}$$

*1: $j \in \mathbf{Nom}$ does not occur in the branch.
*2: This rule can be applied only one time per formula.
*3: The formula above the line is not an accessibility formula. Here, an *accessibility formula* is the formula of the form $@_i\Diamond j$ generated by $[\Diamond]$, where j is a new nominal.

In these rules, the formulas above the line show the formulas that have already occurred in the branch, and the formulas below the line show the formulas that will be added to the branch. The vertical line in the $[\neg\wedge]$ means that the branch splits to the left and right.

Fig. 9.1 The rules of **TAB**

Bolander and Blackburn [8] showed two significant properties of **TAB**. One is the termination property, and the other is completeness.

Theorem 9.1 (Termination) *The tableau calculus* **TAB** *has the termination property. That is, for every tableau in* **TAB**, *all branches of it are finite.*

Theorem 9.2 (Completeness) *The tableau calculus* **TAB** *is complete for the class of all frames.*

9.3 Tableau Calculus for IB(@)

In this paper, we discuss the hybrid logic corresponding to irreflexive and symmetric models. We call the models *orthomodels*, following [14].

Definition 9.9 An *orthoframe* is a Kripke frame $\mathcal{F} = (W, R)$, where R is irreflexive and symmetric. An *orthomodel* is $\mathcal{M} = (\mathcal{F}, V)$, where $\mathcal{F}$ is an orthoframe.

IB(@) is a hybrid logic corresponding to the class of all orthoframes. We can construct it by adding the following two axioms to the hybrid logic **K**(@):

$$(irr) : @_i \neg \Diamond i,$$
$$(sym) : @_i \Box \Diamond i.$$

9.3.1 Constructing Tableau Calculus

Our aim is to construct a tableau calculus corresponding to **IB**(@), keeping the termination property. First, we add the following rule that reflects the symmetry:

$$\frac{\begin{array}{c} @_i \Box \varphi \\ @_j \Diamond i \end{array}}{@_j \varphi} \; [\Box_{sym}] \; .$$

However, this rule allows us to build an infinite branch.

Example 9.1 Consider an example of Fig. 9.2. (Formulas with * are accessibility formulas.) In this branch, nominals are generated infinitely as $i_1, i_3, i_5, \ldots$ and $i_2, i_4, i_6, \ldots$.

In this case, the nominals $i_1, i_3, i_5, \ldots$ play the same role. For any nominal i_{2n+1} ($n \geq 0$) occurring in Θ, we have $@_{i_{2n+1}} \Diamond\Diamond i$, $@_{i_{2n+1}} \Diamond i \in \Theta$. Then we have to prohibit the creation of such redundant nominals.

The solution to this problem has already been proposed in [8, Sect. 5.2] and [9, Sect. 7]. That is to add this restriction ($\mathcal{D}$) to the tableau calculus.

Fig. 9.2 Non-terminating tableau of **TAB** with only $[\Box_{sym}]$

1. $@_{i_0}(\Diamond\Diamond i \wedge @_i\Box\Diamond\Diamond i)$	
2. $@_{i_0}\Diamond\Diamond i$	(1)
3. $@_{i_0}@_i\Box\Diamond\Diamond i$	(1)
4. $@_{i_0}\Diamond i_1{}^*$	(2)
5. $@_{i_1}\Diamond i$	(2)
6. $@_i\Box\Diamond\Diamond i$	(3)
7. $@_{i_1}\Diamond i_2{}^*$	(5)
8. $@_{i_2}i$	(5)
9. $@_{i_1}\Diamond\Diamond i$	$(5, 6, [\Box_{sym}])$
10. $@_i i$	(8)
11. $@_{i_1}\Diamond i_3{}^*$	(9)
12. $@_{i_3}\Diamond i$	(9)
13. $@_{i_3}\Diamond i_4{}^*$	(12)
14. $@_{i_4}i$	(12)
15. $@_{i_3}\Diamond\Diamond i$	$(6, 12, [\Box_{sym}])$
16. $@_{i_3}\Diamond i_5{}^*$	(15)
17. $@_{i_5}\Diamond i$	(15)
18. $@_{i_5}\Diamond i_6{}^*$	(17)
19. $@_{i_6}i$	(17)
20. $@_{i_5}\Diamond\Diamond i$	$(6, 17, [\Box_{sym}])$
⋮	

($\mathcal{D}$) The rule $[\Diamond]$ can only be applied to a formula $@_i\Diamond\varphi$ on a branch Θ if i is a quasi-urfather.

We will give a precise definition of *quasi-urfather* later. Intuitively, the quasi-urfather is the representative of the aforementioned redundant nominals.

The next thing to do is the addition of rules that address irreflexivity. The solution, which is quite similar to that of [9], is straightforward: we add the formula of the form $@_i\neg\Diamond i$ to the branch for every nominal i.

$$\frac{}{@_i\neg\Diamond i}\ [Irr]$$

where i has already occurred in the branch.

Example 9.2 Consider an example of Fig. 9.3. The branch with the root formula $@_i\Diamond(i \wedge p)$ is closed owing to $[Irr]$. Indeed, the formula $@_i\Diamond(i \wedge p)$ never holds in any orthomodel, i.e., $\neg @_i\Diamond(i \wedge p)$ is a theorem of $\mathbf{IB}(@)$.

Fig. 9.3 A closed tableau owing to [*Irr*]

1. $@_i\Diamond(i \wedge p)$	
2. $@_i\neg\Diamond i$	$[Irr]$
3. $@_i\Diamond j^*$	(1)
4. $@_j(i \wedge p)$	(1)
5. $@_j\neg\Diamond j$	$[Irr]$
6. $@_j\neg i$	(2, 3)
7. $@_j i$	(4)
8. $@_j p$	(4)
✗	

Finally, we replace $[\neg]$ with a new rule [*Ref*]. In this rule, $i \in$ **Nom** must have already occurred in the branch. This change makes proving some lemmas, particularly Lemmas 9.8 and 9.12, easier.

$$\frac{}{@_i i}\ [Ref]$$

Now we are ready to construct the tableau calculus for **IB**(@).

Definition 9.10 $\mathbf{TAB}_{\mathbf{IB}(@)}$ (or simply $\mathbf{TAB}_{\mathbf{IB}}$) is the tableau calculus made by adding $[\Box_{sym}]$, $[Irr]$, and $(\mathcal{D})$ to **TAB** and replacing $[\neg]$ with [*Ref*]. Moreover, if there is a closed tableau in $\mathbf{TAB}_{\mathbf{IB}}$ whose root formula is $@_i\neg\varphi$, where $i \in$ **Nom** does not occur in φ, we say that φ is *provable* in $\mathbf{TAB}_{\mathbf{IB}}$.

9.3.2 Termination

A tableau calculus has the *termination property* if, for every tableau, all branches in it are finite. The basic strategy of loop-checking is very close to that of [8, §5]. What we have to do is checking this method still works even after some tuning has been applied with the aim of making it correspond to **IB**(@).

The first thing we do is to insert a relation between nominals occurring in a branch of a tableau. It is the case that some nominals that exist in a branch play similar roles. We need to make them be grouped together as *twins*.

First, we show the quasi-subformula property. $\mathbf{TAB}_{\mathbf{IB}}$ does not have the subformula property that tableau calculus of propositional logic has. However, by extending the concept of subformula, we can prove a similar property.

Definition 9.11 (*quasi-subformula*) Given two formulas of the form $@_i\varphi$ and $@_j\psi$, $@_i\varphi$ is a *quasi-subformula* of $@_j\psi$ if one of the following conditions holds:

- φ is a subformula of ψ.
- φ has the form $\neg\chi$, and χ is a subformula of ψ.

Lemma 9.1 *Let Θ be a branch of a tableau. For every formula of the form $@_i\varphi$ in Θ, one of the following conditions holds.*

- *It is a quasi-subformula of the root formula of Θ.*
- *It is an accessibility formula.*
- *It is a quasi-subformula of $@_i\neg\Diamond i$ for some i occurring in Θ.*

Proof By induction on the number of applied rules. We provide the proof only for the cases $[\Box_{sym}]$, $[\mathit{Irr}]$, and $[\mathit{Ref}]$, since they do not exist in the original tableau system proposed by [8]. The other cases are left to the reader.

$[\Box_{sym}]$ $@_j\varphi$ is a quasi-subformula of $@_i\Box\varphi$. By the induction hypothesis, $@_i\Box\varphi$ is either a quasi-subformula of the root formula or the formula $@_i\neg\Diamond i$ (if $\varphi = \neg i$). Therefore, $@_j\neg\varphi$ is a quasi-subformula of either the root formula or $@_i\neg\Diamond i$.

$[\mathit{Irr}]$ It is straightforward because the formula we add by $[\mathit{Irr}]$ is the very formula of the form $@_i\neg\Diamond i$ for some i occurring in Θ.

$[\mathit{Ref}]$ The lemma straightforwardly follows, because $@_i i$ is a quasi-subformula of $@_i\neg\Diamond i$.

Next, we define the set $T^{\Theta}(i)$, a set of true formulas in the world pointed to by the nominal i.

Definition 9.12 Let Θ be a branch of a tableau. For every nominal i occurring in Θ, the set $T^{\Theta}(i)$ is defined as follows:

$$T^{\Theta}(i) = \{\varphi \mid @_i\varphi \in \Theta \text{ and } @_i\varphi \text{ is a quasi-subformula of the root formula}\}.$$

Lemma 9.2 *For all i occurring in Θ, $T^{\Theta}(i)$ in Definition 9.12 is finite.*

Proof Straightforward from the construction of $T^{\Theta}(i)$.

Using the set $T^{\Theta}(i)$ defined above, we can give a definition of twins.

Definition 9.13 Let Θ be a branch of a tableau. We call nominals i, j *twins* in Θ if $T^{\Theta}(i) = T^{\Theta}(j)$.

We need one more step to understand a quasi-urfather to define a generating relation between nominals.

Definition 9.14 Let Θ be a tableau branch, and let i and j be nominals occurring in Θ. If j is introduced by applying $[\Diamond]$ to a formula $@_i\Diamond\varphi$, we say that j is a *generated nominal* from i (denoted as $i \prec_{\Theta} j$).

Note that the only rule generating a new nominal is [$\Diamond$], even if we add [$\Box_{sym}$], [*Irr*], or [*Ref*]. Since all the accessibility formulas emerge in a branch only when [$\Diamond$] is applied to some formula in the branch, $i \prec_\Theta j$ is equivalent to that there is an accessibility formula $@_i \Diamond j$ in Θ.

Based on these preparations, we can give a precise definition of quasi-urfather.

Definition 9.15 (*Quasi-urfather*) We call a nominal i *quasi-urfather* on Θ if there are no twins j, k such that $j \neq k$ and $j, k \prec^*_\Theta i$, where $\prec^*_\Theta$ denotes a reflexive and transitive closure of $\prec_\Theta$.

Before proving the termination property, we see an example that ($\mathcal{D}$) plays an important role in stopping extending a branch.

Example 9.3 With ($\mathcal{D}$), the tableau in Fig. 9.2 stops before adding the 16th formula $@_{i_3} \Diamond i_5$. In this branch, up to the 15th formula $@_{i_3} \Diamond\Diamond i$, nominals i_1 and i_3 are twins, since $T_\Theta(i_1) = T_\Theta(i_3) = \{\Diamond i, \Diamond\Diamond i\}$. Then i_3 is not a quasi-urfather, so we can no longer apply [$\Diamond$] to the 15th formula $@_{i_3} \Diamond\Diamond i$.

Now we move on to prove the termination property.

Lemma 9.3 *Given a branch Θ of a tableau, we define a structure $G = (N^\Theta, \prec_\Theta)$ where*

- *N^Θ is the set of nominals occurring in Θ.*
- *$\prec_\Theta$ is the binary relation defined in Definition 9.14.*

Then G is a finite set of well-founded and finitely branching trees.

Proof See the proof of [8, Lemma 6.4].

Lemma 9.4 *Let Θ be a branch of a tableau. Then Θ is infinite if and only if we have the following infinite sequence:*

$$i_0 \prec_\Theta i_1 \prec_\Theta i_2 \prec_\Theta \cdots .$$

Proof See the proof of [8, Lemma 6.5].

Theorem 9.3 *The tableau calculus* **TAB$_{\mathbf{IB}}$** *has the termination property.*

Proof This can be proved in almost the same way as [8, Theorem 5.8].

9.3.3 Completeness

Next, we prove the completeness. The basic strategy is the same as in [8, Sect. 5.2]. That is, define an *identity urfather* and use it to create a model from an open saturated branch of tableau. When constructing the model, make sure that the relation R^Θ is symmetric.

Definition 9.16 A branch Θ of a tableau is *saturated* if every new formula generated by applying some rules already exists in Θ. We call a tableau *saturated* if all branches of it are saturated.

The following are some properties that a saturated branch Θ has.

- If $@_i(\varphi \wedge \psi) \in \Theta$, then $@_i\varphi, @_i\psi \in \Theta$.
- If $@_i\Diamond\varphi \in \Theta$ which is not an accessibility formula and i is a quasi-urfather, then there exists some $j \in \mathbf{Nom}$ such that $@_i\Diamond j, @_j\varphi \in \Theta$.
- If $@_i\Box\varphi$ and $@_j\Diamond i \in \Theta$, then $@_j\varphi \in \Theta$.
- For all i occurring in Θ, $@_i\neg\Diamond i \in \Theta$.

Definition 9.17 Let Θ be a tableau branch and i a nominal occurring in Θ. The *identity urfather* of i on Θ (denoted as $\upsilon_\Theta(i)$) is the earliest introduced nominal j satisfying the following conditions:

1. j is a twin of i.
2. j is a quasi-urfather.

Note that there may be no identity urfather of some i. If $\upsilon_\Theta(i)$ exists for given i, we write $i \in \mathrm{dom}(\upsilon_\Theta)$, which is a widely used notation in mathematics.

Now we show some properties of $\mathrm{dom}(\upsilon_\Theta)$.

Lemma 9.5 *Let Θ be a branch of a tableau. If i occurs in the root formula of Θ, then $i \in \mathrm{dom}(\upsilon_\Theta)$.*

Proof Suppose i occurs in the root formula of Θ. Then there exists no nominal j such that $j \prec_\Theta i$, since when we use [$\Diamond$], we must introduce a new nominal. Then i is a quasi-urfather. Therefore, $\upsilon_\Theta(i)$ is definable.

Lemma 9.6 *Let Θ be a saturated branch and i be a quasi-urfather on Θ. If $i \prec_\Theta j$, then $j \in \mathrm{dom}(\upsilon_\Theta)$.*

Proof See [8, Lemma 5.12].

Next, we show some properties of υ_Θ.

Lemma 9.7 *Let Θ be a branch. Suppose $@_i\varphi \in \Theta$ is a quasi-subformula of the root formula of Θ and $i \in \mathrm{dom}(\upsilon_\Theta)$. Then $@_{\upsilon_\Theta(i)}\varphi \in \Theta$.*

Proof If $v_\Theta(i) = i$, there is nothing to prove. Even if else, $v_\Theta(i)$ and i are twins. Then the lemma holds by the definition of twins.

Lemma 9.8 *Let Θ be a saturated branch. If $@_i j \in \Theta$ and $i, j \in \mathrm{dom}(v_\Theta)$, then $v_\Theta(i) = v_\Theta(j)$.*

Proof Suppose $@_i j \in \Theta$. Since Θ is saturated, we obtain $@_i i \in \Theta$ from [*Ref*]. Then by [*Id*], we have $@_j i \in \Theta$. They mean that we can add all the formulas of the form $@_j \varphi$ to Θ where $@_i \varphi \in \Theta$, and vice versa. Thus, we have $T^\Theta(i) = T^\Theta(j)$. Therefore, by definition, $v_\Theta(i) = v_\Theta(j)$.

Lemma 9.9 *Let Θ be a branch. A nominal i is the identity urfather if and only if $v_\Theta(i) = i$.*

Proof Straightforward from the definition of identity urfathers.

Definition 9.18 Given an open saturated branch Θ with a root formula $@_{i_0}\varphi_0$ of a tableau in **TAB$_{\mathbf{IB}}$**, a model $\mathcal{M}^\Theta = (W^\Theta, R^\Theta, V^\Theta)$ is defined as follows:

$$\begin{aligned}
W^\Theta &= \{i \mid i \text{ is an identity urfather on } \Theta\};\\
R^\Theta &= \{(v_\Theta(i), v_\Theta(j)) \mid @_i\Diamond j \in \Theta \text{ and } i, j \in \mathrm{dom}(v_\Theta)\}\\
&\quad \cup \{(v_\Theta(j), v_\Theta(i)) \mid @_i\Diamond j \in \Theta \text{ and } i, j \in \mathrm{dom}(v_\Theta)\};\\
V^\Theta(p) &= \{v_\Theta(i) \mid @_i p \in \Theta\}, \text{ where } p \in \mathbf{Prop};\\
V^\Theta(i) &= \begin{cases} \{v_\Theta(i)\} & \text{if } i \in \mathrm{dom}(v_\Theta),\\ \{i_0\} & \text{otherwise,} \end{cases} \quad \text{where } i \in \mathbf{Nom}.
\end{aligned}$$

Then we obtain the model existence lemma, which ensures that if we have a tableau with a root formula $@_{i_0}\varphi_0$ one of whose branches are open and saturated, there exists a model which falsifies φ_0.

Lemma 9.10 (Model existence lemma) *Let Θ be an open saturated branch of a tableau in* **TAB$_{\mathbf{IB}}$** *and $@_i\varphi$ be a quasi-subformula of the root formula $@_{i_0}\varphi_0$ of Θ, where i denotes an identity urfather. Then we have the following proposition:*

$$\textit{if } @_i\varphi \in \Theta, \textit{ then } \mathcal{M}^\Theta, v_\Theta(i) \models \varphi.$$

In particular, $\mathcal{M}^\Theta, v_\Theta(i_0) \models \varphi_0$.

Proof By induction on the complexity of φ.

$[\varphi = p, \neg p]$ Straightforward.
$[\varphi = j]$ Suppose $@_i j \in \Theta$, where $j \in \mathbf{Nom}$. Since i is an identity urfather and $j \in \mathrm{dom}(v_\Theta)$ (Lemma 9.5), we have $v_\Theta(i) = v_\Theta(j)$ by Lemma 9.8. Therefore, $\mathcal{M}^\Theta, v_\Theta(i) \models j$.

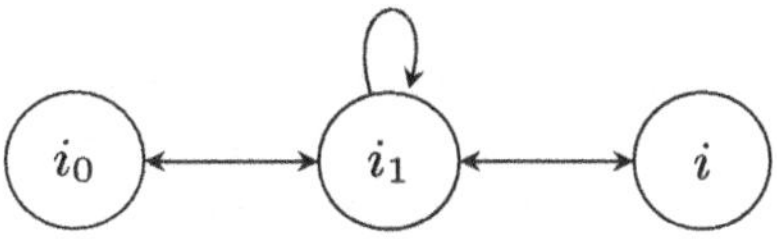

Fig. 9.4 The model $\mathcal{M}_\Theta$ constructed from a branch Θ in Fig. 9.2

[$\varphi = \neg j$] Suppose $@_i \neg j \in \Theta$, where $j \in$ **Nom**. First, by Lemma 9.7, we have $@_{v_\Theta(i)} \neg j \in \Theta$. Second, since Θ is saturated, by [*Ref*], we have $@_j j \in \Theta$. Since j is also a quasi-subformula of the root formula, by Lemma 9.7, we have $@_{v_\Theta(j)} j \in \Theta$. Hence, $@_{v_\Theta(i)} \neg j$, $@_{v_\Theta(j)} j \in \Theta$. However, since Θ is open, $v_\Theta(i) \neq v_\Theta(j)$. Then by the definition of V^Θ, it follows that $\mathcal{M}^\Theta, v_\Theta(i) \models \neg j$.

[$\varphi = (\psi_1 \wedge \psi_2), \neg(\psi_1 \wedge \psi_2)$] In these cases, the proofs are simple. Then we leave them to the reader.

[$\varphi = \Diamond\psi$] Suppose $@_i \Diamond\psi \in \Theta$. From the facts that Θ is saturated, i is an identity urfather, and $@_i \Diamond\psi$ is not an accessibility formula, by [$\Diamond$] there exists a nominal $j \in$ **Nom** such that $@_i \Diamond j$, $@_j \psi \in \Theta$. By Lemma 9.6, there exists a $v_\Theta(j)$. Hence, we have $v_\Theta(i) R^\Theta v_\Theta(j)$ and $\mathcal{M}^\Theta, v_\Theta(j) \models \psi$ from the definition of R^Θ and the induction hypothesis respectively. Therefore, $\mathcal{M}^\Theta, v_\Theta(i) \models \Diamond\psi$.

[$\varphi = \neg\Diamond\psi$] Suppose $@_i \neg\Diamond\psi \in \Theta$, and take a nominal $j \in$ **Nom** such that $v_\Theta(i) R^\Theta v_\Theta(j)$. (If nothing, $\mathcal{M}^\Theta, v_\Theta(i) \models \neg\Diamond\psi$ is straightforward.) Then by the definition of R^Θ, either $@_i \Diamond j$ or $@_j \Diamond i$ is in Θ. If the former exists, by [$\neg\Diamond$] and the saturation of Θ we have $@_j \neg\psi$. If else, we can obtain it using [$\Box_{sym}$]. Then in both cases, $\mathcal{M}^\Theta, v_\Theta(j) \models \neg\psi$ holds by the induction hypothesis. This holds for all j, so we have $\mathcal{M}^\Theta, v_\Theta(i) \models \neg\Diamond\psi$.

[$\varphi = @_j \psi$] Suppose $@_i @_j \psi \in \Theta$. Since Θ is saturated, by [@] we have $@_j \psi \in \Theta$. By the induction hypothesis, $\mathcal{M}^\Theta, v_\Theta(j) \models \psi$ holds. Therefore, $\mathcal{M}^\Theta, v_\Theta(i) \models @_j \psi$.

[$\varphi = \neg @_j \psi$] We can prove similarly to the case of $\varphi = @_j \psi$.

To show the completeness theorem for **TAB$_{\mathbf{IB}}$**, we have to show the symmetry and irreflexivity of $\mathcal{M}^\Theta$. The symmetry is straightforward from the construction; however, it is not the case that $\mathcal{M}^\Theta$ is always irreflexive. The following is an example of $\mathcal{M}^\Theta$ that is not irreflexive.

Example 9.4 Consider the branch Θ in Fig. 9.2. Note that in **TAB$_{\mathbf{IB}}$** this branch is stopped up to the 15th formula $@_{i_3} \Diamond\Diamond i$. From Θ, we create a model as Definition 9.18. Since i_1 and i_3 are twins and i_2, i_4, and i are all twins, we have $v_\Theta(i_3) = i_1$ and $v_\Theta(i_2) = v_\Theta(i_4) = i$. Then $\mathcal{M}^\Theta$ must be as follows (Fig. 9.4 illustrates this model):

$$W^\Theta = \{i_0, i_1, i\},$$
$$R^\Theta = \{(i_0, i_1), (i_1, i_0), (i_1, i_1), (i_1, i), (i, i_1)\},$$
$$V^\Theta(i) = \{i\}.$$

This model has the reflexive world i_1.

To solve this problem, we *bulldoze* all the reflexive points and turn a frame into an irreflexive one. For modal logic, the bulldozing method was originally proposed in Segerberg [19]. Before giving a formal definition, we cite the explanation of bulldozing by Hughes and Cresswell [15].

> ... For if we take any reflexive world in any model, i.e., any world which can see itself, and replace it by a pair of worlds each able to see the other but neither able to see itself, *and* we give each variable the same value in each world in the new pair as it had in the original world, ... [15, p. 176]

For hybrid logic, Blackburn [4] showed that bulldozing works well in proving the completeness of Hilbert-style axiomatization of irreflexive hybrid logic. Our following lemmas claim that the method also works in the tableau calculus.

Definition 9.19 Given a model $\mathcal{M} = (W, R, V)$, the *bulldozed model* $\mathcal{M}_B = (W_B, R_B, V_B)$ of $\mathcal{M}$ is defined as follows:

1. Put $W_r = \{w \in W \mid wRw\}$.
2. Define $W_B = W^- \cup \{(w, n) \mid w \in W_r, n = 0, 1\}$, where $W^- = W \setminus W_r$.
3. Define $\alpha : W_B \to W$ by $\alpha(w) = w$ if $w \in W^-$, and $\alpha((w, n)) = w$ otherwise.
4. Define R_B by wR_Bv if one of the following conditions holds:
   - $w \in W^-$ or $v \in W^-$, and $\alpha(w)R\alpha(v)$,
   - $w = (w', m)$, $v = (v', n)$, $w' \neq v'$, and $w'Rv'$,
   - $w \neq v$ and $\alpha(w) = \alpha(v)$.
5. Define V_B : **Prop** $\cup$ **Nom** $\to \mathcal{P}(W_B)$ as follows:
   - If $p \in$ **Prop**, then $w \in V_B(p)$ if and only if $\alpha(w) \in V(p)$.
   - If $i \in$ **Nom**, then $V_B(i) = \{(w, 0)\}$, where $w = i^V$ for $w \in W_r$, and $V_B(i) = V(i)$ otherwise.

Lemma 9.11 *Let* $\mathcal{M}_B = (W_B, R_B, V_B)$ *be the bulldozed model of* $\mathcal{M} = (W, R, V)$. *If* R *is symmetric, then* $\mathcal{F}_B = (W_B, R_B)$ *is an orthoframe.*

Proof We show that R_B is irreflexive and symmetric.

Irreflexivity By the definition of W_B, it suffices to show that every $w \in \{(w', n) \mid w' \in W_r, n = 0, 1\}$ does not satisfy wR_Bw. We prove it by *reductio ad absurdum*. Suppose that wR_Bw for some $w \in \{(w', n) \mid w' \in W_r, n = 0, 1\}$. By the definition of R_B, there is only one possible case (the first and third case in Condition 4 of Definition 9.19 can never happen): $w = (w', m) = (w'', n)$, $w' \neq w''$, and $w'Rw''$. It follows from $w' \neq w''$ that $(w', m) \neq (w'', n)$, which leads to a contradiction with $w = (w', m) = (w'', n)$.

Symmetry Suppose $(w, v) \in R_B$. By the definition of R_B, there are three possible cases. In each case, we show $(v, w) \in R_B$.

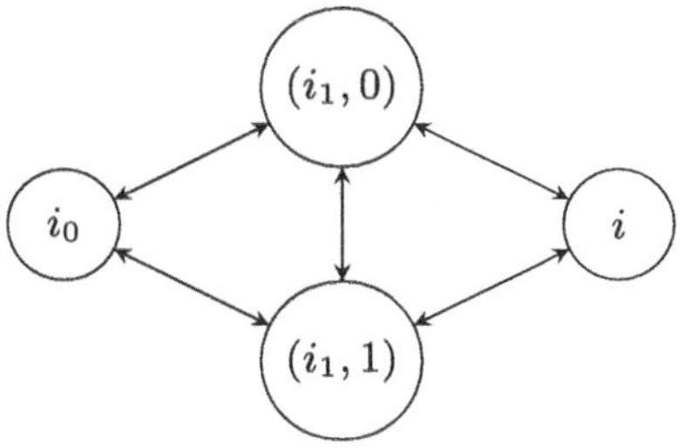

Fig. 9.5 The bulldozed model $\mathcal{M}_{\Theta}^{B}$ constructed from a branch Θ in Fig. 9.2

Case 1 $w \in W^-$ or $v \in W^-$, and $\alpha(w) R \alpha(v)$. Because R is symmetric, $v \in W^-$ or $w \in W^-$, and $\alpha(v) R \alpha(w)$. Thus, $(v, w) \in R_B$.

Case 2 $w = (w', m)$, $v = (v', n)$, $w' \neq v'$, and $w' R v'$. Because R is symmetric, $v = (v', n)$, $w = (w', m)$, $v' \neq w'$, and $v' R w'$. Thus, $(v, w) \in R_B$.

Case 3 $w \neq v$ and $\alpha(w) = \alpha(v)$. Then $v \neq w$ and $\alpha(v) = \alpha(w)$. Thus, $(v, w) \in R_B$.

Example 9.5 Let $\mathcal{M}$ be the model in Fig. 9.4. The bulldozed model $\mathcal{M}_B$ is illustrated in Fig. 9.5. Note that the proposition i_1 (and also i_3) only holds at $(i_1, 0)$ and fails in $(i_1, 1)$.

Unlike in the case of modal logic, this construction cannot completely preserve satisfaction. In the model in Fig. 9.4, we have $\mathcal{M}, i_1 \models \Diamond i_1$, but the bulldozed model in Fig. 9.5, $\mathcal{M}_B, (i_1, 0) \models \Diamond i_1$ does not hold.

However, this method works well if we consider only the quasi-subformulas of the root formula. What is important is that the reflexive worlds in a model from a tableau branch have no information about nominals in the root formula.

Definition 9.20 Let Θ be an open saturated branch and $\mathcal{M}^{\Theta} = (W^{\Theta}, R^{\Theta}, V^{\Theta})$ be a model constructed in the way of Definition 9.18. A world $i \in W^{\Theta}$ is *named by the root formula of* Θ if there exists a nominal j such that $j \in T^{\Theta}(i)$.

Lemma 9.12 *Let* $W_r^{\Theta} = \{w \in W^{\Theta} \mid w R^{\Theta} w\}$. *If* $i \in W_r^{\Theta}$, *then* i *is not named by the root formula of* Θ.

Proof By *reductio ad absurdum*.

Suppose i is named by the root formula of Θ. From $i \in W_r^{\Theta}$, we have $i R^{\Theta} i$, which means that there are nominals j and k such that $v_{\Theta}(j) = v_{\Theta}(k) = i$ and $@_j \Diamond k \in \Theta$. (They do not have to be different; for example, it is possible that $j = k$.) By the assumption, there exists another nominal l such that $l \in T^{\Theta}(i)$. By the definition of identity urfather, all of i, j, k are twins. Thus, we have another nominal l such that $l \in T^{\Theta}(j)$ and $l \in T^{\Theta}(k)$, which means that both $@_j l$ and $@_k l$ are in Θ. Therefore, we can show that Θ is closed as Fig. 9.6, which is a contradiction.

Fig. 9.6 A closed branch Θ if $i \in W_r^{\Theta}$ is named in Θ

1. $@_j \Diamond k$	
2. $@_j l$	
3. $@_k l$	
4. $@_j \neg \Diamond j$	[*Irr*]
5. $@_j j$	[*Ref*]
6. $@_k \neg j$	$(1, 4, [\neg\Diamond])$
7. $@_l j$	$(2, 5, [Id])$
8. $@_l \neg j$	$(3, 6, [Id])$
✗	

Lemma 9.13 *Let Θ be an open saturated branch, and $@_{i_0}\varphi_0$ be a root formula. For any nominal $i \in W_r^{\Theta}$ and formula φ such that $@_i\varphi$ is a quasi-subformula of $@_{i_0}\varphi_0$, we have*

$$\mathcal{M}_B^{\Theta}, (i, 0) \models \varphi \iff \mathcal{M}_B^{\Theta}, (i, 1) \models \varphi.$$

Proof By induction on the complexity of φ. Note that we do not have to consider the case $\varphi = j$ because, by Lemma 9.12, φ cannot be any nominal i occurring in φ_0.

We give the proof only for the cases $\varphi = p, \Diamond\psi$, and $@_j\psi$ and only for left to right. The other cases are left to the reader.

$[\varphi = p]$ If $\mathcal{M}_B^{\Theta}, (i, 0) \models p$, we have $(i, 0) \in V_B^{\Theta}(p)$. By the construction of V_B^{Θ}, it follows that $(i, 1) \in V_B^{\Theta}(p)$. Therefore, $\mathcal{M}_B^{\Theta}, (i, 1) \models p$.

$[\varphi = \Diamond\psi]$ Suppose $\mathcal{M}_B^{\Theta}, (i, 0) \models \Diamond\psi$. Then there exists a world $w \in W_B^{\Theta}$ such that $(i, 0) R_B^{\Theta} w$ and $\mathcal{M}_B^{\Theta}, w \models \psi$. If $w = (i, 1)$, by the induction hypothesis, we have $\mathcal{M}_B^{\Theta}, (i, 0) \models \psi$. Thus, we acquire that $\mathcal{M}_B^{\Theta}, (i, 1) \models \Diamond\psi$. If $w \neq (i, 1)$, we have $(i, 1) R_B^{\Theta} w$ by the construction of R_B^{Θ}. Therefore, we have $\mathcal{M}_B^{\Theta}, (i, 1) \models \Diamond\psi$.

$[\varphi = @_j\psi]$ Suppose $\mathcal{M}_B^{\Theta}, (i, 0) \models @_j\psi$. Then we have $\mathcal{M}_B^{\Theta}, j^{V_B^{\Theta}} \models \psi$. Therefore, we acquire that $\mathcal{M}_B^{\Theta}, (i, 1) \models @_j\psi$.

Lemma 9.14 (Bulldozing lemma) *Let Θ be an open saturated branch. Also, let i be an identity urfather of Θ and φ be a formula such that $@_i\varphi$ be a quasi-subformula of the root formula. Then we have*

$$\mathcal{M}^{\Theta}, i \models \varphi \iff \mathcal{M}_B^{\Theta}, i_B \models \varphi,$$

where

$$i_B = \begin{cases} (i, 0) & \text{if } i \in W_r, \\ i & \text{otherwise.} \end{cases}$$

Proof By induction on the complexity of φ.

$[\varphi = p, \neg p]$ Straightforward from the definition of V_B^Θ.

$[\varphi = j]$ Note that i is not in W_r^Θ owing to Lemma 9.12. Then we can show the equivalence in the following way.

$$\begin{aligned}\mathcal{M}^\Theta, i \models j &\iff V^\Theta(j) = \{i\}\\ &\iff V_B^\Theta(j) = \{i_B\}\\ &\iff \mathcal{M}_B^\Theta, i_B \models j.\end{aligned}$$

$[\varphi = \neg j]$ Similar to the case $\varphi = \neg p$.

$[\varphi = \psi_1 \wedge \psi_2, \neg(\psi_1 \wedge \psi_2)]$ Straightforward.

$[\varphi = \Diamond\psi]$ First, we show the left-to-right direction.

Suppose $\mathcal{M}^\Theta, i \models \Diamond\psi$. Then there exists a world $j \in W^\Theta$ such that $i R^\Theta j$ and $\mathcal{M}^\Theta, j \models \psi$. If $j \neq i$, we have $i_B R^\Theta j_B$ by the definition of R_B^Θ, and $\mathcal{M}_B^\Theta, j_B \models \psi$ by the induction hypothesis. Therefore, $\mathcal{M}_B^\Theta, v_\Theta(i)_B \models \Diamond\psi$.

The case to watch out for is $j = i$. In this case, we have $(i, 0) R^\Theta (i, 1)$, and $\mathcal{M}_B^\Theta, (i, 0) \models \psi$ by the induction hypothesis. However, by Lemma 9.13, we have $\mathcal{M}_B^\Theta, (i, 1) \models \psi$. Therefore, $\mathcal{M}_B^\Theta, (i, 0) \models \Diamond\psi$.

The proof of the other direction is straightforward since for every $w, v \in \Theta$, $w R_B^\Theta v$ implies $\alpha(w) R^\Theta \alpha(v)$.

$[\varphi = \neg\Diamond\psi]$ Similar to the case $\varphi = \Diamond\psi$.

$[\varphi = @_j\psi]$ We only show from right to left, and the other is left to the reader.

Suppose $\mathcal{M}_B^\Theta, i_B \models @_j\psi$. Take a world $w \in W_B^\Theta$ such that $V_B^\Theta(j) = \{w\}$, and we have $\mathcal{M}_B^\Theta, w \models \psi$. Considering that $V^\Theta(j) = \{v_\Theta(j)\}$, it follows that $w = (v_\Theta(j))_B$. Thus, by induction hypothesis, we have $\mathcal{M}^\Theta, v_\Theta(j) \models \psi$. Therefore, $\mathcal{M}^\Theta, i \models @_j\psi$ holds.

$[\varphi = \neg @_j\psi]$ Similar to the case $\varphi = @_j\psi$.

Finally, we reach our goal.

Theorem 9.4 (Completeness) *The tableau calculus* $\mathbf{TAB_{IB}}$ *is complete for the class of all orthoframes.*

Proof We show the contraposition. Suppose φ is not provable in $\mathbf{TAB_{IB}}$. Then we can find a tableau in $\mathbf{TAB_{IB}}$, whose root formula is $@_i\neg\varphi$ where i does not occur in φ, and which has an open and saturated branch Θ. Then by Lemma 9.10, we have $\mathcal{M}^\Theta, v_\Theta(i) \models \neg\varphi$. By Lemma 9.14, we have $\mathcal{M}_B^\Theta, v_\Theta(i)_B \models \neg\varphi$. Moreover, by Lemma 9.11, $\mathcal{M}_B^\Theta$ is an orthomodel. Therefore, we can find an orthoframe which falsifies φ.

Corollary 9.1 $\mathbf{IB}(@)$ *is decidable.*

Proof From the termination property of $\mathbf{TAB_{IB}}$ and the fact that $\mathcal{M}_B^\Theta$ is finite.

Note that Takeuti and Sano [22] also showed the decidability of $\mathbf{IB}(@)$.

9.4 Conclusion and Future Work

In this paper, we constructed a tableau calculus that has the termination property and is complete with respect to all frames of undirected graphs. To prove completeness, the bulldozing method worked well in dealing with tableau calculus as well as Hilbert-style axiomatization.

The first task that comes to our mind as a continuation of this paper—but with difficulty—is to create a tableau calculus corresponding to **I4**(@). The bulldozing method for a transitive model was proposed by [5, Sect. 4.5]. Furthermore, a method to realize transitivity was also proposed in [9]. Combining these methods with the method described in this paper, it should be possible to construct a tableau calculus for the class of irreflexive and transitive frames.

Moreover, there is much remaining work to formulate tableau calculi corresponding to frames with various conditions. Many axioms characterizing some frame conditions were proposed in [9], such as anti-symmetry, trichotomy, and tree-like. If there is a relational structure that is widely studied, creating tableau calculus corresponding to it is worthwhile.

Another direction of future work is to add other operators. Besides the satisfaction operator @, hybrid logic can have more operators, such as the existential operator E and the downarrow operator $\downarrow$. Moreover, hybrid graph logic in [2] contains other modal operators $\Box^+$ and $\Diamond^+$, and [22] introduced new modal operators $\Box^*$ and $\Diamond^*$ to express the planarity.

Acknowledgements We would like to thank Prof. Ryo Kashima and Prof. Katsuhiko Sano for their invaluable advice in writing this paper. The research of the first author was supported by JST SPRING, Grant Number JPMJSP2106. The research of the second author was supported by Grant-in-Aid for JSPS Research Fellow Grant Number JP22KJ1483.

References

1. Areces, C., Ten Cate, B.: 14 hybrid logics. In: Handbook of Modal Logic, pp. 821–868. Elsevier (2007)
2. Benevides, M.R.F., Schechter, L.M.: Using modal logics to express and check global graph properties. Logic J. IGPL **17**(5), 559–587 (2009)
3. Benevides, M.R.F., Schechter, L.M.: A study on multi-dimensional products of graphs and hybrid logic. Theoret. Comput. Sci. **412**, 4946–4966 (2011)
4. Blackburn, P.: Nominal Tense Logic and Other Sorted Intensional Frameworks. PhD thesis, The University of Edinburgh (1990)
5. Blackburn, P., de Rijke, M., de Venema, Y.: Modal Logic. Cambridge University Press (2002)
6. Blackburn, Patrick, ten Cate, Balder: Pure extensions, proof rules, and hybrid axiomatics. Stud. Logica. **84**(2), 277–322 (2006)
7. Blackburn, Patrick, Tzakova, Miroslava: Hybrid languages and temporal logic. Log. J. IGPL **7**(1), 27–54 (1999)
8. Bolander, Thomas, Blackburn, Patrick: Termination for hybrid tableaus. J. Log. Comput. **17**(3), 517–554 (2007)

9. Bolander, Thomas, Blackburn, Patrick: Terminating tableau calculi for hybrid logics extending K. Electron. Notes Theoret. Comput. Sci. **231**, 21–39 (2009)
10. Braüner, T.: Hybrid Logic and its Proof-Theory, vol. 37. Springer Science & Business Media (2011)
11. Fitting, Melvin: Proof Methods for Modal and Intuitionistic Logics. Springer, Dordrecht (1983)
12. Gate, J.S.: Applications of Finite Model Theory: Optimisation Problems, Hybrid Modal Logics and Games. PhD thesis, Durham University (2013)
13. Goldblatt, R.: Logics of Time and Computation. Center for the Study of Language and Information (1987)
14. Goldblatt, R.I.: Semantic analysis of orthologic. J. Philos. Logic **3**, 19–35 (1974)
15. Hughes, G.E., Cresswell, M.J.: A New Introduction to Modal Logic. Routledge (1996)
16. Indrzejczak, Andrzej: Modal hybrid logic. Logic Log. Philos. **16**(2–3), 147–257 (2007)
17. Prior, Arthur N.: Past, Present and Future. Oxford University Press (1967)
18. Prior, A.N.: Papers on Time and Tense. Oxford University Press (1968)
19. Segerberg, K.K.: An Essay in Classical Modal Logic. PhD thesis, Stanford University (1971)
20. Seligman, J., Liu, F., Girard, P.: Logic in the community. In: Proceedings of the 4th Indian Conference on Logic and Its Applications, pp. 178–188. Springer, Berlin (2011)
21. Seligman, J., Liu, F., Girard, P.: Facebook and the epistemic logic of friendship. In: Proceedings of the 14th Conference on Theoretical Aspects of Rationality and Knowledge, pp. 229–238 (2013)
22. Takeuti, I., Sano, K.: Modal logic and planarity of graphs. In: Context, Conflict and Reasoning: Proceedings of the Fifth Asian Workshop on Philosophical Logic, pp. 115–126. Springer Singapore (2020)
23. Ten Cate, B.: Model Theory for Extended Modal Languages. PhD thesis, University of Amsterdam (2005)
24. Tzakova, M.: Tableau calculi for hybrid logics. In: International Conference on Automated Reasoning with Analytic Tableaux and Related Methods, pp. 278–292. Springer (1999)

Chapter 10
Strict Truth, Tolerant Truth, and Generalized Strict-Tolerant Logics

Yuecong Zhang and Xuefeng Wen

Abstract We give a new generalization of Strict-tolerant logic. The main idea is that strict truth is the standard of truth accepted by everyone, and tolerant truth the standard accepted by someone. To formalize the idea, we use generalized matrices in abstract algebraic logic, defining strict truth as the intersection of the closure system in the given generalized matrix, and tolerant truth the union of it. We connect our generalized strict-tolerant logics with p-consequences and propose conditions for such logics to be classical. Compared to existing generalizations of Strict-tolerant logic, our generalization is both intuitive and general enough.

Keywords Strict-tolerant logic · Abstract algebraic logic · Many-valued logic · Classical logic · Non-transitive logic

10.1 Introduction

There are mainly two strategies to obtain nonclassical logics. One is defining some usual logical operators, like negation or implication, in a nonclassical way, either from a semantic approach or a proof-theoretical one. The other is maintaining the semantics or rules of logical operators but redefine the notion of validity or logical consequence. The first strategy had been the main stream in nonclassical logic for a long time. The latter strategy is relatively new and has received more and more attention recently. A typical nonclassical logic in the second strategy is strict-tolerant logic (**ST** logic, henceforth) proposed in [5, 16] (see [2, 6] for further developments). A great advantage of **ST** logic is that it can produce the same inferences as classical logic but meanwhile provides a unified framework for solving the liar paradox and the

Corresponding author. This work was supported by National Social Science Foundation of China (Grant No. 22BZX129).

Y. Zhang · X. Wen (✉)
Sun Yat-Sen University, Guangzhou, China
e-mail: wxflogic@gmail.com
URL: https://sites.google.com/view/xuefeng-wen

K. Sano et al. (eds.), *Exploring Negation, Modality and Proof*,
Logic in Asia: Studia Logica Library,
https://doi.org/10.1007/978-981-96-8566-0_10

sorites paradox [4, 13]. Moreover, it deepens our understanding of the identification of a logic as well as the distinction between inferences and metainferences.

The basic idea of **ST** logic is evaluating premises and conclusion by different standards in a three-valued setting. More specifically, for an inference to be valid in **ST** it no longer requires preserving truth but only requires that when premises are (strictly) true the conclusion is tolerantly true, namely, not false, which can be either true, or neither true nor false. Though the logic can produce the same inferences as classical logic (when logical operators are interpreted as in Kleene's three-valued logic), it can have different metainferences. From an abstract point of view, a logic whose validity is defined like **ST** can be non-transitive (see [3, 14]), and hence non-Tarskian. Like non-monotonic logics, the idea of non-transitive logics can be fruitful, as good inferences are arguably non-transitive [11].

As has been noted, the idea behind **ST** logic can be general, without being confined in a three-valued setting. Indeed, the concept of tolerant truth was mentioned as early as 2008, when Zardini [17] pointed out that the tolerant truth set should be a non-empty upper set. Based on this idea, Cobreros et al. [5] attempted to generalize strict truth and tolerant truth using a duality condition: $\neg\varphi$ is tolerantly true iff φ is not strictly true, and $\neg\varphi$ is strictly true iff φ is not tolerantly true. However, the duality condition may be meaningful only in three-valued settings. If we apply it to 4-valued De Morgan algebra $\mathbf{M}_4 = (\{1, 0, \top, \bot\}, \wedge, \vee, \neg)$, a situation arises where $\{\top, 1\}$ and $\{\bot, 1\}$ satisfy the duality condition, but we cannot determine which is the strict truth set and which is the tolerant one by the inclusion relationship between the sets. In fact, whichever of the two sets is defined as strict (or tolerant), the resulting logic will not satisfy inclusiveness ($\Gamma \vdash \varphi$ if $\varphi \in \Gamma$), which is undesirable.

The notion of p-consequence proposed by Frankowski [9] can be regarded as another generalization of **ST** logic. Frankowski aims to formalize the notion of plausible reasoning studied by Ajdukiewicz [1]. In plausible reasoning, the degree of strength of the conclusion can be smaller than that of the premises. Though neither Ajdukiewicz nor Frankowski anticipated **ST** logic, we can see that the ideas behind them are very similar. A p-consequence is just a consequence operation that is inclusive and monotonic, but not necessarily transitive, to formalize the non-transitivity of plausible reasoning. To obtain a concrete p-consequence, Frankowski introduced p-matrices, which are generalization of matrices in algebraic logic. Instead of stipulating one set for designated truth in matrices, two sets are stipulated in p-matrices, one for the truth of premises and the other for conclusion. The former designated truth set is a subset of the second. In terms of the notions of **ST** logic, the former truth set can be regarded as a general notion of strict truth, whereas the second a generalization of tolerant truth. Though the logical consequence of **ST** logic is a special p-consequence, p-consequences induced by p-matrices seem to generalize too much for **ST** logic. According to p-matrices, when $\{1\}$ is the strict truth set, any supersets of $\{1\}$ can be a tolerant truth set, including $\{1, 0\}$ in a two-valued setting, which is rather unintuitive.

A most recent generalization of **ST** logic is given by Fitting [7], where he generalized **ST** logic using bilattices. In addition to the truth order, a bilattice has a knowledge order. Fitting defines three kinds of truth sets by this knowledge order and a conflation operation, one of which (consistent truth set) can be ignored and the other two (anticonsistent truth set and exact truth set) are used to define strict truth and tolerant truth. Fitting's generalization, due to its concrete algebraic structure, avoids the problem of p-consequences. But there are still two shortcomings of it. First, since Fitting needs Morgan algebras to obtain generalized **ST** logic, some typical logics (e.g., the $\{\wedge, \vee\}$-fragment of classical logic) are excluded. Second, Fitting defines strict truth and tolerant truth in a rather abstract way. The conflation operation lacks intuitive meaning, which makes the definitions of strict truth and tolerant truth difficult to access.

The purpose of this paper is to give a new generalization of **ST** logic, with the intuitive meaning of strict truth and tolerant truth easily understood, and meanwhile general enough to incorporate as many logics as possible. The main idea is that, a proper subset of the truth set represents a basic standard of truth. If $F_1, \cdots, F_n$ are all basic standards of truth, then the intersection $F_1 \cap \cdots \cap F_n$ is the strict truth, and the union $F_1 \cup \cdots \cup F_n$ is the tolerant truth. The moral behind is that strict truth is the truth accepted by everyone, whereas tolerant truth is the truth accepted by someone. For example, when $A = \{0, 1/2, 1\}$ is the set of all truth values, and $F_1 = \{1\}$, $F_2 = \{1, 1/2\}$ are all standards of truth, then $F_1 \cap F_2 = \{1\}$ is the strict truth, and $F_1 \cup F_2 = \{1, 1/2\}$ is the corresponding tolerant truth.

To formalize the idea, we use generalized matrices in abstract algebraic logic [8]. Compared to matrices, which are used to define validity by preserving truth, generalized matrices are meant to define validity by preserving degrees of truth, which makes it a proper tool to generalize **ST** logic.

The rest of the paper is organized as follows. Section 10.2 defines our generalized **ST** logics, connecting them with p-consequences induced by p-matrices. Section 10.3 gives the conditions for generalized **ST** logics to be classical. Section 10.4 concludes the paper and indicates some future work.

10.2 Generalized ST Logics

Let $\mathcal{L}$ be the propositional language to be discussed, Fm its set of well-formed formulas, and $\mathbf{Fm}$ the formula algebra of $\mathcal{L}$. Whenever an abstract algebra $\mathbf{A}$ is given, we assume A is its underlying domain.

Definition 10.1 A generalized matrix for $\mathcal{L}$ is a pair $\mathcal{M} = (\mathbf{A}, \mathcal{C})$, where $\mathbf{A}$ is an $\mathcal{L}$-algebra and $\mathcal{C}$ is a closure system on A, meaning that $\mathcal{C}$ is a collection of subsets of A satisfying that $A \in \mathcal{C}$ and if $\mathcal{B} \subseteq \mathcal{C}$ then $\bigcap \mathcal{B} \in \mathcal{C}$. A generalized matrix $\mathcal{M} = (\mathbf{A}, \mathcal{C})$ is proper, if $\mathcal{C} \neq \{A\}$.

Elements in a closure system $\mathcal{C}$ represent different degrees or standards of truth. According to our idea, $\bigcap \mathcal{C}$ is the strict truth and $\bigcup \{F \in \mathcal{C} \mid F \neq A\}$[1] is the cor-

[1] We certainly do not want acknowledging all propositions as a plausible truth standard. So A is excluded in the union.

responding tolerant truth. Now generalized **ST** logics can be defined as follows.

Definition 10.2 Let $\mathcal{M} = (\mathbf{A}, \mathcal{C})$ be a proper generalized matrix for $\mathcal{L}$. The generalized **ST** logic $\vdash^{\mathbf{ST}}_{\mathcal{M}}$ induced by $\mathcal{M}$ is given below: for all $\Gamma \cup \{\varphi\} \subseteq Fm$,

$$\Gamma \vdash^{\mathbf{ST}}_{\mathcal{M}} \varphi \iff \text{for all } h \in \mathrm{Hom}(\mathbf{Fm}, \mathbf{A}), \text{ if } h[\Gamma] \subseteq \bigcap \mathcal{C}, \text{ then there exists } F \in \mathcal{C} \text{ such that } F \neq A \text{ and } h(\varphi) \in F, \text{ where } \mathrm{Hom}(\mathbf{Fm}, \mathbf{A}) \text{ is the set of all homomorphisms from } \mathbf{Fm} \text{ to } \mathbf{A}.$$

Closure systems and closure operators are in one-to-one correspondence. A closure operator C: $\mathcal{P}(A) \to \mathcal{P}(A)$ on a set A is a function satisfying inclusiveness, monotonicity, and idempotence. For every closure system $\mathcal{C}$ on A, the only closure operator C corresponding to $\mathcal{C}$ satisfies that for all $X \subseteq A$, $C(X) = \bigcap\{F \in \mathcal{C} \mid X \subseteq F\}$. For convenience, we write $C(a)$ for $C(\{a\})$ and $C(X, a)$ for $C(X \cup \{a\})$, respectively.

Proposition 10.1 *Let $\mathcal{C}$ be a closure system on A and C the closure operator corresponding to $\mathcal{C}$. Then*

1. $C(\emptyset) = \bigcap \mathcal{C}$, *and* $\mathcal{C} \neq \{A\} \iff C(\emptyset) \neq A$;
2. *If* $\mathcal{C} \neq \{A\}$, *then for all* $a \in A$, $a \in \bigcup\{F \in \mathcal{C} \mid F \neq A\} \iff C(a) \neq A$.

Proof (1) Let $F \in \mathcal{C}$, then $\emptyset \subseteq F$, which means $C(\emptyset) \subseteq C(F) = F$. It follows by the arbitrariness of F that $C(\emptyset) \subseteq \bigcap \mathcal{C}$. Obviously $C(\emptyset) \in \mathcal{C}$, so we have $\bigcap \mathcal{C} \subseteq C(\emptyset)$. Hence, $C(\emptyset) = \bigcap \mathcal{C}$. Moreover, $\mathcal{C} \neq \{A\} \iff \bigcap \mathcal{C} \neq \{A\} \iff C(\emptyset) \neq A$.

(2) ($\Longrightarrow$) For all $a \in A$, suppose $a \in \bigcup\{F \in \mathcal{C} \mid F \neq A\}$, then there exists $F_1 \in \mathcal{C}$ such that $F_1 \neq A$ and $a \in F_1$. It follows that $C(a) \subseteq C(F_1) = F_1$. Because $F_1 \neq A$, we have $C(a) \neq A$.

($\Longleftarrow$) For all $a \in A$, suppose $C(a) \neq A$, then $C(a) \in \{F \in \mathcal{C} \mid F \neq A\}$, which means $a \in C(a) \subseteq \bigcup\{F \in \mathcal{C} \mid F \neq A\}$. □

Based on the above Proposition, generalized **ST** logics can be redefined as follows.

Definition 10.3 Let $\mathcal{M} = (\mathbf{A}, \mathcal{C})$ be a proper generalized matrix for $\mathcal{L}$. The generalized **ST** logic $\vdash^{\mathbf{ST}}_{\mathcal{M}}$ induced by $\mathcal{M}$ is defined as follows: for all $\Gamma \cup \{\varphi\} \subseteq Fm$,

$$\Gamma \vdash^{\mathbf{ST}}_{\mathcal{M}} \varphi \iff \text{for all } h \in \mathrm{Hom}(\mathbf{Fm}, \mathbf{A}), \text{ if } h[\Gamma] \subseteq C(\emptyset), \text{ then } C(h(\varphi)) \neq A.$$

Using the consequence relation $\vdash^{\mathbf{ST}}_{\mathcal{M}}$, we can define the corresponding consequence operator $C^{\mathbf{ST}}_{\mathcal{M}}$: $\mathcal{P}(Fm) \to \mathcal{P}(Fm)$ as usual: $\varphi \in C^{\mathbf{ST}}_{\mathcal{M}}(\Gamma) \iff \Gamma \vdash^{\mathbf{ST}}_{\mathcal{M}} \varphi$. The following proposition shows that the consequence operator $C^{\mathbf{ST}}_{\mathcal{M}}$ satisfies inclusiveness and monotonicity, meaning that $C^{\mathbf{ST}}_{\mathcal{M}}$ is a p-consequence operation.

Proposition 10.2 *Let $\mathcal{M} = (\mathbf{A}, \mathcal{C})$ be a proper generalized matrix for $\mathcal{L}$. Then*

1. *for all* $X \subseteq Fm$, $X \subseteq C^{\mathbf{ST}}_{\mathcal{M}}(X)$;

2. *for all* $X \cup Y \subseteq Fm$, *if* $X \subseteq Y$, *then* $C^{\mathbf{ST}}_{\mathcal{M}}(X) \subseteq C^{\mathbf{ST}}_{\mathcal{M}}(Y)$.

Proof (1) Let $\varphi \in X$ and $h \in \mathrm{Hom}(\mathbf{Fm}, \mathbf{A})$. Suppose $h[X] \subseteq C(\emptyset)$, then $C(h(\varphi)) \subseteq C(h[X]) \subseteq C(C(\emptyset)) = C(\emptyset)$. It follows by $C(\emptyset) \neq A$ that $C(h(\varphi)) \neq A$. Hence, by the arbitrariness of h and φ, we have $X \subseteq C^{\mathbf{ST}}_{\mathcal{M}}(X)$.

(2) Let $\varphi \in C^{\mathbf{ST}}_{\mathcal{M}}(X)$ and $h \in \mathrm{Hom}(\mathbf{Fm}, \mathbf{A})$. Suppose $X \subseteq Y$ and $h[Y] \subseteq C(\emptyset)$, then $h[X] \subseteq h[Y] \subseteq C(\emptyset)$. Because $\varphi \in C^{\mathbf{ST}}_{\mathcal{M}}(X)$, we have $C(h(\varphi)) \neq A$. Hence, by the arbitrariness of h and φ, we have $C^{\mathbf{ST}}_{\mathcal{M}}(X) \subseteq C^{\mathbf{ST}}_{\mathcal{M}}(Y)$. □

Now we show that p-consequences induced by p-matrices can be defined by generalized **ST** logics, and vice versa. Before giving the results, let us introduce p-matrix and the induced p-consequence first.

Definition 10.4 **([9])** A p-matrix for $\mathcal{L}$ is a structure $\mathcal{M} = (\mathbf{A}, D_1, D_\star)$, where $\mathbf{A}$ is an $\mathcal{L}$-algebra, and $D_1 \subseteq D_\star \subseteq A$.

Definition 10.5 **([9])** Let $\mathcal{M} = (\mathbf{A}, D_1, D_\star)$ be a p-matrix for $\mathcal{L}$. The p-consequence $\vdash^p_{\mathcal{M}}$ induced by $\mathcal{M}$ is defined as follows: for all $\Gamma \cup \{\varphi\} \subseteq Fm$,

$$\Gamma \vdash^p_{\mathcal{M}} \varphi \iff \text{for all } h \in \mathrm{Hom}(\mathbf{Fm}, \mathbf{A}),\ \text{ if } h[\Gamma] \subseteq D_1,\ \text{ then } h(\varphi) \in D_\star.$$

The following theorem says that given a generalized **ST** logic $\vdash^{\mathbf{ST}}_{\mathcal{M}}$, there is a p-consequence $\vdash^p_{\mathcal{M}'}$ coinciding with it.

Theorem 10.1 *Let* $\mathcal{M} = (\mathbf{A}, \mathcal{C})$ *be a proper generalized matrix for* $\mathcal{L}$. *Then there exists a p-matrix* $\mathcal{M}' = (\mathbf{A}', D_1, D_\star)$ *for* $\mathcal{L}$ *such that for all* $\Gamma \cup \{\varphi\} \subseteq Fm$,

$$\Gamma \vdash^{\mathbf{ST}}_{\mathcal{M}} \varphi \iff \Gamma \vdash^p_{\mathcal{M}'} \varphi.$$

Proof Let $D_1 = \bigcap \mathcal{C} = C(\emptyset)$, $D_\star = \bigcup \{F \in \mathcal{C} \mid F \neq A\}$, then $D_1 \subseteq D_\star \subseteq A$. Now let $\mathcal{M}' = (\mathbf{A}, D_1, D_\star)$, obviously $\mathcal{M}'$ is a p-matrix, and it can be verified that $\vdash^{\mathbf{ST}}_{\mathcal{M}} = \vdash^p_{\mathcal{M}'}$. □

The following theorem says that given a p-consequence $\vdash^p_{\mathcal{M}}$, there is a generalized **ST** logic $\vdash^{\mathbf{ST}}_{\mathcal{M}'}$ coinciding with it.

Theorem 10.2 *Let* $\mathcal{M} = (\mathbf{A}, D_1, D_\star)$ *be a p-matrix for* $\mathcal{L}$. *Then there exists a proper generalized matrix* $\mathcal{M}' = (\mathbf{A}', \mathcal{C})$ *for* $\mathcal{L}$ *such that for all* $\Gamma \cup \{\varphi\} \subseteq Fm$,

$$\Gamma \vdash^p_{\mathcal{M}} \varphi \iff \Gamma \vdash^{\mathbf{ST}}_{\mathcal{M}'} \varphi.$$

Proof We prove by the following 2 cases.

- If $D_\star \neq A$, let $\mathcal{C} = \{D_1, D_\star, A\}$, then $\mathcal{C}$ is a closure system on A and $C(\emptyset) = \bigcap \mathcal{C} = D_1$. Because $D_1 \subseteq D_\star \subseteq A$ and $D_\star \neq A$, we have $C(\emptyset) = D_1 \neq A$. Now let $\mathcal{M}' = (\mathbf{A}, \mathcal{C})$, then $\mathcal{M}'$ is a generalized matrix, with $C(\emptyset) \neq A$, and for all $\Gamma \cup \{\varphi\} \subseteq Fm$,

$$\begin{aligned}\Gamma \vdash^{p}_{\mathcal{M}} \varphi &\Longleftrightarrow \text{for all } h \in \mathrm{Hom}(\mathbf{Fm}, \mathbf{A}), \text{ if } h[\Gamma] \subseteq D_1, \text{ then } h(\varphi) \in D_\star;\\ &\Longleftrightarrow \text{for all } h \in \mathrm{Hom}(\mathbf{Fm}, \mathbf{A}), \text{ if } h[\Gamma] \subseteq C(\emptyset), \text{ then } C(h(\varphi)) \neq A;\\ &\Longleftrightarrow \Gamma \vdash^{\mathbf{ST}}_{\mathcal{M}'} \varphi.\end{aligned}$$

- If $D_\star = A$, then for all $\varphi \in Fm$ and $h \in \mathrm{Hom}(\mathbf{Fm}, \mathbf{A})$, $h(\varphi) \in A = D_\star$, which means for all $\Gamma \cup \{\varphi\} \subseteq Fm$, $\Gamma \vdash^{p}_{\mathcal{M}} \varphi$. Now let $\mathcal{M}' = (\mathbf{A}', C)$, where $A' = \{0, 1/2, 1\}$, $C = \{\{1\}, \{1, 1/2\}, \{1, 0\}, \{1, 1/2, 0\}\}$. Then $\mathcal{M}'$ is a generalized matrix, and $C(\emptyset) = \{1\} \neq A'$. Obviously for all $\varphi \in Fm$ and $h \in \mathrm{Hom}(\mathbf{Fm}, \mathbf{A}')$, $C(h(\varphi)) \neq A'$. Hence, for all $\Gamma \cup \{\varphi\} \subseteq Fm$, $\Gamma \vdash^{p}_{\mathcal{M}} \varphi \Longleftrightarrow \Gamma \vdash^{\mathbf{ST}}_{\mathcal{M}'} \varphi$.

□

10.3 When Are Generalized ST Logics Classical?

It is well known that **ST** logics on strong Kleene truth tables and weak Kleene ones produce the same inferences as classical logic. A natural question is when a generalized **ST** logic is classical. This question can be regarded as a generalization of one question answered in [12], where the authors asked what truth tables will give classical logic using strict-tolerant consequence in a three-valued setting. We answer the question by giving the conditions on which generalized matrices are models of the axioms and rules of classical logic. In the sequel, we assume that $\mathcal{L}$ is the standard propositional language, containing negation, implication, conjunction, and disjunction. Before giving the conditions for generalized **ST** logics to be classical, we give the conditions for them to be non-trivial first.

When we use a generalized matrix $\mathcal{M} = (\mathbf{A}, C)$ to define Tarskian logics, if $C = \{A\}$ or $C = \{A, \emptyset\}$, then $\vdash_{\mathcal{M}}$ is *trivial* [8], i.e., for all $\varphi, \psi \in Fm$, $\varphi \vdash_{\mathcal{M}} \psi$, which is undesirable. The following proposition gives the conditions for $\vdash^{\mathbf{ST}}_{\mathcal{M}}$ to be non-trivial.

Proposition 10.3 *Let $\mathcal{M} = (\mathbf{A}, C)$ be a proper generalized matrix for $\mathcal{L}$. Then the following conditions are equivalent:*

1. *$C(\emptyset) \neq \emptyset$ and $\mathcal{M}$ satisfies the condition* (NEG1),[2] *i.e., there is an $a_0 \in A$ such that $C(a_0) = A$;*
2. *$\vdash^{\mathbf{ST}}_{\mathcal{M}}$ is non-trivial, in the sense that there exist $\varphi, \psi \in Fm$ such that $\varphi \nvdash^{\mathbf{ST}}_{\mathcal{M}} \psi$.*

Proof (1) $\Longrightarrow$ (2): Suppose $\vdash^{\mathbf{ST}}_{\mathcal{M}}$ is trivial, then for all $\varphi, \psi \in Fm$, we have $\varphi \vdash^{\mathbf{ST}}_{\mathcal{M}} \psi$. On the other hand, because $C(\emptyset) \neq \emptyset$, there exists $t \in A$ such that $t \in C(\emptyset)$. Now let $p, q \in \mathrm{Prop}$ and $h \in \mathrm{Hom}(\mathbf{Fm}, \mathbf{A})$ such that $h(p) = t$ and $h(q) = a_0$, then $h(p) \in C(\emptyset)$ and $C(h(q)) = C(a_0) = A$, which means $p \nvdash^{\mathbf{ST}}_{\mathcal{M}} q$. This contradicts $p \vdash^{\mathbf{ST}}_{\mathcal{M}} q$.

(2) $\Longrightarrow$ (1): Suppose $C(\emptyset) = \emptyset$ or $\mathcal{M}$ does not satisfy condition (NEG1).

- If $C(\emptyset) = \emptyset$, then for all $h \in \mathrm{Hom}(\mathbf{Fm}, \mathbf{A})$, $h(\varphi) \notin \emptyset = C(\emptyset)$. Hence, $\varphi \vdash^{\mathbf{ST}}_{\mathcal{M}} \psi$;

---

[2] In [15] Tarski used it to characterize classical logic.

- If $\mathcal{M}$ doesn't satisfy condition (NEG1), then for all $h \in \text{Hom}(\mathbf{Fm}, \mathbf{A})$, $C(h(\psi)) \neq A$. Hence, $\varphi \vdash^{\mathbf{ST}}_{\mathcal{M}} \psi$. □

In the sequel, we call a proper generalized matrix *regular* if it satisfies $C(\emptyset) \neq \emptyset$ and the condition (NEG1). We denote by **LK** the sequent calculus given by Gentzen [10] for classical logic, i.e., it contains the axiom $(Id)\ \alpha \Rightarrow \alpha$, the structural rules for weakening, contraction, exchange, the Cut rule and the classical operational rules for negation, implication, conjunction, and disjunction. The logic $\vdash_{\mathbf{LK}}$ defined by Gentzen's sequent calculus is just classical logic $\vdash_{\mathbf{CPL}}$.

Definition 10.6 Let $\mathcal{M} = (\mathbf{A}, \mathcal{C})$ be a proper generalized matrix for $\mathcal{L}$ and $h \in \text{Hom}(\mathbf{Fm}, \mathbf{A})$.

1. Sequent $\Gamma \Rightarrow \Delta$ is **ST**-satisfied in $\mathcal{M}$ under h, denoted by $\mathcal{M} \models^{\mathbf{ST}}_h \Gamma \Rightarrow \Delta$, when if $h[\Gamma] \subseteq C(\emptyset)$, then there exists $\varphi \in \Delta$ such that $C(h(\varphi)) \neq A$;
2. $\mathcal{M}$ is an **ST**-model of sequent $\Gamma \Rightarrow \Delta$, denoted by $\mathcal{M} \models^{\mathbf{ST}} \Gamma \Rightarrow \Delta$, if for all $h \in \text{Hom}(\mathbf{Fm}, \mathbf{A})$, $\mathcal{M} \models^{\mathbf{ST}}_h \Gamma \Rightarrow \Delta$;
3. $\mathcal{M}$ is a global **ST**-model of the sequent rule $\dfrac{\{\Gamma_i \Rightarrow \Delta_i \mid i \in I\}}{\Gamma \Rightarrow \Delta}$, when if $\mathcal{M} \models^{\mathbf{ST}} \Gamma_i \Rightarrow \Delta_i$ for all $i \in I$, then $\mathcal{M} \models^{\mathbf{ST}} \Gamma \Rightarrow \Delta$;
4. $\mathcal{M}$ is a local **ST**-model of the sequent rule $\dfrac{\{\Gamma_i \Rightarrow \Delta_i \mid i \in I\}}{\Gamma \Rightarrow \Delta}$, when for all $h \in \text{Hom}(\mathbf{Fm}, \mathbf{A})$, if $\mathcal{M} \models^{\mathbf{ST}}_h \Gamma_i \Rightarrow \Delta_i$ for all $i \in I$, then $\mathcal{M} \models^{\mathbf{ST}}_h \Gamma \Rightarrow \Delta$.

The following proposition is easily verified.

Proposition 10.4 *Let $\mathcal{M} = (\mathbf{A}, \mathcal{C})$ be a proper generalized matrix for $\mathcal{L}$. Then*

1. *$\mathcal{M}$ is an* **ST***-model of (Id);*
2. *$\mathcal{M}$ is a local* **ST***-model of the structural rules of* **LK**.

Proposition 10.5 *Let $\mathcal{M} = (\mathbf{A}, \mathcal{C})$ be a regular generalized matrix for $\mathcal{L}$. Then the following conditions are equivalent.*

1. *For all $a \in A$, if $C(a) \neq A$, then $a \in C(\emptyset)$;*
2. *$\mathcal{M}$ is a local* **ST***-model of (Cut)*

$$\frac{\Gamma \Rightarrow \Delta, \alpha \quad \alpha, \Gamma \Rightarrow \Delta}{\Gamma \Rightarrow \Delta}.$$

Proof (1) $\Longrightarrow$ (2): Suppose there exists $h \in \text{Hom}(\mathbf{Fm}, \mathbf{A})$ such that $\mathcal{M} \models^{\mathbf{ST}}_h \Gamma \Rightarrow \Delta, \alpha$, $\mathcal{M} \models^{\mathbf{ST}}_h \alpha, \Gamma \Rightarrow \Delta$ but $\mathcal{M} \not\models^{\mathbf{ST}}_h \Gamma \Rightarrow \Delta$, then $h[\Gamma] \subseteq C(\emptyset)$ and for all $\psi \in \Delta$, $C(h(\psi)) = A$, which means $C(h(\alpha)) \neq A$. It follows by (1) that $h(\alpha) \in C(\emptyset)$. Hence $\mathcal{M} \not\models^{\mathbf{ST}}_h \alpha, \Gamma \Rightarrow \Delta$, which contradicts $\mathcal{M} \models^{\mathbf{ST}}_h \alpha, \Gamma \Rightarrow \Delta$.

(2) $\Longrightarrow$ (1): Suppose there exist $a \in A$ such that $C(a) \neq A$ and $a \notin C(\emptyset)$. Since $C(\emptyset) \neq \emptyset$ and $\mathcal{M}$ satisfies the condition (NEG1), there exist $a_1, a_0 \in A$ such that $a_1 \in C(\emptyset)$ and $C(a_0) = A$. For $p, q, r \in \text{Prop}$, let $h \in \text{Hom}(\mathbf{Fm}, \mathbf{A})$

such that $h(p) = a_1$, $h(q) = a$ and $h(r) = a_0$, we have $h(p) \in C(\emptyset)$, $h(q) \notin C(\emptyset)$, $C(h(q)) \neq A$ and $C(h(r)) = A$, which means $\mathcal{M} \models_h^{\mathbf{ST}} p \Rightarrow r, q$, $\mathcal{M} \models_h^{\mathbf{ST}} q, p \Rightarrow r$, but $\mathcal{M} \not\models_h^{\mathbf{ST}} p \Rightarrow r$. This contradicts (2). □

It is easily seen that if $\mathcal{M}$ is a local **ST**-model of a sequent rule then it is also a global **ST**-model of it. We will find out the conditions that make a generalized matrix a model of some operational rules under the condition that it is a global **ST**-model of (Cut).

Proposition 10.6 *Let $\mathcal{M} = (\mathbf{A}, \mathcal{C})$ be a regular generalized matrix for $\mathcal{L}$ that is a global* **ST**-*model of (Cut). Then the following conditions are equivalent.*

1. *For all $a \in A$, if $C(\neg a) = A$, then $C(a) \neq A$;*
2. *$\mathcal{M}$ is a global* **ST**-*model of $(\Rightarrow \neg)$*

$$\frac{\Gamma, \alpha \Rightarrow \Delta}{\Gamma \Rightarrow \neg\alpha, \Delta}.$$

Proof (1) $\Longrightarrow$ (2): Suppose $\mathcal{M} \models^{\mathbf{ST}} \Gamma, \alpha \Rightarrow \Delta$. For all $a \in A$, it follows by (1) that if $C(\neg a) = A$, then $C(a) \neq A$, which means $\mathcal{M}$ is an **ST**-model of $\Gamma \Rightarrow \alpha, \neg\alpha$. Then we can find a simplified derivation as follows:

$$\frac{\dfrac{\Gamma \Rightarrow \alpha, \neg\alpha}{\Gamma \Rightarrow \neg\alpha, \Delta, \alpha} \qquad \dfrac{\Gamma, \alpha \Rightarrow \Delta}{\alpha, \Gamma \Rightarrow \neg\alpha, \Delta}}{\Gamma \Rightarrow \neg\alpha, \Delta}$$

Hence, $\mathcal{M}$ is an **ST**-model of $\Gamma \Rightarrow \neg\alpha, \Delta$.

(2) $\Longrightarrow$ (1): Suppose there exists $a \in A$ such that $C(\neg a) = A$ and $C(a) = A$. Since $C(\emptyset) \neq \emptyset$, there exists $a_1 \in A$ such that $a_1 \in C(\emptyset)$. For $p, q \in \text{Prop}$, let $h \in \text{Hom}(\mathbf{Fm}, \mathbf{A})$ such that $h(p) = a$ and $h(q) = a_1$, we have $h(q) \in C(\emptyset)$, $C(h(\neg p)) = C(\neg h(p)) = A$ and $C(h(p)) = A$, which means $\mathcal{M} \not\models_h^{\mathbf{ST}} q \Rightarrow \neg p, p$, i.e., $\mathcal{M} \not\models^{\mathbf{ST}} q \Rightarrow \neg p, p$. On the other hand, $\mathcal{M}$ is an **ST**-model of $q, p \Rightarrow p$. This contradicts (2). □

Proposition 10.7 *Let $\mathcal{M} = (\mathbf{A}, \mathcal{C})$ be a regular generalized matrix for $\mathcal{L}$ that is a global* **ST**-*model of (Cut). Then the following conditions are equivalent.*

1. *For all $a \in A$, if $\neg a \in C(\emptyset)$, then $a \notin C(\emptyset)$;*
2. *$\mathcal{M}$ is a global* **ST**-*model of $(\neg \Rightarrow)$*

$$\frac{\Gamma \Rightarrow \alpha, \Delta}{\Gamma, \neg\alpha \Rightarrow \Delta}.$$

Proof (1) $\Longrightarrow$ (2): Suppose $\mathcal{M} \models^{\mathbf{ST}} \Gamma \Rightarrow \alpha, \Delta$. For all $a \in A$, it follows by (1) that if $\neg a \in C(\emptyset)$, then $a \notin C(\emptyset)$, which means $\mathcal{M}$ is an **ST**-model of $\alpha, \neg\alpha \Rightarrow \Delta$. Then we can find a simplified derivation as follows:

$$\dfrac{\dfrac{\Gamma \Rightarrow \alpha, \Delta}{\Gamma, \neg\alpha \Rightarrow \Delta, \alpha} \qquad \dfrac{\alpha, \neg\alpha \Rightarrow \Delta}{\alpha, \Gamma, \neg\alpha \Rightarrow \Delta}}{\Gamma, \neg\alpha \Rightarrow \Delta}$$

Hence, $\mathcal{M}$ is an **ST**-model of $\Gamma, \neg\alpha \Rightarrow \Delta$.

(2) $\Longrightarrow$ (1): Suppose there exists $a \in A$ such that $\neg a \in C(\emptyset)$ and $a \in C(\emptyset)$. Since $\mathcal{M}$ satisfies the condition (NEG1), there exists $a_0 \in A$ such that $C(a_0) = A$. For $p, q \in$ Prop, let $h \in \mathrm{Hom}(\mathbf{Fm}, \mathbf{A})$ such that $h(p) = a$ and $h(q) = a_0$, we have $h(p) \in C(\emptyset)$, $h(\neg p) = \neg h(p) \in C(\emptyset)$ and $C(h(q)) = A$, which means $\mathcal{M} \not\models_h^{\mathbf{ST}} p, \neg p \Rightarrow q$, i.e., $\mathcal{M} \not\models^{\mathbf{ST}} p, \neg p \Rightarrow q$. On the other hand, $\mathcal{M}$ is an **ST**-model of $p \Rightarrow p, q$. This contradicts (2). □

Proposition 10.8 *Let $\mathcal{M} = (\mathbf{A}, \mathcal{C})$ be a regular generalized matrix for $\mathcal{L}$ that is a global* **ST***-model of (Cut). Then the following conditions are equivalent.*

1. *For all $a, b \in A$, if $C(a \to b) = A$, then $C(a) \neq A$ and $b \notin C(\emptyset)$;*
2. *$\mathcal{M}$ is a global* **ST***-model of $(\Rightarrow\to)$*

$$\frac{\alpha, \Gamma \Rightarrow \beta, \Delta}{\Gamma \Rightarrow \alpha \to \beta, \Delta}.$$

Proof (1) $\Longrightarrow$ (2): Suppose $\mathcal{M} \models^{\mathbf{ST}} \alpha, \Gamma \Rightarrow \beta, \Delta$. For all $a, b \in A$, if $C(a) = A$ or $b \in C(\emptyset)$, then we can obtain $C(a \to b) \neq A$ following by (1), which means $\mathcal{M}$ is an **ST**-model of $\Gamma \Rightarrow \alpha, \alpha \to \beta$ and $\beta \Rightarrow \alpha \to \beta$. Then we can find a simplified derivation as follows:

$$\dfrac{\dfrac{\Gamma \Rightarrow \alpha, \alpha \to \beta}{\Gamma \Rightarrow \alpha \to \beta, \Delta, \alpha} \qquad \dfrac{\dfrac{\alpha, \Gamma \Rightarrow \beta, \Delta}{\alpha, \Gamma \Rightarrow \alpha \to \beta, \Delta, \beta} \qquad \dfrac{\beta \Rightarrow \alpha \to \beta}{\beta, \alpha, \Gamma \Rightarrow \alpha \to \beta, \Delta}}{\alpha, \Gamma \Rightarrow \alpha \to \beta, \Delta}}{\Gamma \Rightarrow \alpha \to \beta, \Delta}$$

Hence, $\mathcal{M}$ is an **ST**-model of $\Gamma \Rightarrow \alpha \to \beta, \Delta$.

(2) $\Longrightarrow$ (1): Suppose there exist $a, b \in A$ such that $C(a \to b) = A$, and $C(a) = A$ or $b \in C(\emptyset)$. Since $C(\emptyset) \neq \emptyset$, there exists $a_1 \in A$ such that $a_1 \in C(\emptyset)$. For $p, q, r \in$ Prop, let $h \in \mathrm{Hom}(\mathbf{Fm}, \mathbf{A})$ such that $h(p) = a$, $h(q) = b$ and $h(r) = a_1$, we have $C(h(p \to q)) = C(h(p) \to h(q)) = A$ and $h(r) \in C(\emptyset)$.

- If $C(a) = A$, then $C(h(p)) = A$, following that $\mathcal{M} \not\models_h^{\mathbf{ST}} r \Rightarrow p \to q, p$, i.e., $\mathcal{M} \not\models^{\mathbf{ST}} r \Rightarrow p \to q, p$. On the other hand, $\mathcal{M}$ is an **ST**-model of $p, r \Rightarrow q, p$. This contradicts (2);
- If $b \in C(\emptyset)$, then $h(q) \in C(\emptyset)$, following that $\mathcal{M} \not\models_h^{\mathbf{ST}} q, r \Rightarrow p \to q$, i.e., $\mathcal{M} \not\models^{\mathbf{ST}} q, r \Rightarrow p \to q$. On the other hand, $\mathcal{M}$ is an **ST**-model of $p, q, r \Rightarrow q$. This contradicts (2). □

Proposition 10.9 *Let $\mathcal{M} = (\mathbf{A}, \mathcal{C})$ be a regular generalized matrix for $\mathcal{L}$ that is a global* **ST**-*model of* (Cut). *Then the following conditions are equivalent.*

1. *For all $a, b \in A$, if $a \to b \in C(\emptyset)$, then $a \notin C(\emptyset)$ or $C(b) \neq A$;*
2. *$\mathcal{M}$ is a global* **ST**-*model of* $(\to\Rightarrow)$

$$\frac{\Gamma \Rightarrow \Delta, \alpha \quad \beta, \Gamma \Rightarrow \Delta}{\alpha \to \beta, \Gamma \Rightarrow \Delta}.$$

Proof (1) $\Longrightarrow$ (2): Suppose $\mathcal{M} \models^{\mathbf{ST}} \Gamma \Rightarrow \Delta, \alpha$ and $\mathcal{M} \models^{\mathbf{ST}} \beta, \Gamma \Rightarrow \Delta$. For all $a, b \in A$, if $a \in C(\emptyset)$ and $C(b) = A$, then we can obtain $a \to b \notin C(\emptyset)$ following by (1), which means $\mathcal{M}$ is an **ST**-model of $\alpha, \alpha \to \beta \Rightarrow \beta$. Then we can find a simplified derivation as follows:

$$\cfrac{\cfrac{\cfrac{\Gamma \Rightarrow \Delta, \alpha}{\alpha \to \beta, \Gamma \Rightarrow \Delta, \beta, \alpha} \quad \cfrac{\alpha, \alpha \to \beta \Rightarrow \beta}{\alpha, \alpha \to \beta, \Gamma \Rightarrow \Delta, \beta}}{\alpha \to \beta, \Gamma \Rightarrow \Delta, \beta} \quad \cfrac{\beta, \Gamma \Rightarrow \Delta}{\beta, \alpha \to \beta, \Gamma \Rightarrow \Delta}}{\alpha \to \beta, \Gamma \Rightarrow \Delta}$$

Hence, $\mathcal{M}$ is an **ST**-model of $\alpha \to \beta, \Gamma \Rightarrow \Delta$.

(2) $\Longrightarrow$ (1): Suppose there exist $a, b \in A$ such that $a \to b \in C(\emptyset), a \in C(\emptyset)$ and $C(b) = A$. For $p, q \in \text{Prop}$, let $h \in \text{Hom}(\mathbf{Fm}, \mathbf{A})$ such that $h(p) = a$ and $h(q) = b$, we have $h(p \to q) = h(p) \to h(q) \in C(\emptyset), h(p) \in C(\emptyset)$ and $C(h(q)) = A$, which means $\mathcal{M} \not\models_h^{\mathbf{ST}} p \to q, p \Rightarrow q$, i.e., $\mathcal{M} \not\models^{\mathbf{ST}} p \to q, p \Rightarrow q$. On the other hand, $\mathcal{M}$ is an **ST**-model of $p \Rightarrow q, p$ and $q, p \Rightarrow q$. This contradicts (2). □

Proposition 10.10 *Let $\mathcal{M} = (\mathbf{A}, \mathcal{C})$ be a regular generalized matrix for $\mathcal{L}$ that is a global* **ST**-*model of* (Cut). *Then the following conditions are equivalent.*

1. *For all $a, b \in A$, if $C(a \wedge b) = A$, then $a \notin C(\emptyset)$ or $b \notin C(\emptyset)$;*
2. *$\mathcal{M}$ is a global* **ST**-*model of* $(\Rightarrow\wedge)$

$$\frac{\Gamma \Rightarrow \alpha, \Delta \quad \Gamma \Rightarrow \beta, \Delta}{\Gamma \Rightarrow \alpha \wedge \beta, \Delta}.$$

Proof (1) $\Longrightarrow$ (2): Suppose $\mathcal{M} \models^{\mathbf{ST}} \Gamma \Rightarrow \alpha, \Delta$ and $\mathcal{M} \models^{\mathbf{ST}} \Gamma \Rightarrow \beta, \Delta$. For all $a, b \in A$, if $a \in C(\emptyset)$ and $b \in C(\emptyset)$, then we can obtain $C(a \wedge b) \neq A$ following by (1), which means $\mathcal{M}$ is an **ST**-model of $\alpha, \beta \Rightarrow \alpha \wedge \beta$. Then we can find a simplified derivation as follows:

$$\cfrac{\cfrac{\Gamma \Rightarrow \beta, \Delta}{\Gamma \Rightarrow \alpha \wedge \beta, \Delta, \beta} \quad \cfrac{\cfrac{\Gamma \Rightarrow \alpha, \Delta}{\beta, \Gamma \Rightarrow \alpha \wedge \beta, \Delta, \alpha} \quad \cfrac{\alpha, \beta \Rightarrow \alpha \wedge \beta}{\alpha, \beta, \Gamma \Rightarrow \alpha \wedge \beta, \Delta}}{\beta, \Gamma \Rightarrow \alpha \wedge \beta, \Delta}}{\Gamma \Rightarrow \alpha \wedge \beta, \Delta}$$

Hence, $\mathcal{M}$ is an **ST**-model of $\Gamma \Rightarrow \alpha \wedge \beta, \Delta$.

(2) $\Longrightarrow$ (1): Suppose there exist $a, b \in A$ such that $C(a \wedge b) = A$, $a \in C(\emptyset)$ and $b \in C(\emptyset)$. For $p, q \in \text{Prop}$, let $h \in \text{Hom}(\mathbf{Fm}, \mathbf{A})$ such that $h(p) = a$ and $h(q) = b$, we have $h(p) \in C(\emptyset)$, $h(q) \in C(\emptyset)$ and $C(h(p \wedge q)) = C(h(p) \wedge h(q)) = A$, which means $\mathcal{M} \not\models_h^{\mathbf{ST}} p, q \Rightarrow p \wedge q$, i.e., $\mathcal{M} \not\models^{\mathbf{ST}} p, q \Rightarrow p \wedge q$. On the other hand, $\mathcal{M}$ is an **ST**-model of $p, q \Rightarrow p$ and $p, q \Rightarrow q$. This contradicts (2). □

Proposition 10.11 *Let $\mathcal{M} = (\mathbf{A}, \mathcal{C})$ be a regular generalized matrix for $\mathcal{L}$ that is a global* **ST***-model of (Cut). Then the following conditions are equivalent.*

1. *For all $a, b \in A$, if $a \wedge b \in C(\emptyset)$, then $C(a) \neq A$ and $C(b) \neq A$;*
2. *$\mathcal{M}$ is a global* **ST***-model of $(\wedge \Rightarrow)$*

$$\frac{\Gamma, \alpha, \beta \Rightarrow \Delta}{\Gamma, \alpha \wedge \beta \Rightarrow \Delta}.$$

Proof (1) $\Longrightarrow$ (2): Suppose $\mathcal{M} \models^{\mathbf{ST}} \Gamma, \alpha, \beta \Rightarrow \Delta$. For all $a, b \in A$, it follows by (1) that if $a \wedge b \in C(\emptyset)$, then $C(a) \neq A$ and $C(b) \neq A$, which means $\mathcal{M}$ is an **ST**-model of $\alpha \wedge \beta \Rightarrow \alpha$ and $\alpha \wedge \beta \Rightarrow \beta$. Then we can find a simplified derivation as follows:

$$\dfrac{\dfrac{\alpha \wedge \beta \Rightarrow \beta}{\Gamma, \alpha \wedge \beta \Rightarrow \Delta, \beta} \qquad \dfrac{\dfrac{\alpha \wedge \beta \Rightarrow \alpha}{\beta, \Gamma, \alpha \wedge \beta \Rightarrow \Delta, \alpha} \qquad \dfrac{\Gamma, \alpha, \beta \Rightarrow \Delta}{\alpha, \beta, \Gamma, \alpha \wedge \beta \Rightarrow \Delta}}{\beta, \Gamma, \alpha \wedge \beta \Rightarrow \Delta}}{\Gamma, \alpha \wedge \beta \Rightarrow \Delta}$$

Hence, $\mathcal{M}$ is an **ST**-model of $\Gamma, \alpha \wedge \beta \Rightarrow \Delta$.

(2) $\Longrightarrow$ (1): Suppose there exist $a, b \in A$ such that $a \wedge b \in C(\emptyset)$, and $C(a) = A$ or $C(b) = A$. Without loss of generality, let $C(a) = A$. For $p, q \in \text{Prop}$, let $h \in \text{Hom}(\mathbf{Fm}, \mathbf{A})$ such that $h(p) = a$ and $h(q) = b$, we have $h(p \wedge q) = h(p) \wedge h(q) \in C(\emptyset)$ and $C(h(p)) = A$, which means $\mathcal{M} \not\models_h^{\mathbf{ST}} p \wedge q \Rightarrow p$, i.e., $\mathcal{M} \not\models^{\mathbf{ST}} p \wedge q \Rightarrow p$. On the other hand, $\mathcal{M}$ is an **ST**-model of $p, q \Rightarrow p$. This contradicts (2). □

Proposition 10.12 *Let $\mathcal{M} = (\mathbf{A}, \mathcal{C})$ be a regular generalized matrix for $\mathcal{L}$ that is a global* **ST***-model of (Cut). Then the following conditions are equivalent.*

1. *For all $a, b \in A$, if $C(a \vee b) = A$, then $a \notin C(\emptyset)$ and $b \notin C(\emptyset)$;*
2. *$\mathcal{M}$ is a global* **ST***-model of $(\Rightarrow \vee)$*

$$\frac{\Gamma \Rightarrow \alpha, \beta, \Delta}{\Gamma \Rightarrow \alpha \vee \beta, \Delta}.$$

Proof (1) $\Longrightarrow$ (2): Suppose $\mathcal{M} \models^{\mathbf{ST}} \Gamma \Rightarrow \alpha, \beta, \Delta$. For all $a, b \in A$, if $a \in C(\emptyset)$ or $b \in C(\emptyset)$, then we can obtain $C(a \vee b) \neq A$ following by (1), which means $\mathcal{M}$ is an **ST**-model of $\alpha \Rightarrow \alpha \vee \beta$ and $\beta \Rightarrow \alpha \vee \beta$. Then we can find a simplified derivation as follows:

$$\dfrac{\dfrac{\dfrac{\Gamma \Rightarrow \alpha, \beta, \Delta}{\Gamma \Rightarrow \alpha \vee \beta, \Delta, \beta, \alpha} \qquad \dfrac{\alpha \Rightarrow \alpha \vee \beta}{\alpha, \Gamma \Rightarrow \alpha \vee \beta, \Delta, \beta}}{\Gamma \Rightarrow \alpha \vee \beta, \Delta, \beta} \qquad \dfrac{\beta \Rightarrow \alpha \vee \beta}{\beta, \Gamma \Rightarrow \alpha \vee \beta, \Delta}}{\Gamma \Rightarrow \alpha \vee \beta, \Delta}$$

Hence, $\mathcal{M}$ is an **ST**-model of $\Gamma \Rightarrow \alpha \vee \beta, \Delta$.

(2) $\Longrightarrow$ (1): Suppose there exist $a, b \in A$ such that $C(a \vee b) = A$, and $a \in C(\emptyset)$ or $b \in C(\emptyset)$. Without loss of generality, let $a \in C(\emptyset)$. For $p, q \in$ Prop, let $h \in \mathrm{Hom}(\mathbf{Fm}, \mathbf{A})$ such that $h(p) = a$ and $h(q) = b$, we have $C(h(p \vee q)) = C(h(p) \vee h(q)) = A$ and $h(p) \in C(\emptyset)$, which means $\mathcal{M} \not\models_h^{\mathbf{ST}} p \Rightarrow p \vee q$, i.e., $\mathcal{M} \not\models^{\mathbf{ST}} p \Rightarrow p \vee q$. On the other hand, $\mathcal{M}$ is an **ST**-model of $p \Rightarrow p, q$. This contradicts (2). □

Proposition 10.13 *Let $\mathcal{M} = (\mathbf{A}, \mathcal{C})$ be a regular generalized matrix for $\mathcal{L}$ that is a global* **ST***-model of (Cut). Then the following conditions are equivalent.*

1. *For all $a, b \in A$, if $a \vee b \in C(\emptyset)$, then $C(a) \neq A$ or $C(b) \neq A$;*
2. *$\mathcal{M}$ is a global* **ST***-model of $(\vee \Rightarrow)$*

$$\frac{\Gamma, \alpha \Rightarrow \Delta \quad \Gamma, \beta \Rightarrow \Delta}{\Gamma, \alpha \vee \beta \Rightarrow \Delta}.$$

Proof (1) $\Longrightarrow$ (2): Suppose $\mathcal{M} \models^{\mathbf{ST}} \Gamma, \alpha \Rightarrow \Delta$ and $\mathcal{M} \models^{\mathbf{ST}} \Gamma, \beta \Rightarrow \Delta$. For all $a, b \in A$, it follows by (1) that if $a \vee b \in C(\emptyset)$, then $C(a) \neq A$ or $C(b) \neq A$, which means $\mathcal{M}$ is an **ST**-model of $\alpha \vee \beta \Rightarrow \alpha, \beta$. Then we can find a simplified derivation as follows:

$$\dfrac{\dfrac{\dfrac{\alpha \vee \beta \Rightarrow \alpha, \beta}{\Gamma, \alpha \vee \beta \Rightarrow \Delta, \beta, \alpha} \qquad \dfrac{\Gamma, \alpha \Rightarrow \Delta}{\alpha, \Gamma, \alpha \vee \beta \Rightarrow \Delta, \beta}}{\Gamma, \alpha \vee \beta \Rightarrow \Delta, \beta} \qquad \dfrac{\Gamma, \beta \Rightarrow \Delta}{\beta, \Gamma, \alpha \vee \beta \Rightarrow \Delta}}{\Gamma, \alpha \vee \beta \Rightarrow \Delta}$$

Hence, $\mathcal{M}$ is an **ST**-model of $\Gamma, \alpha \vee \beta \Rightarrow \Delta$.

(2) $\Longrightarrow$ (1): Suppose there exist $a, b \in A$ such that $a \vee b \in C(\emptyset)$, $C(a) = A$ and $C(b) = A$. For $p, q \in$ Prop, let $h \in \mathrm{Hom}(\mathbf{Fm}, \mathbf{A})$ such that $h(p) = a$ and $h(q) = b$, we have $h(p \vee q) = h(p) \vee h(q) \in C(\emptyset)$, $C(h(p)) = A$ and $C(h(q)) = A$, which means $\mathcal{M} \not\models_h^{\mathbf{ST}} p \vee q \Rightarrow p, q$, i.e., $\mathcal{M} \not\models^{\mathbf{ST}} p \vee q \Rightarrow p, q$. On the other hand, $\mathcal{M}$ is an **ST**-model of $p \Rightarrow p, q$ and $q \Rightarrow p, q$. This contradicts (2). □

Let $\mathcal{M} = (\mathbf{A}, \mathcal{C})$ be a regular generalized matrix for $\mathcal{L}$, we can define that the *classical truth set* of $\mathcal{M}$ is $\mathrm{CT}(\mathcal{M}) = C(\emptyset) \cup \{a \mid C(a) = A\}$. $\mathrm{CT}(\mathcal{M})$ may be a subuniverse of $\mathbf{A}$. If $\mathrm{CT}(\mathcal{M})$ is a subuniverse of $\mathbf{A}$, then $\mathbf{CT}(\mathcal{M})$ is a subalgebra of $\mathbf{A}$, and there exists a regular generalized matrix corresponding to $\mathcal{M}$:

$$\mathcal{M}^{ct} = (\mathbf{CT}(\mathcal{M}), C^{ct}) = (\mathbf{CT}(\mathcal{M}), \mathcal{C}^{ct}),$$

where $\mathcal{C}^{ct} = \{C(\emptyset), \mathrm{CT}(\mathcal{M})\}$.

Lemma 10.1 *Let $\mathcal{M} = (\mathbf{A}, \mathcal{C})$ be a regular generalized matrix, and* $\mathrm{CT}(\mathcal{M})$ *is a subuniverse of* $\mathbf{A}$. *Then for all $\Gamma \cup \Delta \subseteq Fm$, If $\mathcal{M} \models^{\mathbf{ST}} \Gamma \Rightarrow \Delta$, then $\mathcal{M}^{ct} \models^{\mathbf{ST}} \Gamma \Rightarrow \Delta$.*

Proof Let $h \in \mathrm{Hom}(\mathbf{Fm}, \mathbf{CT}(\mathcal{M}))$, then $\mathrm{id} \circ h \in \mathrm{Hom}(\mathbf{Fm}, \mathbf{A})$. Suppose $h[\Gamma] \subseteq C^{ct}(\emptyset)$. Since $C^{ct}(\emptyset) = C(\emptyset)$, we have $\mathrm{id}[h[\Gamma]] = h[\Gamma] \subseteq C(\emptyset)$, following by $\mathcal{M} \models^{\mathbf{ST}} \Gamma \Rightarrow \Delta$ that there exists $\varphi \in \Delta$ such that $C(\mathrm{id}(h(\varphi))) \neq A$. Then $h(\varphi) = \mathrm{id}(h(\varphi)) \notin \{a \in A \mid C(a) = A\}$, which means $C^{ct}(h(\varphi)) \neq \mathrm{CT}(\mathcal{M})$. Hence, $\mathcal{M}^{ct} \models^{\mathbf{ST}} \Gamma \Rightarrow \Delta$. □

If $\mathcal{M}_1 = (\mathbf{A}_1, \mathcal{C}_1)$ and $\mathcal{M}_2 = (\mathbf{A}_2, \mathcal{C}_2)$ are two generalized matrices for $\mathcal{L}$, then we say that h is a *bilogical morphism* from $\mathcal{M}_1$ to $\mathcal{M}_2$ when $h \in \mathrm{Hom}(\mathbf{A}_1, \mathbf{A}_2)$ is surjective and $\mathcal{C}_1 = \left\{h^{-1}[F] \mid F \in \mathcal{C}_2\right\}$. It is easy to verify that two generalized matrices having a bilogical morphism between them will obtain the same generalized **ST** logic. We say that $\mathcal{M}$ is *hyper-classical*, if $\mathrm{CT}(\mathcal{M})$ is a subuniverse of $\mathbf{A}$, and there exists a bilogical morphism from $\mathcal{M}^{ct}$ to $\mathbb{L}_2 = (\mathbf{2}, \{\{1\}, \{1, 0\}\})$[3].

Lemma 10.2 *Let $\mathcal{M} = (\mathbf{A}, \mathcal{C})$ be a regular generalized matrix for $\mathcal{L}$. Then for the following three conditions:*

1. $\vdash^{\mathbf{ST}}_{\mathcal{M}} = \vdash_{\mathbf{CPL}}$
2. *$\mathcal{M}$ is hyper-classical*
3. $\vdash^{\mathbf{ST}}_{\mathcal{M}} \subseteq \vdash_{\mathbf{CPL}}$

(1) *is a sufficient condition of* (2), *and* (3) *is a necessary condition of* (2).

Proof (1) $\Longrightarrow$ (2): We first show that $\mathrm{CT}(\mathcal{M})$ is closed under $\neg$. Let $a \in \mathrm{CT}(\mathcal{M}) = C(\emptyset) \cup \{a \mid C(a) = A\}$, then $a \in C(\emptyset)$ or $C(a) = A$.

- When $a \in C(\emptyset)$, if $C(\neg a) \neq A$, then $\mathcal{M}$ is an **ST**-model of $p \Rightarrow \neg p$, which contradicts (1). Hence, $C(\neg a) = A$, $\neg a \in \mathrm{CT}(\mathcal{M})$;
- When $C(a) = A$, if $\neg a \notin C(\emptyset)$, then $\mathcal{M}$ is an **ST**-model of $\neg p \Rightarrow p$, which contradicts (1). Hence, $\neg a \in C(\emptyset) \subseteq \mathrm{CT}(\mathcal{M})$.

Thus, $\mathrm{CT}(\mathcal{M})$ is closed under $\neg$. In the same way, $\mathrm{CT}(\mathcal{M})$ is closed under $\rightarrow$, $\wedge$, $\vee$. It is similar to the proof of Lemma A.6 in [12], and the proof about $\rightarrow$ needs sequents $\neg p \rightarrow p \Rightarrow \neg p, \neg p \rightarrow \neg p \Rightarrow \neg p, p \rightarrow p \Rightarrow \neg p$ and $p \Rightarrow p \rightarrow \neg p$. Now define $h : \mathrm{CT}(\mathcal{M}) \rightarrow \{1, 0\}$ such that $h(a) = 1$ iff $a \in C(\emptyset)$. It is easy to see that h is a bilogical morphism from $\mathcal{M}^{ct}$ to $\mathbb{L}_2$.

(2) $\Longrightarrow$ (3): Let $\Gamma \cup \{\varphi\} \subseteq Fm$. Suppose $\Gamma \vdash^{\mathbf{ST}}_{\mathcal{M}} \varphi$, then by (2) and Lemma 10.1 we have $\Gamma \vdash^{\mathbf{ST}}_{\mathcal{M}^{ct}} \varphi$, and $\Gamma \vdash^{\mathbf{ST}}_{\mathbb{L}_2} \varphi$, i.e., $\Gamma \vdash_{\mathbf{CPL}} \varphi$. Hence, $\vdash^{\mathbf{ST}}_{\mathcal{M}} \subseteq \vdash_{\mathbf{CPL}}$. □

---

[3] The generalized **ST** logic induced by $\mathbb{L}_2$ is just classical logic $\vdash_{\mathbf{CPL}}$.

Theorem 10.3 *Let $\mathcal{M} = (\mathbf{A}, \mathcal{C})$ be a proper generalized matrix for $\mathcal{L}$. Then $\vdash^{\mathbf{ST}}_{\mathcal{M}} = \vdash_{\mathbf{CPL}}$ if and only if $\mathcal{M}$ satisfies the following conditions.*

1. *$C(\emptyset) \neq \emptyset$ and $\mathcal{M}$ satisfies the condition* (NEG1)*;*
2. *$\mathcal{M}$ is hyper-classical;*
3. *$\mathcal{M}$ is a global* **ST***-model of (Cut); and for all $a, b \in A$,*
4. *if $C(\neg a) = A$, then $C(a) \neq A$;*
5. *if $\neg a \in C(\emptyset)$, then $a \notin C(\emptyset)$;*
6. *if $C(a \to b) = A$, then $C(a) \neq A$ and $b \notin C(\emptyset)$;*
7. *if $a \to b \in C(\emptyset)$, then $a \notin C(\emptyset)$ or $C(b) \neq A$;*
8. *if $C(a \wedge b) = A$, then $a \notin C(\emptyset)$ or $b \notin C(\emptyset)$;*
9. *if $a \wedge b \in C(\emptyset)$, then $C(a) \neq A$ and $C(b) \neq A$;*
10. *if $C(a \vee b) = A$, then $a \notin C(\emptyset)$ and $b \notin C(\emptyset)$;*
11. *if $a \vee b \in C(\emptyset)$, then $C(a) \neq A$ or $C(b) \neq A$.*

Proof ($\Longrightarrow$): Suppose $\vdash^{\mathbf{ST}}_{\mathcal{M}} = \vdash_{\mathbf{CPL}}$, then $\vdash^{\mathbf{ST}}_{\mathcal{M}}$ is non-trivial. It follows by Proposition 10.3 that $C(\emptyset) \neq \emptyset$ and $\mathcal{M}$ satisfies the condition (NEG1), which means $\mathcal{M}$ is hyper-classical. Since $\vdash^{\mathbf{ST}}_{\mathcal{M}} = \vdash_{\mathbf{CPL}} = \vdash_{\mathbf{LK}}$, it is easy to see that $\mathcal{M}$ is a global **ST**-model of (Cut) and all the operational rules of **LK**. Hence, by Propositions 10.6–10.13, $\mathcal{M}$ satisfies the above conditions.

($\Longleftarrow$): Let $\Gamma \cup \{\varphi\} \subseteq Fm$. Suppose $\Gamma \vdash_{\mathbf{CPL}} \varphi$. Then $\Gamma \vdash_{\mathbf{LK}} \varphi$, i.e., there exists a finite $\Gamma_0 \subseteq \Gamma$ such that $\Gamma_0 \Rightarrow \varphi$ is derivable in **LK**. From Propositions 10.4 and 10.6–10.13, it follows that $\mathcal{M}$ is an **ST**-model of the axiom (Id) and a global **ST**-model of all rules of **LK**. Then $\mathcal{M}$ is an **ST**-model of $\Gamma_0 \Rightarrow \varphi$, and $\varphi \in C^{\mathbf{ST}}_{\mathcal{M}}(\Gamma_0) \subseteq C^{\mathbf{ST}}_{\mathcal{M}}(\Gamma)$, i.e., $\Gamma \vdash^{\mathbf{ST}}_{\mathcal{M}} \varphi$. Hence, $\vdash_{\mathbf{CPL}} \subseteq \vdash^{\mathbf{ST}}_{\mathcal{M}}$. It follows by Lemma 10.2 that $\vdash^{\mathbf{ST}}_{\mathcal{M}} = \vdash_{\mathbf{CPL}}$. □

We give two examples illustrating the above theorem.

Example 10.1 Let $\mathcal{M}_{\mathbf{B3}} = ((\{0, 1/2, 1\}, \wedge, \vee, \neg, \to), \{\{1\}, \{1, 1/2, 0\}\})$, and the operators be defined as in Bochvar's 3-valued external logic. It can be verified that $\mathcal{M}_{\mathbf{B3}}$ is a local **ST**-model of (Cut) and all the conditions in Theorem 10.3 are satisfied. Hence, $\vdash^{\mathbf{ST}}_{\mathcal{M}_{\mathbf{B3}}} = \vdash_{\mathbf{CPL}}$.

Example 10.2 Let $\mathcal{M}_{\mathbf{L3}} = ((\{0, 1/2, 1\}, \wedge, \vee, \neg, \to), \{\{1\}, \{1, 1/2\}, \{1, 1/2, 0\}\})$, and the operators be defined as in Łukasiewicz's 3-valued logic. It can be verified that the condition (3) in Theorem 10.3 is not satisfied, since $\mathcal{M}_{\mathbf{L3}} \models^{\mathbf{ST}} \neg p \to p \Rightarrow \neg(p \to \neg p), p$, $\mathcal{M}_{\mathbf{L3}} \models^{\mathbf{ST}} p, \neg p \to p \Rightarrow \neg(p \to \neg p)$, but $\mathcal{M}_{\mathbf{L3}} \not\models^{\mathbf{ST}} \neg p \to p \Rightarrow \neg(p \to \neg p)$. Hence, $\vdash^{\mathbf{ST}}_{\mathcal{M}_{\mathbf{L3}}} \neq \vdash_{\mathbf{CPL}}$.

10.4 Conclusion

We generalize Strict-tolerant logic using generalized matrices. Strict truth is the intersection of the closure system of the given generalized matrix, and tolerant truth is a restricted union of the closure system. We show that generalized **ST** logics induced by generalized matrices are as general as p-consequences induced by p-matrices. We also give the conditions for generalized **ST** logics to be classical. We haven't found the characterization condition for a generalized matrix to be a global **ST**-model of the Cut rule. We leave it for future work, which also includes a detailed comparison with Fitting's work on **ST** logics and exploring the conditions for generalized **ST** logics to coincide with their corresponding Tarskian logics.

References

1. Ajdukiewicz, K.: Pragmatic Logic. D. Reidel Pub. Co, Dordrecht (1974)
2. Barrio, E., Rosenblatt, L., Tajer, D.: The logics of strict-tolerant logic. J. Philos. Log. **44**(5), 551–571 (2015)
3. Chen, L., Wen, X.: On the transitivity of logical consequence without assuming monotonicity. Log. Univers. (2024). https://doi.org/10.1007/s11787-024-00345-3
4. Cobreros, P., Egré, P., Ripley, D., van Rooij, R.: Reaching transparent truth. Mind **122**(488), 841–866 (2013)
5. Cobreros, P., Egré, P., Ripley, D., van Rooij, R.: Tolerant, classical, strict. J. Philos. Log. **41**(2), 347–385 (2012)
6. Cobreros, P., Egré, P., Ripley, D., van Rooij, R.: Inferences and metainferences in ST. J. Philos. Log. **49**(6), 1057–1077 (2020)
7. Fitting, M.: The Strict/Tolerant Idea and Bilattices. In: Arieli, O., Zamansky, A. (eds.) Arnon Avron on Semantics and Proof Theory of Non-Classical Logics, Outstanding Contributions to Logic, pp. 167–191. Springer International Publishing, Cham (2021)
8. Font, J.M.: Abstract Algebraic Logic: An Introductory Textbook. College Publications (2016)
9. Frankowski, S.: Formalization of a plausible inference. Bullet. Sect. Logic **33**(1), 41–52 (2004)
10. Gentzen, G.: Investigations into Logical Deduction. American Philosophical Quarterly **1**(4), 288–306 ([1935] 1964)
11. Nair, S.: Must good reasoning satisfy cumulative transitivity? Philos. Phenomenol. Res. **98**(1), 123–146 (2019)
12. Re, B. D., Szmuc, D., Chemla, E., Egré, P.: On Three-valued preservations of classical logic. Rev. Symb. Logic 1–26 (2023)
13. Ripley, E.: Paradoxes and failures of cut. Australas. J. Philos. **91**(1), 139–164 (2013)
14. Ripley, E.: On the 'transitivity' of consequence relations. J. Log. Comput. **28**(2), 433–450 (2018)
15. Tarski, A.: On some fundamental concepts of metamathematics. In: Logic, Semantics, Metamathematics, Papers from 1923 to 1938 by Alfred Tarski, Translated by J. H. Woodger, pp. 30–37. Hackett Publishing Company (1930)
16. van Rooij, R.: Vagueness, tolerance and non-transitive entailment. In: Understanding Vagueness, pp. 205–221. College Publications (2011)
17. Zardini, E.: A model of tolerance. Studia Logica **90**(3), 337–368 (2008)

Chapter 11
Variations of Axioms K and S in Substructural Logics

Takahiro Seki

Abstract In this paper, we discuss variations of K and S from the syntactical view of substructural logics without assuming both exchange and associativity. K and S are usually adopted as axioms on implication in Hilbert-style calculus for classical logic and intuitionistic logic. Particularly, K and S are related to weakening and contraction-like rules in sequent calculus **LK** and **LJ**. In substructural logics without the exchange rule, several variations of K and S can be considered because two types of implication exist. We present non-commutative and non-associative substructural logics with variations of K and S using Hilbert-style calculus and consider consecution calculus, which is a type of Gentzen-style calculus, algebraic semantics, the derivability of structural rules, such as contraction, exchange, and associativity, and the hierarchy.

Keywords Axioms K and S · Substructural logic · Associativity · Commutativity

11.1 Introduction

Substructural logics are logics obtained from sequent calculus **LK** and **LJ** as a result of removing structural rules such as weakening, contraction, and exchange rules, and sometimes adding further axioms. As an important subclass of substructural logics, Lambek calculus, BCK logic, relevant logics, many-valued logics, fuzzy logics, and linear logic exist. Basic literature on substructural logics includes books by Bimbó [1], Cintula and Noguera [4], Galatos et al. [7], Ono [9], Paoli [10], Restall [11], and Schroder-Heister and Došen [13].

In Hilbert-style calculus for classical logic and intuitionistic logic, the axioms

$$\begin{aligned} &\mathsf{K} : \alpha \to (\beta \to \alpha), \\ &\mathsf{S} : [\alpha \to (\beta \to \delta)] \to [(\alpha \to \beta) \to (\alpha \to \delta)] \end{aligned}$$

T. Seki (✉)
Headquarters for Management Strategy University Evaluation Center, Niigata University, Nishi-ku, Niigata, Japan
e-mail: tseki@adm.niigata-u.ac.jp

K. Sano et al. (eds.), *Exploring Negation, Modality and Proof*, Logic in Asia: Studia Logica Library,
https://doi.org/10.1007/978-981-96-8566-0_11

are included and many theorems on implication can be derived from them with the help of modus ponens. Particularly, K and S are related to weakening and contraction-like rules in sequent calculus **LK** and **LJ**.

In substructural logics, the logical connective $\cdot$, called *fusion* (or *multiplicative conjunction*), is added to the language. It is well known that fusion and implication are residual. Idempotency, commutativity, and associativity of fusion, which are related to structural rules, hold in classical logic and intuitionistic logic because fusion and (additive) conjunction are identifiable; however, some of them do not necessarily hold in substructural logics.

In substructural logics without exchange, two types of implications, denoted by $\backslash$ and $/$ in accordance with [7], must be considered. Therefore, several variations of K and S exist in which each $\rightarrow$ is replaced by $\backslash$ or $/$ in such logics. Then a question arises regarding whether formulas related to exchange and associativity rules are still derivable in substructural logics with such variations.

In this paper, we discuss substructural logics with variations of K and S without assuming both associativity and exchange using Gentzen-style calculus and an algebraic method. Related studies on *non-associative* substructural logics other than [1, 4, 10, 11, 13], which partially discussed such logics, are as follows: Galatos and Ono [8] introduced Gentzen and Hilbert-style calculi and discussed algebraizable logic. Galatos and Jipsen [6] extended the results in [8] and explored the connection between Gentzen-style calculi and relational semantics. Cintula and Noguera et al. [2–4] introduced Hilbert-style calculi different from [8] and studied them using algebraic methods. Seki [12] introduced a Hilbert-style calculus for logics weaker than [2–4, 8] and discussed metacompleteness. In these studies, variations of K and S were not fully considered. As another approach to non-associative substructural logics, Dunn and Meyer [5] introduced structurally free logics, which can be given by Gentzen-style calculus for non-associative substructural logics with combinatory rules involving K and S. The development of structurally free substructural logics was summarized in [1].

This paper is structured as follows: In Sect. 11.2, we present the Hilbert-style formulation of the basic logics **BSL** and **SL** in which we do not assume all structural rules including associativity, and variations of K and S. Roughly, **SL** is equivalent to the basic non-associative substructural logic in both [4, 8], and **BSL**, which originates from [12], is obtained from **SL** by removing left unitness on 1. In Sect. 11.3, we provide a Gentzen-style calculus for **BSL** and **SL**, and variations of K and S using consecution calculus, and discuss their cut elimination. In Sect. 11.4, we provide algebraic semantics for **BSL**, **SL**, and their extensions using variations of K and S. In Sect. 11.5, we consider structural properties such as contractivity, commutativity, and associativity in logics obtained from **BSL** and **SL** by adding variations of K and S. In Sect. 11.6, we consider the hierarchy of such logics. In Sect. 11.7, we consider the admissibility of the identity axiom $\alpha\backslash\alpha$ in logics obtained from **BSL** and **SL** by adding variations of K and S and by removing the identity axiom.

11.2 Logics and Variations of K and S

The language of (propositional) substructural logic consists of (i) individual variables; (ii) binary logical connectives $\wedge$ (*and*), $\vee$ (*or*), $\cdot$ (*fusion*), $\backslash$ (*left division*), and / (*right division*); and (iii) constant 1. Formulas are defined in the usual manner and denoted, for example, by $\alpha, \beta, \gamma, \delta$. We adopt the convention of writing $\alpha\beta$ for $\alpha \cdot \beta$, and in the absence of parentheses, we assume that $\cdot$ is performed first. Hence, for example, $(\alpha\beta)\gamma\backslash\delta$ is a simplification of $[(\alpha \cdot \beta) \cdot \gamma]\backslash\delta$.

The logic **BSL** is defined as follows:

(a) Axioms

(A1) $\alpha\backslash\alpha$
(A2) $(\alpha \wedge \beta)\backslash\alpha$
(A3) $(\alpha \wedge \beta)\backslash\beta$
(A4) $[(\alpha\backslash\beta) \wedge (\alpha\backslash\delta)]\backslash[\alpha\backslash(\beta \wedge \delta)]$
(A5) $\alpha\backslash(\alpha \vee \beta)$
(A6) $\beta\backslash(\alpha \vee \beta)$
(A7) $[(\alpha\backslash\delta) \wedge (\beta\backslash\delta)]\backslash[(\alpha \vee \beta)\backslash\delta]$
(A8) 1

(b) Rules of inference

$$\text{(R1)}\ \frac{\alpha \quad \alpha\backslash\beta}{\beta} \quad \text{(R2)}\ \frac{\alpha \quad \beta}{\alpha \wedge \beta} \quad \text{(R3)}\ \frac{\alpha\backslash\beta \quad \beta\backslash\delta}{\alpha\backslash\delta} \quad \text{(R4)}\ \frac{\alpha\beta\backslash\delta}{\beta\backslash(\alpha\backslash\delta)}$$

$$\text{(R5)}\ \frac{\beta\backslash(\alpha\backslash\delta)}{\alpha\backslash(\delta/\beta)} \quad \text{(R6)}\ \frac{\alpha\backslash(\delta/\beta)}{\alpha\beta\backslash\delta} \quad \text{(R7)}\ \frac{\alpha}{1\backslash\alpha}.$$

Note that **BSL** is the propositional fragment of **BSL** in [12]. Additionally, formulas (T1) and (T2) are theorems of **BSL**, and rules of inference (Q1)–(Q10) are derivable in **BSL**:

$$\text{(T1)}\ \alpha\backslash[(\beta/\alpha)\backslash\beta] \quad \text{(T2)}\ \alpha\backslash[\beta/(\alpha\backslash\beta)] \quad \text{(Q1)}\ \frac{\beta/\alpha \quad \alpha}{\beta} \quad \text{(Q2)}\ \frac{\beta\backslash(\alpha\backslash\delta)}{\alpha\beta\backslash\delta}$$

$$\text{(Q3)}\ \frac{\alpha\backslash(\delta/\beta)}{\beta\backslash(\alpha\backslash\delta)} \quad \text{(Q4)}\ \frac{\alpha\beta\backslash\delta}{\alpha\backslash(\delta/\beta)} \quad \text{(Q5)}\ \frac{\alpha\backslash\beta}{\alpha\delta\backslash\beta\delta} \quad \text{(Q6)}\ \frac{\alpha\backslash\beta}{\delta\alpha\backslash\delta\beta}$$

$$\text{(Q7)}\ \frac{\alpha\backslash\beta}{(\delta\backslash\alpha)\backslash(\delta\backslash\beta)} \quad \text{(Q8)}\ \frac{\alpha\backslash\beta}{(\beta\backslash\delta)\backslash(\alpha\backslash\delta)} \quad \text{(Q9)}\ \frac{\alpha\backslash\beta}{(\alpha/\delta)\backslash(\beta/\delta)} \quad \text{(Q10)}\ \frac{\alpha\backslash\beta}{(\delta/\beta)\backslash(\delta/\alpha)}.$$

We obtain the logic **SL** from **BSL** by adding the rules of inference:

$$\text{(R8)}\ \frac{\alpha\backslash\beta}{\beta/\alpha} \quad \text{(R9)}\ \frac{\beta/\alpha}{\alpha\backslash\beta}.$$

Note that **SL** is equivalent to **sHL** in [8] and **SL** in [2–4] restricted to our language.

We consider extensions of **BSL** and **SL** by adding variations of K and S as axioms. Let **L** be either **BSL** or **SL**, and X be a set that consists of variations of K and S. Then $\mathbf{L} + X$ denotes the logic we obtain from **L** by adding all elements of X as axioms. In particular, if X is a singleton set, we write $\mathbf{SL} + (\mathrm{K1})$ instead of $\mathbf{SL} + \{(\mathrm{K1})\}$, for example. For logics **L1** and **L2**, $\mathbf{L1} = \mathbf{L2}$ indicates that α is a theorem of **L1** if and only if it is a theorem of **L2**.

Because all axioms except (A8) are of the form $\alpha\backslash\beta$, such variations are expected to have the form $\alpha\backslash\beta$. The *implicational variations* of K and S are obtained as follows: the outermost implication is replaced by $\backslash$ and the other implications are replaced by either $\backslash$ or $/$. Then implicational variations (K1) and (K2) of K, and (S1)–(S16) and (S1)′–(S16)′ of S exist as presented in Table 11.1. We remark that (K1) and (K2) can be replaced by $\beta\backslash(\alpha/\alpha)$ and $\beta\backslash(\alpha\backslash\alpha)$, respectively, which have the same form as K', given (R5) and (Q3).

However, the implicational variations of S converge to (S1)–(S16) because the following holds:

Lemma 11.1 *Let* **L** *be either* **BSL** *or* **SL**. *For* $i = 1, \cdots, 16$, $\mathbf{L} + (\mathrm{S}i) = \mathbf{L} + (\mathrm{S}i)'$.

Proof Let γ/β be β in $(\mathrm{S}i)$. Then we can see that $(\mathrm{S}i)'$ is a theorem of $\mathbf{L} + (\mathrm{S}i)$. Additionally, let $\beta\backslash\gamma$ be β in $(\mathrm{S}i)'$. Then we can see that $(\mathrm{S}i)$ is a theorem of $\mathbf{L} + (\mathrm{S}i)'$. We remark that we use the fact that (T1) and (T2) are theorems of **BSL** in this proof. □

Table 11.1 Implicational variations of K and S

Name	Implicational variation	Name	Implicational variation
(K1)	$\alpha\backslash(\beta\backslash\alpha)$	(K2)	$\alpha\backslash(\alpha/\beta)$
(S1)	$[\alpha\backslash(\beta\backslash\gamma)]\backslash[(\alpha\backslash\beta)\backslash(\alpha\backslash\gamma)]$	(S1)′	$[\alpha\backslash(\gamma/\beta)]\backslash[(\alpha\backslash\gamma)/(\alpha\backslash\beta)]$
(S2)	$[\alpha\backslash(\beta\backslash\gamma)]\backslash[(\alpha\backslash\gamma)/(\alpha\backslash\beta)]$	(S2)′	$[\alpha\backslash(\gamma/\beta)]\backslash[(\alpha\backslash\beta)\backslash(\alpha\backslash\gamma)]$
(S3)	$[\alpha\backslash(\beta\backslash\gamma)]\backslash[(\alpha\backslash\beta)\backslash(\gamma/\alpha)]$	(S3)′	$[\alpha\backslash(\gamma/\beta)]\backslash[(\gamma/\alpha)/(\alpha\backslash\beta)]$
(S4)	$[\alpha\backslash(\beta\backslash\gamma)]\backslash[(\gamma/\alpha)/(\alpha\backslash\beta)]$	(S4)′	$[\alpha\backslash(\gamma/\beta)]\backslash[(\alpha\backslash\beta)\backslash(\gamma/\alpha)]$
(S5)	$[\alpha\backslash(\beta\backslash\gamma)]\backslash[(\beta/\alpha)\backslash(\alpha\backslash\gamma)]$	(S5)′	$[(\gamma/\beta)/\alpha]\backslash[(\alpha\backslash\gamma)/(\alpha\backslash\beta)]$
(S6)	$[\alpha\backslash(\beta\backslash\gamma)]\backslash[(\alpha\backslash\gamma)/(\beta/\alpha)]$	(S6)′	$[(\gamma/\beta)/\alpha]\backslash[(\alpha\backslash\beta)\backslash(\alpha\backslash\gamma)]$
(S7)	$[\alpha\backslash(\beta\backslash\gamma)]\backslash[(\beta/\alpha)\backslash(\gamma/\alpha)]$	(S7)′	$[(\gamma/\beta)/\alpha]\backslash[(\gamma/\alpha)/(\alpha\backslash\beta)]$
(S8)	$[\alpha\backslash(\beta\backslash\gamma)]\backslash[(\gamma/\alpha)/(\beta/\alpha)]$	(S8)′	$[(\gamma/\beta)/\alpha]\backslash[(\alpha\backslash\beta)\backslash(\gamma/\alpha)]$
(S9)	$[(\beta\backslash\gamma)/\alpha]\backslash[(\alpha\backslash\beta)\backslash(\alpha\backslash\gamma)]$	(S9)′	$[\alpha\backslash(\gamma/\beta)]\backslash[(\alpha\backslash\gamma)/(\beta/\alpha)]$
(S10)	$[(\beta\backslash\gamma)/\alpha]\backslash[(\alpha\backslash\gamma)/(\alpha\backslash\beta)]$	(S10)′	$[\alpha\backslash(\gamma/\beta)]\backslash[(\beta/\alpha)\backslash(\alpha\backslash\gamma)]$
(S11)	$[(\beta\backslash\gamma)/\alpha]\backslash[(\alpha\backslash\beta)\backslash(\gamma/\alpha)]$	(S11)′	$[\alpha\backslash(\gamma/\beta)]\backslash[(\gamma/\alpha)/(\beta/\alpha)]$
(S12)	$[(\beta\backslash\gamma)/\alpha]\backslash[(\gamma/\alpha)/(\alpha\backslash\beta)]$	(S12)′	$[\alpha\backslash(\gamma/\beta)]\backslash[(\beta/\alpha)\backslash(\gamma/\alpha)]$
(S13)	$[(\beta\backslash\gamma)/\alpha]\backslash[(\beta/\alpha)\backslash(\alpha\backslash\gamma)]$	(S13)′	$[(\gamma/\beta)/\alpha]\backslash[(\alpha\backslash\gamma)/(\beta/\alpha)]$
(S14)	$[(\beta\backslash\gamma)/\alpha]\backslash[(\alpha\backslash\gamma)/(\beta/\alpha)]$	(S14)′	$[(\gamma/\beta)/\alpha]\backslash[(\beta/\alpha)\backslash(\alpha\backslash\gamma)]$
(S15)	$[(\beta\backslash\gamma)/\alpha]\backslash[(\beta/\alpha)\backslash(\gamma/\alpha)]$	(S15)′	$[(\gamma/\beta)/\alpha]\backslash[(\gamma/\alpha)/(\beta/\alpha)]$
(S16)	$[(\beta\backslash\gamma)/\alpha]\backslash[(\gamma/\alpha)/(\beta/\alpha)]$	(S16)′	$[(\gamma/\beta)/\alpha]\backslash[(\beta/\alpha)\backslash(\gamma/\alpha)]$

Table 11.2 Fusion variations of K and S

Name	Alter. name	Fusion variation	Name	Alter. name	Fusion variation
(KF1)	(LW)	$\beta\alpha\backslash\alpha$	(KF2)	(RW)	$\alpha\beta\backslash\alpha$
(SF1)	(SAE)	$\alpha(\beta\gamma)\backslash(\alpha\beta)(\alpha\gamma)$	(SF9)	(SAG)	$\alpha(\beta\gamma)\backslash(\alpha\beta)(\gamma\alpha)$
(SF2)	(SBE)	$\alpha(\gamma\beta)\backslash(\alpha\beta)(\alpha\gamma)$	(SF10)	(SBG)	$\alpha(\gamma\beta)\backslash(\alpha\beta)(\gamma\alpha)$
(SF3)	(SCE)	$(\beta\gamma)\alpha\backslash(\alpha\beta)(\alpha\gamma)$	(SF11)	(SCG)	$(\beta\gamma)\alpha\backslash(\alpha\beta)(\gamma\alpha)$
(SF4)	(SDE)	$(\gamma\beta)\alpha\backslash(\alpha\beta)(\alpha\gamma)$	(SF12)	(SDG)	$(\gamma\beta)\alpha\backslash(\alpha\beta)(\gamma\alpha)$
(SF5)	(SAF)	$\alpha(\beta\gamma)\backslash(\beta\alpha)(\alpha\gamma)$	(SF13)	(SAH)	$\alpha(\beta\gamma)\backslash(\beta\alpha)(\gamma\alpha)$
(SF6)	(SBF)	$\alpha(\gamma\beta)\backslash(\beta\alpha)(\alpha\gamma)$	(SF14)	(SBH)	$\alpha(\gamma\beta)\backslash(\beta\alpha)(\gamma\alpha)$
(SF7)	(SCF)	$(\beta\gamma)\alpha\backslash(\beta\alpha)(\alpha\gamma)$	(SF15)	(SCH)	$(\beta\gamma)\alpha\backslash(\beta\alpha)(\gamma\alpha)$
(SF8)	(SDF)	$(\gamma\beta)\alpha\backslash(\beta\alpha)(\alpha\gamma)$	(SF16)	(SDH)	$(\gamma\beta)\alpha\backslash(\beta\alpha)(\gamma\alpha)$

We can also consider the *fusion variations*, which are of the form $\alpha\backslash\beta$, where α and β contain only fusion. Then fusion variations (KF1) and (KF2) of K, and (SF1)–(SF16) of S are as presented in Table 11.2. The alternative names (LW) and (RW) of (KFi) depend on whether β is a left or right component of fusion in the antecedent, and (SXY) of (SFi) is given as follows: (a) X represents A, B, C, and D if the antecedent is $\alpha(\beta\gamma)$, $\alpha(\gamma\beta)$, $(\beta\gamma)\alpha$, and $(\gamma\beta)\alpha$, respectively, and (b) Y represents E, F, G, and H if the succedent is $(\alpha\beta)(\alpha\gamma)$, $(\beta\alpha)(\alpha\gamma)$, $(\alpha\beta)(\gamma\alpha)$, and $(\beta\alpha)(\gamma\alpha)$, respectively.

Theorem 11.1 *Let* **L** *be either* **BSL** *or* **SL**.

(1) *For* $i = 1, 2$, $\mathbf{L} + (\mathrm{K}i) = \mathbf{L} + (\mathrm{KF}i)$.
(2) *For* $i = 1, \cdots, 16$, $\mathbf{L} + (\mathrm{S}i) = \mathbf{L} + (\mathrm{SF}i)$.

Proof (1) is easily proved using (Q3), (R4), (R6), and (Q5). We present a proof of (2). Let $\mathrm{A} = \{1, 5, 9, 13\}$, $\mathrm{B} = \{2, 6, 10, 14\}$, $\mathrm{C} = \{3, 7, 11, 15\}$, $\mathrm{D} = \{4, 8, 12, 16\}$, $\mathrm{E} = \{1, 2, 3, 4\}$, $\mathrm{F} = \{5, 6, 7, 8\}$, $\mathrm{G} = \{9, 10, 11, 12\}$, and $\mathrm{H} = \{13, 14, 15, 16\}$.

First, we consider the logic $\mathbf{L} + (\mathrm{S}i)$. Let β_0 be $\alpha\beta$ for $i \in \mathrm{E} \cup \mathrm{G}$ and $\beta\alpha$ for $i \in \mathrm{F} \cup \mathrm{H}$; and γ_0 be $(\alpha\beta)(\alpha\gamma)$ for $i \in \mathrm{E}$, $(\beta\alpha)(\alpha\gamma)$ for $i \in \mathrm{F}$, $(\alpha\beta)(\gamma\alpha)$ for $i \in \mathrm{G}$, and $(\beta\alpha)(\gamma\alpha)$ for $i \in \mathrm{H}$. Then $\gamma\backslash[\alpha\backslash(\beta_0\backslash\gamma_0)]$ and $\gamma\backslash[(\beta_0\backslash\gamma_0)/\alpha]$ are theorems for $i \in \mathrm{E} \cup \mathrm{F}$ and $i \in \mathrm{G} \cup \mathrm{H}$, respectively. Moreover, $\beta\backslash(\alpha\backslash\beta_0)$ and $\beta\backslash(\beta_0/\alpha)$ are theorems for $i \in \mathrm{E} \cup \mathrm{G}$ and $i \in \mathrm{F} \cup \mathrm{H}$, respectively. Let β_0 and γ_0 be β and γ, respectively, in (Si). Then we can show that $\gamma\backslash[\beta\backslash(\alpha\backslash\gamma_0)]$, $\gamma\backslash[(\alpha\backslash\gamma_0)/\beta]$, $\gamma\backslash[\beta\backslash(\gamma_0/\alpha)]$, and $\gamma\backslash[(\gamma_0/\alpha)/\beta]$ are theorems for $i \in \mathrm{A}$, $i \in \mathrm{B}$, $i \in \mathrm{C}$, and $i \in \mathrm{D}$, respectively. By (R6) and (Q3), (SFi) is a theorem of $\mathbf{L} + (\mathrm{S}i)$.

Next, we consider the logic $\mathbf{L} + (\mathrm{SF}i)$. Let β_1 be $\alpha\backslash\beta$ for $i \in \mathrm{E} \cup \mathrm{G}$ and β/α for $i \in \mathrm{F} \cup \mathrm{H}$; and γ_1 be $\alpha\backslash(\beta\backslash\gamma)$ for $i \in \mathrm{E} \cup \mathrm{F}$ and $(\beta\backslash\gamma)/\alpha$ for $i \in \mathrm{G} \cup \mathrm{H}$. Then $(\alpha\beta_1)(\alpha\gamma_1)\backslash\gamma$, $(\beta_1\alpha)(\alpha\gamma_1)\backslash\gamma$, $(\alpha\beta_1)(\gamma_1\alpha)\backslash\gamma$, and $(\beta_1\alpha)(\gamma_1\alpha)\backslash\gamma$ are theorems for $i \in \mathrm{E}$, $i \in \mathrm{F}$, $i \in \mathrm{G}$, and $i \in \mathrm{H}$, respectively. Let β_1 and γ_1 be β and γ, respectively, in (SFi). Then we can show that $\alpha(\beta_1\gamma_1)\backslash\gamma$, $\alpha(\gamma_1\beta_1)\backslash\gamma$, $(\beta_1\gamma_1)\alpha\backslash\gamma$, and $(\gamma_1\beta_1)\alpha\backslash\gamma$

are theorems for $i \in$ A, $i \in$ B, $i \in$ C, and $i \in$ D, respectively. By (R4) and (Q5), (Si) is a theorem of $\mathbf{L} + (\mathrm{SF}i)$. □

Therefore, it is sufficient for us to consider only logics obtained from **BSL** and **SL** by adding fusion variations as axioms.

11.3 Consecution Calculus

Gentzen-style calculus for non-associative substructural logics can be given by consecution calculus instead of the usual sequent calculus, such as **LJ** and **FL**, that is, the full Lambek calculus. For general information on consecution calculus, see [1].

To introduce consecution calculus, we prepare some notions and notation. *Structures* are defined as follows:

(i) An empty sequence is a structure, called an *empty* structure.
(ii) A formula α is a structure.
(iii) If A and B are structures, then $(A;\ B)$ is also a structure.

When A is a structure and ε is an empty structure, structures $(A; \varepsilon)$ and $(\varepsilon; A)$ are defined by A. We remark that empty structures are allowed. For a structure A and formula α, an expression of the form '$A \Rightarrow \alpha$' is called a *consecution.*

The notation $\Gamma[A]$ indicates one particular occurrence of the non-empty structure A in the structure Γ; however, in (cut) defined below, $\Gamma[A]$ in the lower consecution indicates that the result replaces γ of $\Gamma[\gamma]$ in the upper consecution by A, which may be empty in **LSL**.

The consecution calculus **LBSL** is defined as follows, where Γ, A, and B denote structures:

(a) Axiom (Id) $\alpha \Rightarrow \alpha$ (Unit) $\Rightarrow 1$
(b) Cut rules

$$\frac{A \Rightarrow \gamma \quad \Gamma[\gamma] \Rightarrow \delta}{\Gamma[A] \Rightarrow \delta}(\text{cut}) \qquad \frac{\Rightarrow \gamma \quad \Gamma[(A;\gamma)] \Rightarrow \delta}{\Gamma[A] \Rightarrow \delta}(\text{cut0})$$

(c) Rules for logical connectives and the constant

$$\frac{\Gamma[\alpha] \Rightarrow \delta}{\Gamma[\alpha \wedge \beta] \Rightarrow \delta}(\wedge \Rightarrow) \qquad \frac{\Gamma[\beta] \Rightarrow \delta}{\Gamma[\alpha \wedge \beta] \Rightarrow \delta}(\wedge \Rightarrow) \qquad \frac{A \Rightarrow \alpha \quad A \Rightarrow \beta}{A \Rightarrow \alpha \wedge \beta}(\Rightarrow \wedge)$$

$$\frac{\Gamma[\alpha] \Rightarrow \delta \quad \Gamma[\beta] \Rightarrow \delta}{\Gamma[\alpha \vee \beta] \Rightarrow \delta}(\vee \Rightarrow) \qquad \frac{A \Rightarrow \alpha}{A \Rightarrow \alpha \vee \beta}(\Rightarrow \vee) \qquad \frac{A \Rightarrow \beta}{A \Rightarrow \alpha \vee \beta}(\Rightarrow \vee)$$

$$\frac{A \Rightarrow \alpha \quad \Gamma[\beta] \Rightarrow \delta}{\Gamma[(A;\alpha\backslash\beta)] \Rightarrow \delta}(\backslash \Rightarrow) \qquad \frac{(\alpha; A) \Rightarrow \beta}{A \Rightarrow \alpha\backslash\beta}(\Rightarrow \backslash)$$

$$\frac{A \Rightarrow \alpha \quad \Gamma[\beta] \Rightarrow \delta}{\Gamma[(\beta/\alpha; A)] \Rightarrow \delta}(/ \Rightarrow) \qquad \frac{(A; \alpha) \Rightarrow \beta}{A \Rightarrow \beta/\alpha}(\Rightarrow /)$$

$$\frac{\Gamma[(\alpha;\beta)] \Rightarrow \delta}{\Gamma[\alpha\beta] \Rightarrow \delta}(\cdot \Rightarrow) \qquad \frac{A \Rightarrow \alpha \quad B \Rightarrow \beta}{(A; B) \Rightarrow \alpha\beta}(\Rightarrow \cdot) \qquad \frac{\Gamma[A] \Rightarrow \delta}{\Gamma[(A; 1)] \Rightarrow \delta}(1\mathrm{R}).$$

Right Unit Condition. In (cut), (cut0), ($\backslash \Rightarrow$), and ($\Rightarrow /$), A must be non-empty, and in ($\Rightarrow \cdot$), if A is empty, then B must be empty.

The right unit condition is necessary for avoiding the following undesired scenarios: (1) $1\backslash\alpha \Rightarrow \alpha$ and $\alpha \Rightarrow 1\alpha$ are provable, and (2) $\Rightarrow \alpha\backslash\beta$ is provable if and only if $\Rightarrow \beta/\alpha$ is provable.

The consecution calculus **LSL** is obtained from **LBSL** by removing (cut0) and the right unit condition, and by adding the following:

$$\frac{\Gamma[A] \Rightarrow \delta}{\Gamma[(1; A)] \Rightarrow \delta}(1\mathrm{L}).$$

Note that **LSL** is equivalent to **GL** in [6, 8].

Lemma 11.2 *If a formula α is a theorem of* **BSL** *and* **SL**, *then a consecution $\Rightarrow \alpha$ is provable in* **LBSL** *and* **LSL**, *respectively.*

Proof We prove the lemma by induction on the length of the proof in **BSL** and **SL**. As an example, we present the case (R6) in **BSL**.

$$\dfrac{\dfrac{\dfrac{\Rightarrow \alpha\backslash(\delta/\beta) \qquad \dfrac{\alpha \Rightarrow \alpha \qquad \dfrac{\delta \Rightarrow \delta \qquad \beta \Rightarrow \beta}{(\delta/\beta;\beta) \Rightarrow \delta}(/ \Rightarrow)}{((\alpha;\alpha\backslash(\delta/\beta));\beta) \Rightarrow \delta}(\backslash \Rightarrow)}{(\alpha;\beta) \Rightarrow \delta}(\mathrm{cut0})}{\alpha\beta \Rightarrow \delta}(\cdot \Rightarrow)}{\Rightarrow \alpha\beta\backslash\delta}(\Rightarrow \backslash)$$

□

To relate consecutions to formulas, we provide *interpretations* ι for consecutions in terms of formulas as follows:

(i) For empty structures ε, $\iota(\varepsilon) = 1$.
(ii) For formulas α, $\iota(\alpha) = \alpha$.
(iii) For structures $(A; B)$, $\iota((A; B)) = \iota(A) \cdot \iota(B)$.
(iv) For structures A and formulas α, $\iota(A \Rightarrow \alpha) = \iota(A)\backslash\iota(\alpha)$.

Then we can prove the following by induction on the length of the proof in **LBSL** and **LSL**.

Lemma 11.3 *If a consecution $A \Rightarrow \alpha$ is provable in* **LBSL** *and* **LSL**, *then $\iota(A \Rightarrow \alpha)$ is a theorem of* **BSL** *and* **SL**, *respectively.*

Theorem 11.2 *A formula α is a theorem of* **BSL** *and* **SL** *if and only if a consecution $\Rightarrow \alpha$ is provable in* **LBSL** *and* **LSL**, *respectively.*

Proof The only-if part follows from Lemma 11.2. The if part is proved as follows: Suppose $\Rightarrow \alpha$ is provable in **LBSL** and **LSL**. By Lemma 11.3, $\iota(\Rightarrow \alpha)$, that is, $1\backslash\alpha$, is a theorem of **BSL** and **SL**, respectively. Therefore, α is a theorem of **BSL** and **SL**, respectively. □

We consider corresponding Gentzen-style rules for variations of K and S. To obtain them, we consider a proof of **LBSL** that ends with a consecution $\Rightarrow \alpha$, where α is any fusion variation.

For (LW) and (RW), we identify the following *weakening rules*, abbreviated as *W-rules*:

$$\frac{\Gamma[A] \Rightarrow \delta}{\Gamma[(B;A)] \Rightarrow \delta}\ (\text{Lw}) \qquad \frac{\Gamma[A] \Rightarrow \delta}{\Gamma[(A;B)] \Rightarrow \delta}\ (\text{Rw}).$$

For (SAE)–(SDH), we identify the following *S-rules*:

$$\frac{\Gamma[((A;B);(A;C))] \Rightarrow \delta}{\Gamma[(A;(B;C))] \Rightarrow \delta}\ (\text{sAE}) \qquad \frac{\Gamma[((A;B);(A;C))] \Rightarrow \delta}{\Gamma[(A;(C;B))] \Rightarrow \delta}\ (\text{sBE})$$

$$\frac{\Gamma[((A;B);(A;C))] \Rightarrow \delta}{\Gamma[((B;C);A)] \Rightarrow \delta}\ (\text{sCE}) \qquad \frac{\Gamma[((A;B);(A;C))] \Rightarrow \delta}{\Gamma[((C;B);A)] \Rightarrow \delta}\ (\text{sDE})$$

$$\frac{\Gamma[((B;A);(A;C))] \Rightarrow \delta}{\Gamma[(A;(B;C))] \Rightarrow \delta}\ (\text{sAF}) \qquad \frac{\Gamma[((B;A);(A;C))] \Rightarrow \delta}{\Gamma[(A;(C;B))] \Rightarrow \delta}\ (\text{sBF})$$

$$\frac{\Gamma[((B;A);(A;C))] \Rightarrow \delta}{\Gamma[((B;C);A)] \Rightarrow \delta}\ (\text{sCF}) \qquad \frac{\Gamma[((B;A);(A;C))] \Rightarrow \delta}{\Gamma[((C;B);A)] \Rightarrow \delta}\ (\text{sDF})$$

$$\frac{\Gamma[((A;B);(C;A))] \Rightarrow \delta}{\Gamma[(A;(B;C))] \Rightarrow \delta}\ (\text{sAG}) \qquad \frac{\Gamma[((A;B);(C;A))] \Rightarrow \delta}{\Gamma[(A;(C;B))] \Rightarrow \delta}\ (\text{sBG})$$

$$\frac{\Gamma[((A;B);(C;A))] \Rightarrow \delta}{\Gamma[((B;C);A)] \Rightarrow \delta}\ (\text{sCG}) \qquad \frac{\Gamma[((A;B);(C;A))] \Rightarrow \delta}{\Gamma[((C;B);A)] \Rightarrow \delta}\ (\text{sDG})$$

$$\frac{\Gamma[((B;A);(C;A))] \Rightarrow \delta}{\Gamma[(A;(B;C))] \Rightarrow \delta}\ (\text{sAH}) \qquad \frac{\Gamma[((B;A);(C;A))] \Rightarrow \delta}{\Gamma[(A;(C;B))] \Rightarrow \delta}\ (\text{sBH})$$

$$\frac{\Gamma[((B;A);(C;A))] \Rightarrow \delta}{\Gamma[((B;C);A)] \Rightarrow \delta}\ (\text{sCH}) \qquad \frac{\Gamma[((B;A);(C;A))] \Rightarrow \delta}{\Gamma[((C;B);A)] \Rightarrow \delta}\ (\text{sDH}).$$

Let **L** be either **LBSL** or **LSL** and R be a set of rules $\{(\text{r}_1), \cdots, (\text{r}_n)\}$. Then $\mathbf{L} + R$ denotes the calculus obtained from **L** by adding all rules that belong to R. In particular, if R is a singleton $\{(\text{r})\}$, we write **LSL**+(r) instead of **LSL** + $\{(\text{r})\}$.

Theorem 11.3 (1) *For each* $\text{Z} \in \{\text{L}, \text{R}\}$, *a formula α is a theorem of* **BSL**+(ZW) *and* **SL**+(ZW) *if and only if a consecution $\Rightarrow \alpha$ is provable in* **LBSL**+(Zw) *and* **LSL**+(Zw), *respectively.*

(2) *For each* X ∈ {A, B, C, D} *and* Y ∈ {E, F, G, H}, *a formula* α *is a theorem of* **BSL**+(SXY) *and* **SL**+(SXY) *if and only if a consecution* $\Rightarrow \alpha$ *is provable in* **LBSL**+(sXY) *and* **LSL**+(sXY), *respectively.*

Proof We only show the correspondence between (LW) and (Lw). Other cases can be proved similarly. The left-hand side shows the essential part of 'from (LW) to (Lw)' and the right-hand side shows 'from (Lw) to (LW)':

$$\dfrac{\dfrac{}{\iota(B)\cdot\iota(A)\backslash\iota(A)}\,(\mathrm{LW}) \qquad \iota(A)\backslash\delta}{\iota(B)\cdot\iota(A)\backslash\delta}\,(\mathrm{Q1}). \qquad\qquad \dfrac{\dfrac{\dfrac{\alpha\Rightarrow\alpha}{(\beta;\alpha)\Rightarrow\alpha}\,(\mathrm{Lw})}{\beta\alpha\Rightarrow\alpha}\,(\cdot\Rightarrow)}{\Rightarrow\beta\alpha\backslash\alpha}\,(\Rightarrow\backslash)$$

□

The cut elimination theorem of **LSL** has already been proved because it is effectively the same calculus as **GL** in [6, 8]. In a similar manner, we can prove the cut elimination theorem of **LBSL**. Furthermore, because W-rules and S-rules are *simple rules in* **LBSL** and **LSL** in the sense of [6], we obtain the following:

Theorem 11.4 *For any subset X of W-rules and S-rules, the cut elimination theorem holds in* **LBSL** + *X and* **LSL** + *X.*

We remark that Theorem 11.4 can be also proved using the proof-theoretical method described in [1, 7, 9], that is, by double induction on the degree and rank. For the calculus not containing S-rules, it is not necessary to introduce alternative rules, such as the mix rule. For the calculus containing S-rules, the theorem is proved by eliminating the multiple cut rule, which is essentially the same as the cut rule.

11.4 Algebraic Semantics

A structure $\langle A, \wedge, \vee, \cdot, \backslash, /, 1\rangle$ is called a **BSL**-*algebra* if it satisfies the following postulates. For all $a, b, c \in A$:

(g1) $\langle A, \wedge, \vee\rangle$ is a lattice;
(g2) $a(b \vee c) = ab \vee ac$;
(g3) $(b \vee c)a = ba \vee ca$;
(g4) $ab \leq c$ iff $b \leq a\backslash c$ iff $a \leq c/b$;
(g5) $a1 = a$,

where $\leq$ denotes the lattice order. Furthermore, we adopt the convention of writing ab for $a \cdot b$, and in the absence of parentheses, we assume that $\cdot$ is performed first.

BSL-algebras satisfying the following postulate are called **SL**-*algebras*. For all $a \in A$:

(g6) $1a = a$.

Note that **BSL**-algebras are **BSL**-algebras in [12], removing the postulate that 0 is an arbitrary element of A. We remark that **BSL**-algebras and **SL**-algebras are regarded as right-unital residuated lattice-ordered groupoids and unital residuated lattice-ordered groupoids, respectively.

Table 11.3 Algebraic postulates for variations of K and S

Variation	Name	Algebraic postulate	Variation	Name	Algebraic postulate
(LW)	(lw)	$ba \leq a$	(RW)	(rw)	$ab \leq a$
(SAE)	(sae)	$a(bc) \leq (ab)(ac)$	(SAG)	(sag)	$a(bc) \leq (ab)(ca)$
(SBE)	(sbe)	$a(cb) \leq (ab)(ac)$	(SBG)	(sbg)	$a(cb) \leq (ab)(ca)$
(SCE)	(sce)	$(bc)a \leq (ab)(ac)$	(SCG)	(scg)	$(bc)a \leq (ab)(ca)$
(SDE)	(sde)	$(cb)a \leq (ab)(ac)$	(SDG)	(sdg)	$(cb)a \leq (ab)(ca)$
(SAF)	(saf)	$a(bc) \leq (ba)(ac)$	(SAH)	(sah)	$a(bc) \leq (ba)(ca)$
(SBF)	(sbf)	$a(cb) \leq (ba)(ac)$	(SBH)	(sbh)	$a(cb) \leq (ba)(ca)$
(SCF)	(scf)	$(bc)a \leq (ba)(ac)$	(SCH)	(sch)	$(bc)a \leq (ba)(ca)$
(SDF)	(sdf)	$(cb)a \leq (ba)(ac)$	(SDH)	(sdh)	$(cb)a \leq (ba)(ca)$

Theorem 11.5 *Let* $\langle A, \wedge, \vee, \cdot, \backslash, /, 1\rangle$ *be a* **BSL**-*algebra or* **SL**-*algebra. For any* $a, b \in A$, $a\backslash b = \max\{x \in A \mid ax \leq b\}$ *and* $b/a = \max\{x \in A \mid xa \leq b\}$.

Proof We present a proof of $a\backslash b = \max\{x \in A \mid ax \leq b\}$. For any $a, b \in A$, let M be the set $\{x \in A \mid ax \leq b\}$. Because $a\backslash b \leq a\backslash b$, $a(a\backslash b) \leq b$ by (g4); hence, $a\backslash b \in M$. For any $c \in M$, $ac \leq b$; hence, $c \leq a\backslash b$ by (g4). Therefore, $a\backslash b = \max M$. □

Let **L** be either **BSL** or **SL**. Valuation v on **L**-algebras and interpretations I associated with v are defined in the usual manner. Then a formula α is *valid in* an **L**-algebra if $1 \leq I(\alpha)$ for any valuation v on the **L**-algebra.

Let X be a set of variations of K and S. **L**-algebras in which all elements of X are valid are called **L** + X*-algebras*. Then given a variation (V) of K and S, **L**+(V)-algebras satisfy the postulates presented in Table 11.3.

Then the following holds:

Theorem 11.6 *Let* **L** *be either* **BSL** *or* **SL**, *and* (V) *be a variation of* K *and* S. *Then* α *is a theorem of* **L** + (V) *if and only if* α *is valid in all* **L**+(V)*-algebras.*

11.5 Associativity, Commutativity, and Contractivity

Substructural logics are called *associative* if both (Ass1) $(\alpha\beta)\gamma\backslash\alpha(\beta\gamma)$ and (Ass2) $\alpha(\beta\gamma)\backslash(\alpha\beta)\gamma$ are theorems, *commutative* if (Comm) $\alpha\beta\backslash\beta\alpha$ is a theorem, and *contractive* if (Cont) $\alpha\backslash\alpha\alpha$ is a theorem. In consecution calculus, associativity, commutativity, and contractivity are formalized as the following rules of inference, that is, both (a1) and (a2), (e), and (c), respectively:

$$\frac{\Gamma[(A; (B; C))] \Rightarrow \delta}{\Gamma[((A; B); C)] \Rightarrow \delta}\,(\mathrm{a1}) \qquad \frac{\Gamma[((A; B); C)] \Rightarrow \delta}{\Gamma[(A; (B; C))] \Rightarrow \delta}\,(\mathrm{a2})$$

$$\frac{\Gamma[(A; B)] \Rightarrow \delta}{\Gamma[(B; A)] \Rightarrow \delta}\,(\mathrm{e}) \qquad \frac{\Gamma[(A; A)] \Rightarrow \delta}{\Gamma[A] \Rightarrow \delta}\,(\mathrm{c}).$$

BSL-algebras and **SL**-algebras $\langle A, \wedge, \vee, \cdot, \backslash, /, 1\rangle$, hereafter denoted by **A**, are called *associative*, *commutative*, and *contractive* if they satisfy (ass) $a(bc) = (ab)c$, (comm) $ab \leq ba$, and (cont) $a \leq aa$, respectively, for any $a, b, c \in A$.

The following results can be proved using both consecution calculus and an algebraic method; we present a proof using the latter.

Theorem 11.7 (1) **BSL** + (LW) = **BSL** + {(LW), (RW)}.
(2) **SL** + (LW) = **SL** + (RW) = **SL** + {(LW), (RW)}.

Proof (1) Let **A** be a **BSL** + (LW)-algebra. For any $a, b \in A$, because $b1 \leq 1$, $ab = a(b1) \leq a1 = a$.
(2) By (1), **SL** + (LW) = **SL** + {(LW), (RW)}. To prove that **SL** + (RW) = **SL** + {(LW), (RW)}, let **A** be a **SL** + (RW)-algebra. For any $a, b \in A$, because $1b \leq 1$, $ba = (1b)a \leq 1a = a$. □

Theorem 11.8 (1) *For* $X \in$ {AE, BE, CE, AF, BF, AH}, **BSL** + (SX) *is contractive.*
(2) *For* $X \in$ {DF, DH}, **BSL** + (SX) *is both commutative and contractive.*
(3) *For* X ∈ {A, C} *and* Y ∈ {E, F, G, H}, **SL** + (SXY) *is contractive.*
(4) *For* X ∈ {B, D} *and* Y ∈ {E, F, G, H}, **SL** + (SXY) *is both commutative and contractive.*

Proof (1) For $X \in$ {AE, BE, CE, AF, BF, AH}, let **A** be a **BSL** + (SX)-algebra.
If $X \in$ {AE, BE}, for any $a \in A$, $a = a(11) \leq (a1)(a1) = aa$.
If $X =$ CE, for any $a \in A$, $a = (a1)1 \leq (1a)(11) = 1a = (11)a \leq (a1)(a1) = aa$.
If $X =$ AF, for any $a \in A$, $a = a(11) \leq (1a)(a1) = (1a)a = (1(a1))a \leq ((a1)(11))a = aa$.
If $X =$ BF, for any $a \in A$, $a = a(11) \leq (1a)(a1) = (1a)a = (1(a1))a \leq ((11)(1a))a = (1(1a))a \leq ((a1)(11))a = aa$.
If $X =$ AH, for any $a \in A$, $a = a(11) \leq (1a)(1a) = (1(a1))(1(a1)) \leq ((a1)(11))((a1)(11)) = aa$.
(2) For $X \in$ {DF, DH}, let **A** be a **BSL** + (SX)-algebra.
If $X =$ DF, for any $a, b \in A$, $ab = (ab)1 \leq (b1)(1a) = b(1a) = b((1a)1) \leq b((a1)(11)) = ba$, and $a = a(11) \leq (11)a \leq (1a)(a1) \leq (a1)(a1) = aa$.
If $X =$ DH, for any $a, b \in A$, $ab = (ab)1 \leq (b1)(a1) = ba$, and $a = a(11) \leq (11)a \leq (1a)(1a) \leq (a1)(a1) = aa$.
(3) For X ∈ {A, C} and Y ∈ {E, F, G, H}, if $b = 1$ and $c = 1$ in algebraic postulates of **SL** + (SXY)-algebra, we can easily prove $a \leq aa$ with the help of (g5) and (g6).
(4) For X ∈ {B, D} and Y ∈ {E, F, G, H}, in algebraic postulates of **SL** + (SXY)-algebra, we can easily prove $cb \leq bc$ and $a \leq aa$, with the help of (g5) and (g6), if $a = 1$ and $b = c = 1$, respectively. □

Lemma 11.4 (Comm) *is a theorem of* **BSL** + {(LW), (Cont)}, **SL** + {(LW), (Cont)}, *and* **SL** + {(RW), (Cont)}.

Proof We present a proof for $\mathbf{BSL} + \{\text{(LW)}, \text{(Cont)}\}$ only. Let $\mathbf{A}$ be a $\mathbf{BSL} + \{\text{(LW)}, \text{(Cont)}\}$-algebra. With the help of Theorem 11.7, for any $a, b \in A$, $ab \leq (ab)(ab) \leq ba$. □

Lemma 11.5 (1) (Ass2) *is a theorem of* $\mathbf{BSL} + \{\text{(Comm)}, \text{(Ass1)}\}$ *and* $\mathbf{SL} + \{\text{(Comm)}, \text{(Ass1)}\}$.
(2) (Ass1) *is a theorem of* $\mathbf{BSL} + \{\text{(Comm)}, \text{(Ass2)}\}$ *and* $\mathbf{SL} + \{\text{(Comm)}, \text{(Ass2)}\}$.

Proof (1) We present a proof for $\mathbf{BSL} + \{\text{(Comm)}, \text{(Ass1)}\}$ only. Let $\mathbf{A}$ be a $\mathbf{BSL} + \{\text{(Comm)}, \text{(Ass1)}\}$-algebra. Because $(ab)c \leq a(bc)$, it is sufficient to prove that $a(bc) \leq (ab)c$ for any $a, b, c \in A$, which is proved as follows: $a(bc) \leq (bc)a \leq (cb)a \leq c(ba) \leq (ba)c \leq (ab)c$.
(2) We present a proof for $\mathbf{BSL} + \{\text{(Comm)}, \text{(Ass2)}\}$ only. Let $\mathbf{A}$ be a $\mathbf{BSL} + \{\text{(Comm)}, \text{(Ass2)}\}$-algebra. Because $a(bc) \leq (ab)c$, it is sufficient to prove that $(ab)c \leq a(bc)$ for any $a, b, c \in A$, which is proved as follows: $(ab)c \leq c(ab) \leq c(ba) \leq (cb)a \leq (bc)a \leq a(bc)$. □

The following theorems show whether associativity, commutativity, and contractivity hold for each substructural logic with variations of both K and S.

Theorem 11.9 *For* $X \in \{\text{AE}, \text{BE}\}$, *logics* $\mathbf{BSL} + \{\text{(RW)}, (\mathsf{S}X)\}$ *are contractive, but neither associative nor commutative.*

Proof (i) Contractivity of $\mathbf{BSL} + \{\text{(RW)}, (\mathsf{S}X)\}$ for $X \in \{\text{AE}, \text{BE}\}$ is trivial by (1) of Theorem 11.8.

(ii) Let $\mathbf{B}$ be the $\mathbf{BSL}$-algebra $\langle\{0, a, b, c, 1\}, \wedge, \vee, \cdot, \backslash, /, 1\rangle$, where the lattice order of $\{0, a, b, c, 1\}$ and $\cdot$ are defined as follows (we remark that $\backslash$ and $/$ can be obtained from $\cdot$ using Theorem 11.5; the same applies in the following):

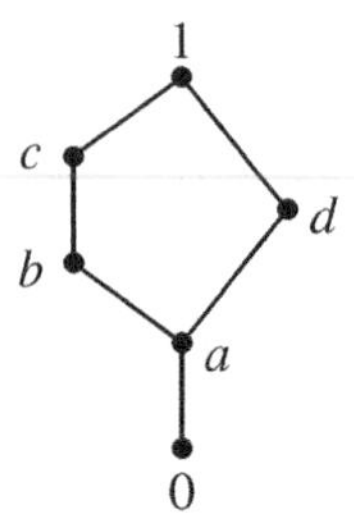

$\cdot$	0	a	b	c	d	1
0	0	0	0	0	0	0
a	0	a	a	a	a	a
b	0	b	b	b	b	b
c	0	b	b	c	c	c
d	0	a	a	c	d	d
1	0	b	b	c	1	1

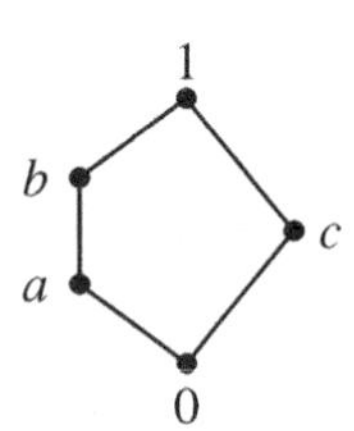

$\cdot$	0	a	b	c	1
0	0	0	0	0	0
a	0	a	a	a	a
b	0	a	b	b	b
c	0	0	c	c	c
1	0	a	1	1	1

Then **B** is a **BSL** + {(RW), (SX)}-algebra. The failure of associativity and commutativity is proved as follows: $(1c)a = 1a = a \nleq 0 = 10 = 1(ca)$ and $1b = 1 \nleq b = b1$, respectively.
The logic **BSL** + {(RW), (SBG)} is contractive, but it is unclear whether it is associative and commutative. Contractivity is proved as follows: let A be a **BSL** + {(RW), (SBG)}-algebra. Then $a = a(11) \leq (a1)(1a) = a(1a) \leq (aa)(1a) \leq aa$, for any a $\in A$. □

Theorem 11.10 *The logic* **BSL** + {(RW), (SAG)} *is associative and contractive, but not commutative.*

Proof (i) Let **A** be a **BSL** + {(RW), (SAG)}-algebra. Contractivity is proved as follows: for any $a \in A$, $a = a(11) \leq (a1)(1a) = a(1a) \leq (a1)(aa) \leq aa$. Associativity is proved, with the help of contractivity, as follows: $a(bc) \leq (ab)(ca) \leq (ab)c$ and $(ab)c \leq (ab)(cc) \leq ((ab)c)(c(ab)) \leq a(c(ab)) \leq a((ca)(bc)) \leq (a(ca))((bc)a) \leq a(bc)$, for any $a, b, c \in A$.

(ii) Let **C** be the **BSL**-algebra $\langle\{0, a, 1\}, \wedge, \vee, \cdot, \backslash, /, 1\rangle$, where $0 < a < 1$ and $\cdot$ is defined as follows:

$\cdot$	0	a	1
0	0	0	0
a	0	a	a
1	0	1	1

Then **C** is a **BSL** + {(RW), (SAG)}-algebra. The failure of commutativity is proved as follows: $1a = 1 \nleq a = a1$. □

Theorem 11.11 *For $X \in$* {CE, DE, AF, BF, DF, CG, DG, AH, BH, DH}, *logics* **BSL** + {(RW), (SX)} *are associative, commutative, and contractive.*

Proof For $X \in$ {CE, AF, BF, AH}, contractivity follows from (1) of Theorem 11.8, and for $X \in$ {DF, DH}, contractivity and commutativity follow from (2) of Theorem 11.8.

For $X \in$ {CE, DE, AF, BF, DF, CG, DG, AH, BH, DH}, let **A** be a **BSL** + {(RW), (SX)}-algebra.

(1) If $X =$ CE, commutativity is proved as follows: for any $a, b \in A$, $ab = (a1)b \leq (ba)(b1) \leq ba$; and associativity follows from Lemma 11.5, commutativity, and the following: $(ab)c \leq (ca)(cb) \leq (ac)(bc) \leq a(bc)$, for any $a, b, c \in A$.
(2) If $X =$ DE, commutativity is proved as follows: for any $a, b \in A$, $ab = (a1)b \leq (b1)(ba) \leq ((ba)1)((ba)b) \leq (ba)1 = ba$; contractivity is proved, with the help of commutativity, as follows: for any $a \in A$, $a = a(11) \leq (11)a \leq (a1)(a1) = aa$; and associativity follows from Lemma 11.5, commutativity, and the following: $(ab)c \leq (cb)(ca) \leq (bc)(ac) \leq (bc)a \leq a(bc)$, for any $a, b, c \in A$.
(3) If $X =$ AF, commutativity is proved as follows: for any $a, b \in A$, $ab = a(b1) \leq (ba)(a1) \leq ba$; and associativity follows from Lemma 11.5, commutativity, and the following: $a(bc) \leq (ba)(ac) \leq (ab)(ca) \leq (ab)c$, for any $a, b, c \in A$.

(4) If $X = \mathrm{BF}$, commutativity is proved as follows: for any $a, b \in A, ab = (ab)(11) \le (1(ab))((ab)1) \le (1(ab))a \le ((b1)(1a))a \le ba$; and associativity follows from Lemma 11.5, commutativity, and the following: $a(bc) \le (ca)(ab) \le c(ab) \le (ab)c$, for any $a, b, c \in A$.
(5) If $X = \mathrm{DF}$, associativity follows from Lemma 11.5, commutativity, and the following: $a(bc) \le (bc)a \le (ca)(ab) \le c(ab) \le (ab)c$, for any $a, b, c \in A$.
(6) If $X = \mathrm{CG}$, commutativity is proved as follows: for any $a, b \in A, ab = (a1)b \le (ba)(1b) \le ba$; contractivity is proved, with the help of commutativity, as follows: for any $a \in A$, $a = a(11) \le (11)a \le (a1)(1a) \le (a1)(a1) = aa$; and associativity follows from Lemma 11.5, commutativity, and the following: $(ab)c \le (ca)(bc) \le (ac)(bc) \le a(bc)$, for any $a, b, c \in A$.
(7) If $X = \mathrm{DG}$, commutativity is proved as follows: for any $a, b \in A, ab = (a1)b \le (b1)(ab) \le ba$; contractivity is proved, with the help of commutativity, as follows: for any $a \in A$, $a = a(11) \le (11)a \le (a1)(1a) \le (a1)(a1) = aa$; and associativity follows from Lemma 11.5, commutativity, and the following: $(ab)c \le (cb)(ac) \le (cb)a \le (bc)a \le a(bc)$, for any $a, b, c \in A$.
(8) If $X = \mathrm{AH}$, commutativity is proved as follows: for any $a, b \in A, ab = a(b1) \le (ba)(1a) \le ba$; and associativity follows from Lemma 11.5, commutativity, and the following: $a(bc) \le (ba)(ca) \le (ba)c \le (ab)c$, for any $a, b, c \in A$.
(9) If $X = \mathrm{BH}$, commutativity is proved as follows: for any $a, b \in A, ab = (ab)(11) \le (1(ab))(1(ab)) \le 1(ab) \le (b1)(a1) = ba$; contractivity is proved, with the help of commutativity, as follows: for any $a \in A$, $a = a(11) \le (1a)(1a) \le (a1)(a1) = aa$; and associativity follows from Lemma 11.5, commutativity, and the following: $a(bc) \le (ca)(ba) \le c(ba) \le c(ab) \le (ab)c$, for any $a, b, c \in A$.
(10) If $X = \mathrm{DH}$, associativity follows from Lemma 11.5, commutativity, and the following: $(ab)c \le (bc)(ac) \le (bc)a \le a(bc)$, for any $a, b, c \in A$. □

Theorem 11.12 *For $X \in \{\mathrm{CF}, \mathrm{CH}\}$, logics* **BSL** + {(RW), (S*X*)} *are not associative, commutative, nor contractive.*

Proof Let **D** be the **BSL**-algebra $\langle \{0, a, b, 1\}, \wedge, \vee, \cdot, \backslash, /, 1 \rangle$, where $0 < a < b < 1$ and $\cdot$ is defined as follows:

$\cdot$	0	a	b	1
0	0	0	0	0
a	0	a	a	a
b	0	a	a	b
1	0	1	1	1

Then **D** is a **BSL** + {(RW), (S*X*)}-algebra. The failure of associativity, commutativity, and contractivity is proved as follows: $b(1b) = b1 = b \neq a = bb = (b1)b$, $1a = 1 \not\le a = a1$, and $b \not\le a = bb$, respectively. □

Theorem 11.13 *For $X \in \{\mathrm{AE}, \mathrm{BE}, \cdots, \mathrm{DH}\} - \{\mathrm{CF}, \mathrm{CH}\}$, logics* **BSL** + {(LW), (S*X*)} *are associative, commutative, and contractive.*

Proof For $X \in \{\text{CE, DE, AF, BF, DF, CG, DG, AH, BH, DH}\}$, associativity, commutativity, and contractivity follow from Theorem 11.11; for $X \in \{\text{AE, BE, BG}\}$, contractivity and commutativity follow from Theorem 11.9 and Lemma 11.4; and for $X \in \{\text{AG}\}$, associativity, contractivity, and commutativity follow from Theorem 11.10 and Lemma 11.4.

To show associativity for $X \in \{\text{AE, BE, BG}\}$, let $\mathbf{A}$ be a $\mathbf{BSL} + \{(\text{LW}), (\text{S}X)\}$-algebra. Associativity follows from Lemma 11.5, commutativity, and the following. For any $a, b, c \in A$:

(1) If $X = \text{AE}$, $a(bc) \leq (ab)(ac) \leq (ab)c$.
(2) If $X = \text{BE}$, $a(bc) \leq a(cb) \leq (ab)(ac) \leq (ab)c$.
(3) If $X = \text{BG}$, $a(bc) \leq a(cb) \leq (ab)(ca) \leq (ab)c$. □

Theorem 11.14 *For $X \in \{\text{CF, CH}\}$, logics $\mathbf{BSL} + \{(\text{LW}), (\text{S}X)\}$ are associative, but neither commutative nor contractive.*

Proof (i) For $X \in \{\text{CF, CH}\}$, let $\mathbf{A}$ be a $\mathbf{BSL} + \{(\text{LW}), (\text{S}X)\}$-algebra. We remark that $\mathbf{A}$ satisfies not only (lw) but also (rw) by Theorem 11.7. Associativity is proved as follows: For any $a, b, c \in A$:

If $X = \text{CF}, a(bc) = (a1)(bc) \leq (a(bc))((bc)1) \leq (ab)(bc) \leq (ab)c$ and $(ab)c \leq (ac)(cb) \leq (ac)b \leq (ab)(bc) \leq a(bc)$.

If $X = \text{CH}, a(bc) = (a1)(bc) \leq (a(bc))(1(bc)) \leq (ab)(1(bc)) \leq (ab)c$ and $(ab)c \leq (ac)(bc) \leq a(bc)$.

(ii) Let $\mathbf{E}$ be the $\mathbf{BSL}$-algebra $\langle\{0, a, 1\}, \wedge, \vee, \cdot, \backslash, /, 1\rangle$, where $0 < a < 1$ and $\cdot$ is defined as follows:

$\cdot$	0	a	1
0	0	0	0
a	0	0	a
1	0	0	1

Then for $X \in \{\text{CF, CH}\}$, $\mathbf{E}$ is a $\mathbf{BSL} + \{(\text{LW}), (\text{S}X)\}$-algebra. The failure of commutativity and contractivity is proved as follows: $a1 = a \not\leq 0 = 1a$ and $a \not\leq 0 = aa$, respectively. □

Theorem 11.15 *For* $\text{Z} \in \{\text{L, R}\}$, $\text{X} \in \{\text{A, B, C, D}\}$ *and* $\text{Y} \in \{\text{E, F, G, H}\}$, *logics* $\mathbf{SL} + \{(\text{ZW}), (\text{SXY})\}$ *are associative, commutative, and contractive.*

Proof Contractivity and commutativity follow from (3) and (4) of Theorem 11.8 and Lemma 11.4. Associativity follows from Theorems 11.13, 11.14, and 11.7. □

11.6 Hierarchy of Logics with Variations of K and S

In the previous section, we considered the associativity, commutativity, and contractivity of substructural logics with both a variation of K and a variation of S. Associative, commutative, and contractive substructural logics with both a variation of K and a variation of S, which include the logics mentioned in Theorems 11.11, 11.13, and 11.15, are equivalent to the positive fragment of intuitionistic logic. By contrast, there are non-associative, non-commutative, or non-contractive substructural logics with both a variation of K and a variation of S. We consider the interrelationships among such logics using an algebraic method.

First, we consider logics obtained from **BSL** + (LW) by adding any non-empty subset of {(SCF), (SCH)}. We remark that all such logics are neither contractive nor commutative.

Theorem 11.16 **BSL** + {(LW), (SCF)} = **BSL** + {(LW), (SCH)} = **BSL**+ {(LW), (SCF), (SCH)}.

Proof For $X \in \{\text{CF}, \text{CH}\}$, let **A** be a **BSL** + {(LW), (SX)}-algebra. We remark that **A** satisfies not only (lw) but also (rw) by Theorem 11.7.

If $X = \text{CF}$, (sch) can be derived as follows: for any $a, b, c \in A$, $(bc)a \leq (ba)(ac) \leq (b(ac))((ac)a) \leq (ba)((ac)a) \leq (ba)(ca)$.

If $X = \text{CH}$, (scf) can be derived as follows: for any $a, b, c \in A$, $(bc)a \leq (ba)(ca) \leq (b(ca))(a(ca)) \leq (ba)(a(ca)) \leq (ba)(ac)$. □

Therefore, the logics obtained from **BSL** + (LW) by adding any non-empty subset of the set of all variations of S are collapsed to two logics: the logic containing only (SCF) and (SCH), and the logic containing all variations of S. We remark that the latter is commutative and contractive, but the former is not.

Next, we consider logics obtained from **BSL** + (RW) by adding any non-empty subset of {(SAE), (SBE), (SCF), (SAG), (SBG), (SCH)}. We remark that all such logics are not commutative.

Contractive logics and non-contractive logics can be clearly distinguished as follows:

Lemma 11.6 *For any non-empty subset S of* {(SCF), (SCH)} *and any non-empty subset T of* {(SAE), (SBE), (SBG), (SAG)}, **BSL** + (RW) + $S \neq$ **BSL** + (RW) + T.

Proof The algebra **D** in the proof of Theorem 11.12 is a **BSL** + (RW) + S-algebra, but not a **BSL** + (RW) + T-algebra. For the latter, because $b(1a) = b1 = b$, $(b1)(ba) = ba = a$, $(ba)(b1) = ab = a$, $(ba)(1b) = a1 = a$, and $(b1)(ab) = ba = a$, the failure of (sae), (sbe), (sbg), and (sag) is proved as follows: $b(1a) \nleq (b1)(ba)$, $b(1a) \nleq (ba)(b1)$, $b(1a) \nleq (ba)(1b)$, and $b(1a) \nleq (b1)(ab)$, respectively. □

For logics **L1** and **L2**, **L1** $\subseteq$ **L2** indicates that all theorems of **L1** are theorems of **L2**, and **L1** $\subset$ **L2** indicates that both **L1** $\subseteq$ **L2** and **L1** $\neq$ **L2**.

For contractive logics, we have the following:

Lemma 11.7 $\mathbf{BSL} + \{(\text{RW}), (\text{SAE})\} \subseteq \mathbf{BSL} + \{(\text{RW}), (\text{SBE})\} \subseteq \mathbf{BSL} + \{(\text{RW}), (\text{SBG})\} \subseteq \mathbf{BSL} + \{(\text{RW}), (\text{SAG})\}$.

Proof For $X \in \{\text{BE}, \text{BG}, \text{AG}\}$, let $\mathbf{A}$ be a $\mathbf{BSL} + \{(\text{RW}), (\text{S}X)\}$-algebra.

(1) If $X = \text{BE}$, (sae) can be derived as follows: for any $a, b, c \in A$, $a(bc) \leq (ac)(ab) \leq ((ac)b)((ac)a) \leq (ab)(ac)$.
(2) If $X = \text{BG}$, (sbe) can be derived as follows: for any $a, b, c \in A$, $a(cb) \leq (ab)(ca) \leq ((ab)a)(c(ab)) \leq (ab)(c(ab)) \leq (ab)((cb)(ac)) \leq ((ab)(ac))((cb)(ab)) \leq (ab)(ac)$.
(3) If $X = \text{AG}$, (sbg) can be derived as follows: for any $a, b, c \in A$, $a(cb) \leq (ac)(ba) \leq ((ac)b)(a(ac)) \leq (ab)(a(ac)) \leq (ab)((aa)(ca)) \leq ((ab)(aa))((ca)(ab)) \leq (ab)(ca)$. □

Lemma 11.8 (1) $\mathbf{BSL} + \{(\text{RW}), (\text{SAE})\} \neq \mathbf{BSL} + \{(\text{RW}), (\text{SBE})\}$.
(2) $\mathbf{BSL} + \{(\text{RW}), (\text{SBE})\} \neq \mathbf{BSL} + \{(\text{RW}), (\text{SBG})\}$.

Proof (1) Let $\mathbf{F}$ be the $\mathbf{BSL}$-algebra $\langle\{0, a, b, 1\}, \wedge, \vee, \cdot, \backslash, /, 1\rangle$, where $0 < a < b < 1$ and $\cdot$ is defined as follows:

$\cdot$	0	a	b	1
0	0	0	0	0
a	0	a	a	a
b	0	b	b	b
1	0	b	1	1

Then $\mathbf{F}$ is a $\mathbf{BSL} + \{(\text{RW}), (\text{SAE})\}$-algebra, but not a $\mathbf{BSL} + \{(\text{RW}), (\text{SBE})\}$-algebra. For the latter, the failure of (sbe) is proved as follows: $1(1a) = 1b = 1 \not\leq b = b1 = (1a)(11)$.
(2) The algebra $\mathbf{B}$ in the proof of Theorem 11.17 is a $\mathbf{BSL} + \{(\text{RW}), (\text{SBE})\}$-algebra, but not a $\mathbf{BSL} + \{(\text{RW}), (\text{SBG})\}$-algebra. For the latter, the failure of (sbg) is proved as follows: $a(cc) = ac = a \not\leq 0 = a0 = (ac)(ca)$.
(3) $\mathbf{BSL} + \{(\text{RW}), (\text{SAG})\}$ is associative by Theorem 11.10, but $\mathbf{BSL} + \{(\text{RW}), (\text{SBG})\}$ is not by Theorem 11.9. □

By Lemmas 11.7 and 11.8, we have the following:

Theorem 11.17 $\mathbf{BSL} + \{(\text{RW}), (\text{SAE})\} \subset \mathbf{BSL} + \{(\text{RW}), (\text{SBE})\} \subset \mathbf{BSL} + \{(\text{RW}), (\text{SBG})\} \subset \mathbf{BSL} + \{(\text{RW}), (\text{SAG})\}$.

For non-contractive logics, we have the following:

Lemma 11.9 *For any non-empty subset S of* $\{(\text{SCF}), (\text{SCH})\}$, $\mathbf{BSL} + (\text{RW}) + S \subseteq \mathbf{BSL} + \{(\text{RW}), (\text{SAG})\}$.

Proof Let **A** be a **BSL** + {(RW), (SAG)}-algebra. We remark that **A** is associative and contractive. (scf) can be derived as follows: for any $a, b, c \in A$, $(bc)a \leq (bc)(aa) \leq ((bc)a)(a(bc)) \leq (ba)(a(bc)) = (ba)((ab)c) \leq (ba)(ac)$. (sch) can be derived as follows: for any $a, b, c \in A$, $(bc)a \leq (bc)(aa) \leq ((bc)a)(a(bc)) \leq (ba)(a(bc)) \leq (ba)((ab)(ca)) = ((ba)(ab))(ca) \leq (ba)(ca)$. □

Lemma 11.10 **BSL** + {(RW), (SCH)} $\not\subseteq$ **BSL** + {(RW), (SCF)}.

Proof (1) Let **G** be the **BSL**-algebra $\langle\{0, a, b, c, 1\}, \wedge, \vee, \cdot, \backslash, /, 1\rangle$, where $0 < a < b < c < 1$ and $\cdot$ is defined as follows:

$\cdot$	0	a	b	c	1
0	0	0	0	0	0
a	0	a	a	a	a
b	0	b	b	b	b
c	0	b	c	c	c
1	0	c	1	1	1

Then **G** is a **BSL** + {(RW), (SCH)}-algebra, but not a **BSL** + {(RW), (SCF)}-algebra. For the latter, the failure of (scf) is proved as follows: $(11)a = 1a = c \not\leq b = ca = (1a)(a1)$. □

By Lemmas 11.6, 11.9 and 11.10, we have the following:

Theorem 11.18 (1) **BSL** + {(RW), (SCH)} $\subset$ **BSL** + {(RW), (SCF), (SCH)} $\subset$ **BSL** + {(RW), (SAG)}.

We have at least the following results for the logics obtained from **BSL** + (RW) by adding a non-empty subset of {(SAE), (SBE), (SBG)} and that of {(SCF), (SCH)}.

Lemma 11.11 (1) **BSL** + {(RW), (SBE), (SCF)} = **BSL** + {(RW), (SBG), (SCF)}.
(2) **BSL** + {(RW), (SBG), (SCF)} = **BSL** + {(RW), (SAG)}.
(3) **BSL** + {(RW), (SBE), (SCH)} = **BSL** + {(RW), (SAG)}.

Proof **BSL** + {(RW), (SBE), (SCF)}, **BSL** + {(RW), (SBG), (SCF)}, **BSL** + {(RW), (SAG)}, and **BSL** + {(RW), (SBE), (SCH)} are denoted by **M1**, **M2**, **M3**, and **M4**, respectively.

(1) Because **M1** $\subseteq$ **M2** by Theorem 11.17, it is sufficient to prove that **M2** $\subseteq$ **M1**. Let **A** be a **M1**-algebra. (sbg) can be derived with the help of contractivity as follows: for any $a, b, c \in A$, $a(cb) \leq (aa)(cb) \leq (a(cb))((cb)a) \leq (a(cb))(ca) \leq ((ab)(ac))(ca) \leq (ab)(ca)$.
(2) Because **M2** $\subseteq$ **M3** by Theorems 11.17 and 11.18, it is sufficient to prove that **M3** $\subseteq$ **M2**. Let **A** be a **M2**-algebra. (sag) can be derived as follows: for any $a, b, c \in A, a(bc) \leq (ac)(ba) \leq (a(ba))((ba)c) \leq (ab)(bc) \leq ((ab)c)(b(ab)) \leq (ab)c \leq (ac)(cb) \leq ((ac)b)(c(ac)) \leq (ab)(ca)$.
(3) Because **M4** $\subseteq$ **M3** by Theorems 11.17 and 11.18, it is sufficient to prove that **M3** $\subseteq$ **M4**. Let **A** be a **M4**-algebra. (sag) can be derived as follows: for any $a, b, c \in A, a(bc) \leq (ac)(ab) \leq (a(ab))(c(ab)) \leq (a(ab))(ca) \leq ((ab)(aa))(ca) \leq (ab)(ca)$. □

Theorem 11.19 (1) *For any non-empty subset* S *of* {(SBE), (SBG)} *and non-empty subset* T *of* {(SCF), (SCH)}, **BSL** + (RW) + $S + T$ = **BSL** + {(RW), (SAG)}.
(2) **BSL** + {(RW), (SAE), (SCH)} $\subset$ **BSL** + {(RW), (SAE), (SCF), (SCH)} $\subset$ **BSL** + {(RW), (SAG)}.

Proof (1) Given Theorems 11.17 and 11.18, **BSL** + {(RW), (SBE), (SCF)} $\subseteq$ **BSL** + (RW) + $S + T \subseteq$ **BSL** + {(RW), (SAG)} or **BSL** + {(RW), (SBE), (SCH)} $\subseteq$ **BSL** + (RW) + $S + T \subseteq$ **BSL** + {(RW), (SAG)}. By Lemma 11.11, **BSL** + (RW) + $S + T$ = **BSL** + {(RW), (SAG)}.
(2) **BSL** + {(RW), (SAE), (SCH)}, **BSL** + {(RW), (SAE), (SCF), (SCH)}, and **BSL** + {(RW), (SAG)} are denoted by **M5**, **M6**, and **M3**, respectively. It is obvious that **M5** $\subseteq$ **M6** $\subseteq$ **M3**. The algebra **G** in the proof of Lemma 11.10 is indeed an **M5**-algebra, but not an **M6**-algebra. Therefore, **M5** $\neq$ **M6**. The algebra **F** in the proof of Lemma 11.8 is indeed an **M6**-algebra, but not an **M3**-algebra. Therefore, **M6** $\neq$ **M3**. □

It remains to be clarified whether **BSL** + {(RW), (SCF)} $\subset$ **BSL** + {(RW), (SCF), (SCH)} or **BSL** + {(RW), (SCF)} = **BSL**+ {(RW), (SCF), (SCH)} and (2)

11.7 Admissibility of the Axiom (A1)

In classical logic and intuitionistic logic, it is well known that the axiom

$$\mathsf{I} : \alpha \to \alpha$$

can be derived from K and S with the help of modus ponens. In substructural logics with variations of both K and S, the axiom (A1), $\alpha\backslash\alpha$, of the basic logics **BSL** and **SL** may be superfluous. Finally, we consider the admissibility of (A1).

BSL$^-$ and **SL**$^-$ denote the logics obtained from **BSL** and **SL**, respectively, by removing the axiom (A1). In **BSL**$^-$ and **SL**$^-$, (T1) and (T2) are not theorems and (Q5)–(Q10) are not derivable; hence, the analogues of Lemma 11.1 and Theorem 11.1 do not hold. Therefore, if **BSL**$^-$ and **SL**$^-$ are taken as basic logics, the implicational variations and fusion variations of S cannot be identified. Moreover, we can consider **BSL**$^-$-algebras and **SL**$^-$-algebras, which are no longer lattice-ordered groupoids.

Theorem 11.20 (A1) *is admissible in* **BSL**$^-$ + (K2), **SL**$^-$ + (K2), *and* **SL**$^-$ + (K1), *that is,* **BSL**$^-$ + (K2) = **BSL** + (K2), **SL**$^-$ + (K2) = **SL** + (K2), *and* **SL**$^-$ + (K1) = **SL** + (K1). *Therefore,* (A1) *is admissible in all logics obtained from any of* **BSL**$^-$ + (K2), **SL**$^-$ + (K2), *or* **SL**$^-$ + (K1) *by adding any subset of the set of implicational variations of* S.

Proof Because both 1 and $1\backslash(\alpha\backslash\alpha)$ are theorems of **BSL**$^-$ + (K2) and **SL**$^-$ + (K2), $\alpha\backslash\alpha$ is also a theorem. Because both 1 and $1\backslash(\alpha/\alpha)$ are theorems of **SL**$^-$ + (K1), α/α is a theorem; hence $\alpha\backslash\alpha$ is a theorem. □

For $\mathbf{BSL}^- + (\mathrm{K1})$ with implicational variations of S, there are still problems, but at least we obtain the following positive results.

Theorem 11.21 *For $S \in \{(\mathrm{S1}), (\mathrm{S2}), (\mathrm{S5}), (\mathrm{S6}), (\mathrm{S9}), (\mathrm{S10}), (\mathrm{S13}), (\mathrm{S14}), (\mathrm{S9})', (\mathrm{S10})', (\mathrm{S13})', (\mathrm{S14})'\}$, (A1) is admissible in $\mathbf{BSL}^- + \{(\mathrm{K1}), S\}$, that is, $\mathbf{BSL}^- + \{(\mathrm{K1}), S\} = \mathbf{BSL} + \{(\mathrm{K1}), S\}$.*

Proof We remark that α/α is a theorem of $\mathbf{BSL}^- + (\mathrm{K1})$ and the rules of inference (Q1) and (R8) are still derivable in $\mathbf{BSL}^- + (\mathrm{K1})$.

For $S \in \{(\mathrm{S1}), (\mathrm{S2}), (\mathrm{S9}), (\mathrm{S10})\}$, let $\alpha\backslash\alpha$ and α be β and γ, respectively, in S. Then we can see that (A1) is a theorem of $\mathbf{BSL}^- + \{(\mathrm{K1}), S\}$.

For $S \in \{(\mathrm{S5}), (\mathrm{S6}), (\mathrm{S13}), (\mathrm{S14}), (\mathrm{S9})', (\mathrm{S10})', (\mathrm{S13})', (\mathrm{S14})'\}$, let α be β and γ in S. Then we can see that (A1) is a theorem of $\mathbf{BSL}^- + \{(\mathrm{K1}), S\}$. □

Acknowledgements We thank the anonymous referees and participants of AWPL2024 for many valuable suggestions, comments, and corrections. This work was supported by JSPS KAKENHI Grant Number 19K03600 and the Research Institute for Mathematical Sciences, an International Joint Usage/Research Center located in Kyoto University. We thank Edanz (https://jp.edanz.com/ac) for editing a draft of this manuscript.

References

1. Bimbó, K.: Proof Theory: Sequent Calculi and Related Formalisms. Chapman & Hall (2014)
2. Cintula, P., Noguera, C.: A general framework for mathematical fuzzy logic. In: Cintula, C., Hájek, P., Noguera, C. (ed.), Handbook of Mathematical Fuzzy Logic, vol. 1, pp. 103–207 (2011)
3. Cintula, P., Horčík, P., Noguera, C.: Non-associative substructural logics and their semilinear extensions: axiomatization and completeness properties. Rev. Symb. Logic **6**, 394–423 (2011)
4. Cintula, P., Noguera, C.: Logic and Implication: An Introduction to the General Algebraic Study of Non-classical Logics, Trends in Logic, vol. 57. Springer (2021)
5. Dunn, J.M., Meyer, R.K.: Combinators and structurally free logic. Logic J. IGPL **5**, 505–537 (1997)
6. Galatos, N., Jipsen, P.: Residuated frames with applications to decidability. Trans. Am. Math. Soc. **365**, 1219–1249 (2013)
7. Galatos, N., Jipsen, P., Kowalski, T., Ono, H.: Residuated Lattices: An Algebraic Glimpse at Substructural Logics. Elsevier (2007)
8. Galatos, N., Ono, H.: Cut elimination and strong separation for substructural logics: an algebraic approach. Ann. Pure Appl. Logic **161**, 1097–1133 (2010)
9. Ono, H.: Proof Theory and Algebra in Logic. Springer (2019)
10. Paoli, F.: Substructural Logics: A Primer. Kluwer Academic Publishers (2002)
11. Restall, G.: An Introduction to Substructural Logics. Routledge (2000)
12. Seki, T.: Metacompleteness of substructural logics. Stud. Logica. **100**, 1175–1199 (2012)
13. Schroeder-Heister, P., Došen, K. (ed.): Substructural Logics. Oxford University Press (1993)

Chapter 12
Leśniewski's Ontology Satisfies Interpolation

Andrzej Indrzejczak

Abstract The interpolation theorem of Craig is one of the most important results in formal logic. In the paper it is shown that it holds for elementary ontology of Leśniewski (LO), commonly regarded as the most comprehensive calculus of names. LO is a theoretical basis for mereology, an alternative foundation of mathematics, and can be also used as an efficient tool for direct formalisation of reasoning in natural languages. Showing that it satisfies interpolation increases its credibility of a well-behaved logical system. A cut-free analytic sequent calculus GO is used for this aim and a Maehara-style constructive proof of this result is provided.

Keywords Leśniewski's ontology · Craig · Maehara · Mereology · Sequent calculus · Interpolation theorem

12.1 Introduction

The aim of the paper is to provide a constructive proof of the Craig's interpolation theorem for the ontology of Leśniewski (LO), a general calculus of names proposed as a formalisation of logic alternative to the Fregean paradigm. In a nutshell, LO is a theory of the binary predicate ε understood as a formalisation of the Greek 'esti' or Latin 'est'. LO in the original form was introduced as a formal basis for developing another, better known theory of Leśniewski—mereology [24]. Thus LO was introduced as an alternative to Frege's construction of logic and as a theoretical basis for mereology which provides a foundation for mathematics, alternative to set theory. LO is often examined in restricted form, as a kind of first-order theory of ε

Funded by the European Union (ERC, ExtenDD, project number: 101054714). Views and opinions expressed are however those of the author(s) only and do not necessarily reflect those of the European Union or the European Research Council. Neither the European Union nor the granting authority can be held responsible for them.

A. Indrzejczak (✉)
Department of Logic, University of Lodz, Lodz, Poland
e-mail: andrzej.indrzejczak@filhist.uni.lodz.pl

K. Sano et al. (eds.), *Exploring Negation, Modality and Proof*,
Logic in Asia: Studia Logica Library,
https://doi.org/10.1007/978-981-96-8566-0_12

based on classical first-order logic (FOL) and called elementary ontology (see, e.g. Słupecki [32], Iwanuś [20], Urbaniak [36]), and we follow this line here.

The interpolation theorems, as well as their consequences like Beth definability theorem or Robinson's joint consistency theorem, belong to the most important results in logic, originally proved by Craig for classical first-order logic (FOL). Satisfaction of interpolation property is usually treated as a witness of a well-behaved logic. Moreover, Beth definability shows that the syntax and semantics of a logic are well-balanced; what is implicitly determined by semantics, can be explicitly defined syntactically. Additionally, there is a close relationship of interpolation with important algebraic properties, like amalgamation or filter extension (see, e.g. Czelakowski [4, 5]). Also other, more general forms of interpolation theorem, like uniform interpolation, are intensely studied. Hence, it is not surprising that several attempts were made to prove their counterparts for the extensions of classical logic and several non-classical logics (e.g. Fitting [7], Gherardi, Maffezzioli and Orlandelli [9], Maffezzioli and Orlandelli [26], Muskens and Wintein [38], Indrzejczak and Petrukhin [15]). Also in computer science the interpolation theorems found important applications in restricting the proof-search space in automated deduction and verification (cf. D'Silva, Kroening and Wessenbacher [6], McMillan [27] or Bonacina and Johansson [2]). The failure of interpolation for several non-classical logics is a sign of their problems with expressivity, but sometimes it is possible to fix the situation by extending the language (see, e.g. Areces, Blackburn and Marx [1], Indrzejczak and Zawidzki [17] or Indrzejczak [12]).

Interpolation theorems are proven either by means of model theoretic methods, or constructively, by means of proof-theoretic methods. Even if we restrict attention to constructive proof-theoretic approaches one may find several techniques which were applied for this aim (a useful comparison may be found in [10]). However, one of the most popular method, in particular in the setting of sequent calculus, is due to Maehara [25] (see Takeuti [34], Ono [29] or Troelstra and Schwichtenberg [35]). In this paper we follow this tradition and apply Maehara's method to obtain the Craig's interpolation theorem for LO.

In spite of the importance and originality of LO the problem of the interpolation property was not examined so far in this setting. It seems that non-constructive semantic methods of proving interpolation cannot be easily adapted to LO because of the problems concerning the interpretation of quantifiers in ontology (cf. Küng and Canty [21] or Rickey [31]). One may think about using a constructive proof-theoretic approach instead. However, for such kind of proof we need suitable tableau or sequent calculus satisfying the subformula property and it was only recently that some proof-theoretic approach to LO was provided which allows us to prove the interpolation theorem in a constructive way. LO was originally formulated as an axiomatic system but Ishimoto and Kobayashi [19] introduced also a tableau system for some part of (a quantifier-free) ontology. In [11] we introduced a sequent calculus for elementary ontology and proved the cut elimination theorem for it which makes possible the application of Maehara's strategy to obtain the interpolation theorem.

More specifically, in Sect. 12.2 we briefly characterise elementary ontology and in Sect. 12.3 we present an adequate cut-free sequent calculus GO which is equivalent to the axiomatic formulation. Then we prove the interpolation theorem for GO in Sect. 12.4. In the next section we extend this result to the cut-free calculus GOP with rules for several predicate constants. Some remarks on further extensions and open problems close the paper.

12.2 Elementary Ontology

The ontology of Leśniewski (LO) is a general calculus of names proposed as a formalisation of logic alternative to the Fregean paradigm. Despite of the great success of the latter approach, it is often difficult to apply it in a direct and satisfactory way to formalisation of natural languages. The following two features of natural languages are usually discussed in this context: (1) the subject-predicate structure of atomic sentences, characteristic not only for traditional logic but also for modern linguistics with its NP+VP model of sentences applied in generative grammar; (2) the wide class of naming expressions which are used not only to refer to x, but also to convey information about x, and even if they refer to something it is not necessarily the singular reference.

No wonder that several approaches alternative to FOL (first-order logic) were proposed, attempting to obtain a formalisation of arguments in natural languages which is closer to their original structure. It seems that one of the oldest approaches of this kind is Leśniewski's ontology. It satisfies both features mentioned above: the subject-predicate structure of atomic sentences and a wide understanding of names, including empty (like 'Hercules') and general names (like 'an emperor' or 'a dog').

Let us briefly recall the basic information concerning LO, before we proceed with the technical result.[1] In its most basic form it is a theory of the binary predicate ε understood as a formalisation of the Greek 'esti' or Latin 'est'. Atomic sentences have the form $a\varepsilon b$, where a, b denote names. Such formulae are to be read as 'the a is (a/the) b', so in order to be true a must be an individual name whereas b can be individual or general name. In the original formulation of Leśniewski, ontology is the middle part of the hierarchical structure involving also the protothetics and mereology (see the presentation in Urbaniak [36]). Protothetics, a very general form of propositional logic, is the basis of the overall construction. Its generality follows from the fact that, in addition to sentence variables, arbitrary sentence-functors (connectives) are allowed as variables, and quantifiers binding all these kinds of variables are involved. Similarly in Leśniewski's ontology, we have a quantification over name variables but also over arbitrary name-functors creating complex names. In consequence we obtain a very expressive logic which is then extended to mereology. The latter, which is the most well-known ingredient of Leśniewski's construction, is a theory of parthood relation, which provides an alternative formalisation of the theory of classes and foundations of mathematics.

[1] For more detailed general presentation of Leśniewski's systems one may consult Urbaniak [36] and for a detailed study of Leśniewski's ontology see Iwanuś [20] or Słupecki [32].

Despite of the dependance of the Leśniewski's ontology on his protothetics, it is often examined in separation, as a kind of first-order theory of ε based on classical first-order logic (FOL) and called elementary ontology. Such an elementary ontology was investigated, among others, by Słupecki [32] and Iwanuś [20], and we follow this line here. The expressive power of such an approach is strongly reduced, in particular, quantifiers apply only to name variables. One should note however that, despite of the appearances, it is not just another elementary theory in the standard sense, since in LO terms in quantifier rules are restricted to variables ranging over arbitrary names (including empty and general ones). This means, in particular, that quantifiers do not have an existential import, in contrast to standard FOL.

LO was originally formulated as an axiomatic system, but also a form of natural deduction proof system was applied by many authors following the way of presenting proofs by Leśniewski (see, e.g. his [22], [23] where the original Leśniewski's natural deduction proofs are employed—more on the construction of Leśniewski's natural deduction in [13]). Ishimoto and Kobayashi [19] introduced also a tableau system for some part of (a quantifier-free) ontology. However these attempts can hardly be treated as a proof-theoretic study of Leśniewski's ontology but only as a convenient way of simplifying a presentation of axiomatic proofs. In particular, it is not possible to use them as a basis for constructive proof of the interpolation theorem. In [11] we introduced a sequent calculus for elementary ontology and proved the cut elimination theorem for it. This kind of formalisation of LO makes possible the application of Maehara's strategy to obtain the proof of the interpolation theorem.

In this article, by Leśniewski's elementary ontology LO we mean the system which obtains by addition of the Leśniewski's axiom LA to arbitrary axiomatisation of classical pure FOL (i.e. with no terms other than variables). However, we should remember that, because of the aforementioned reasons, such a theory is not an elementary theory in the standard sense.

Following [11] we consider two languages for ontology. In both we assume a denumerable set of name variables, applying a graphical distinction between the bound variables (denoted by $x, y, z, \ldots$, possibly with subscripts), and the free variables usually called parameters (denoted by $a, b, c, \ldots$). It makes easier an elaboration of some technical issues concerning substitution and proof transformations. These are the only terms we admit, and both kinds will be called simply name variables. φ, ψ, χ denote formulae and $\Gamma, \Delta, \Pi, \Sigma$ their multisets. The basic language L_o consists of the propositional constants: $\top, \bot$, connectives: $\neg, \wedge, \vee, \rightarrow$, first-order quantifiers: $\forall, \exists$, and binary predicate ε. The differences with the version of the basic language from [11] is the addition of $\top$ and $\bot$, and deletion of $\leftrightarrow$ to simplify a proof of the interpolation theorem. In addition to the standard logical vocabulary of FOL, the only specific constant is a predicate ε with the formation rule: $t\varepsilon t'$ is an atomic formula, for any terms t, t'. In what follows we will use a convention: instead of $t\varepsilon t'$ we will write tt'. In fact, this is in accordance with several natural languages, including Japanese, where such special copula like 'is' is not used at all. $\varphi[t/t']$ is used for the operation of correct substitution of a term t' for all occurrences of term t in φ. This convention extends to multisets in the natural way—$\Gamma[t/t']$.

The only specific axiom of elementary ontology is Leśniewski's axiom:

$$\forall xy(xy \leftrightarrow \exists z(zx) \wedge \forall z(zx \rightarrow zy) \wedge \forall zv(zx \wedge vx \rightarrow zv)) \qquad (LA)$$

which expresses that for xy to be true x must have a unique designate which is also a designate of y.

This basic form of ontology is usually enriched with several kinds of predicates and term functors by means of definitions. We restrict our attention only to the most popular predicates. The language L_p adds to the vocabulary of L_o a number of unary and binary predicates. D, V, S, G are unary predicates informing that a is denoting, empty (or void), singular or general. D and S are Leśniewski's *ex* and *ob* respectively. He preferred also to apply $sol(a)$ which we symbolise with U (for unique). Here are the definitions of these predicates:

$Da := \exists x(xa)$	$Va := \neg\exists x(xa)$
$Sa := \exists x(ax)$	$Ga := \exists xy(xa \wedge ya \wedge \neg xy)$
$Ua := \forall xy(xa \wedge ya \rightarrow xy)$	[or simply $\neg Ga$]

The binary predicates $=$, $\equiv$, $\approx$, $\bar{\varepsilon}$, $\subset$, $\not\subset$, A, E, I, O of identity (weak and strong), coextensiveness, separation, subsumption and antysubsumption and of four predicates corresponding to Aristotelian categorical sentences are defined in the following way:

$a = b := ab \wedge ba$	$a\bar{\varepsilon}b := aa \wedge \neg ab$
$a \equiv b := \forall x(xa \leftrightarrow xb)$	$a \subset b := \forall x(xa \rightarrow xb)$
$a \approx b := a \equiv b \wedge Da$	$a \not\subset b := \forall x(xa \rightarrow \neg xb)$
$aAb := a \subset b \wedge Da$	$aEb := a \not\subset b \wedge Da$
$aIb := \exists x(xa \wedge xb)$	$aOb := \exists x(xa \wedge \neg xb)$

Note that different predicates characterised above may be added to LO independently of each other. Thus we can obtain several sublanguages of L_p and, accordingly, several subsystems of the full LO with respective predicates.

Note also that in L_o and L_p we deal only with simple terms. However Leśniewski considered also complex terms built with the help of specific term-forming functors. We have discussed briefly in [11] such extensions and the problems generated by them in the setting of sequent calculus. In Sect. 12.6 we add some comments to this issue and sketch a possible solution leading to more satisfactory proof-theoretic treatment.

12.3 Sequent Calculus GO

Elementary ontology will be formalised as a sequent calculus with sequents $\Gamma \Rightarrow \Delta$ which are ordered pairs of finite multisets of formulae called the antecedent and the succedent, respectively. We will use the calculus GO introduced in [11] which consists of the rules from Fig. 12.1. It is essentially the calculus G1 of Troelstra and Schwichtenberg [35] with added four specific rules characterising ε. The names of these rules come from reflexivity, transitivity, symmetry and extensionality. In case of (R) and (S) it is a kind of prefixed, or bounded, reflexivity and symmetry $(ab \rightarrow aa, bb \rightarrow (ab \rightarrow ba))$. These two rules, as well as (T) are similar to the rules introduced by Ishimoto and Kobayashi [19] in their quantifier-free tableau formalisation of the fragment of LO. Roughly speaking, these three rules allow us to prove the left-right implication of LA. (E) is responsible for proving the right-left implication of LA and in fact expresses the weak form of extensionality for LO.

All necessary structural rules, including cut, weakening and contraction are primitive. The only difference with the version examined in [11] is that for technical reasons we have to add $\bot$ and $\top$ to the language and suitable axioms $(\top)$ and $(\bot)$ to the calculus. We have also deleted the rules for $\leftrightarrow$ which were introduced in [11] only to simplify proofs of equivalence with axiomatic formulation of LO. These small changes do not affect the results proved for GO in [11].

Let us recall that formulae displayed in the schemata are active, whereas the remaining ones are parametric, or form a context. In particular, all active formulae in the premisses are called side formulae, and the ones in the conclusion are the principal formulae of the respective rule application. Note that only two rules (T) and (S) have two principal formulae. Also the four specific rules characterising

$$(AX)\ \varphi \Rightarrow \varphi \qquad (\top)\ \Rightarrow \top \qquad (\bot)\ \bot \Rightarrow$$

$$(W\Rightarrow)\ \frac{\Gamma \Rightarrow \Delta}{\varphi, \Gamma \Rightarrow \Delta} \qquad (\neg\Rightarrow)\ \frac{\Gamma \Rightarrow \Delta, \varphi}{\neg\varphi, \Gamma \Rightarrow \Delta} \qquad (\Rightarrow\neg)\ \frac{\varphi, \Gamma \Rightarrow \Delta}{\Gamma \Rightarrow \Delta, \neg\varphi}$$

$$(\Rightarrow W)\ \frac{\Gamma \Rightarrow \Delta}{\Gamma \Rightarrow \Delta, \varphi} \qquad (\wedge\Rightarrow)\ \frac{\varphi, \psi, \Gamma \Rightarrow \Delta}{\varphi \wedge \psi, \Gamma \Rightarrow \Delta} \qquad (\Rightarrow\wedge)\ \frac{\Gamma \Rightarrow \Delta, \varphi \qquad \Gamma \Rightarrow \Delta, \psi}{\Gamma \Rightarrow \Delta, \varphi \wedge \psi}$$

$$(C\Rightarrow)\ \frac{\varphi, \varphi, \Gamma \Rightarrow \Delta}{\varphi, \Gamma \Rightarrow \Delta} \qquad (\Rightarrow\vee)\ \frac{\Gamma \Rightarrow \Delta, \varphi, \psi}{\Gamma \Rightarrow \Delta, \varphi \vee \psi} \qquad (\vee\Rightarrow)\ \frac{\varphi, \Gamma \Rightarrow \Delta \qquad \psi, \Gamma \Rightarrow \Delta}{\varphi \vee \psi, \Gamma \Rightarrow \Delta}$$

$$(\Rightarrow C)\ \frac{\Gamma \Rightarrow \Delta, \varphi, \varphi}{\Gamma \Rightarrow \Delta, \varphi} \qquad (\Rightarrow\rightarrow)\ \frac{\varphi, \Gamma \Rightarrow \Delta, \psi}{\Gamma \Rightarrow \Delta, \varphi \rightarrow \psi} \qquad (\rightarrow\Rightarrow)\ \frac{\Gamma \Rightarrow \Delta, \varphi \qquad \psi, \Gamma \Rightarrow \Delta}{\varphi \rightarrow \psi, \Gamma \Rightarrow \Delta}$$

$$(\forall\Rightarrow)\ \frac{\varphi[x/t], \Gamma \Rightarrow \Delta}{\forall x\varphi, \Gamma \Rightarrow \Delta} \qquad (\Rightarrow\exists)\ \frac{\Gamma \Rightarrow \Delta, \varphi[x/t]}{\Gamma \Rightarrow \Delta, \exists x\varphi} \qquad (Cut)\ \frac{\Gamma \Rightarrow \Delta, \varphi \qquad \varphi, \Pi \Rightarrow \Sigma}{\Gamma, \Pi \Rightarrow \Delta, \Sigma}$$

$$(\Rightarrow\forall)\ \frac{\Gamma \Rightarrow \Delta, \varphi[x/d]}{\Gamma \Rightarrow \Delta, \forall x\varphi} \qquad (\exists\Rightarrow)\ \frac{\varphi[x/d], \Gamma \Rightarrow \Delta}{\exists x\varphi, \Gamma \Rightarrow \Delta} \qquad (E)\ \frac{da, \Gamma \Rightarrow \Delta, dc \quad dc, \Gamma \Rightarrow \Delta, da \quad ab, \Gamma \Rightarrow \Delta}{cb, \Gamma \Rightarrow \Delta}$$

$$(R)\ \frac{aa, \Gamma \Rightarrow \Delta}{ab, \Gamma \Rightarrow \Delta} \qquad (T)\ \frac{ac, \Gamma \Rightarrow \Delta}{ab, bc, \Gamma \Rightarrow \Delta} \qquad (S)\ \frac{ba, \Gamma \Rightarrow \Delta}{ab, bb, \Gamma \Rightarrow \Delta}$$

where in rules $(\Rightarrow \forall)$, $(\exists \Rightarrow)$ and (E), d is a fresh parameter (eigenvariable)

Fig. 12.1 Calculus GO

jointly ε have different character than the rules characterising logical constants. On the other hand, in all cases the principal formulae are placed only in the antecedents of the sequents which is necessary to prove the cut elimination theorem. In this respect they represent a similar approach like the one presented in Negri and von Plato [28] providing cut-free SC formalisations of several formal theories. The only difference is that (E) is not purely one-sided in the sense that some side formulae are also present in the succedents. But this does not do any harm to proofs of metalogical properties.

Proofs in GO are defined in a standard way as finite trees with nodes labelled by sequents. The height of a proof $\mathcal{D}$ of $\Gamma \Rightarrow \Delta$ is defined as the number of nodes of the longest branch in $\mathcal{D}$. $\vdash_k \Gamma \Rightarrow \Delta$ means that $\Gamma \Rightarrow \Delta$ has a proof of the height at most k. In [11] it was also proved that GO satisfies:

Lemma 12.1 (Substitution) *If* $\vdash_k \Gamma \Rightarrow \Delta$, *then* $\vdash_k \Gamma[a/b] \Rightarrow \Delta[a/b]$.

This result allows us to assume that all proofs are regular, i.e. that every eigenvariable a which is fresh by side condition on the respective rule must be fresh in the entire proof, not only on the branch where this rule was applied. Thus no eigenvariable is used more than once in the whole regular proof which simplifies the cut elimination proof but also the proof of the interpolation theorem. In what follows we assume that we deal only with such proofs; there is no loss of generality since every proof may be systematically transformed into a regular one by the substitution lemma. The central result of [11] is:

Theorem 12.1 (Cut elimination) *If* $\vdash \Gamma \Rightarrow \Delta$ *in GO, then* $\vdash \Gamma \Rightarrow \Delta$ *in GO with no application of* (Cut).

In what follows we assume that we are dealing only with cut-free and regular proofs. It may be easily observed that such proofs are analytic in the sense that all applied rules satisfy the subformula property.

12.4 Interpolation in GO

Let us recall that a logic L satisfies the *Craig interpolation property* if, for all formulae φ and ψ having common non-logical symbols, if $\vdash \varphi \rightarrow \psi$, then there is an interpolant formula, that is, a formula χ such that $\vdash \varphi \rightarrow \chi$, $\vdash \chi \rightarrow \psi$ and χ contains only non-logical symbols which are common to both φ and ψ. Moreover, if φ and ψ have no non-logical symbols in common, then either $\neg\varphi$ or ψ is a thesis.

We have mentioned that the most popular method of proving interpolation via sequent calculi is due to Maehara [25] and we follow this route. In fact, the Maehara's method justifies a slightly generalised version of the interpolation theorem, sometimes called Maehara's interpolation. Therefore we will be using with no risk of ambiguity two phrases: Maehara's method of proof, and Maehara's interpolation theorem, to the effect that if $\vdash \Gamma \Rightarrow \Delta$, then for any partition $((\Gamma_1, \Delta_1), (\Gamma_2, \Delta_2))$ of Γ, Δ we can construct an interpolant φ, such that:

1. $\vdash \Gamma_1 \Rightarrow \Delta_1, \varphi$
2. $\vdash \varphi, \Gamma_2 \Rightarrow \Delta_2$
3. $L(\varphi) \subseteq L(\Gamma_1 \sqcup \Delta_1) \cap L(\Gamma_2 \sqcup \Delta_2)$

where $\sqcup$ denotes a union of multisets which is like ordinary set union $\cup$ but with additional counting of the number of occurrences of the same formula. $L(\Gamma)$ is the set of non-logical symbols in Γ which, in the case of elementary LO as presented here, is simply the set of parameters. Conditions 1, 2 can be called provability criteria, and condition 3, a language condition for our interpolant.

It is obvious that Maehara's theorem implies Craig's theorem. Since by taking a partition $\Gamma_1 := [\varphi],\ \Delta_2 := [\psi],\ \Gamma_2 = \Delta_1 = \varnothing$ we derive from Maehara's theorem an interpolant χ, such that $\vdash \varphi \Rightarrow \chi$ and $\vdash \chi \Rightarrow \psi$.

A necessary condition for the successful application of Maehara's procedure is that the sequent calculus satisfies the subformula property. Since cut-free GO satisfies this condition we are free to examine if it satisfies the Maehara's interpolation theorem.

Theorem 12.2 *Maehera's interpolation theorem holds for GO.*

Proof It goes by induction on the height of a proof of the sequent. All the details of the proof concerning the axioms and rules for connectives and quantifiers are like in other presentations of this proof for FOL (see, e.g. Takeuti [34]), it remains only to show that the specific rules for GO preserve the feature that the interpolant for the conclusion can be computed on the basis of interpolants for premisses.

In the case of (R) the situation is clear. For arbitrary partition of the conclusion, the same partition of the premiss has an interpolant by the induction hypothesis (IH), and this is also the interpolant for the conclusion, since the set of parameters is intact.

Let us examine (T). All possible partitions of the conclusion belong to one of the four cases:

(a) $(ab, bc, \Gamma_1, \Delta_1), (\Gamma_2, \Delta_2)$
(b) $(\Gamma_1, \Delta_1), (ab, bc, \Gamma_2, \Delta_2)$
(c) $(ab, \Gamma_1, \Delta_1), (bc, \Gamma_2, \Delta_2)$
(d) $(bc, \Gamma_1, \Delta_1), (ab, \Gamma_2, \Delta_2)$.

By the IH we have interpolants φ_1, φ_2 such that:

1.1. $\vdash ac, \Gamma_1 \Rightarrow \Delta_1, \varphi_1$
1.2. $\vdash \varphi_1, \Gamma_2 \Rightarrow \Delta_2$
1.3. $L(\varphi_1) \subseteq L(\{ac\} \sqcup \Gamma_1 \sqcup \Delta_1) \cap L(\Gamma_2 \sqcup \Delta_2)$

and

2.1. $\vdash \Gamma_1 \Rightarrow \Delta_1, \varphi_2$
2.2. $\vdash \varphi_2, ac, \Gamma_2 \Rightarrow \Delta_2$
2.3. $L(\varphi_2) \subseteq L(\Gamma_1 \sqcup \Delta_1) \cap L(\{ac\} \sqcup \Gamma_2 \sqcup \Delta_2)$

Let us consider (a). From 1.1 we derive $\vdash ab, bc, \Gamma_1 \Rightarrow \Delta_1, \varphi_1$ by (T), and 1.3. obviously implies $L(\varphi_1) \subseteq L(\{ab, bc\} \sqcup \Gamma_1 \sqcup \Delta_1) \cap L(\Gamma_2 \sqcup \Delta_2)$, hence φ_1 remains the interpolant and we are done.

By symmetric argument but on the basis of 2.1., 2.2. and 2.3. we prove that in the case (b) the interpolant φ_2 is also inherited from the respective partition of the premiss.

Consider the case (c). By the IH, 1.1., 1.2., 1.3. hold and either $c \in L(\Gamma_1 \sqcup \Delta_1)$ or not. Assume the former, then the interpolant is $bc \rightarrow \varphi_1$ and the provability conditions are satisfied by the following arguments on the basis of 1.1. and 1.2.:

$$\dfrac{\dfrac{ac, \Gamma_1 \Rightarrow \Delta_1, \varphi_1}{ab, bc, \Gamma_1 \Rightarrow \Delta_1, \varphi_1}\,(T)}{ab, \Gamma_1 \Rightarrow \Delta_1, bc \rightarrow \varphi_1}\,(\Rightarrow\rightarrow)$$

$$(\rightarrow\Rightarrow)\,\dfrac{bc, \Gamma_2 \Rightarrow \Delta_2, bc \qquad \varphi_1, \Gamma_2 \Rightarrow \Delta_2}{bc \rightarrow \varphi_1, bc, \Gamma_2 \Rightarrow \Delta_2}$$

To simplify proof figures we omit explicit application of structural rules like weakening or contraction; this convention applies also to proofs displayed below. Note that $L(bc \rightarrow \varphi_1) \subseteq L(\{ab\} \sqcup \Gamma_1 \sqcup \Delta_1) \cap L(\{bc\} \sqcup \Gamma_2 \sqcup \Delta_2)$ follows by 1.3. and the fact that b is in both parts and c as well by the assumption. Now assume the latter, i.e. $c \notin L(\Gamma_1 \sqcup \Delta_1)$. In this case if c does not occur in φ_1 we can keep the interpolant intact, but if c occurs in φ_1, an interpolant must take the form $\forall x(bx \rightarrow \varphi_1[c/x])$. Provability conditions are satisfied by the application of $(\Rightarrow \forall)$ to the interpolant in the first proof figure (correct since c occurs only in the interpolant by assumption) and by $(\forall \Rightarrow)$ in the second.

Now consider the fourth case, i.e. (d). By the IH, 2.1., 2.2., 2.3. hold and again, either $c \in L(\Gamma_2 \sqcup \Delta_2)$ or not. In the first case the interpolant is $\varphi_2 \wedge bc$ and in the second $\exists x(\varphi_2[c/x] \wedge bx)$ (provided that c is also in φ_2). The respective proof figures are the following:

$$\dfrac{\dfrac{bc, \Gamma_1 \Rightarrow \Delta_1, \varphi_2 \qquad bc, \Gamma_1 \Rightarrow \Delta_1, bc}{bc, \Gamma_1 \Rightarrow \Delta_1, \varphi_2 \wedge bc}\,(\Rightarrow \wedge)}{bc, \Gamma_1 \Rightarrow \Delta_1, \exists x(\varphi_2[c/x] \wedge bx)}\,(\Rightarrow \exists)$$

$$(\exists \Rightarrow)\,\dfrac{(\wedge \Rightarrow)\,\dfrac{(T)\,\dfrac{\varphi_2, ac, \Gamma_2 \Rightarrow \Delta_2}{\varphi_2, ab, bc, \Gamma_2 \Rightarrow \Delta_2}}{\varphi_2 \wedge bc, ab, \Gamma_2 \Rightarrow \Delta_2}}{\exists x(\varphi_2[c/x] \wedge bx), ab, \Gamma_2 \Rightarrow \Delta_2}$$

with the last applications of rules for $\exists$ not required if $c \in L(\Gamma_2 \sqcup \Delta_2)$ or does not occur in φ_2. In the latter case, i.e. if $c \notin L(\Gamma_2 \sqcup \Delta_2)$ but c occurs in φ_2, the quantification of c is necessary to satisfy the language condition.

In the case of (S) the situation is similar but simpler. Again there are four possible classes of partitions of the conclusion:

(a) $(ab, bb, \Gamma_1, \Delta_1), (\Gamma_2, \Delta_2)$
(b) $(\Gamma_1, \Delta_1), (ab, bb, \Gamma_2, \Delta_2)$
(c) $(ab, \Gamma_1, \Delta_1), (bb, \Gamma_2, \Delta_2)$
(d) $(bb, \Gamma_1, \Delta_1), (ab, \Gamma_2, \Delta_2)$.

The cases (a) and (b) allow us to inherit the interpolant from the premiss. Cases (c) and (d) require similar proofs like for (T), In the first case the interpolant is $bb \to \varphi_1$ in the second $\varphi_2 \wedge bb$. In both cases the language condition is satisfied without further inference steps.

Let us consider the case of (E). There are only two possible classes of partition of the conclusion:

(a) $(cb, \Gamma_1, \Delta_1), (\Gamma_2, \Delta_2)$
(b) $(\Gamma_1, \Delta_1), (cb, \Gamma_2, \Delta_2)$

Let us consider (a). By the IH we have:

1. $\vdash da, \Gamma_1 \Rightarrow \Delta_1, dc, \varphi_1$
2. $\vdash \varphi_1, \Gamma_2 \Rightarrow \Delta_2$
3. $L(\varphi_1) \subseteq L(\{da, dc\} \sqcup \Gamma_1 \sqcup \Delta_1) \cap L(\Gamma_2 \sqcup \Delta_2)$
4. $\vdash dc, \Gamma_1 \Rightarrow \Delta_1, da, \varphi_2$
5. $\vdash \varphi_2, \Gamma_2 \Rightarrow \Delta_2$
6. $L(\varphi_2) \subseteq L(\{da, dc\} \sqcup \Gamma_1 \sqcup \Delta_1) \cap L(\Gamma_2 \sqcup \Delta_2)$
7. $\vdash ab, \Gamma_1 \Rightarrow \Delta_1, \varphi_3$
8. $\vdash \varphi_3, \Gamma_2 \Rightarrow \Delta_2$
9. $L(\varphi_3) \subseteq L(\{ab\} \sqcup \Gamma_1 \sqcup \Delta_1) \cap L(\Gamma_2 \sqcup \Delta_2)$

Either $a \in L(\Gamma_1 \sqcup \Delta_1)$ or not. Assume the former, then the interpolant is $\varphi_1 \vee \varphi_2 \vee \varphi_3$. The following proofs justify provability conditions on the basis of 1, 4, 7, and 2, 5, 8:

$$\dfrac{\dfrac{da, \Gamma_1 \Rightarrow \Delta_1, dc, \varphi_1 \qquad dc, \Gamma_1 \Rightarrow \Delta_1, da, \varphi_2 \qquad ab, \Gamma_1 \Rightarrow \Delta_1, \varphi_3}{cb, \Gamma_1 \Rightarrow \Delta_1, \varphi_1, \varphi_2, \varphi_3}\,(E)}{cb, \Gamma_1 \Rightarrow \Delta_1, \varphi_1 \vee \varphi_2 \vee \varphi_3}\,(\Rightarrow \vee)$$

$$(\vee \Rightarrow)\,\dfrac{(\vee \Rightarrow)\,\dfrac{\varphi_1, \Gamma_2 \Rightarrow \Delta_2 \qquad \varphi_2, \Gamma_2 \Rightarrow \Delta_2}{\varphi_1 \vee \varphi_2, \Gamma_2 \Rightarrow \Delta_2} \qquad \varphi_3, \Gamma_2 \Rightarrow \Delta_2}{\varphi_1 \vee \varphi_2 \vee \varphi_3, \Gamma_2 \Rightarrow \Delta_2}$$

$L(\varphi_1 \vee \varphi_2 \vee \varphi_3) \subseteq L(\{cb\} \sqcup \Gamma_1 \sqcup \Delta_1) \cap L(\Gamma_2 \sqcup \Delta_2)$ follows from 3, 6, 9. d is fresh in the premisses so it cannot occur in interpolants (otherwise it has to be in $\Gamma_2 \sqcup \Delta_2$ as well). Since, by assumption $a \in L(\Gamma_1 \sqcup \Delta_1)$ it is not possible that there is some other parameter which is in $\varphi_1 \vee \varphi_2 \vee \varphi_3$ but not in $L(\{cb\} \sqcup \Gamma_1 \sqcup \Delta_1) \cap L(\Gamma_2 \sqcup \Delta_2)$. If $a \notin L(\Gamma_1 \sqcup \Delta_1)$ but also $a \notin L(\Gamma_2 \sqcup \Delta_2)$ the same interpolant holds. However, if $a \in L(\Gamma_2 \sqcup \Delta_2)$ it may appear also in $\varphi_1 \vee \varphi_2 \vee \varphi_3$ so we must quantify its occurrences. In this case the interpolant is $\forall x(\varphi_1 \vee \varphi_2 \vee \varphi_3)[a/x]$ which is derivable by $(\Rightarrow \forall)$ in the first and $(\forall \Rightarrow)$ in the second proof schema.

The case of partition (b) is treated in the symmetric way and the interpolation is either of the form $\varphi_1 \wedge \varphi_2 \wedge \varphi_3$, if $a \in L(\Gamma_2 \sqcup \Delta_2)$, or $\exists x(\varphi_1 \wedge \varphi_2 \wedge \varphi_3)[a/x]$ if $a \notin L(\Gamma_2 \sqcup \Delta_2)$ but $a \in L(\Gamma_1 \sqcup \Delta_1)$ and $a \in L(\varphi_1 \wedge \varphi_2 \wedge \varphi_3)$. □

12.5 Interpolation in GOP

As we mentioned in Sect. 12.2, an elementary ontology is usually considered as extended with several predicates introduced by means of definitions. In [11] we have provided an alternative proof-theoretic characterisation which allows us to enrich GO with arbitrary selection of rules for predicates. In contrast to the four rules characterising ε, these additional rules come in pairs introducing formulae with respective predicates either to antecedents or to succedents of sequents and in this respect are similar to the rules characterising logical constants. In particular:

1. All rules for predicates are explicit, separate and symmetric, which are usual requirements for well-behaved rules in sequent calculi (see, e.g. [10]).
2. All these new rules satisfy the subformula property.
3. The substitution lemma holds for GO with any combination of these rules.
4. All rules are pairwise reductive, in the sense of Ciabattoni [3], modulo substitution of terms.

The system with added rules for all predicates listed in Sect. 12.2 is called GOP. In [11] it was shown that:

Theorem 12.3 (Cut elimination) *Every proof in GOP can be transformed into a cut-free proof.*

As a consequence of this theorem and the subformula property we have:

Theorem 12.4 *Maehera's interpolation theorem holds for GOP.*

Proof Since the rules for predicates are modular it is enough to extend the proof of theorem 2 by consideration of the situation when the last step of the proof is obtained by additional rule. Due to modularity this result holds for every subsystem based on a selection of predicates and rules characterising them.

In all cases there is only one principal formula in the conclusion, so we have always only two possible partitions: with the principal formula occurring either with Γ_1, Δ_1, or with Γ_2, Δ_2. We will call them respectively the left and the right division. Since we are interested in the compact presentation of the proof, we will not present the rules in pairs characterising suitable predicates one by one, but rather divide all the rules into groups representing similar structure, so that in each group the construction of the interpolant for the conclusion is identical in all cases.

Group A:

$$(D\Rightarrow)\ \frac{ca,\Gamma\Rightarrow\Delta}{Da,\Gamma\Rightarrow\Delta}\qquad (S\Rightarrow)\ \frac{ac,\Gamma\Rightarrow\Delta}{Sa,\Gamma\Rightarrow\Delta}\qquad (\Rightarrow V)\ \frac{ca,\Gamma\Rightarrow\Delta}{\Gamma\Rightarrow\Delta,Va}$$

$$(G\Rightarrow)\ \frac{da,ca,\Gamma\Rightarrow\Delta,dc}{Ga,\Gamma\Rightarrow\Delta}\qquad (\Rightarrow U)\ \frac{da,ca,\Gamma\Rightarrow\Delta,dc}{\Gamma\Rightarrow\Delta,Ua}\qquad (=\Rightarrow)\ \frac{ab,ba,\Gamma\Rightarrow\Delta}{a=b,\Gamma\Rightarrow\Delta}$$

$$(\bar{\varepsilon}\Rightarrow)\ \frac{aa,\Gamma\Rightarrow\Delta,ab}{a\bar{\varepsilon}b,\Gamma\Rightarrow\Delta}\qquad (\Rightarrow\subset)\ \frac{ca,\Gamma\Rightarrow\Delta,cb}{\Gamma\Rightarrow\Delta,a\subset b}\qquad (\Rightarrow\not\subset)\ \frac{ca,cb,\Gamma\Rightarrow\Delta}{\Gamma\Rightarrow\Delta,a\not\subset b}$$

$$(I\Rightarrow)\ \frac{ca,cb,\Gamma\Rightarrow\Delta}{aIb,\Gamma\Rightarrow\Delta}\qquad (O\Rightarrow)\ \frac{ca,\Gamma\Rightarrow\Delta,cb}{aOb,\Gamma\Rightarrow\Delta}$$

where c and d are new in all schemata.

For this class of rules the situation is unproblematic like in the case of (R). For arbitrary partition of the conclusion if φ is an interpolant of the same partition of the premiss (in the sense that all side formulae are in the same class), then it holds for the conclusion.

Group B:

$$(\Rightarrow D)\ \frac{\Gamma\Rightarrow\Delta,ba}{\Gamma\Rightarrow\Delta,Da}\qquad (\Rightarrow S)\ \frac{\Gamma\Rightarrow\Delta,ab}{\Gamma\Rightarrow\Delta,Sa}\qquad (V\Rightarrow)\ \frac{\Gamma\Rightarrow\Delta,ba}{Va,\Gamma\Rightarrow\Delta}$$

In this group for the left division if b is identical to a or occurs in Γ_1, Δ_1, then the interpolant φ is the same as in the respective partition of the premiss. Otherwise, it is of the form $\forall x\varphi[b/x]$ to satisfy language condition. In the case of the right division it is either φ (the same as in the premiss) or $\exists x\varphi[b/x]$.

Group C:

$$(\Rightarrow=)\ \frac{\Gamma\Rightarrow\Delta,ab\quad \Gamma\Rightarrow\Delta,ba}{\Gamma\Rightarrow\Delta,a=b}\qquad (\Rightarrow\bar{\varepsilon})\ \frac{\Gamma\Rightarrow\Delta,aa\quad ab,\Gamma\Rightarrow\Delta}{\Gamma\Rightarrow\Delta,a\bar{\varepsilon}b}$$

$$(\Rightarrow\equiv)\ \frac{ca,\Gamma\Rightarrow\Delta,cb\quad cb,\Gamma\Rightarrow\Delta,ca}{\Gamma\Rightarrow\Delta,a\equiv b}$$

where c is new in the last schema.

Note that in the first two rules there are no other terms present in the premisses, and in the third one, the only term not in the conclusion is fresh. Hence, the language conditions are automatically satisfied and the interpolant is simply deduced from interpolants of the premisses. In case of the left partition it is $\varphi_1 \vee \varphi_2$, in case of the right partition it is $\varphi_1 \wedge \varphi_2$, where φ_1 is an interpolant of (suitable partition) of the first premiss, and φ_2 of the second.

Group D:

$$(\equiv\Rightarrow)\ \frac{\Gamma \Rightarrow \Delta, ca, cb \qquad ca, cb, \Gamma \Rightarrow \Delta}{a \equiv b, \Gamma \Rightarrow \Delta} \qquad (\subset\Rightarrow)\ \frac{\Gamma \Rightarrow \Delta, ca \qquad cb, \Gamma \Rightarrow \Delta}{a \subset b, \Gamma \Rightarrow \Delta}$$

$$(\approx\Rightarrow)\ \frac{da, \Gamma \Rightarrow \Delta, ca, cb \qquad ca, cb, da, \Gamma \Rightarrow \Delta}{a \approx b, \Gamma \Rightarrow \Delta} \qquad (\not\subset\Rightarrow)\ \frac{\Gamma \Rightarrow \Delta, ca \qquad \Gamma \Rightarrow \Delta, cb}{a \not\subset b, \Gamma \Rightarrow \Delta}$$

$$(A \Rightarrow)\ \frac{da, \Gamma \Rightarrow \Delta, ca \qquad cb, da, \Gamma \Rightarrow \Delta}{aAb, \Gamma \Rightarrow \Delta} \qquad (\Rightarrow A)\ \frac{da, \Gamma \Rightarrow \Delta, db \qquad \Gamma \Rightarrow \Delta, ca}{\Gamma \Rightarrow \Delta, aAb}$$

$$(E \Rightarrow)\ \frac{da, \Gamma \Rightarrow \Delta, ca \qquad da, \Gamma \Rightarrow \Delta, cb}{aEb, \Gamma \Rightarrow \Delta} \qquad (\Rightarrow E)\ \frac{da, db, \Gamma \Rightarrow \Delta \qquad \Gamma \Rightarrow \Delta, ca}{\Gamma \Rightarrow \Delta, aEb}$$

$$(\Rightarrow I)\ \frac{\Gamma \Rightarrow \Delta, ca \qquad \Gamma \Rightarrow \Delta, cb}{\Gamma \Rightarrow \Delta, aIb} \qquad (\Rightarrow O)\ \frac{\Gamma \Rightarrow \Delta, ca \qquad cb, \Gamma \Rightarrow \Delta}{\Gamma \Rightarrow \Delta, aOb}$$

where d is new and c arbitrary (but c can be identical to d in rules for $\approx$, A, E).

For this group we have also always two interpolants of (respective partitions of) two premisses φ_1 and φ_2. In all rules there is also one arbitrary term c in the premisses. Hence for the left partition the interpolant of the conclusion is either $\varphi_1 \vee \varphi_2$ or $\forall x(\varphi_1 \vee \varphi_2)[c/x]$ (if c is not in Γ_1, Δ_1 but in φ_1 or φ_2), and the interpolant for the right partition is either $\varphi_1 \wedge \varphi_2$ or $\exists x(\varphi_1 \wedge \varphi_2)$ (if c is not in Γ_2, Δ_2 but in φ_1 or φ_2).

Group E:

$$(\Rightarrow G)\ \frac{\Gamma \Rightarrow \Delta, ba \qquad \Gamma \Rightarrow \Delta, ca \qquad bc, \Gamma \Rightarrow \Delta}{\Gamma \Rightarrow \Delta, Ga}$$

$$(U \Rightarrow)\ \frac{\Gamma \Rightarrow \Delta, ba \qquad \Gamma \Rightarrow \Delta, ca \qquad bc, \Gamma \Rightarrow \Delta}{Ua, \Gamma \Rightarrow \Delta}$$

$$(\Rightarrow\approx)\ \frac{da, \Gamma \Rightarrow \Delta, db \qquad db, \Gamma \Rightarrow \Delta, da \qquad \Gamma \Rightarrow \Delta, ca}{\Gamma \Rightarrow \Delta, a \approx b}$$

where d is new and b, c arbitrary (but c can be identical to d in the last rule).

The main differences with the former cases are: three premisses and accordingly three interpolants $\varphi_1, \varphi_2, \varphi_3$ and two arbitrary terms b, c which may lead to problems with language conditions. Accordingly, the interpolants for the left partition may be of the form: $\varphi_1 \vee \varphi_2 \vee \varphi_3$ or $\forall x(\varphi_1 \vee \varphi_2 \vee \varphi_3)[x/b]$ or $\forall x(\varphi_1 \vee \varphi_2 \vee \varphi_3)[x/c]$ or $\forall xy(\varphi_1 \vee \varphi_2 \vee \varphi_3)[x/b, y/c]$. And for the right partition: $\varphi_1 \wedge \varphi_2 \wedge \varphi_3$ or $\exists x(\varphi_1 \wedge \varphi_2 \wedge \varphi_3)[x/b]$ or $\exists x(\varphi_1 \wedge \varphi_2 \wedge \varphi_3)[x/c]$ or $\exists xy(\varphi_1 \wedge \varphi_2 \wedge \varphi_3)[x/b, y/c]$. The reader can easily establish which kind of interpolant holds by checking if b or

c occurs in any interpolants of the premisses but is not present in Γ_1, Δ_1 or Γ_2, Δ_2 respectively. □

12.6 Conclusion

We have considered the classical version of LO and proved the interpolation theorem for it using the fact that it can be characterised by means of a well-behaved sequent calculus. However, the advantages of using sequent calculus go further. Note that we can easily obtain an intuitionistic version of GO by making only minor changes. It is sufficient to change the rules for their variants defined on intuitionistic sequents with at most one formula in the succedent. As a result we restrict Δ in all schemata either to empty or to singular multiset, get rid of $(\Rightarrow C)$ and instead of $(\Rightarrow \vee)$ introduce its intuitionistic variants:

$$(\Rightarrow\vee_1)\ \frac{\Gamma \Rightarrow \varphi}{\Gamma \Rightarrow \varphi \vee \psi} \qquad\qquad (\Rightarrow\vee_2)\ \frac{\Gamma \Rightarrow \psi}{\Gamma \Rightarrow \varphi \vee \psi}$$

In a similar way we can obtain an intuitionistic version of GOP replacing $(\equiv\Rightarrow)$ and $(\approx\Rightarrow)$ with:

$$(\equiv\Rightarrow_1)\ \frac{\Gamma \Rightarrow ca \quad cb, \Gamma \Rightarrow \Delta}{a \equiv b, \Gamma \Rightarrow \Delta} \qquad (\equiv\Rightarrow_2)\ \frac{\Gamma \Rightarrow cb \quad ca, \Gamma \Rightarrow \Delta}{a \equiv b, \Gamma \Rightarrow \Delta}$$

$$(\approx\Rightarrow_1)\ \frac{da, \Gamma \Rightarrow ca \qquad cb, da, \Gamma \Rightarrow \Delta}{a \approx b, \Gamma \Rightarrow \Delta} \qquad (\approx\Rightarrow_2)\ \frac{da, \Gamma \Rightarrow cb \qquad ca, da, \Gamma \Rightarrow \Delta}{a \approx b, \Gamma \Rightarrow \Delta}$$

and restricting the remaining rules to variants with Δ empty or singular. One may easily obtain a proof of the interpolation theorem for intuitionistic versions of GO and GOP by slightly modifying the proofs from Sects. 12.4 and 12.5 (cf. the details of proof for intuitionistic logic in Ono [29]).

Since LO was shown capable of being investigated by means of proof-theoretic methods it opens the possibility to investigate proof-search strategies and derive semi-decision procedures (or decision procedures for quantifier-free subsystems) on the basis of the theoretic tools already provided. In particular, due to modularity of provided rules, it is possible to obtain in this way decision procedures for several quantifier-free subsystems investigated in Pietruszczak [30], or the interesting subsystem investigated in the works of Ishimoto [18], Ishimoto and Kobayashi [19], and Takano [33]. Automatisation of proof-search in such systems is another task worth exploration. Taking into account that LO is a very convenient tool for expressing arguments in natural languages such an application may enrich significantly the toolkit of automated deduction.

Closely connected to the problem of the applicability of LO to efficient processing of natural language is the issue of treating complex names. In L_o and L_p we introduced only name variables ranging over all names (individual, general and empty), but without taking care of their internal structure. Leśniewski considered also complex terms built with the help of specific term-forming functors in ontology. In [11] we have discussed briefly his way of dealing with the problem and the difficulties it generates in the setting of sequent calculus (creativity of his definitions, the loss of the subformula property).

Instead of following the original Leśniewski's approach it seems more promising to apply a different approach to systematic construction of complex names in LO. One possibility can be based on the approach described in Waragai [37], another one on the use of lambda operator. The latter seems to be very useful, at least from the standpoint of proof theory and automated deduction. One could extend the language of LO with lambda operator used to form complex terms. If we additionally admit in the language also predicate parameters of arbitrary arity we can obtain a very expressive system allowing us to deal in a natural and direct way with complex term structures which are common in natural language communication. The advantages of using lambda operator to build complex predicates and to combine them with iota operator building complex (singular) terms (definite descriptions) were already shown in several works (cf. Fitting and Mendelsohn [8] or Indrzejczak and Zawidzki [16, 17]).

The specific features of LO, in particular the facts that names are taken in the broad sense of the word, and the subject-predicate structure is taken as the basic form of the sentence, make such an application of lambda operator even more suitable here. In this approach the atomic sentence is of the form $t\varepsilon t'$, where each term may be either a name parameter or a lambda term expressing complex name of any sort. Since terms in LO are taken in a wide sense there is no need of using iota operator to formalise individual complex names, like definite descriptions. In this setting lambda operator is a sufficient device to cover the role of (proper and improper) definite and indefinite descriptions. This kind of extended Leśniewski's ontology is under investigation now and the first results in this area were presented in [14].

Acknowledgements The author would like to express his gratitude to the reviewer for his comments and suggestions.

References

1. Areces, C., Blackburn, P., Marx, M.: Hybrid logics: characterization, interpolation and complexity. J. Symb. Log. **66**(3), 977–1010 (2001)
2. Bonacina, M.P., Johansson, M.: Interpolation systems for ground proofs in automated deduction: a survey. J. Automat. Reason. **54** 353–390 (2015)
3. Ciabattoni, A.: Automated generation of analytic calculi for logics with linearity. In: Marcinkowski, J., Tarlecki, A. (eds.), CSL 2004, LNCS, vol. 3210, pp. 503–517. Springer, Heidelberg (2004)
4. Czelakowski, J.: Logical matrices and the amalgamation property. Stud. Logica **41**(4), 329–342 (1982)
5. Czelakowski, J.: Sentential logics and Maehara interpolation property. Stud. Logica **44**(3), 265–284 (1985)

6. D'Silva, V., Kroening, D., Weissenbacher G.: A survey of automated techniques for formal software verification. IEEE Trans. Comput.-Aided Design Integr. Circuits Syst. **27** 1165–1178 (2008)
7. Fitting, M.: Proof Methods for Modal and Intuitionistic Logics. Reidel, Dordrecht (1983)
8. Fitting, M., Mendelsohn, R.L.: First-Order Modal Logic. Synthese Library, vol. 277. Springer, Dordrecht (1998)
9. Gherardi, G., Maffezioli, P., Orlandelli, E.: Interpolation in extensions of first-order logic. Stud. Logica **108**(3), 619–648 (2020)
10. Indrzejczak, A.: Sequents and Trees. An Introduction to the Theory and Applications of Propositional Sequent Calculi, Birkhäuser (2021)
11. Indrzejczak, A.: Leśniewski's ontology – proof-theoretic characterization. In: Blanchette, J., Kovacs, L., Pattinson, D. (eds.), Automated Reasoning, IJCAR 2022, LNAI vol. 13385, pp. 541–558. Springer, Heidelberg (2022)
12. Indrzejczak, A.: Bisequent calculus for four-valued quasi-relevant logics; cut elimination and interpolation. J. Autom. Reason. **67**(37) (2023)
13. Indrzejczak, A.: The problem of natural representation of reasoning in the Lvov-Warsaw school. Hist. Philos. Logic **45**(2), 142–160 (2024)
14. Indrzejczak, A.: When Epsilon meets Lambda: Extended Leśniewski's Ontology. In: Nancy (ed.), CEUR Workshop Proceedings, vol. 3875, ARQNL 2024, pp. 62–79 (2024). https://iltp.de/ARQNL 2024/CEUR/proceedings_tmp/ARQNL2024_paper6.pdf
15. Indrzejczak, A., Petrukhin, Y.: A uniform formalisation of three-valued logics in bisequent calculus. In: International Conference on Automated Deduction. CADE 2023: Automated Deduction – CADE 29, pp. 325–343. Springer, Rome (2023)
16. Indrzejczak, A., Zawidzki, M.: When Iota meets Lambda. Synthese 201/72 (2023). https://doi.org/10.1007/s11229-023-04048-y.
17. Indrzejczak, A., Zawidzki, M.: Definite Descriptions and Hybrid Tense Logic. Synthese 202/98 (2023). https://doi.org/10.1007/s11229-023-04319-8
18. Ishimoto, A.: A propositional fragment of Leśniewski's ontology. Stud. Logica **36**(4), 285–299 (1977)
19. Ishimoto, A., Kobayashi, M.: A propositional fragment of Leśniewski's ontology and its formulation by the Tableau method. Stud. Logica **41**(2–3), 181–196 (1982)
20. Iwanuś, B.: On Leśniewski's elementary ontology. Stud. Logica **31**(1), 73–119 (1973)
21. Küng, G., Canty, J.T.: Substitutional quantification and Leśniewskian quantifiers. Theoria **36**, 165–182 (1970)
22. Leśniewski, S.: Über Funktionen, deren Felder Gruppen mit Rücksicht auf diese Funktionen sind. Fundamenta Mathematicae **13**, 319–332 (1929). [English translation in: Leśniewski [24]]
23. Leśniewski, S.: Über Funktionen, deren Felder Abelsche Gruppen in Bezug auf diese Funktionen sind. Fund. Math. **14**, 242–251 (1929). [English translation in: Leśniewski [24]]
24. Leśniewski, S.: Collected Works, vol. II. S., Srzednicki, J., Barnett, D.I. Kluwer/PWN, Surma (1992)
25. Maehara, S.: On the interpolation theorem of Craig. Suugaku **12**, 235–237 (1960). [in Japanese]
26. Maffezioli, P., Orlandelli, E.: Full cut elimination and interpolation for intuitionistic logic with existence predicate. Bullet. Sect. Logic **48**(2), 137–158 (2019)
27. McMillan, K.L.: Interpolation and SAT-based model checking. In: Hunt W.A., Jr., Somenzi, F., (Eds.), Computer Aided Verification, 15th International Conference, CAV 2003, LNCS, vol. 2725, pp. 1–13. Springer (2003)
28. Negri, S., von Plato, J.: Structural Proof Theory. Cambridge University Press, Cambridge (2001)
29. Ono, H.: Proof-theoretic methods in nonclassical logic – an introduction. In: M. Takahashi (ed.), Theories of Types and Proofs. MSJ-Memoir 2, pp. 207–254. Mathematical Society of Japan (1998)
30. Pietruszczak, A.: Quantifier-free Calculus of Names. Systems and Metatheory, Toruń (1991). [in Polish]
31. Rickey, F.: Interpretations of Leśniewski's ontology. Dialectica **39**(3), 181–192 (1985)

32. Słupecki, J.: S. Leśniewski's Calculus of Names. Studia Logica **3**(1), 7–72 (1955)
33. Takano, M.: A semantical investigation into Leśniewski's axiom and his ontology. Stud. Logica. **44**(1), 71–78 (1985)
34. Takeuti, G.: Proof Theory. North-Holland, Amsterdam (1987)
35. Troelstra, A.S., Schwichtenberg, H.: Basic Proof Theory. Oxford University Press, Oxford (1996)
36. Urbaniak, R.: Leśniewski's Systems of Logic and Foundations of Mathematics. Springer, Cham (2014)
37. Waragai, T.: Ontology as a natural extension of predicate calculus with identity equipped with description. Ann. Japan Assoc. Philos. Sci. **7**(5), 233–250 (1990)
38. Wintein, S., Muskens, R.: Interpolation methods for Dunn logics and their extensions. Stud. Logica. **105**(6), 1319–1348 (2017)

Chapter 13
Proof Construction Style Representation of Cut-Elimination

Ryo Takemura

Abstract We formalize a sequent calculus mG3 for multi-sequents. Our multi-sequent is a sequence of the usual sequents representing successive applications of the cut-rule. Proof-reduction rules for eliminating the cut-rule are represented as proof-construction rules in mG3. Cut-eliminability is then characterized by provability in mG3.

13.1 Introduction

We propose a sequent calculus system, in which a proof represents a cut-elimination process. The cut-elimination theorem is typically proved by reducing a cut-formula to atoms, then by deleting the atomic cut-formulas. (Cf. [11, 12]) For example, by principal $\wedge$-reduction, *cut* on $A \wedge B$ on the following left, where we consider *cut* in the context sharing form, is typically reduced to *cut* on A and on B on the following right.

$$\frac{\Gamma \vdash A \wedge B \quad A \wedge B, \Gamma \vdash}{\Gamma \vdash} (A \wedge B) \quad \rhd \quad \frac{\Gamma \vdash B \quad \dfrac{B, \Gamma \vdash A \quad A, B, \Gamma \vdash}{B, \Gamma \vdash} (A)}{\Gamma \vdash} (B)$$

Such a reduction step may be regarded as the decomposition step of a given cut-formula, and then, we can see the similarities between cut reductions and proof-construction (proof-search) operations. By the above reduction, focusing on upper sequents, $A \wedge B, \Gamma \vdash$ on the left changes to $A, B, \Gamma \vdash$ on the right, which is analogous to the $\wedge l$-rule; and $\Gamma \vdash A \wedge B$ changes to $\Gamma \vdash B$ and $B, \Gamma \vdash A$, which is analogous to the $\wedge r$-rule.

To track the change of upper sequents of *cut* in the reduction, we introduce a *multi-sequent*, which is a sequence of the usual sequents, and we express the above *cut* on $A \wedge B$ on the left as $(\Gamma \vdash A \wedge B \mid A \wedge B, \Gamma \vdash)$, where $\mid$ expresses an application of

R. Takemura (✉)
Nihon University, Chiyoda, Japan
e-mail: takemura.ryo@nihon-u.ac.jp

K. Sano et al. (eds.), *Exploring Negation, Modality and Proof*,
Logic in Asia: Studia Logica Library,
https://doi.org/10.1007/978-981-96-8566-0_13

cut. We can then express the above principal $\wedge$-reduction in the proof construction style as follows, where the lower sequent represents the original *cut* on $A \wedge B$, and the upper sequent represents the resultant *cut* on A and on B.

$$\frac{(\Gamma \vdash B \mid (B, \Gamma \vdash A \mid A, B, \Gamma \vdash))}{(\Gamma \vdash A \wedge B \mid A \wedge B, \Gamma \vdash)}$$

However, auxiliary reductions are problematic. Let R be an inference from Π to Γ such as $\wedge l$. Then the following proof on the left is reduced to the proof on the right.

$$\dfrac{\Gamma \vdash B \quad \dfrac{B, \Gamma \vdash A \quad \dfrac{A, B, \Pi \vdash}{A, B, \Gamma \vdash}\,R}{B, \Gamma \vdash}\,(A)}{\Gamma \vdash}\,(B) \quad \rhd \quad \dfrac{\Gamma \vdash B \quad \dfrac{\dfrac{B, \Pi \vdash A \quad A, B, \Pi \vdash}{B, \Pi \vdash}\,(A)}{B, \Gamma \vdash}\,R}{\Gamma \vdash}\,(B)$$

The topmost $B, \Pi \vdash A$ in the resultant proof is obtained by applying the inversion lemma to $B, \Gamma \vdash A$. If we try to express the above reduction by using our multi-sequents, we find that $\mid$ no longer represents a simple *cut*, because contexts Γ and Π would no longer match in the upper sequent.

$$\frac{(\Gamma \vdash B \mid (B, \Pi \vdash A \mid A, B, \Pi \vdash))}{(\Gamma \vdash B \mid (B, \Gamma \vdash A \mid A, B, \Gamma \vdash))}$$

To avoid this difficulty, maintaining the context sharing form of *cut*, we regard the applications of *cut* on A and on B as one lump called *cut-string*, and reduce such a sequence of *cut* as a whole.

$$\dfrac{\Gamma \vdash B \quad \dfrac{B, \Gamma \vdash A \quad \dfrac{A, B, \Pi \vdash}{A, B, \Gamma \vdash}\,R}{B, \Gamma \vdash}\,(A)}{\Gamma \vdash}\,(B) \quad \rhd \quad \dfrac{\dfrac{\Pi \vdash B \quad \dfrac{B, \Pi \vdash A \quad A, B, \Pi \vdash}{B, \Pi \vdash}\,(A)}{\Pi \vdash}\,(B)}{\Gamma \vdash}\,R$$

Then, the above auxiliary reduction is expressed as follows.

$$\frac{(\Pi \vdash B \mid (B, \Pi \vdash A \mid A, B, \Pi \vdash))}{(\Gamma \vdash B \mid (B, \Gamma \vdash A \mid A, B, \Gamma \vdash))}\,R$$

In Sect. 13.2, we introduce cut-strings and prove our cut-string elimination theorem in essentially the same way as the usual proof of cut-elimination. Then, in Sect. 13.3, by introducing multi-sequents as a sequent representation of cut-strings, we define sequent calculus mG3 for multi-sequents. We show that every cut-string reduction is represented by an mG3-rule, and that cut-eliminability in G3 is characterized by provability in mG3.

13.2 Sequent Calculus and Cut-Elimination

In Sect. 13.2.1, we review sequent calculus G3 of [12]. In Sect. 13.2.2, we prove our cut-elimination theorem by introducing cut-strings.

13.2.1 Sequent Calculus G3

We introduce the sequent calculus G3 of [4, 11, 12]. Following [4], we introduce the negation $\neg$ as a primitive connective instead of $\bot$. We consider only classical logic, whose propositional connectives are $\neg$, $\wedge$, $\rightarrow$, and $\vee$; hence, we refer to our system simply as G3 instead of G3cp by omitting c for classical and p for propositional.

Formulas, denoted by $A, B, C, \ldots$, are defined inductively as usual. Atoms are denoted by $Q, R, \ldots$. The **weight** $w(A)$ of a formula A is inductively defined as follows. $w(Q) = 0$; $w(\neg A) = w(A) + 1$; $w(A \wedge B) = w(A \rightarrow B) = w(A \vee B) = w(A) + w(B) + 1$. A sequent is an expression $\Gamma \vdash \Delta$, where Γ and Δ are finite multisets of formulas. Γ is called the antecedent of the sequent, while Δ is called the succedent of the sequent.

Definition 1 (G3-*rules*) The inference rules of G3 are defined as follows.

$$\frac{\Gamma \vdash \Delta, A}{\neg A, \Gamma \vdash \Delta}\neg l \qquad \frac{A, \Gamma \vdash \Delta}{\Gamma \vdash \Delta, \neg A}\neg r \qquad \frac{A, B, \Gamma \vdash \Delta}{A \wedge B, \Gamma \vdash \Delta}\wedge l \qquad \frac{\Gamma \vdash \Delta, A \quad \Gamma \vdash \Delta, B}{\Gamma \vdash \Delta, A \wedge B}\wedge r$$

$$\frac{A, \Gamma \vdash \Delta \quad B, \Gamma \vdash \Delta}{A \vee B, \Gamma \vdash \Delta}\vee l \qquad \frac{\Gamma \vdash \Delta, A, B}{\Gamma \vdash \Delta, A \vee B}\vee r$$

$$\frac{\Gamma \vdash \Delta, A \quad B, \Gamma \vdash \Delta}{A \rightarrow B, \Gamma \vdash \Delta}\rightarrow l \qquad \frac{A, \Gamma \vdash \Delta, B}{\Gamma \vdash \Delta, A \rightarrow B}\rightarrow r$$

An axiom is a sequent of the form $\Gamma, Q \vdash Q, \Delta$, where Q is an atom.

In each rule, Γ and Δ are collectively called the context. Contexts are denoted by the Greek capital letters for $\Gamma, \Delta, \Pi, \Lambda, \ldots$. In the lower sequent of each rule, the formula not in the context is called the main formula. The abovementioned form of a rule whose main formula is C is referred by an **inferences on** C.

We investigate G3 + *cut* by adding the following rule of *cut* to G3. We regard the following two expressions of *cut* equivalent.

$$\frac{\Gamma \vdash \Delta, C \quad C, \Gamma \vdash \Delta}{\Gamma \vdash \Delta}\,(C) \qquad\qquad \frac{C, \Gamma \vdash \Delta \quad \Gamma \vdash \Delta, C}{\Gamma \vdash \Delta}\,(C)$$

The set of the above two occurrences of formula C in the *cut*-rule is called the **cut-formula**. When a cut-formula is an atom, it is called a **cut-atom**. In an application of *cut*, we indicate the cut-formula in place of the name of the rule. In the following text, by (C) with parentheses, we refer to an application of *cut* on cut-formula C.

A proof, which is denoted by $\pi, \pi_1, \pi_2, \ldots$, is a finite tree constructed using inference rules whose leaves are axioms. The height of a proof denotes the greatest number of successive applications of rules in it, where a proof consisting only of an axiom has the height 0. See [11, 12] for a formal definition.

The following inversion, weakening, and contraction lemmas hold in G3 [11, 12], and they are height-preserving.

Lemma 1 (Inversion in G3) *For every* G3*-rule, if the lower sequent of the rule is provable with a height of proof at most n, then so is the upper sequent(s) thereof.*

We write $\Rightarrow_n \Gamma \vdash \Delta$, when $\Gamma \vdash \Delta$ is provable with a height of proof at most n.

Lemma 2 (Weakening and contraction in G3) *If* $\Rightarrow_n \Gamma \vdash \Delta$, *then* $\Rightarrow_n \Pi, \Gamma \vdash \Delta, \Lambda$ *for any* Π, Λ. *If* $\Rightarrow_n A, A, \Gamma \vdash \Delta$, *then* $\Rightarrow_n A, \Gamma \vdash \Delta$. *If* $\Rightarrow_n \Gamma \vdash \Delta, A, A$, *then* $\Rightarrow_n \Gamma \vdash \Delta, A$.

13.2.2 Cut-Elimination of G3 + *cut*

We regard the successive applications of *cut* in a proof as one lump called a **cut-string**, and we define cut-string reduction rules.

Definition 2 (*Cut-string*) Beginning with an application of *cut* (C_n) in a given proof, by tracing back successive applications of *cut* to the top without branching, let an application of *cut* (C_1) be the endpoint whose two upper sequents are not derived from *cut*. The resulting path of successive applications of *cut* on $C_1, \ldots, C_n$, from the top of the proof, is called a **cut-string** and denoted by $(C_n(\cdots(C_2(C_1))\cdots))$, which is abbreviated by $(\vec{C})$.

When we focus on a successive part $C_i, C_{i+1}, \ldots, C_k$ in a cut-string $(\vec{C})$, we denote the cut-string as $(\overrightarrow{C_{k+1}}(\vec{C_i}(\vec{C_1})))$, where $(\vec{C_1})$ is $(C_{i-1}(\cdots(C_1))\cdots)$.

In a given proof, two cut-strings may overlap, and a cut-string may be a part of another cut-string.

Remark 1 Although our notation $(\overrightarrow{C})$ focuses only on cut-formulas, a cut-string is defined as successive applications of the *cut*-rule; hence, it includes not only cut-formulas but also contexts. Thus, even if two cut-strings are referred by the same $(\overrightarrow{C})$, that is, they consist of the same cut-formulas, they may be different as cut-strings in their contexts.

The following notion of minor upper sequents is an extension of the usual two upper sequents of an application of *cut*.

Definition 3 Let $(\overrightarrow{C})$ be a cut-string in a given proof. For every *cut* (C_i) in the cut-string, one of the upper sequents, which is not derived from the *cut* (C_{i-1}) $(i \geq 2)$, is called the **minor upper sequent** of (C_i). In particular, two upper sequents of the topmost *cut* (C_1) are minor upper sequents of (C_1). The sub-proof of a given proof, which derives the minor upper sequent of (C_i), is called the **minor sub-proof** of (C_i).

Minor upper sequents of $(\overrightarrow{C})$ are all minor upper sequents of (C_i) for all i.

Note that the minor upper sequent of (C_i) $(i \geq 2)$ may be derived from an application of *cut*, which belongs to another cut-string apart from $(\overrightarrow{C})$.

We now define our cut-string reduction rules.

In our reduction rules, we significantly rewrite a given proof, using the inversion lemma. Thus, we define our reduction rules only to the topmost cut-string. We describe our rules for $\land$ and $\neg$, and see, Appendix for other rules.

Definition 4 (*Reduction rules*) Assume that the given proof π contains a cut-string $(\overrightarrow{C}) \equiv (C_n(\cdots(C_2(C_1))\cdots))$, where the minor upper sequents of (C_1) are $\Gamma \vdash \Delta, C_1$ and $C_1, \Gamma \vdash \Delta$, and no other *cut* appears in π. For any non atomic $A \in \Gamma \cup \Delta \cup \{C_1\}$, (1) when $A \in \overrightarrow{C}$, the following **principal reduction** is applied; (2) when $A \notin \overrightarrow{C}$ and a minor sub-proof of (C_i) exists, whose last is an inference on A, then the following **auxiliary reduction** is applied; (3) otherwise, any reduction cannot be applied.

(Principal reduction) When $A \equiv C_i$ of $\overrightarrow{C}$, the following cases occur.

1-1. (Principal $\land l$**)** When $A \equiv D_1 \land D_2$, let the given cut-string be the following form. $(\overrightarrow{C_2})$, as well as $(\overrightarrow{C_{i+1}})$, is the successive applications of *cut* on $C_2, \ldots, C_{i-1}$ (resp. $C_{i+1}, \ldots, C_n$), which may be empty.

$$
\begin{array}{c}
\dfrac{\overset{\vdots\ \pi_3}{\Pi \vdash \Lambda, D_1 \land D_2} \qquad \begin{array}{c} \dfrac{\overset{\vdots\ \pi_1}{D_1 \land D_2, \Gamma \vdash \Delta, C_1} \quad \overset{\vdots\ \pi_2}{C_1, D_1 \land D_2, \Gamma \vdash \Delta}}{D_1 \land D_2, \Gamma \vdash \Delta}\,(C_1) \\ \vdots\ (\overrightarrow{C_2}) \\ D_1 \land D_2, \Pi \vdash \Lambda \end{array}}{\Pi \vdash \Lambda}\,(D_1 \land D_2) \\
\vdots\ (\overrightarrow{C_{i+1}}) \\
\Sigma \vdash \Phi
\end{array}
$$

The above cut-string is transformed as follows. We call $\Pi \vdash \Lambda, D_2$ of $\pi_3^{\circ'}$ the **additional sequent**.

$$
\dfrac{\begin{array}{c}\vdots\ \pi_3^{\circ'}\\ \Pi \vdash \Lambda, D_2\end{array} \qquad \dfrac{\begin{array}{c}\vdots\ \pi_3^{\circ}\\ D_2^{*}, \Pi \vdash \Lambda, D_1\end{array} \qquad \begin{array}{c}\dfrac{\begin{array}{c}\vdots\ \pi_1^{\circ}\\ D_1, D_2, \Gamma \vdash \Delta, C_1\end{array} \quad \begin{array}{c}\vdots\ \pi_2^{\circ}\\ C_1, D_1, D_2, \Gamma \vdash \Delta\end{array}}{D_1, D_2, \Gamma \vdash \Delta}\,(C_1)\\ \vdots\ (\overrightarrow{C_2}^{D_1 D_2})\\ D_1, D_2, \Pi \vdash \Lambda\end{array}}{D_2, \Pi \vdash \Lambda}\,(D_1)}{\begin{array}{c}\Pi \vdash \Lambda\\ \vdots\ (\overrightarrow{C_{i+1}})\\ \Sigma \vdash \Phi\end{array}}\,(D_2)
$$

π_i° (and $\pi_i^{\circ'}$) is obtained by rewriting the given minor sub-proof π_i using the inversion lemma on $D_1 \wedge D_2$. $(\overrightarrow{C_2}^{D_1 D_2})$ expresses that the single occurrence of cut-formula $D_1 \wedge D_2$ is replaced by D_1, D_2 in every sequent that constitutes the given cut-string $(\overrightarrow{C_2})$. D_2^{*} (of π_3°) is obtained by applying the weakening lemma. In the following text, by $^{\circ}$ we express a sub-proof obtained using the inversion lemma, and by * a formula obtained using the weakening lemma. Note that the cut-formula $D_1 \wedge D_2$ occurs in all minor upper sequents of $(C_{i-1}(\cdots(C_1)\cdots))$, and they are treated in the same manner as π_1 and π_2. The same applies to the other principal reductions in the following text.

1-2. (Principal $\wedge r$) When $A \equiv D_1 \wedge D_2$, let the given cut-string be the following form.

$$
\dfrac{\begin{array}{c}\dfrac{\begin{array}{c}\vdots\ \pi_1\\ \Gamma \vdash \Delta, D_1 \wedge D_2, C_1\end{array} \quad \begin{array}{c}\vdots\ \pi_2\\ C_1, \Gamma \vdash \Delta, D_1 \wedge D_2\end{array}}{\Gamma \vdash \Delta, D_1 \wedge D_2}\,(C_1)\\ \vdots\ (\overrightarrow{C_2})\\ \Pi \vdash \Lambda, D_1 \wedge D_2\end{array} \qquad \begin{array}{c}\vdots\ \pi_3\\ D_1 \wedge D_2, \Pi \vdash \Lambda\end{array}}{\begin{array}{c}\Pi \vdash \Lambda\\ \vdots\ (\overrightarrow{C_{i+1}})\\ \Sigma \vdash \Phi\end{array}}\,(D_1 \wedge D_2)
$$

The above cut-string is transformed as follows.

$$
\dfrac{
\begin{array}{c}
\dfrac{\overset{\vdots\ \pi_1^{\circ\prime}}{\Gamma \vdash \Delta, D_2, C_1} \quad \overset{\vdots\ \pi_2^{\circ\prime}}{C_1, \Gamma \vdash \Delta, D_2}}{\Gamma \vdash \Delta, D_2}\,(C_1) \\
\vdots\ (\overrightarrow{C_2}^{D_2}) \\
\Pi \vdash \Lambda, D_2
\end{array}
\qquad
\dfrac{
\begin{array}{c}
\dfrac{\overset{\vdots\ \pi_1^{\circ}}{D_2^*, \Gamma \vdash \Delta, D_1, C_1} \quad \overset{\vdots\ \pi_2^{\circ}}{C_1, D_2^*, \Gamma \vdash \Delta, D_1}}{D_2, \Gamma \vdash \Delta, D_1}\,(C_1) \\
\vdots\ (\overrightarrow{C_2}^{D_1}) \\
D_2, \Pi \vdash \Lambda, D_1
\end{array}
\quad
\overset{\vdots\ \pi_3^{\circ}}{D_1, D_2, \Pi \vdash \Lambda}
}{D_2, \Pi \vdash \Lambda}\,(D1)
}{
\begin{array}{c}
\Pi \vdash \Lambda \\
\vdots\ (\overrightarrow{C_{i+1}}) \\
\Sigma \vdash \Phi
\end{array}
}\,(D_2)
$$

$\Pi \vdash \Lambda, D_2$ is the additional sequent. We call the resulting $(\overrightarrow{C_{i+1}}(D_2(D_1(\overrightarrow{C_1}))))$ the **main string**, and $(C_{i-1}(\cdots(C_1)\cdots))$ of the additional sequent $\Pi \vdash \Lambda, D_2$ the **additional string**.

Note that when $(C_{i-1}(\cdots(C_1)\cdots))$ is empty, $\wedge l$ and $\wedge r$ reductions coincide.

1-3. (Principal $\neg l$) When $A \equiv \neg D$ and the given cut-string is the following form on the left, it is transformed into the following proof on the right.

$$
\dfrac{
\overset{\vdots\ \pi_3}{\Pi \vdash \Lambda, \neg D}
\quad
\begin{array}{c}
\dfrac{\overset{\vdots\ \pi_1}{\neg D, \Gamma \vdash \Delta, C_1} \quad \overset{\vdots\ \pi_2}{C_1, \neg D, \Gamma \vdash \Delta}}{\neg D, \Gamma \vdash \Delta}\,(C_1) \\
\vdots\ (\overrightarrow{C_2}) \\
\neg D, \Pi \vdash \Lambda
\end{array}
}{
\begin{array}{c}
\Pi \vdash \Lambda \\
\vdots\ (\overrightarrow{C_{i+1}}) \\
\Sigma \vdash \Phi
\end{array}
}\,(\neg D)
\qquad
\dfrac{
\begin{array}{c}
\dfrac{\overset{\vdots\ \pi_1^{\circ}}{\Gamma \vdash \Delta, D, C_1} \quad \overset{\vdots\ \pi_2^{\circ}}{C_1, \Gamma \vdash \Delta, D}}{\Gamma \vdash \Delta, D}\,(C_1) \\
\vdots\ (\overrightarrow{C_2}^{D}) \\
\Pi \vdash \Lambda, D
\end{array}
\quad
\overset{\vdots\ \pi_3^{\circ}}{D, \Pi \vdash \Lambda}
}{
\begin{array}{c}
\Pi \vdash \Lambda \\
\vdots\ (\overrightarrow{C_{i+1}}) \\
\Sigma \vdash \Phi
\end{array}
}\,(D)
$$

1-4. (Principal $\neg r$) When $A \equiv \neg D$ and the given cut-string is the following form on the left, it is transformed into the following proof on the right.

$$
\dfrac{
\begin{array}{c}
\dfrac{\overset{\vdots\ \pi_1}{\Gamma \vdash \Delta, \neg D, C_1} \quad \overset{\vdots\ \pi_2}{C_1, \Gamma \vdash \Delta, \neg D}}{\Gamma \vdash \Delta, \neg D}\,(C_1) \\
\vdots\ (\overrightarrow{C_2}) \\
\Pi \vdash \Lambda, \neg D
\end{array}
\quad
\overset{\vdots\ \pi_3}{\neg D, \Pi \vdash \Lambda}
}{
\begin{array}{c}
\Pi \vdash \Lambda \\
\vdots\ (\overrightarrow{C_{i+1}}) \\
\Sigma \vdash \Phi
\end{array}
}\,(\neg D)
\qquad
\dfrac{
\overset{\vdots\ \pi_3^{\circ}}{\Pi \vdash \Lambda, D}
\quad
\begin{array}{c}
\dfrac{\overset{\vdots\ \pi_1^{\circ}}{D, \Gamma \vdash \Delta, C_1} \quad \overset{\vdots\ \pi_2^{\circ}}{C_1, D, \Gamma \vdash \Delta}}{D, \Gamma \vdash \Delta}\,(C_1) \\
\vdots\ (\overrightarrow{C_2}^{D}) \\
D, \Pi \vdash \Lambda
\end{array}
}{
\begin{array}{c}
\Pi \vdash \Lambda \\
\vdots\ (\overrightarrow{C_{i+1}}) \\
\Sigma \vdash \Phi
\end{array}
}\,(D)
$$

When $(C_{i-1}(\cdots(C_1)\cdots))$ is empty, $\neg l$ and $\neg r$ reductions coincide.

1-5. (Principal $\vee\, l$) Similar to $\wedge r$.
1-6. (Principal $\vee\, r$) Similar to $\wedge l$.
1-7. (Principal $\rightarrow l$) Similar to $\wedge r$.
1-8. (Principal $\rightarrow r$) Similar to $\wedge l$.

(Auxiliary reduction) When $A \notin \overrightarrow{C}$, and a minor sub-proof of C_i exists, whose last is an inference on A, the following cases occur.
2-1. **(Auxiliary** $\wedge\, l$) When $A \equiv B_1 \wedge B_2$, let the given cut-string be the following form, where cut-formulas $\overrightarrow{C}$ are not explicitly represented.

$$\begin{matrix}\dfrac{\dfrac{\begin{matrix}\vdots\ \pi_3\\ B_1, B_2, \Pi_2 \vdash \Lambda_2\end{matrix}}{B_1\wedge B_2, \Pi_2\vdash \Lambda_2}\wedge l \qquad \begin{matrix}\dfrac{\begin{matrix}\vdots\ \pi_1\\ B_1\wedge B_2, \Gamma_1\vdash \Delta_1\end{matrix}\quad \begin{matrix}\vdots\ \pi_2\\ B_1\wedge B_2, \Gamma_2\vdash \Delta_2\end{matrix}}{B_1\wedge B_2, \Gamma_3\vdash \Delta_3}(C_1)\\ \vdots\ (\overrightarrow{C_2})\\ B_1\wedge B_2, \Pi_1\vdash \Lambda_1\end{matrix}}{B_1\wedge B_2, \Pi_3\vdash \Lambda_3}(C_i)\\ \vdots\ (\overrightarrow{C_{i+1}})\\ B_1\wedge B_2, \Sigma\vdash \Phi\end{matrix}$$

The above cut-string is transformed as follows.

$$\begin{matrix}\dfrac{\begin{matrix}\vdots\ \pi_3\\ B_1, B_2, \Pi_2 \vdash \Lambda_2\end{matrix} \qquad \begin{matrix}\dfrac{\begin{matrix}\vdots\ \pi_1^\circ\\ B_1, B_2, \Gamma_1\vdash \Delta_1\end{matrix}\quad \begin{matrix}\vdots\ \pi_2^\circ\\ B_1, B_2, \Gamma_2\vdash \Delta_2\end{matrix}}{B_1, B_2, \Gamma_3\vdash \Delta_3}(C_1)\\ \vdots\ (\overrightarrow{C_2}^{B_1 B_2})\\ B_1, B_2, \Pi_1\vdash \Lambda_1\end{matrix}}{B_1, B_2, \Pi_3\vdash \Lambda_3}(C_i)\\ \vdots\ (\overrightarrow{C_{i+1}}^{B_1 B_2})\\ \dfrac{B_1, B_2, \Sigma\vdash \Phi}{B_1\wedge B_2, \Sigma\vdash \Phi}\wedge l\end{matrix}$$

Note that, apart from π_3, there are applications of $\wedge l$ on $B_1 \wedge B_2$ in minor sub-proofs of ($\overrightarrow{C}$). Furthermore, a number of minor sub-proofs may not end with $\wedge l$ on $B_1 \wedge B_2$. They are treated in the same manner as π_3 or π_1° and π_2°, by applying the inversion lemma if necessary. The same applies to the other auxiliary reductions that follows.
2-2. **(Auxiliary** $\wedge\, r$) When $A \equiv B_1 \wedge B_2$, let the given cut-string be the following form.

$$\begin{matrix}\dfrac{\dfrac{\begin{matrix}\vdots\ \pi_{31}\\ \Pi_2 \vdash \Lambda_2, B_1\end{matrix}\quad \begin{matrix}\vdots\ \pi_{32}\\ \Pi_2 \vdash \Lambda_2, B_2\end{matrix}}{\Pi_2\vdash \Lambda_2, B_1\wedge B_2}\wedge r \qquad \begin{matrix}\dfrac{\begin{matrix}\vdots\ \pi_1\\ \Gamma_1\vdash \Delta_1, B_1\wedge B_2\end{matrix}\quad \begin{matrix}\vdots\ \pi_2\\ \Gamma_2\vdash \Delta_2, B_1\wedge B_2\end{matrix}}{\Gamma_3\vdash \Delta_3, B_1\wedge B_2}(C_1)\\ \vdots\ (\overrightarrow{C_2})\\ \Pi_1\vdash \Lambda_1, B_1\wedge B_2\end{matrix}}{\Pi_3\vdash \Lambda_3, B_1\wedge B_2}(C_i)\\ \vdots\ (\overrightarrow{C_{i+1}})\\ \Sigma\vdash \Phi, B_1\wedge B_2\end{matrix}$$

The above cut-string is transformed as follows.

$$
\dfrac{
\begin{matrix}
\dfrac{\begin{matrix}\vdots\ \pi_{31}\\ \Pi_2 \vdash \Lambda_2, B_1\end{matrix} \quad \begin{matrix}\dfrac{\begin{matrix}\vdots\ \pi_1^\circ \\ \Gamma_1\vdash\Delta_1,B_1\end{matrix}\quad\begin{matrix}\vdots\ \pi_2^\circ\\ \Gamma_2\vdash\Delta_2,B_1\end{matrix}}{\Gamma_3\vdash\Delta_3,B_1}\,(C_1)\\ \vdots\ (\overrightarrow{C_2}^{B_1}) \\ \Pi_1\vdash\Lambda_1,B_1\end{matrix}}{\Pi_3\vdash\Lambda_3,B_1}\,(C_i)\\
\vdots\ (\overrightarrow{C_{i+1}}^{B_1})\\
\Sigma\vdash\Phi,B_1
\end{matrix}
\quad
\begin{matrix}
\dfrac{\begin{matrix}\vdots\ \pi_{32}\\ \Pi_2 \vdash \Lambda_2, B_2\end{matrix} \quad \begin{matrix}\dfrac{\begin{matrix}\vdots\ \pi_1^{\circ'} \\ \Gamma_1\vdash\Delta_1,B_2\end{matrix}\quad\begin{matrix}\vdots\ \pi_2^{\circ'}\\ \Gamma_2\vdash\Delta_2,B_2\end{matrix}}{\Gamma_3\vdash\Delta_3,B_2}\,(C_1)\\ \vdots\ (\overrightarrow{C_2}^{B_2}) \\ \Pi_1\vdash\Lambda_1,B_2\end{matrix}}{\Pi_3\vdash\Lambda_3,B_2}\,(C_i)\\
\vdots\ (\overrightarrow{C_{i+1}}^{B_2})\\
\Sigma\vdash\Phi,B_2
\end{matrix}
}{\Sigma\vdash\Phi,B_1\wedge B_2}\,\wedge r
$$

2-3. (**Auxiliary** $\neg l$) When $A \equiv \neg B$ and the given cut-string is the following form on the left, it is transformed into the following proof on the right.

$$
\begin{matrix}
\dfrac{\dfrac{\begin{matrix}\vdots\ \pi_3\\ \Pi_2\vdash\Lambda_2,B\end{matrix}}{\neg B,\Pi_2\vdash\Lambda_2}\,\neg l \quad \begin{matrix}\dfrac{\begin{matrix}\vdots\ \pi_1\\ \neg B,\Gamma_1\vdash\Delta_1\end{matrix}\quad\begin{matrix}\vdots\ \pi_2\\ \neg B,\Gamma_2\vdash\Delta_2\end{matrix}}{\neg B,\Gamma_3\vdash\Delta_3}\,(C_1)\\ \vdots\ (\overrightarrow{C_2})\\ \neg B,\Pi_1\vdash\Lambda_1\end{matrix}}{\neg B,\Pi_3\vdash\Lambda_3}\,(C_i)\\
\vdots\ (\overrightarrow{C_{i+1}})\\
\neg B,\Sigma\vdash\Phi
\end{matrix}
\qquad
\dfrac{\begin{matrix}
\dfrac{\begin{matrix}\vdots\ \pi_3\\ \Pi_2\vdash\Lambda_2,B\end{matrix}\quad\begin{matrix}\dfrac{\begin{matrix}\vdots\ \pi_1^\circ\\ \Gamma_1\vdash\Delta_1,B\end{matrix}\quad\begin{matrix}\vdots\ \pi_2^\circ\\ \Gamma_2\vdash\Delta_2,B\end{matrix}}{\Gamma_3\vdash\Delta_3,B}\,(C_1)\\ \vdots\ (\overrightarrow{C_2}^{B})\\ \Pi_1\vdash\Lambda_1,B\end{matrix}}{\Pi_3\vdash\Lambda_3,B}\,(C_i)\\
\vdots\ (\overrightarrow{C_{i+1}}^{B})\\
\Sigma\vdash\Phi,B
\end{matrix}}{\neg B,\Sigma\vdash\Phi}\,\neg l
$$

2-4. (**Auxiliary** $\neg r$) When $A \equiv \neg B$ and the given cut-string is the following form on the left, it is transformed into the following proof on the right.

$$
\begin{matrix}
\dfrac{\dfrac{\begin{matrix}\vdots\ \pi_3\\ B,\Pi_2\vdash\Lambda_2\end{matrix}}{\Pi_2\vdash\Lambda_2,\neg B}\,\neg r \quad \begin{matrix}\dfrac{\begin{matrix}\vdots\ \pi_1\\ \Gamma_1\vdash\Delta_1,\neg B\end{matrix}\quad\begin{matrix}\vdots\ \pi_2\\ \Gamma_2\vdash\Delta_2,\neg B\end{matrix}}{\Gamma_3\vdash\Delta_3,\neg B}\,(C_1)\\ \vdots\ (\overrightarrow{C_2})\\ \Pi_1\vdash\Lambda_1,\neg B\end{matrix}}{\Pi_3\vdash\Lambda_3,\neg B}\,(C_i)\\
\vdots\ (\overrightarrow{C_{i+1}})\\
\Sigma\vdash\Phi,\neg B
\end{matrix}
\qquad
\dfrac{\begin{matrix}
\dfrac{\begin{matrix}\vdots\ \pi_3\\ B,\Pi_2\vdash\Lambda_2\end{matrix}\quad\begin{matrix}\dfrac{\begin{matrix}\vdots\ \pi_1^\circ\\ B,\Gamma_1\vdash\Delta_1\end{matrix}\quad\begin{matrix}\vdots\ \pi_2^\circ\\ B,\Gamma_2\vdash\Delta_2\end{matrix}}{B,\Gamma_3\vdash\Delta_3}\,(C_1)\\ \vdots\ (\overrightarrow{C_2}^{B})\\ B,\Pi_1\vdash\Lambda_1\end{matrix}}{B,\Pi_3\vdash\Lambda_3}\,(C_i)\\
\vdots\ (\overrightarrow{C_{i+1}}^{B})\\
B,\Sigma\vdash\Phi
\end{matrix}}{\Sigma\vdash\Phi,\neg B}\,\neg r
$$

2-5. (**Auxiliary** $\vee l$) Similar to $\wedge r$.
2-6. (**Auxiliary** $\vee r$) Similar to $\wedge l$.
2-7. (**Auxiliary** $\rightarrow l$) Similar to $\wedge r$.
2-8. (**Auxiliary** $\rightarrow r$) Similar to $\wedge l$.

To prove our cut-elimination theorem, we define the complexity of a cut-string in terms of cut-weight and cut-height.

Definition 5 Let $(\overrightarrow{C}) \equiv (C_n(\cdots(C_2(C_1))\cdots))$ be a cut-string in a given proof, and let $1 \leq i \leq n$. The **cut-weight** $w(\overrightarrow{C})$ is the sum of the weight $w(C_i)$ of every C_i. The **cut-height** $h(\overrightarrow{C})$ is the sum of heights of the minor sub-proof of every (C_i). The complexity of $(\overrightarrow{C})$ is the pair $(w(\overrightarrow{C}), h(\overrightarrow{C}))$.

Thus, $h(\overrightarrow{C}) = 0$ when every minor sub-proof of $(\overrightarrow{C})$ consists of an axiom, and $w(\overrightarrow{C}) = 0$ when every cut-formula of $(\overrightarrow{C})$ is an atom.

Theorem 1 (Cut-elimination) *If $\Gamma \vdash \Delta$ is provable in* G3 + *cut, then it is provable in* G3 *without cut.*

Proof Assume that given proof π ends with *cut* between $\Gamma \vdash \Delta, C$ and $C, \Gamma \vdash \Delta$, and no other *cut* is applied in π. To prove the theorem, it is sufficient to show that (C) is eliminable (cf. [12]). If we apply our reduction to "any" cut-string in a given proof, then, in an application of the principal reduction, new and complex cut-strings may be produced. Thus, we first define **proper cut-strings**, according to our reduction rules Definition 4 as follows.

(1) The given first *cut* (C) is the proper cut-string.

(2) In the principal $\neg l$ and $\neg r$ reductions of Definition 4, if the original $(\overrightarrow{C_{i+1}}(\neg D(\overrightarrow{C_1})))$ is proper, then the resulting $(\overrightarrow{C_{i+1}}(D(\overrightarrow{C_1})))$ is also proper.

(3) In the principal $\star l$- and $\star r$-reductions, where $\star$ is $\wedge$, $\vee$, or $\rightarrow$, of Definition 4, if the original $(\overrightarrow{C_{i+1}}(D_1 \star D_2(\overrightarrow{C_1})))$ is proper, then the resulting $(\overrightarrow{C_{i+1}}(D_2(D_1(\overrightarrow{C_1}))))$ is also proper. Furthermore, the additional string $(C_{i-1}(\cdots(C_1)\cdots))$ is another proper cut-string, if any.

(4) In every auxiliary reduction, if the original cut-string is proper, then the resulting one is also proper. In the auxiliary $\wedge r$, $\vee l$, and $\rightarrow l$ reductions, both cut-strings in the resulting proof are proper.

We call a cut-string in a given proof as **topmost**, when no *cut* appears in any minor sub-proof thereof. Let us consider the following general form, which is topmost. The usual *cut* is a particular case, where $(\overrightarrow{C_2})$ is empty.

$$\dfrac{\begin{array}{c}\vdots\ \pi_1\\ \Gamma \vdash \Delta, C_1\end{array} \quad \begin{array}{c}\vdots\ \pi_2\\ C_1, \Gamma \vdash \Delta\end{array}}{\begin{array}{c}\Gamma \vdash \Delta\\ \vdots\ (\overrightarrow{C_2})\\ \Sigma \vdash \Phi\end{array}}\ (C_1)$$

We prove our claim by induction on the complexity $(w(\overrightarrow{C}), h(\overrightarrow{C}))$ of the given topmost proper cut-string.

(Induction step) When $w(\vec{C}) > 0$ or $h(\vec{C}) > 0$, let non atomic $A \in \Gamma \cup \Delta \cup \{C_1\}$.

• If $A \in \vec{C}$, then by applying a principal reduction, $w(\vec{C})$ is reduced. Even if the number of proper cut-strings is increased, their complexities are smaller than that of the original cut-string; hence, the induction hypothesis is applied.

• If $A \notin \vec{C}$ and a minor sub-proof of C_i of $\vec{C}$ exists, whose last is an inference on A, then we apply an auxiliary reduction. In this case, $h(\vec{C})$ is reduced (under the assumption of the height-preserving inversion lemma).

• If no $A \in \Gamma \cup \Delta \cup \{C_1\}$ exists, to which any reduction rule is applied, then $(\vec{C})$ consists only of atoms, i.e., $w(\vec{C}) = 0$, because no principal reduction is applied. Furthermore, we find $h(\vec{C}) = 0$ by examining the end of every minor sub-proof of $(\vec{C})$ as follows. (i) It is not a rule on a cut-formula, because $(\vec{C})$ consists of atoms; (ii) It is not a rule on non cut-formula, because no auxiliary reduction is applied; (iii) It is not *cut*, since the given cut-string is topmost. Therefore, every minor sub-proof of $(\vec{C})$ consists of an axiom.

(Base step) When $w(\vec{C}) = 0$ and $h(\vec{C}) = 0$, the given cut-string has the following form on the left.

$$\dfrac{\Gamma \vdash \Delta, Q_1 \quad Q_1, \Gamma \vdash \Delta}{\Gamma \vdash \Delta} (Q_1) \qquad\qquad \begin{array}{c}\Gamma \vdash \Delta \\ \vdots\ (\vec{Q_2}) \\ \Pi \vdash \Lambda\end{array}$$
$$\begin{array}{c}\vdots\ (\vec{Q_2}) \\ \Pi \vdash \Lambda\end{array}$$

In this case, the cut-string $(\vec{Q})$ is eliminated in order from Q_1. We find that an atom $R \in \Gamma \cap \Delta$. To the contrary, assume that no $R \in \Gamma \cap \Delta$ exists. Then, non cut-atom $Q_1 \in \Gamma$ since $\Gamma \vdash \Delta, Q_1$ is an axiom, and similarly, non cut-atom $Q_1 \in \Delta$ since $Q_1, \Gamma \vdash \Delta$ is an axiom. However, $Q_1 \in \Gamma \cap \Delta$ contradicts the assumption. Thus, we have $R \in \Gamma \cap \Delta$, which means $\Gamma \vdash \Delta$ is an axiom by itself. Thus, by eliminating (Q_1), we obtain the above cut-string on the right. All minor sub-proofs are axioms in $(\vec{Q_2})$, because $h(\vec{Q}) = 0$; hence, by the same manner, all $(\vec{Q_2})$ are eliminated. ■

In terms of the abovementioned base step, in every axiom $\Gamma, R \vdash R, \Delta$ of G3 + *cut*, we can assume that R is not a cut-atom.

Lemma 3 *Let* $(\vec{Q})$ *be a cut-string that consists only of cut-atoms and* $h(\vec{Q}) = 0$. *For every minor upper sequent* $\Gamma \vdash \Delta$ *of* $(\vec{Q})$, *an atom* R *exists, such that* R *is not a cut-atom and* $R \in \Gamma \cap \Delta$.

Example 1 The following is an example of our cut-string reduction.

Let a given proof be the following form, where the cut-formula is indicated by $\Box$.

$$\dfrac{\begin{matrix}\vdots & & \vdots\\ A, B\wedge C \vdash A\wedge B, \boxed{\neg(A\wedge\neg B)} & & \boxed{\neg(A\wedge \neg B)}, A, B\wedge C\vdash A\wedge B\end{matrix}}{A, B\wedge C\vdash A\wedge B}\ (\neg(A\wedge\neg B))$$

By the principal $\neg l$-reduction, we obtain the following proof.

$$\dfrac{\begin{matrix}\vdots & & \vdots\\ \boxed{A\wedge\neg B}, A, B\wedge C \vdash A\wedge B & & A, B\wedge C\vdash A\wedge B, \boxed{A\wedge \neg B}\end{matrix}}{A, B\wedge C\vdash A\wedge B}\ (A\wedge\neg B)$$

By the principal $\wedge r$-reduction, we obtain the following proof.

$$\dfrac{\begin{matrix}\vdots\\ A, B\wedge C\vdash A\wedge B, \boxed{\neg B}\end{matrix} \qquad \dfrac{\begin{matrix}\vdots & & \vdots\\ \boxed{A}, \neg B, A, B\wedge C\vdash A\wedge B & & \neg B^{*}, A, B\wedge C\vdash A\wedge B, \boxed{A}\end{matrix}}{\boxed{\neg B}, A, B\wedge C\vdash A\wedge B}\ (A)}{A, B\wedge C\vdash A\wedge B}\ (\not\angle\, gB)$$

By the principal $\neg l$-reduction, we obtain the following proof, where we assume the application of $\wedge l$.

$$\dfrac{\dfrac{\begin{matrix}\vdots\\ B, A, B, C\vdash A\wedge B\end{matrix}}{\boxed{B}, A, B\wedge C\vdash A\wedge B}\ \wedge l \qquad \dfrac{\begin{matrix}\vdots & & \vdots\\ \boxed{A}, A, B\wedge C\vdash B, A\wedge B & & A, B\wedge C\vdash A\wedge B, B, \boxed{A}\end{matrix}}{A, B\wedge C\vdash A\wedge B, \boxed{B}}\ (A)}{A, B\wedge C\vdash A\wedge B}\ (B)$$

By the auxiliary $\wedge l$-reduction, we obtain the following proof, where we assume the application of $\wedge r$.

$$\dfrac{\dfrac{\dfrac{B, A, B, C\vdash A \quad B, A, B, C\vdash B}{\boxed{B}, A, B, C\vdash A\wedge B}\ \wedge r \qquad \dfrac{\begin{matrix}\vdots & & \vdots\\ \boxed{A}, A, B, C\vdash B, A\wedge B & & A, B, C\vdash A\wedge B, B, \boxed{A}\end{matrix}}{A, B, C\vdash A\wedge B, \boxed{B}}\ (A)}{A, B, C\vdash A\wedge B}\ (B)}{A, B\wedge C\vdash A\wedge B}\ \wedge l$$

By the auxiliary $\wedge r$-reduction, we obtain the following proof, where S_1 is $\boxed{A}$, $A, B, C \vdash B, B$ and S_2 is $A, B, C \vdash B, B, \boxed{A}$.

$$\dfrac{\dfrac{\dfrac{\boxed{B}, A, B, C\vdash A \qquad \dfrac{\boxed{A}, A, B, C\vdash B, A \quad A, B, C\vdash A, B, \boxed{A}}{A, B, C\vdash A, \boxed{B}}\ (A)}{A, B, C\vdash A}\ (B) \qquad \dfrac{\boxed{B}, A, B, C\vdash B \quad \dfrac{S_1 \quad S_2}{A, B, C\vdash B, \boxed{B}}\ (A)}{A, B, C\vdash B}\ (B)}{A, B, C\vdash A\wedge B}\ \wedge r}{A, B\wedge C\vdash A\wedge B}\ \wedge l$$

13.3 Multi-Sequent Representation of Cut-Elimination

We define our sequent calculus for multi-sequents mG3 in Sect. 13.3.1. We show that the cut-string reduction in G3 is represented by the proof construction in mG3 in Sect. 13.3.2.

13.3.1 Sequent Calculus for Multi-Sequents mG3

We extend the usual sequents to multi-sequents. Although our multi-sequent is similar to a hypersequent [1, 2], our multi-sequent is specialized to represent a cut-string, and its interpretation differs from the standard one. When we distinguish the usual sequent from a multi-sequent, we call it a **single sequent**.

We first define the intersection among sequents as follows. For any n,

$$(\Gamma_1 \vdash \Delta_1) \cap \cdots \cap (\Gamma_n \vdash \Delta_n) = \Gamma_1 \cap \cdots \cap \Gamma_n \vdash \Delta_1 \cap \cdots \cap \Delta_n$$

where $\cap$ in the right-hand side is the multiset intersection operation. For example, $\{A, A, B, B, C\} \cap \{A, A, B, D\} = \{A, A, B\}$. We sometimes write the above intersection as $\bigcap(\Gamma_i \vdash \Delta_i)$.

Our *cut* has the context sharing form; thus, *cut* can be specified by the intersection of its upper sequents. This observation is extended to any cut-string as follows.

Lemma 4 *Let a cut-string* $(\vec{C})$ *in a given proof be the following form.*

$$\dfrac{\Pi \vdash \Lambda, C_n \qquad \dfrac{C_n, \Pi \vdash \Lambda, C_{n-1} \qquad \begin{array}{c}\dfrac{\Gamma \vdash \Delta, C_1 \quad C_1, \Gamma \vdash \Delta}{\Gamma \vdash \Delta}\,(C_1)\\ \vdots\,(\vec{C_2})\\ C_{n-1}, C_n, \Pi \vdash \Lambda\end{array}}{C_n, \Pi \vdash \Lambda}\,(C_{n-1})}{\Pi \vdash \Lambda}\,(C_n)$$

where $(\vec{C_2})$ *is the successive applications of cut on* $C_2, \ldots, C_{n-2}$. *The intersection of all minor upper sequents of* $(\vec{C})$ *is then the end sequent* $\Pi \vdash \Lambda$ *of* $(\vec{C})$.

Proof By induction on the length n of the given cut-string, where the length of a cut-string is the length of the corresponding path in a given proof.
(Base step) When $n = 1$, we obtain $(\Gamma \vdash \Delta, C_1) \cap (C_1, \Gamma \vdash \Delta) = \Gamma \vdash \Delta$.
(Induction step) When $n > 1$, by the induction hypothesis, we obtain $(\Gamma \vdash \Delta, C_1) \cap (C_1, \Gamma \vdash \Delta) \cap \cdots \cap (C_n, \Pi \vdash \Lambda, C_{n-1}) = C_n, \Pi \vdash \Lambda$. Hence, we have $(\Pi \vdash \Lambda, C_n) \cap (C_n, \Pi \vdash \Lambda) = \Pi \vdash \Lambda$. ■

Our multi-sequent is obtained, which begins from a single sequent, by connecting another single sequent individually.

Definition 6 A sequence of single sequents called a **multi-sequent** $\mathcal{M}, \mathcal{M}_1, \mathcal{M}_2, \ldots$, and cut-formulas thereof are inductively defined as follows.

- A single sequent $\Gamma \vdash \Delta$ is a multi-sequent, which has no cut-formula.
- Let $\mathcal{M}$ be a multi-sequent that consists of single sequents $\Pi_1 \vdash \Lambda_1, \ldots, \Pi_n \vdash \Lambda_n$.
- $(\Gamma \vdash \Delta, C \mid \mathcal{M})_C$ is a multi-sequent if $(\Pi_1 \vdash \Lambda_1) \cap \cdots \cap (\Pi_n \vdash \Lambda_n) = C, \Gamma \vdash \Delta$.
- $(C, \Gamma \vdash \Delta \mid \mathcal{M})_C$ is a multi-sequent if $(\Pi_1 \vdash \Lambda_1) \cap \cdots \cap (\Pi_n \vdash \Lambda_n) = \Gamma \vdash \Delta, C$.

The **cut-formulas** of the multi-sequent $(\Gamma \vdash \Delta, C \mid \mathcal{M})_C$ are, in addition to the cut-formulas of $\mathcal{M}$, the set of single occurrences of C in $\Gamma \vdash \Delta, C$ and in every $C, \Gamma \vdash \Delta$ shared by all single sequents of $\mathcal{M}$. Similarly for $(C, \Gamma \vdash \Delta \mid \mathcal{M})_C$. We assume $(\Gamma \vdash \Delta \mid \mathcal{M})_C = (\mathcal{M} \mid \Gamma \vdash \Delta)_C$ for any multi-sequent $\mathcal{M}$ and any sequent $\Gamma \vdash \Delta$. Each single sequent that constitutes a multi-sequent $\mathcal{M}$ is called as a **component sequent** of $\mathcal{M}$. By $\bigcap \mathcal{M}$, we mean $\bigcap(\Gamma_i \vdash \Delta_i)$ for all component sequents $\Gamma_i \vdash \Delta_i$ of $\mathcal{M}$.

We sometimes write a multi-sequent $(\Gamma \vdash \Delta \mid \mathcal{M})_C$ as $(\Gamma \vdash \Delta \mid \mathcal{M})$ without the subscription of its cut-formula when it is unnecessary. In what follows, when we write $(\mathcal{M}_1 \mid \mathcal{M}_2)$, either $\mathcal{M}_1$ or $\mathcal{M}_2$ is a single sequent.

Remark 2 We may omit the parentheses of our multi-sequent by assuming that its component sequents are arranged so that they are aligned from right to left. However, the subscription of the cut-formula is sometimes useful, and especially in the description of our mG3-rules, it is convenient not to fix the order of the component sequents; hence, we will keep our description with brackets here.

When $\mathcal{M} \equiv (\Gamma \vdash \Delta \mid \mathcal{M}')_C$, the difference between $\Gamma \vdash \Delta$ and $\bigcap \mathcal{M}'$ is only in the (position of) cut-formula C. The same applies to every component sequent of $\mathcal{M}'$ by assuming $\mathcal{M}' = (\Pi \vdash \Lambda \mid \mathcal{M}'')_D$. Thus, $\bigcap \mathcal{M}$ is the single sequent obtained from any component sequent of $\mathcal{M}$ by deleting all cut-formulas of $\mathcal{M}$. See also Example 2 below.

Example 2 In the following multi-sequent, the underlined C_1, C_2, C_3 are the cut-formulas of this multi-sequent.

$$(\underline{C_3}, A \vdash C_1 \mid (A \vdash C_1, \underline{C_2}, \underline{C_3} \mid (\underline{C_2}, A \vdash C_1, \underline{C_1}, \underline{C_3} \mid \underline{C_2}, \underline{C_1}, A \vdash C_1, \underline{C_3})_{C_1})_{C_2})_{C_3}$$

Note that cut-formulas are occurrences; hence, the underlined occurrences of C_1 are cut-formulas, and the other occurrences of C_1 are not.

Given a multi-sequent $\mathcal{M}$, by indicating a multi-sequent $\mathcal{M}'$, which is a part of $\mathcal{M}$, and by abbreviating the other parts of $\mathcal{M}$ other than $\mathcal{M}'$, we denote $\mathcal{M}$ as $\mathcal{C}(\mathcal{M}')$. For example, the multi-sequent in Example 2 can be expressed as $\mathcal{C}(A \vdash C_1, C_2, C_3 \mid \mathcal{M}')_{C_2}$ or $\mathcal{C}'(\mathcal{M}')$, where $\mathcal{M}' \equiv (C_2, A \vdash C_1, C_1, C_3 \mid C_2, C_1, A \vdash C_1, C_3)_{C_1}$.

Our calculus is formalized in the proof-search style. In other words, given a multi-sequent, we apply our rule from bottom to top. The following mG3-rules are read from bottom to top. The lower sequent contains an expression like $\mathcal{M}_i\langle A \vdash B\rangle$ that describes the condition in which A occurs in the antecedent and B occurs in the succedent of “all” component sequents of $\mathcal{M}_i$. If this condition is not satisfied, then the rule cannot be applied. The upper sequent contains an expression like $\mathcal{M}_i\{C, D \vdash E, F\}$ that expresses the multi-sequent obtained by replacing the occurrence of A, indicated in the lower sequent $\mathcal{M}_i\langle A \vdash B\rangle$, with C, D and the occurrence of B with E, F for all component sequents of $\mathcal{M}_i$. The abovementioned A, B, C, D, E, or F may be missing. When A (as well as B) is missing, C, D (resp. E, F), which may also be missing, is added to the designated place. When both C and D (resp. E and F) are missing, A (resp. B) is deleted from the designated place.

Definition 7 **mG3-rules** consist of the following principal and auxiliary rules.

Principal rules: Let $\mathcal{M}_1$ and $\mathcal{M}_2$ be multi-sequents such that $\mathcal{M}_1 \cap \mathcal{M}_2 = \Gamma \vdash \Delta$, and one of them is single.

$$\frac{\mathcal{C}(\mathcal{M}_1\{D \vdash\} \mid \mathcal{M}_2\{\vdash D\})_D}{\mathcal{C}(\mathcal{M}_1\langle\vdash \neg D\rangle \mid \mathcal{M}_2\langle\neg D \vdash\rangle)_{\neg D}}\ \neg$$

$$\frac{\mathcal{M}_1\{\vdash D_2\} \quad \mathcal{C}(\Gamma \vdash \Delta, D_2 \mid (\mathcal{M}_1\{D_2 \vdash D_1\} \mid \mathcal{M}_2\{D_1, D_2 \vdash\})_{D_1})_{D_2}}{\mathcal{C}(\mathcal{M}_1\langle\vdash D_1 \wedge D_2\rangle \mid \mathcal{M}_2\langle D_1 \wedge D_2 \vdash\rangle)_{D_1 \wedge D_2}}\ \wedge$$

$$\frac{\mathcal{M}_2\{D_2 \vdash\} \quad \mathcal{C}(D_2, \Gamma \vdash \Delta \mid (\mathcal{M}_1\{\vdash D_1, D_2\} \mid \mathcal{M}_2\{D_1 \vdash D_2\})_{D_1})_{D_2}}{\mathcal{C}(\mathcal{M}_1\langle\vdash D_1 \vee D_2\rangle \mid \mathcal{M}_2\langle D_1 \vee D_2 \vdash\rangle)_{D_1 \vee D_2}}\ \vee$$

$$\frac{\mathcal{M}_2\{D_2 \vdash\} \quad \mathcal{C}(D_2, \Gamma \vdash \Delta \mid (\mathcal{M}_1\{D_1 \vdash D_2\} \mid \mathcal{M}_2\{\vdash D_2, D_1\})_{D_1})_{D_2}}{\mathcal{C}(\mathcal{M}_1\langle\vdash \mathcal{D}_1 \to D_2\rangle \mid \mathcal{M}_2\langle D_1 \to D_2 \vdash\rangle)_{D_1 \to D_2}}\ \to$$

$\mathcal{M}_1\{\vdash D_2\}$ of the $\wedge$-rule and $\mathcal{M}_2\{D_2 \vdash\}$ of $\vee$ and $\to$ rules are called **additional sequents**.

Auxiliary rules:

$$\frac{\mathcal{M}\{\vdash B\}}{\mathcal{M}\langle\neg B \vdash\rangle}\ \neg l \qquad \frac{\mathcal{M}\{B \vdash\}}{\mathcal{M}\langle\vdash \neg B\rangle}\ \neg r \qquad \frac{\mathcal{M}\{B_1, B_2 \vdash\}}{\mathcal{M}\langle B_1 \wedge B_2 \vdash\rangle}\ \wedge l \qquad \frac{\mathcal{M}\{\vdash B_1\} \quad \mathcal{M}\{\vdash B_2\}}{\mathcal{M}\langle\vdash B_1 \wedge B_2\rangle}\ \wedge r$$

$$\frac{\mathcal{M}\{B_1 \vdash\} \quad \mathcal{M}\{B_2 \vdash\}}{\mathcal{M}\langle B_1 \vee B_2 \vdash\rangle}\ \vee l \qquad \frac{\mathcal{M}\{\vdash B_1, B_2\}}{\mathcal{M}\langle\vdash B_1 \vee B_2\rangle}\ \vee r$$

$$\frac{\mathcal{M}\{\vdash B_1\} \quad \mathcal{M}\{B_2 \vdash\}}{\mathcal{M}\langle B_1 \to B_2 \vdash\rangle}\ \to l \qquad \frac{\mathcal{M}\{B_1 \vdash B_2\}}{\mathcal{M}\langle\vdash B_1 \to B_2\rangle}\ \to r$$

An axiom of mG3 is a multi-sequent such that each of its component sequents is an axiom of G3 on an atom that is not a cut-formula.

A non-single additional sequent corresponds to an additional string in the corresponding cut-string reduction rule. In the principal $\wedge$-rule, when $\mathcal{M}_1$ is single, the additional sequent coincides with $\Gamma \vdash \Delta, D_2$; in principal $\vee$ and $\rightarrow$ rules, when $\mathcal{M}_2$ is single, the additional sequent coincides with $D_2, \Gamma \vdash \Delta$.

Every auxiliary rule has a form in which the usual G3-rule is applied to every component sequent of the given $\mathcal{M}$ simultaneously. In particular, when $\mathcal{M}$ in the rule is a single sequent, it is the usual G3-rule. The main formula of every auxiliary rule is not a cut-formula. The reason is that, by the condition described in the lower sequent, the main formula, such as $B_1 \wedge B_2$ of the $\wedge l$-rule, should appear at the designated place of every component sequent of the lower sequent $\mathcal{M}$. This condition is not satisfied by any cut-formula. In contrast, the main formula of every principal rule is a cut-formula, which is indicated by the subscription of the given lower multi-sequent.

Definition 8 An **m-proof** of $\mathcal{M}$ is a finite tree constructed using rules of mG3, whose root is $\mathcal{M}$, and leaves are axioms of mG3.

Example 3 The following m-proof corresponds to the cut-string reduction in Example 1, where cut-formulas are indicated by $\boxed{}$, and
$\mathcal{M}_1 \equiv (\boxed{B}, A, B, C \vdash A \mid (A, B, C \vdash A, \boxed{A}, \boxed{B} \mid \boxed{A}, A, B, C \vdash A, \boxed{B}))$;
$\mathcal{M}_2 \equiv (\boxed{B}, A, B, C \vdash B \mid (A, B, C \vdash B, \boxed{A}, \boxed{B} \mid \boxed{A}, A, B, C \vdash B, \boxed{B}))$, and
we omit the sub-proof π of the additional sequent $A, B \wedge C \vdash A \wedge B, \neg B$.

$$
\dfrac{\dfrac{\begin{matrix}\vdots\ \pi\\ A, B \wedge C \vdash A \wedge B, \neg B\end{matrix} \quad \dfrac{\dfrac{\dfrac{\mathcal{M}_1 \qquad \mathcal{M}_2}{(\boxed{B}, A, B, C \vdash A \wedge B \mid (A, B, C \vdash A \wedge B, \boxed{A}, \boxed{B} \mid \boxed{A}, A, B, C \vdash A \wedge B, \boxed{B}))}\wedge r}{(\boxed{B}, A, B \wedge C \vdash A \wedge B \mid (A, B \wedge C \vdash A \wedge B, \boxed{A}, \boxed{B} \mid \boxed{A}, A, B \wedge C \vdash A \wedge B, \boxed{B}))}\wedge l}{(A, B \wedge C \vdash A \wedge B, \boxed{\neg B} \mid (\boxed{\neg B}, A, B \wedge C \vdash A \wedge B, \boxed{A} \mid \boxed{A}, \boxed{\neg B}, A, B \wedge C \vdash A \wedge B \mid))}\neg}{(\boxed{A \wedge \neg B}, A, B \wedge C \vdash A \wedge B \mid A, B \wedge C \vdash A \wedge B, \boxed{A \wedge \neg B})}\wedge}{(A, B \wedge C \vdash A \wedge B, \boxed{\neg(A \wedge \neg B)} \mid \boxed{\neg(A \wedge \neg B)}, A, B \wedge C \vdash A \wedge B)}\neg
$$

As in G3, we have the following weakening and inversion lemmas in mG3.

Lemma 5 (Weakening in mG3) *If $\mathcal{M}$ is provable in mG3, then $\mathcal{M}'$, which is obtained from $\mathcal{M}$ by replacing every component sequent $\Gamma \vdash \Delta$ of $\mathcal{M}$ with $\Pi, \Gamma \vdash \Delta, \Lambda$ is also provable for any Π, Λ.*

Lemma 6 (Inversion in mG3) *For every mG3-rule, if the lower sequent of the rule is provable, then the upper sequents thereof are also provable.*

Proof We show the lemma by induction on the height of a given m-proof of the lower sequent of each rule. Since our auxiliary rule is defined as the simultaneous applications of the corresponding G3-rule, the lemma holds for auxiliary rules in the same way as in G3 [12].

For principal rules, we show only the case of principal $\wedge$. The cases of $\neg$, $\vee$ and $\rightarrow$ are similar. Let $\mathcal{C}(\mathcal{M}_1\langle\vdash D_1 \wedge D_2\rangle \mid \mathcal{M}_2\langle D_1 \wedge D_2 \vdash\rangle)$ be provable, where $\mathcal{M}_1 \cap \mathcal{M}_2 = \Gamma \vdash \Delta$. We show that $\mathcal{M}_1\{\vdash D_2\}$ and $\mathcal{C}(\Gamma \vdash \Delta, D_2 \mid (\mathcal{M}_1\{D_2 \vdash D_1\} \mid \mathcal{M}_2\{D_1, D_2 \vdash\}))$ are provable. We divide the following cases according to the last rule applied in the given m-proof of $\mathcal{C}(\mathcal{M}_1\langle\vdash D_1 \wedge D_2\rangle \mid \mathcal{M}_2\langle D_1 \wedge D_2 \vdash\rangle)$.

- When the last rule is the principal $\wedge$-rule on $D_1 \wedge D_2$, the lemma is immediate.
- When the last rule is an axiom, that is, $\mathcal{C}(\mathcal{M}_1\langle\vdash D_1 \wedge D_2\rangle \mid \mathcal{M}_2\langle D_1 \wedge D_2 \vdash\rangle)$ is an axiom of mG3, every component sequent of $\mathcal{M}_1$ and $\mathcal{M}_2$ is a G3-axiom. Then, $\Gamma \vdash \Delta$ is also a G3-axiom, since the difference between $\mathcal{M}_1$ and $\mathcal{M}_2$ is in the cut-formula $D_1 \wedge D_2$. Therefore, both $\mathcal{M}_1\{\vdash D_2\}$ and $\mathcal{C}(\Gamma \vdash \Delta, D_2 \mid (\mathcal{M}_1\{D_2 \vdash D_1\} \mid \mathcal{M}_2\{D_1, D_2 \vdash\}))$ are mG3-axioms.
- When the last rule is an auxiliary rule, the lemma is shown in the same way as in G3.
- When the last rule is the principal $\wedge$ on $C_1 \wedge C_2$ different from $D_1 \wedge D_2$. We examine the following case, where $\mathcal{M}_1$ is not single. (The other cases are simpler.)

$$\frac{\mathcal{M}_3\{\vdash C_2\} \quad \mathcal{C}(\mathcal{C}_1(\Pi \vdash \Lambda, C_2 \mid (\mathcal{M}_3\{C_2 \vdash C_1\} \mid \mathcal{M}_4\{C_1, C_2 \vdash\})_{C_1})_{C_2} \mid \mathcal{M}_2)_{D_1 \wedge D_2}}{\mathcal{C}(\mathcal{C}_1(\mathcal{M}_3\langle\vdash C_1 \wedge C_2\rangle \mid \mathcal{M}_4\langle C_1 \wedge C_2 \vdash\rangle)_{C_1 \wedge C_2} \mid \mathcal{M}_2)_{D_1 \wedge D_2}} \wedge$$

The following (1), which is the additional sequent of $\wedge$ on $D_1 \wedge D_2$, and (2) are provable by the induction hypothesis.

$$\mathcal{C}_1(\Pi \vdash \Lambda, C_2 \mid \mathcal{M}_3\{C_2 \vdash C_1\} \mid \mathcal{M}_4\{C_1, C_2 \vdash\})\{\vdash D_2\} \tag{13.1}$$

$$\mathcal{C}(\Gamma \vdash \Delta, D_2 \mid \mathcal{C}_1(\Pi \vdash \Lambda, C_2 \mid \mathcal{M}_3\{C_2 \vdash C_1\} \mid \mathcal{M}_4\{C_1, C_2 \vdash\})\{D_2 \vdash D_1\} \mid \mathcal{M}_2\{D_1, D_2 \vdash\}) \tag{13.2}$$

From $\mathcal{M}_3\{\vdash C_2\}\langle\vdash D_1 \wedge D_2\rangle$ by applying the inversion lemma and weakening lemma to $\vdash D_1 \wedge D_2$, which is not a cut-formula in this additional sequent, the following (3) and (4) are provable.

$$\mathcal{M}_3\{\vdash C_2\}\{D_2 \vdash D_1\} \tag{13.3}$$

$$\mathcal{M}_3\{\vdash C_2\}\{\vdash D_2\} \tag{13.4}$$

By applying the $\wedge$-rule on $C_1 \wedge C_2$ to the above (3) and (2), as well as to (4) and (1), the following (5) and (6) are provable, which are required sequents.

$$\mathcal{C}(\Gamma \vdash \Delta, D_2 \mid \mathcal{C}_1(\mathcal{M}_3\langle\vdash C_1 \wedge C_2\rangle \mid \mathcal{M}_4\langle C_1 \wedge C_2 \vdash\rangle)\{D_2 \vdash D_1\} \mid \mathcal{M}_2\{D_1, D_2 \vdash\}) \tag{13.5}$$

$$\mathcal{C}_1(\mathcal{M}_3\langle\vdash C_1 \wedge C_2\rangle \mid \mathcal{M}_4\langle C_1 \wedge C_2 \vdash\rangle)\{\vdash D_2\} \tag{13.6}$$

■

13.3.2 *Cut-Elimination in* G3 *and* mG3

We investigate cut-elimination of G3 in terms of mG3. We first show that cut-strings are represented by our multi-sequents.

Lemma 7 *For any cut-string* $(\vec{C})$ *in a given* G3*-proof, there exists a multi-sequent that consists of all minor upper sequents of* $(\vec{C})$*.*

Proof By induction on the length n of the given cut-string.
(Base step) When $n = 1$, let the given cut-string be the following form on the left. Then, the required multi-sequent is that on the right.

$$\frac{\Gamma \vdash \Delta, C_1 \quad C_1, \Gamma \vdash \Delta}{\Gamma \vdash \Delta}\,(C_1) \qquad\qquad (\Gamma \vdash \Delta, C_1 \mid C_1, \Gamma \vdash \Delta)_{C_1}$$

(Induction step) When $n > 1$, let the given cut-string be the following form.

$$\dfrac{\Pi \vdash \Lambda, C_n \qquad \dfrac{C_n, \Pi \vdash \Lambda, C_{n-1} \qquad \begin{array}{c}\dfrac{\Gamma \vdash \Delta, C_1 \quad C_1, \Gamma \vdash \Delta}{\Gamma \vdash \Delta}\,(C_1)\\ \vdots\ (\vec{C_2})\\ C_{n-1}, C_n, \Pi \vdash \Lambda\end{array}}{C_n, \Pi \vdash \Lambda}\,(C_{n-1})}{\Pi \vdash \Lambda}\,(C_n)$$

By the induction hypothesis, for the cut-string $(C_{n-1}(\cdots(C_1)\cdots))$, there exists a multi-sequent $\mathcal{M}'$ which consists of all minor upper sequents of the cut-string, that is, $\mathcal{M}' = (C_n, \Pi \vdash \Lambda, C_{n-1} \mid \vec{C_2}(\Gamma \vdash \Delta, C_1 \mid C_1, \Gamma \vdash \Delta)_{C_1})_{C_{n-1}}$, where $\vec{C_2}$ expresses the combination of minor upper sequents of $(\vec{C_2})$. $\bigcap \mathcal{M}' = C_n, \Pi \vdash \Lambda$ by Lemma 4; thus, we have the required multi-sequent $(\Pi \vdash \Lambda, C_n \mid \mathcal{M}')_{C_n}$. ■

Definition 9 For a cut-string $(\vec{C})$, the corresponding multi-sequent that consists of all minor upper sequents of $(\vec{C})$ is denoted by $m(\vec{C})$.

We next investigate the relationship between cut-eliminability in G3 and provability in mG3. We first show that cut-string reduction rules are represented by mG3-rules.

Lemma 8 *Let* $(\vec{C})$ *be the topmost cut-string in a* G3*-proof.*

1. *By the principal* $\wedge r$*,* $\vee l$*, or* $\rightarrow l$ *reduction, if* $(\vec{C})$ *is reduced to the main string* $(\vec{D})$ *and the additional string* $(\vec{E})$*, then the following is an instance of an* mG3*-rule.*

$$\frac{m(\vec{E}) \quad m(\vec{D})}{m(\vec{C})}$$

2. *By the principal $\wedge l$, $\vee r$, or $\rightarrow r$ reduction, if $(\vec{C})$ is reduced to $(\vec{D})$, and $\Gamma \vdash \Delta$ is its additional sequent, then the following is an instance of an* mG3*-rule.*

$$\frac{\Gamma \vdash \Delta \quad m(\vec{D})}{m(\vec{C})}$$

3. *By the auxiliary $\wedge r$, $\vee l$, or $\rightarrow l$ reduction, if $(\vec{C})$ is reduced to $(\vec{D})$ and $(\vec{E})$, then the following is an instance of an* mG3*-rule.*

$$\frac{m(\vec{D}) \quad m(\vec{E})}{m(\vec{C})}$$

4. *By a reduction rule other than (1), (2), or (3), if $(\vec{C})$ is reduced to $(\vec{D})$, then the following is an instance of an* mG3*-rule.*

$$\frac{m(\vec{D})}{m(\vec{C})}$$

Proof We examine every reduction rule of G3 described in Definition 4. We show principal $\wedge l$ and $\wedge r$ reductions, as well as auxiliary $\wedge l$ and $\wedge r$ reductions. The other reductions are similar.

- The principal $\wedge l$-reduction is represented by the $\wedge$-rule in mG3 as follows.

$$\frac{\Pi \vdash \Lambda, D_2 \quad \overrightarrow{C_{i+1}}(\Pi \vdash \Lambda, D_2 \mid (D_2, \Pi \vdash \Lambda, D_1 \mid \vec{C_2}^{D_1 D_2}(D_1, D_2, \Gamma \vdash \Delta, C_1 \mid C_1, D_1, D_2, \Gamma \vdash \Delta)_{C_1})_{D_1})_{D_2}}{\overrightarrow{C_{i+1}}(\Pi \vdash \Lambda, D_1 \wedge D_2 \mid \vec{C_2}(D_1 \wedge D_2, \Gamma \vdash \Delta, C_1 \mid C_1, D_1 \wedge D_2, \Gamma \vdash \Delta)_{C_1})_{D_1 \wedge D_2}} \wedge$$

The lower multi-sequent represents the given cut-string, where $\vec{C_2}$, as well as $\overrightarrow{C_{i+1}}$, expresses the combination of minor upper sequents of $(\vec{C_2})$, as well as $(\overrightarrow{C_{i+1}})$. The upper multi-sequent represents the resultant cut-string by applying the reduction rule.

- The principal $\wedge r$-reduction is represented by the $\wedge$-rule in mG3 as follows.

$$\frac{\mathcal{M} \quad \overrightarrow{C_{i+1}}(\Pi \vdash \Lambda, D_2 \mid (\vec{C_2}^{D_1}(D_2, \Gamma \vdash \Delta, D_1, C_1 \mid C_1, D_2, \Gamma \vdash \Delta, D_1)_{C_1} \mid D_1, D_2, \Pi \vdash \Lambda)_{D_1})_{D_2}}{\overrightarrow{C_{i+1}}(\vec{C_2}(\Gamma \vdash \Delta, D_1 \wedge D_2, C_1 \mid C_1, \Gamma \vdash \Delta, D_1 \wedge D_2)_{C_1} \mid D_1 \wedge D_2, \Pi \vdash \Lambda)_{D_1 \wedge D_2}} \wedge$$

where the additional sequent $\mathcal{M} \equiv \vec{C_2}^{D_2}(\Gamma \vdash \Delta, D_2, C_1 \mid C_1, \Gamma \vdash \Delta, D_2)_{C_1}$.

- Auxiliary $\wedge l$- and $\wedge r$-reductions are represented by $\wedge l$- and $\wedge r$-rules, respectively, in mG3 as follows.

$$\frac{\overrightarrow{C_{i+1}}^{B_1B_2}(B_1, B_2, \Pi_2 \vdash \Lambda_2 \mid \overrightarrow{C_2}^{B_1B_2}(B_1, B_2, \Gamma_1 \vdash \Delta_1 \mid B_1, B_2, \Gamma_2 \vdash \Delta_2)_{C_1})_{C_i}}{\overrightarrow{C_{i+1}}(B_1 \wedge B_2, \Pi_2 \vdash \Lambda_2 \mid \overrightarrow{C_2}(B_1 \wedge B_2, \Gamma_1 \vdash \Delta_1 \mid B_1 \wedge B_2, \Gamma_2 \vdash \Delta_2)_{C_1})_{C_i}} \wedge l$$

$$\frac{\mathcal{M}_1 \qquad \mathcal{M}_2}{\overrightarrow{C_{i+1}}(\Pi_2 \vdash \Lambda_2, B_1 \wedge B_2 \mid \overrightarrow{C_2}(\Gamma_1 \vdash \Delta_1, B_1 \wedge B_2 \mid \Gamma_2 \vdash \Delta_2, B_1 \wedge B_2)_{C_1})_{C_i}} \wedge r$$

where $\mathcal{M}_1 \equiv \overrightarrow{C_{i+1}}^{B_1}(\Pi_2 \vdash \Lambda_2, B_1 \mid \overrightarrow{C_2}^{B_1}(\Gamma_1 \vdash \Delta_1, B_1 \mid \Gamma_2 \vdash \Delta_2, B_1)_{C_1})_{C_i}$ and $\mathcal{M}_2 \equiv \overrightarrow{C_{i+1}}^{B_2}(\Pi_2 \vdash \Lambda_2, B_2 \mid \overrightarrow{C_2}^{B_2}(\Gamma_1 \vdash \Delta_1, B_2 \mid \Gamma_2 \vdash \Delta_2, B_2)_{C_1})_{C_i}$. ■

As for axioms in mG3, the following holds.

Lemma 9 *Let* $(\overrightarrow{C})$ *be the topmost cut-string in a given* G3*-proof, where* $w(\overrightarrow{C}) = 0$ *and* $h(\overrightarrow{C}) = 0$. *Then, the multi-sequent* $m(\overrightarrow{C})$ *is an* mG3*-axiom.*

Proof Since $h(\overrightarrow{C}) = 0$, every minor upper sequent of $(\overrightarrow{C})$ is an axiom in G3. Furthermore, by Lemma 3 with $w(\overrightarrow{C}) = 0$, they are axioms of G3 on non cut-formulas. Thus, $m(\overrightarrow{C})$ is an axiom of mG3. ■

By summing up Lemmas 8 and 9, we obtain the following result.

Theorem 2 *If there exists a* G3*-proof* π *whose last is cut between* $\Gamma \vdash \Delta, C$ *and* $C, \Gamma \vdash \Delta$, *and no other cut is applied in* π, *then there exists an m-proof of* $(\Gamma \vdash \Delta, C \mid C, \Gamma \vdash \Delta)_C$ *in* mG3.

Proof In our proof of the cut-elimination of G3, there exists a reduction process that, by reducing the topmost cut-string, transforms the given cut-string so that $w(\overrightarrow{Q}) = 0$ and $h(\overrightarrow{Q}) = 0$. By Lemma 8, every reduction step is represented by an mG3-rule, and by Lemma 9, $m(\overrightarrow{Q})$ is an axiom of mG3. Thus, there exists an m-proof of $(\Gamma \vdash \Delta, C \mid C, \Gamma \vdash \Delta)_C$. ■

We next show that by deleting cut-formulas from an m-proof of mG3, we obtain a cut-free G3-proof. We first define a normal form of m-proofs in mG3.

Definition 10 A **regular m-proof** is an m-proof, in which the cut-formulas of every axiom are atoms, and for every branch of the m-proof, there exists a unique multi-sequent $\mathcal{M}$ such that only auxiliary rules are applied above $\mathcal{M}$, and only principal rules are applied below $\mathcal{M}$.

Thus, a regular m-proof consists, from the bottom, of applications of principal rules followed by applications of auxiliary rules. From the viewpoint of cut-elimination in G3, a regular m-proof represents a reduction strategy of a cut-string by applying principal reductions followed by auxiliary reductions. Given that a multi-sequent $\mathcal{M}$ is provable, by using the inversion Lemma 6 of mG3, we are able to construct a regular m-proof of $\mathcal{M}$.

Lemma 10 *If $\mathcal{M}$ is provable in* mG3, *then there exists a regular m-proof of $\mathcal{M}$.*

Definition 11 In any multi-sequent $\mathcal{M}$, the concatenation part of two single sequents of the form $(\Gamma \vdash \Delta, C \mid C, \Gamma \vdash \Delta)_C$ is uniquely determined. The cut-formula C of that part is called the **innermost cut-formula** of $\mathcal{M}$. An atomic formula that is the innermost cut-formula is called the **innermost atom**.
Let $\mathcal{C}(\Gamma \vdash \Delta, Q \mid Q, \Gamma \vdash \Delta)_Q$ be a multi-sequent whose innermost atom is Q. By deleting the two occurrences of Q and by unifying the remaining Γ and Δ, a multi-sequent $\mathcal{C}(\Gamma \vdash \Delta)$ is obtained. We call this operation the **innermost atom deletion**.

By the innermost atom deletion, we will eliminate cut-atoms one by one. However, removing cut-atoms from an additional sequent may render an application of a principal rule invalid. Let us examine the following m-proof.

$$\dfrac{\begin{matrix}\vdots\ \pi_3\\ \vdash \neg(B \wedge C)\end{matrix} \qquad \dfrac{\dfrac{\begin{matrix}\vdots\ \pi_1\\ (\vdash C, Q \mid Q \vdash C)_Q\end{matrix} \quad \begin{matrix}\vdots\ \pi_2\\ ((\vdash C, Q \mid Q \vdash C)_Q \mid ((C \vdash B, Q \mid Q, C \vdash B)_Q \mid B, C \vdash)_B)_C\end{matrix}}{((\vdash B \wedge C, Q \mid Q \vdash B \wedge C)_Q \mid B \wedge C \vdash)_{B \wedge C}}\ \boxed{\wedge}}{(\vdash \neg(B \wedge C) \mid (\neg(B \wedge C) \vdash Q \mid Q, \neg(B \wedge C) \vdash)_Q)_{\neg(B \wedge C)}}\ \neg}{(\vdash Q \wedge \neg(B \wedge C) \mid Q \wedge \neg(B \wedge C) \vdash)_{Q \wedge \neg(B \wedge C)}}\ \wedge$$

Assume that no additional sequents appear in π_1 and π_2. Then, the additional sequent $(\vdash C, Q \mid Q \vdash C)_Q$ of π_1 is the topmost, which corresponds to the topmost cut-string. By eliminating the innermost atom Q from π_1, the above $\boxed{\wedge}$ is transformed as follows, where π_1^{-Q} is an m-proof obtained by eliminating the innermost atom Q from π_1.

$$\dfrac{\begin{matrix}\vdots\ \pi_1^{-Q}\\ \vdash C\end{matrix} \qquad \begin{matrix}\vdots\ \pi_2\\ ((\vdash C, Q \mid Q \vdash C)_Q \mid ((C \vdash B, Q \mid Q, C \vdash B)_Q \mid B, C \vdash)_B)_C\end{matrix}}{((\vdash B \wedge C, Q \mid Q \vdash B \wedge C)_Q \mid B \wedge C \vdash)_{B \wedge C}}$$

The structure above is not a valid instance of $\wedge$-rule. This is because $\mathcal{M}_1 \equiv (\vdash B \wedge C, Q \mid Q \vdash B \wedge C)_Q$, and $\vdash C$ of π_1^{-Q} is not in the form of $\mathcal{M}_1\{\vdash C\}$.

We can overcome this difficulty by eliminating the innermost atom Q not only from π_1 but from the entire m-proof, including π_2. However, to remain consistent with the topmost reduction strategy in our cut-string elimination, we will eliminate the innermost atom only from the topmost additional sequent. Thus, we introduce the above invalid instance of an mG3-rule as a *quasi rule*. These quasi rules appear only during the process of innermost atom deletion, and they are removed in the finally obtained cut-free proof.

Definition 12 (*Quasi rules*) Let the following structure be the principal $\star$ rule, where $\star$ is $\wedge$, $\vee$, or $\rightarrow$. $\mathcal{M}'$ is obtained by appropriate substitution for $\mathcal{M}$, and $i = 1$ for $\wedge$ and $i = 2$ for $\vee$ and $\rightarrow$. Let $\mathcal{M}'_i$ be a non-single multi-sequent.

$$\frac{\mathcal{M}_i' \quad \mathcal{C}(\Pi \vdash \Lambda \mid (\mathcal{M}_1' \mid \mathcal{M}_2')_{D_1})_{D_2}}{\mathcal{C}(\mathcal{M}_1 \mid \mathcal{M}_2)_{D_1 \star D_2}} \star$$

The following structure obtained by replacing the additional sequent $\mathcal{M}_i'$ with another single additional sequent $\Pi \vdash \Lambda$ is called the **quasi $\star$ rule**.

$$\frac{\Pi \vdash \Lambda \quad \mathcal{C}(\Pi \vdash \Lambda \mid (\mathcal{M}_1' \mid \mathcal{M}_2')_{D_1})_{D_2}}{\mathcal{C}(\mathcal{M}_1 \mid \mathcal{M}_2)_{D_1 \star D_2}} \; quasi \star$$

Note that we assume $\mathcal{M}_i'$ to be a non-single multi-sequent. If $\mathcal{M}_i'$ is single, then the structure of quasi $\star$ is the valid principal rule.

Lemma 11 (Innermost atom deletion) *Let π be a regular m-proof of non-single $\mathcal{M}$, where all additional sequents are single sequents, and applications of quasi rules may appear. For every multi-sequent in π,*

1. *apply the innermost atom deletion;*
2. *delete all non-single multi-sequents that do not contain any innermost atom;*
3. *delete every single additional sequent, as well as the sub-proof thereof, that is associated with the deleted multi-sequent by (2).*

The resulting structure π' is then a regular m-proof of $\mathcal{M}'$ such that $\bigcap \mathcal{M} = \bigcap \mathcal{M}'$.

Proof By definition of regular m-proofs, all cut-formulas are decomposed into cut-atoms, and there exists an innermost atom. We show that mG3-rules are preserved after the above operations by induction on the height h of π.

(Base step) When $h = 0$, π consists of the following axiom on the left. By applying (1) to Q, we obtain the multi-sequent on the right.

$$\mathcal{C}(\Gamma \vdash \Delta, Q \mid Q, \Gamma \vdash \Delta)_Q \qquad\qquad \mathcal{C}(\Gamma \vdash \Delta)$$

By definition of the regular m-proof, Γ and Δ share a common atom, and the other component sequents remain the same. Hence, $\mathcal{C}(\Gamma \vdash \Delta)$ is also an axiom. Furthermore, we have $\bigcap \mathcal{C}(\Gamma \vdash \Delta, Q \mid Q, \Gamma \vdash \Delta)_Q = \bigcap \mathcal{C}(\Gamma \vdash \Delta)$, because $\bigcap(\Gamma \vdash \Delta, Q \mid Q, \Gamma \vdash \Delta)_Q = \Gamma \vdash \Delta$. The same applies to the other cases below.

(Induction step) When $h > 0$, we examine each rule applied in the given π. We first examine the auxiliary rules.

• $(\wedge l)$ When the last rule of π is the $\wedge l$-rule, π is the following form on the left. By applying the innermost atom deletion, we obtain the structure on the right.

$$\frac{\mathcal{C}(B_1, B_2, \Gamma \vdash \Delta, Q \mid Q, B_1, B_2, \Gamma \vdash \Delta)_Q}{\mathcal{C}(B_1 \wedge B_2, \Gamma \vdash \Delta, Q \mid Q, B_1 \wedge B_2, \Gamma \vdash \Delta)_Q} \wedge l \qquad\qquad \frac{\mathcal{C}(B_1, B_2, \Gamma \vdash \Delta)}{\mathcal{C}(B_1 \wedge B_2, \Gamma \vdash \Delta)}$$

By the induction hypothesis, $\mathcal{C}(B_1, B_2, \Gamma \vdash \Delta)$ has a regular m-proof, and the above structure is an application of the $\wedge l$-rule. Thus, we obtain a regular m-proof.

• ($\wedge r$) When the last rule of π is $\wedge r$-rule, π is the following form on the left. By applying the innermost atom deletion, we obtain the structure on the right, which is an application of $\wedge r$-rule.

$$\frac{\mathcal{C}(\Gamma \vdash \Delta, B_1, Q \mid Q, \Gamma \vdash \Delta, B_1) \quad \mathcal{C}(\Gamma \vdash \Delta, B_2, Q \mid Q, \Gamma \vdash \Delta, B_2)}{\mathcal{C}(\Gamma \vdash \Delta, B_1 \wedge B_2, Q \mid Q, \Gamma \vdash \Delta, B_1 \wedge B_2)} \wedge r \qquad \frac{\mathcal{C}(\Gamma \vdash \Delta, B_1) \quad \mathcal{C}(\Gamma \vdash \Delta, B_2)}{\mathcal{C}(\Gamma \vdash \Delta, B_1 \wedge B_2)}$$

Similarly for the other auxiliary rules.

For principal rules, we divide the following cases dependent on whether or not the rule is related to the innermost atom Q.

• ($\wedge$ on $Q \wedge D_2$) When the last rule of π is the $\wedge$-rule on $Q \wedge D_2$, π is the following form.

$$\frac{\Gamma \vdash \Delta, D_2 \quad \mathcal{C}(\Gamma \vdash \Delta, D_2 \mid (D_2, \Gamma \vdash \Delta, Q \mid Q, D_2, \Gamma \vdash \Delta)_Q)_{D_2}}{\mathcal{C}(\Gamma \vdash \Delta, Q \wedge D_2 \mid Q \wedge D_2, \Gamma \vdash \Delta)_{Q \wedge D_2}} \wedge$$

Because the lower sequent does not contain any innermost atom, the entire multi-sequent is deleted by (2), and the additional sequent $\Gamma \vdash \Delta, D_2$ is also deleted by (3); thus, we obtain the following multi-sequent by applying (1): $\mathcal{C}(\Gamma \vdash \Delta, D_2 \mid D_2, \Gamma \vdash \Delta)_{D_2}$. We obtain a regular m-proof by the induction hypothesis.

• ($\wedge$ on $D_1 \wedge D_2$) When the last rule of π is the $\wedge$-rule on $D_1 \wedge D_2$, π is the following form, where $\mathcal{M}_1$ of the rule is a single sequent $\Gamma \vdash \Delta, D_1 \wedge D_2$. Since we assume that all additional sequents are single, $\mathcal{M}_1$ should be a single sequent.

$$\frac{\Gamma \vdash \Delta, D_2 \quad \mathcal{C}(\Gamma \vdash \Delta, D_2 \mid (D_2, \Gamma \vdash \Delta, D_1 \mid \mathcal{C}_2(D_1, D_2, \Gamma \vdash \Delta, Q \mid Q, D_1, D_2, \Gamma \vdash \Delta)_Q)_{D_1})_{D_2}}{\mathcal{C}(\Gamma \vdash \Delta, D_1 \wedge D_2 \mid \mathcal{C}_2(D_1 \wedge D_2, \Gamma \vdash \Delta, Q \mid Q, D_1 \wedge D_2, \Gamma \vdash \Delta)_Q)_{D_1 \wedge D_2}} \wedge$$

By applying the innermost atom deletion, we obtain the following structure, which is an application of the $\wedge$-rule.

$$\frac{\Gamma \vdash \Delta, D_2 \quad \mathcal{C}(\Gamma \vdash \Delta, D_2 \mid (D_2, \Gamma \vdash \Delta, D_1 \mid \mathcal{C}_2(D_1, D_2, \Gamma \vdash \Delta))_{D_1})_{D_2}}{\mathcal{C}(\Gamma \vdash \Delta, D_1 \wedge D_2 \mid \mathcal{C}_2(D_1 \wedge D_2, \Gamma \vdash \Delta))_{D_1 \wedge D_2}}$$

• (quasi $\wedge$ on $D_1 \wedge D_2$) When the last rule of π is the quasi $\wedge$-rule on $D_1 \wedge D_2$, π is the following form.

$$\frac{\Gamma \vdash \Delta, D_2 \quad \mathcal{C}(\Gamma \vdash \Delta, D_2 \mid (\mathcal{C}_1(D_2, \Gamma \vdash \Delta, D_1, Q \mid Q, D_2, \Gamma \vdash \Delta, D_1)_Q \mid D_1, D_2, \Gamma \vdash \Delta)_{D_1})_{D_2}}{\mathcal{C}(\mathcal{C}_1(\Gamma \vdash \Delta, D_1 \wedge D_2, Q \mid Q, \Gamma \vdash \Delta, D_1 \wedge D_2)_Q \mid D_1 \wedge D_2, \Gamma \vdash \Delta)_{D_1 \wedge D_2}} \text{quasi} \wedge$$

By applying the innermost atom deletion, we obtain the following structure, which is an application of the quasi $\wedge$-rule.

$$\frac{\Gamma \vdash \Delta, D_2 \quad \mathcal{C}(\Gamma \vdash \Delta, D_2 \mid (\mathcal{C}_1(D_2, \Gamma \vdash \Delta, D_1) \mid D_1, D_2, \Gamma \vdash \Delta)_{D_1})_{D_2}}{\mathcal{C}(\mathcal{C}_1(\Gamma \vdash \Delta, D_1 \wedge D_2) \mid D_1 \wedge D_2, \Gamma \vdash \Delta)_{D_1 \wedge D_2}}$$

Similarly for the other principal rules. ■

Since the innermost atom deletion does not provide any new non-single additional sequent, by applying Lemma 11 successively, we obtain a cut-free G3-proof.

Lemma 12 *Let π be a regular m-proof of non-single $\mathcal{M}$, where all additional sequents are single sequents, and applications of quasi rules may appear. π is transformed into a cut-free* G3*-proof π' of $\bigcap \mathcal{M}$.*

Proof Since all additional sequents are single, no essential branching occurs with the application of a principal rule. Additionally, in the application of an auxiliary rule which makes branching, the cut-formulas in its lower sequent are preserved in its upper sequents. Therefore, in π, cut-formulas are uniformly decomposed into cut-atoms equally across all non-single top sequents. Hence, we show the lemma by induction on the number n of cut-atoms contained in a non-single top sequent of the highest branch in π.
(Induction step) When $n > 1$, by Lemma 11, π is transformed into a regular m-proof whose highest top sequent contains cut-atoms less than those in π. Thus, the induction hypothesis is applied.
(Base step) When $n = 1$, the top sequent takes the form $(\Gamma \vdash \Delta, Q \mid Q, \Gamma \vdash \Delta)_Q$. Because the upper sequent of the $\wedge$, $\vee$, or $\rightarrow$ rule contains at least two cut-formulas, the possible rules under the above top sequent are the principal $\neg$-rule and auxiliary rules. By an application of these rules, the number of cut-formulas is unchanged. Thus, the resulting regular m-proof of $\bigcap \mathcal{M}$, which is obtained by the innermost atom deletion in Lemma 11, consists only of single sequents. Any regular m-proof that consists only of single sequents is a cut-free G3-proof. ■

Lemma 12 corresponds to the elimination of the topmost cut-string. Thus, by applying Lemma 12 to the leftmost topmost non-single additional sequent successively in a given m-proof, we finally obtain a cut-free G3-proof.

Theorem 3 *Any m-proof π of $\mathcal{M}$ is transformed into a cut-free* G3*-proof π' of $\bigcap \mathcal{M}$.*

Proof By Lemma 10, we assume π is a regular m-proof. We show the theorem by induction on the number k of non-single additional sequents in π.
(Base step) When $k = 0$, we obtain a cut-free G3-proof π' by Lemma 12.
(Induction step) When $k > 0$, let λ be a sub-proof of a non-single additional sequent in π, where all additional sequents are single. Then, by applying Lemma 12 to λ, the given π is transformed into an m-proof that contains $k - 1$ non-single additional sequents; hence, we can apply the induction hypothesis. (Note that by applying Lemma 12 to λ, an application of a principal rule becomes an application of the quasi rule.) ■

In particular, we have the following corollary.

Corollary 1 *If there exists an m-proof of* $(\Gamma \vdash \Delta, C \mid C, \Gamma \vdash \Delta)_C$ *in* mG3, *then there exists a cut-free* G3*-proof of* $\Gamma \vdash \Delta$.

Example 4 Let us apply the innermost atom deletion to the m-proof given in Example 3, assuming that A and B are atoms. By eliminating the innermost atom $\boxed{A}$, as well as the lower sequents of the last $\neg$ and of $\wedge$ above it, we obtain the following m-proof.

$$\dfrac{\dfrac{\dfrac{(\boxed{B}, A, B, C \vdash A \mid A, B, C \vdash A, \boxed{B}) \quad (\boxed{B}, A, B, C \vdash B \mid A, B, C \vdash B, \boxed{B})}{(\boxed{B}, A, B, C \vdash A \wedge B \mid A, B, C \vdash A \wedge B, \boxed{B})}\wedge r}{(\boxed{B}, A, B \wedge C \vdash A \wedge B \mid A, B \wedge C \vdash A \wedge B, \boxed{B})}\wedge l}{(A, B \wedge C \vdash \boxed{\neg B} \mid \boxed{\neg B}, A, B \wedge C \vdash A \wedge B)}\neg$$

Then, by eliminating the innermost atom $\boxed{B}$, as well as the lower sequent of the last $\neg$, we obtain the following cut-free proof.

$$\dfrac{\dfrac{A, B, C \vdash A \quad A, B, C \vdash B}{A, B, C \vdash A \wedge B}\wedge r}{A, B \wedge C \vdash A \wedge B}\wedge l$$

13.4 Conclusion and Future Work

By introducing cut-strings, which are sequences of applications of the *cut*-rule, we proved cut-string elimination (Theorem 1). By introducing multi-sequents, which correspond to cut-strings, we introduced sequent calculus mG3 for multi-sequents. We then investigated cut-elimination in terms of mG3. We showed that every cut-string reduction rule is represented as a proof construction rule in mG3 (Theorem 2). We further proved that, by deleting cut-formulas from a given regular m-proof in mG3, we obtain a cut-free G3-proof (Theorem 3).

Our cut-string elimination, as well as mG3, heavily depends on the inversion lemma of G3. Let us consider the following cut-string $(C \wedge D(A \wedge B))$.

$$\dfrac{\dfrac{\dfrac{\Gamma \vdash C \wedge D, A \quad \Gamma \vdash C \wedge D, B}{\Gamma \vdash C \wedge D, A \wedge B}\wedge r \quad \dfrac{}{A \wedge B, \Gamma \vdash C \wedge D}R}{\Gamma \vdash C \wedge D}(A \wedge B) \quad C \wedge D, \Gamma \vdash}{\Gamma \vdash}(C \wedge D)$$

By applying the inversion lemma of G3, we assume that the inference rule R is $\wedge l$ on $A \wedge B$, and apply $\wedge$-reduction. However, without the inversion lemma, when R is $\wedge r$ on $C \wedge D$, it seems difficult to rewrite the above proof while retaining the cut-string. The same difficulty arises in systems where the inversion lemma does not hold.

Although we may consider *cut*-rule in the non-context-sharing from in G3, for a multi-sequent with different contexts in its component sequents, the description of auxiliary rules becomes more complicated. Furthermore, non-context-sharing *cut* introduces another issue related to *contraction*. Let us examine the following principal reduction.

$$\dfrac{\dfrac{\Gamma \vdash \Delta, A \quad \Gamma \vdash \Delta, B}{\Gamma \vdash \Delta, A \wedge B}\wedge r \quad \dfrac{A, B, \Pi \vdash \Lambda}{A \wedge B, \Pi \vdash \Lambda}\wedge l}{\Gamma, \Pi \vdash \Delta, \Lambda}(A \wedge B) \quad \rhd \quad \dfrac{\Gamma \vdash \Delta, B \quad \dfrac{\Gamma \vdash \Delta, A \quad A, B, \Pi \vdash \Lambda}{B, \Gamma, \Pi \vdash \Delta, \Lambda}(A)}{\Gamma, \Gamma, \Pi \vdash \Delta, \Delta, \Lambda}(B)$$

At the end of the above transformation, we need *contraction* to match the context of the resulting end sequent to the original one. However, it is difficult to represent such applications of *contraction* by our m-proof. These applications of *contraction* need to be managed at the meta level.

If we investigate mG3 as an independent system, separate from cut-elimination, we may consider variants of our calculus for multi-sequents. For example, we could explicitly introduce *contraction* and *weakening* as rules for multi-sequents. We leave the investigation of such systems for future work.

As another future research, we will investigate a dialogical interpretation of cut-elimination. After the seminal work of Lorenzen [8, 9] in the 1950s, it has been well-established that logical rules can be interpreted as a dialogue between a Proponent and an Opponent. Among others, Felscher [5] proved a correspondence between the existence of proofs in classical/intuitionistic logic and that of winning strategies. Herbelin [7] extended the correspondence to that between the winning strategies of the Lorenzen dialogue and the cut-free proofs of a variant of sequent calculus. Okada and Takemura [10] provided essentially the same result as [7] by introducing a sequent calculus representation of the Lorenzen dialogue. However, with a few exceptions by Coquand [3] and Herbelin [6], not much research has been devoted to the Lorenzen dialogical counterpart of *cut* and cut-elimination processes. Our proof construction style representation of cut-elimination, combined with the sequent calculus representation of [10], enables the investigation of a Lorenzen dialogical interpretation of cut-elimination.

Appendix: Principal Reductions for ∨ and →

(Principal $\vee l$**)** When $A \equiv D_1 \vee D_2$, let the given cut-string be the following form.

$$\begin{array}{c}\dfrac{\begin{matrix}\vdots\ \pi_3\\ \Pi \vdash \Lambda, D_1 \vee D_2\end{matrix} \qquad \begin{matrix}\dfrac{\begin{matrix}\vdots\ \pi_1\\ D_1 \vee D_2, \Gamma \vdash \Delta, C_1\end{matrix} \quad \begin{matrix}\vdots\ \pi_2\\ C_1, D_1 \vee D_2, \Gamma \vdash \Delta\end{matrix}}{D_1 \vee D_2, \Gamma \vdash \Delta}\,(C_1)\\ \vdots\ (\overrightarrow{C_2})\\ D_1 \vee D_2, \Pi \vdash \Lambda\end{matrix}}{\Pi \vdash \Lambda}\,(D_1 \vee D_2)\\ \vdots\ (\overrightarrow{C_{i+1}})\\ \Sigma \vdash \Phi\end{array}$$

The above cut-string is transformed as follows.

$$\begin{array}{c}\dfrac{\dfrac{\begin{matrix}\vdots\ \pi_3^\circ\\ \Pi \vdash \Lambda, D_1, D_2\end{matrix} \qquad \begin{matrix}\dfrac{\begin{matrix}\vdots\ \pi_1^\circ\\ D_1, \Gamma \vdash \Delta, D_2^*, C_1\end{matrix} \quad \begin{matrix}\vdots\ \pi_2^\circ\\ C_1, D_1, \Gamma \vdash D_2^*, \Delta\end{matrix}}{D_1, \Gamma \vdash \Delta, D_2}\,(C_1)\\ \vdots\ (\overrightarrow{C_2}^{D_1})\\ D_1, \Pi \vdash \Lambda, D_2\end{matrix}}{\Pi \vdash \Lambda, D_2}\,(D_1) \qquad \begin{matrix}\dfrac{\begin{matrix}\vdots\ \pi_1^{\circ\prime}\\ D_2, \Gamma \vdash \Delta, C_1\end{matrix} \quad \begin{matrix}\vdots\ \pi_2^{\circ\prime}\\ C_1, D_2, \Gamma \vdash \Delta\end{matrix}}{D_2, \Gamma \vdash \Delta}\,(C_1)\\ \vdots\ (\overrightarrow{C_2}^{D_2})\\ D_2, \Pi \vdash \Lambda\end{matrix}}{\Pi \vdash \Lambda}\,(D_2)\\ \vdots\ (\overrightarrow{C_{i+1}})\\ \Sigma \vdash \Phi\end{array}$$

$(\overrightarrow{C_{i+1}}(D_2(D_1(\overrightarrow{C_1}))))$ is the main string. $D_2, \Pi \vdash \Lambda$ is the additional sequent, and $(\overrightarrow{C_1})$ of the additional sequent is the additional string.

(Principal $\vee r$**)** When $A \equiv D_1 \vee D_2$, let the given cut-string be the following form.

$$\begin{array}{c}\dfrac{\begin{matrix}\dfrac{\begin{matrix}\vdots\ \pi_1\\ \Gamma \vdash \Delta, D_1 \vee D_2, C_1\end{matrix} \quad \begin{matrix}\vdots\ \pi_2\\ C_1, \Gamma \vdash \Delta, D_1 \vee D_2\end{matrix}}{\Gamma \vdash \Delta, D_1 \vee D_2}\,(C_1)\\ \vdots\ (\overrightarrow{C_2})\\ \Pi \vdash \Lambda, D_1 \vee D_2\end{matrix} \qquad \begin{matrix}\vdots\ \pi_3\\ D_1 \vee D_2, \Pi \vdash \Lambda\end{matrix}}{\Pi \vdash \Lambda}\,(D_1 \vee D_2)\\ \vdots\ (\overrightarrow{C_{i+1}})\\ \Sigma \vdash \Phi\end{array}$$

The above cut-string is transformed as follows.

$$
\begin{array}{c}
\dfrac{\dfrac{\begin{array}{c}\dfrac{\begin{array}{c}\vdots\ \pi_1^\circ\\ \Gamma\vdash\Delta,D_1,D_2,C_1\end{array}\quad\begin{array}{c}\vdots\ \pi_2^\circ\\ C_1,\Gamma\vdash\Delta,D_1,D_2\end{array}}{\Gamma\vdash\Delta,D_1,D_2}\,(C_1)\\ \vdots\ (\overrightarrow{C_2}^{D_1D_2})\\ \Pi\vdash\Lambda,D_1,D_2\end{array}\quad\begin{array}{c}\vdots\ \pi_3^\circ\\ D_1,\Pi\vdash\Lambda,D_2^*\end{array}}{\Pi\vdash\Lambda,D_2}\,(D_1)\quad\begin{array}{c}\vdots\ \pi_3^{\circ\prime}\\ D_2,\Pi\vdash\Lambda\end{array}}{\Pi\vdash\Lambda}\,(D_2)\\
\vdots\ (\overrightarrow{C_{i+1}})\\
\Sigma\vdash\Phi
\end{array}
$$

$D_2, \Pi \vdash \Lambda$ is the additional sequent.

(Principal $\rightarrow l$**)** When $A \equiv D_1 \rightarrow D_2$, let the given cut-string be the following form.

$$
\begin{array}{c}
\dfrac{\begin{array}{c}\vdots\ \pi_3\\ \Pi\vdash\Lambda,D_1\rightarrow D_2\end{array}\quad\begin{array}{c}\dfrac{\begin{array}{c}\vdots\ \pi_1\\ D_1\rightarrow D_2,\Gamma\vdash\Delta,C_1\end{array}\quad\begin{array}{c}\vdots\ \pi_2\\ C_1,D_1\rightarrow D_2,\Gamma\vdash\Delta\end{array}}{D_1\rightarrow D_2,\Gamma\vdash\Delta}\,(C_1)\\ \vdots\ (\overrightarrow{C_2})\\ D_1\rightarrow D_2,\Pi\vdash\Lambda\end{array}}{\Pi\vdash\Lambda}\,(D_1\rightarrow D_2)\\
\vdots\ (\overrightarrow{C_{i+1}})\\
\Sigma\vdash\Phi
\end{array}
$$

The above cut-string is transformed as follows.

$$
\begin{array}{c}
\dfrac{\dfrac{\begin{array}{c}\dfrac{\begin{array}{c}\vdots\ \pi_1^\circ\\ \Gamma\vdash\Delta,D_1,D_2^*,C_1\end{array}\quad\begin{array}{c}\vdots\ \pi_2^\circ\\ C_1,\Gamma\vdash\Delta,D_1,D_2^*\end{array}}{\Gamma\vdash\Delta,D_2,D_1}\,(C_1)\\ \vdots\ (\overrightarrow{C_2})^{D_1}\\ \Pi\vdash\Lambda,D_2,D_1\end{array}\quad\begin{array}{c}\vdots\ \pi_3^\circ\\ D_1,\Pi\vdash\Lambda,D_2\end{array}}{\Pi\vdash\Lambda,D_2}\,(D_1)\quad\begin{array}{c}\dfrac{\begin{array}{c}\vdots\ \pi_1^{\circ\prime}\\ D_2,\Gamma\vdash\Delta,C_1\end{array}\quad\begin{array}{c}\vdots\ \pi_2^{\circ\prime}\\ C_1,D_2,\Gamma\vdash\Delta\end{array}}{D_2,\Gamma\vdash\Delta}\,(C_1)\\ \vdots\ (\overrightarrow{C_2})^{D_2}\\ D_2,\Pi\vdash\Lambda\end{array}}{\Pi\vdash\Lambda}\,(D_2)\\
\vdots\ (\overrightarrow{C_{i+1}})\\
\Sigma\vdash\Phi
\end{array}
$$

$D_2, \Pi \vdash \Lambda$ is the additional sequent, and the cut-string $(C_{i-1}(\cdots(C_1)\cdots))$ of the additional sequent is the additional string, and $(\overrightarrow{C_{i+1}}(D_2(D_1(\overrightarrow{C_1}))))$ is the main string.

(Principal $\rightarrow r$**)** When $A \equiv D_1 \rightarrow D_2$, let the given cut-string be the following form.

$$
\dfrac{\dfrac{\dfrac{\vdots\,\pi_1 \qquad\qquad \vdots\,\pi_2}{\Gamma \vdash \Delta, D_1 \to D_2, C_1 \quad C_1, \Gamma \vdash \Delta, D_1 \to D_2}}{\Gamma \vdash \Delta, D_1 \to D_2}\,(C_1) \atop {\vdots\,(\overrightarrow{C_2}) \atop \Pi \vdash \Lambda, D_1 \to D_2} \qquad {\vdots\,\pi_3 \atop D_1 \to D_2, \Pi \vdash \Lambda}}{\Pi \vdash \Lambda}\,(D_1 \to D_2)
$$
$$
\vdots\,(\overrightarrow{C_{i+1}})
$$
$$
\Sigma \vdash \Phi
$$

The above cut-string is transformed as follows.

$$
\dfrac{\dfrac{{\vdots\,\pi_3^\circ \atop \Pi \vdash \Lambda, D_2^*, D_1} \qquad {\dfrac{{\vdots\,\pi_1^\circ \atop D_1, \Gamma \vdash \Delta, D_2, C_1} \quad {\vdots\,\pi_2^\circ \atop C_1, D_1, \Gamma \vdash \Delta, D_2}}{D_1, \Gamma \vdash \Delta, D_2}\,(C_1) \atop {\vdots\,(\overrightarrow{C_2})^{D_1 D_2} \atop D_1, \Pi \vdash \Lambda, D_2}}}{\Pi \vdash \Lambda, D_2}\,(D_1) \qquad {\vdots\,\pi_3^{\circ\prime} \atop D_2, \Pi \vdash \Lambda}}{\Pi \vdash \Lambda}\,(D_2)
$$
$$
\vdots\,(\overrightarrow{C_{i+1}})
$$
$$
\Sigma \vdash \Phi
$$

$D_2, \Pi \vdash \Lambda$ is the additional sequent.

References

1. Avron, A.: A constructive analysis of RM. J. Symb. Log. **52**(4), 939–951 (1987)
2. Avron, A.: The method of hypersequents in the proof theory of propositional non-classical logics. In: Hodges, W. (ed.), Logic: Foundations to Applications, pp. 1–32. Oxford (1996)
3. Coquand, T.: A semantics of evidence for classical arithmetic. J. Symb. Log. **60**(1), 325–337 (1995)
4. Dragalin, A.G.: Mathematical Intuitionism: Introduction to Proof Theory. American Mathematical Society, Translations of Mathematical Monographs (1987)
5. Felscher, W.: Dialogues, strategies and intuitionistic provability. Ann. Pure Appl. Logic **28**, 217–254 (1985)
6. Herbelin, H.: Games and weak-head reduction for classical PCF. In: Proceedings of Typed Lambda Calculus and Applications (TLCA'97), LNCS 1210, pp. 214–230
7. Herbelin, H.: Séquents qu'on calcule: de l'interprétation du calcul des séquents comme calcul de lambda-termes et comme calcul de stratégies gagnantes, PhD thesis (1995)
8. Lorenz, K., Lorenzen, P.: Dialogische Logik. Darmstadt (1978)
9. Lorenzen, P.: Ein dialogisches Konstruktivitätskriterium, in Infinitistic Methods, pp. 193–200. Pergamon Press, Oxford (1961)
10. Okada, M., Takemura, R.: A new proof-theoretical view on an old "dialogue logic". In: Essays in the Foundations of Logical and Phenomenological Studies, pp. 153–167. Keio University Press (2007)
11. Negri, S., von Plato, J.: Structural Proof Theory. Cambridge University Press (2001)
12. Troelstra, A.S., Schwichtenberg, H.: Basic Proof Theory, 2nd edn. Cambridge University Press (2000)

Chapter 14
Invitation to Constructive Nonreflexive and Nontransitive Logics

Rohan French

Abstract Rejecting structural principles, like the reflexivity or transitivity of the entailment relation, provides an attractive solution to the semantic paradoxes, allowing us to give a uniform analysis and solution to the paradoxes of self-reference without having to engage in piecemeal rejection of inferential principles governing the sentential connectives. To date all of the proposed nonreflexive and nontransitive logics have been substructural companions of classical logic. In this paper we correct this by investigating the substructural companions of intuitionistic logic, and their relation to their classical counterparts.

Keywords Substructural logics · Metainferences · Intuitionistic logic

14.1 Motivation and Preliminaries

According to substructural approaches to the paradoxes of self-reference, rather than addressing paradoxes like the Liar or Curry's Paradox by rejecting inferences involving negation or the conditional, we should instead 'grapple with the paradoxes where they live: in the basic features of argumentation' [16, p. 310], by rejecting *structural rules*—schematic rules of inference which encode basic structural features of arguments themselves. Our focus in the present paper will be nonreflexive and nontransitive substructural theories, which advise us to reject the claim that all sentences entail themselves (as encoded in the principle $[Id]$), or that arguments can be chained together (as encoded in the principle $[Cut]$).[1,2]

---

[1] Nonreflexive approaches to paradox have been argued for in [6, 7], and nontransitive approaches in [3, 14, 15].

[2] We will deal solely with propositional languages. Given a propositional language $\mathcal{L}$ a *sequent* is a pair $\langle \Gamma, \Delta \rangle$ of finite sets of formulas taken from $\mathcal{L}$, which we will typically write as $[\Gamma \succ \Delta]$. We will

---

R. French (✉)
University of California, Davis, Davis, CA, USA
e-mail: rfrench@ucdavis.edu
URL: http://rohan-french.github.io

K. Sano et al. (eds.), *Exploring Negation, Modality and Proof*,
Logic in Asia: Studia Logica Library,
https://doi.org/10.1007/978-981-96-8566-0_14

$$\frac{}{[A \succ A]}\,[\text{Id}] \qquad \frac{[\Gamma \succ \Delta, A] \quad [A, \Gamma' \succ \Delta']}{[\Gamma, \Gamma' \succ \Delta, \Delta']}\,[\text{Cut}]$$

The most prominent nonreflexive and nontransitive propositional logics are the logics dubbed TS and ST in [3]. Throughout we assume that our language $\mathcal{L}$ is an absolutely free-algebra of formulas with a set Atom of countably many generators $p_1, p_2, p_3 \ldots$ (the first three of which we abbreviate p, q, r) and the primitive singulary connective $\neg$, and primitive binary connectives $\wedge, \vee, \rightarrow$. Following [3] we can characterize these logics semantically in terms of sets of trivaluations, functions from our propositional language $\mathcal{L}$ to the space of values $\{1, i, 0\}$ which interpret the sentential connectives as indicated by the matrices in Fig. 14.1.[3]

Say that a trivaluation is a Strong Kleene valuation whenever it agrees with the matrices in Fig. 14.1. Then we can then characterize the sequents which are valid according to both TS and ST as follows.

Definition 14.1 Let $T = \{1\}$ and $S = \{1, i\}$. Then a Strong Kleene valuation v is an XY-counterexample to the sequent $[\Gamma \succ \Delta]$ iff $v[\Gamma] \subseteq X$ and $v[\Delta] \cap Y = \emptyset$, where $X, Y \in \{S, T\}$. A sequent $[\Gamma \succ \Delta]$ which has no XY-counterexamples is XY-valid.

As noted in [3], the notions of SS and TT validity determine Strong Kleene logic and the Logic of Paradox. The more interesting notions are the 'mixed' ones. As sets of arguments the TS- and ST-valid arguments are not that distinctive—the ST-valid arguments are all and only the classically valid arguments, and set of TS-valid arguments is empty. As is pointed out in [7] in the context of TS, though, the best way to understand these logics is in terms of their valid *metainferences*—inferences which have inferences as both premises and conclusions. Say that a Strong Kleene valuation v is an XY-counterexample to a metainference $\Sigma \Rightarrow \sigma$ iff it is not an XY-counterexample to any sequent in Σ, but is an XY-counterexample to σ. Then we can say that a metainference is XY-(locally) valid iff it has no XY-counterexamples. It is routine to see that there are ST counterexamples to the metainference corresponding to $[Cut]$, making the set of ST-valid metainferences distinct from those for classical

$\neg$	
1	0
i	i
0	1

$\wedge$	1	i	0
1	1	i	0
i	i	i	0
0	0	0	0

$\vee$	1	i	0
1	1	1	1
i	1	i	i
0	1	i	0

$\rightarrow$	1	i	0
1	1	i	0
i	1	i	i
0	1	1	1

Fig. 14.1 The Strong Kleene Matrices

also be concerned throughout with metainferences, with metasequents standing to metainferences in the same way sequents stand to inferences, with a *metasequent* being defined as an ordered pair $\langle \Sigma, \sigma \rangle$ of a finite set of sequents Σ and a sequent σ, which we write as $\Sigma \Rightarrow \sigma$.

[3] For a thorough investigation of the valuational model theory of nonreflexive and nontransitive logics the interested reader should consult [10].

logic, while there is no TS-counterexample to the metainference $[p \succ p], [q \succ q] \Rightarrow [p, q \succ p \wedge q]$, distinguishing the TS-valid metainferences from the empty set.

Of particular interest to us here is the fact that TS and ST are, in a very precisely specifiable way, nonreflexive and nontransitive versions of classical logic—in fact [15] has even argued that in the present signature ST is nothing more than classical logic. To make this more precise, let us temporarily identify a logic L with a collection of valuations V_L paired with a counterexample relation $\divideontimes_L$ on sequents. Then, when R is a metasequent, we can say that a logic L_1 is the R-companion of a logic L_2 whenever (i) R is not locally valid according to $\divideontimes_{L_2}$, and (ii) the set of metasequents which are valid according to $\divideontimes_{L_2}$ when restricting to valuations in V which are not $\divideontimes_{L_2}$-counterexamples to R is identical to those metasequents which are valid according to $\divideontimes_{L_1}$.

Theorem 14.1 *Classical logic is the* $[Id]$*-companion of* TS, *and the* $[Cut]$*-companion of* ST.

Proof We prove the first of these. As noted above, $[Id]$ is not TS-valid. To show that the restriction to Strong Kleene valuations which TS-validate $[Id]$ determines classical logic, consider that for any trivaluation v, whenever $v(A) = i$ it follows that v is a TS-counterexample to $[A \succ A]$. So the trivaluations which TS-validate all instances of $[Id]$ are those valuations which never assign a formula i, which are precisely the Boolean valuations. □

This result points to one of the key advantages of substructural approaches to the paradoxes. Substructural solutions let us very clearly point out both what the mistake involved in paradoxical reasoning is, as well as what the correct canons of reasoning are when that mistake is not relevant, in a relatively modular way. For example, according to the nonreflexive logician the mistake in paradoxical reasoning is thinking that paradoxical sentences entail themselves. Given the above result, the nonreflexive logician who thinks that, outside of the context of paradox, classical logic gives the canons of correct reasoning, should accept TS as the correct logic.

What about the nonreflexive and nontransitive logicians who don't wish to hew as closely to classical logic outside of paradoxical contexts, though? The purpose of the present paper is to act as the first steps in the investigation of this question in the context of intuitionistic logic, by investigating the nonreflexive and nontransitive counterparts of intuitionistic logic. The structure of the remainder of the paper is as follows. In Sect. 14.2 we begin by outlining a modification of the Kripke semantics for intuitionistic logic over which we are able to define analogues of the nonreflexive and nontransitive logics TS and ST. In §14.3 I then outline a labelled metasequent calculus in the sense of [8] which in Sect. 14.4 we prove to be sound & complete for the models defined in Sect. 14.2. Finally in Sect. 14.5 we look at the relations between the nonreflexive and nontransitive companions of classical logic and intuitionistic logic, before concluding in Sect. 14.6.

14.2 Semantics

In the present section we will define a general semantic framework for reasoning about nonreflexive and nontransitive intuitionistic logic. Our present framework is essentially the Kripke semantic version of the topological semantics which Girard used in [11] in his discussion of the semantics of Cut elimination in intuitionistic logic.[4] A model for substructural intuitionistic logic (henceforth simply 'a model') is a structure $\langle W, \leq, V_s, V_t \rangle$ where W is a non-empty set, $\leq$ a partial order on W, and V_s and V_t are functions from Atom to subsets of W which satisfy the following three conditions:

- CONTAINMENT: $V_s(p_i) \subseteq V_t(p_i)$
- T- PERSISTENCE: If $w \in V_t(p_i)$ and $w \leq v$ then $v \in V_t(p_i)$.
- S- PERSISTENCE: If $w \in V_s(p_i)$ and $w \leq v$ then $v \in V_s(p_i)$.

It can help to have some way of interpreting what's going on in this semantics. Following the interpretation given in [13], we will think of the members of W as representing information states, and $\leq$ as representing the relation of one information state extending another (with possibly further information). The main departure here is to think of information states as providing two different kinds of warrant for statements. The first kind of warrant they can provide is *strict* (or strong) warrant, which we should think of as the kind of warrant for a statement which supports us accepting it. The second kind of warrant a state can provide is *tolerant* (or tentative) warrant, which is sufficient warrant to not reject a statement, but not necessarily sufficient warrant to accept it. On this way of interpreting our semantics CONTAINMENT represents the reasonable thought that a warrant to accept a statement is also a warrant against rejecting it, while S- PERSISTENCE and T- PERSISTENCE record the idea that warrants are preserved under the gaining of new information—one consequence of which is that the tentativeness of tolerant warrant should be thought of as simply a matter of their strength, not whether they can later be overruled.

The distinction between strict and tolerant warrant is reminiscent of the distinction drawn in Artemov and Protopoescu's Intuitionistic Epistemic Logic [2] between the intuitionistic conceptions of truth and knowledge. Following the BHK-interpretation of the intuitionistic connectives Artemov and Protopoescu identify intuitionistic truth with proof, and characterize intuitionistic knowledge as verification that a proof exists, where proofs are understood to be a species of verification. This results in a novel epistemic logic IEL^- which validates the converse of the T axiom ($A \rightarrow \mathbf{K}A$, where K is the knowledge operator), while invalidating the $\mathbf{T}$ axiom itself (in the form $\mathsf{K}A \rightarrow A$) which is usually taken as characterizing the factivity for knowledge. This failure of intuitionistic knowledge to be factive in this sense is due to the fact

[4] Another similar semantic treatment of nontransitive intuitionistic logic is given in [20]. The main departure from the present approach is that their clause for the evaluation of the conditional is *partial*, only including the 'only if' direction of our clause for a conditional s-satisfied relative to an information state. We will leave further consideration of the connections between their approach and the present approach for future work.

that, on the intuitionistic conception advanced by Artemov and Protopoescu, having a verification that a proof exists (as is required for $\mathsf{K}A$ to hold) doesn't necessarily mean that one has a proof in hand (as is required for A to hold).[5] This is similar to our situation, where any statement which has a strict warrant also has a tolerant warrant, but not all statements which have tolerant warrants necessarily have strict warrants. The logics we get by understanding strict warrant as intuitionistic truth, and tolerant warrant as intuitionistic knowledge are significantly weaker than the logics **ILts** and **ILst** we characterize below. We leave further investigation of this parallel to future work.

The recursive clauses for evaluating formulas in our language over these models is relativized to the two different sorts of warrants states can provide, with x-satisfaction of a formula A at a point w in a model $\mathcal{M}$ ($=\mathcal{M}, w \Vdash_{\mathsf{x}} A$), where $\mathsf{x} \in \{s, t\}$, being defined inductively as follows:

- $\mathcal{M}, w \Vdash_t p_i$ iff $w \in V_t(p_i)$.
- $\mathcal{M}, w \Vdash_s p_i$ iff $w \in V_s(p_i)$.
- $\mathcal{M}, w \Vdash_t A \wedge B$ iff $\mathcal{M}, w \Vdash_t A$ and $\mathcal{M}, w \Vdash_t B$
- $\mathcal{M}, w \Vdash_s A \wedge B$ iff $\mathcal{M}, w \Vdash_s A$ and $\mathcal{M}, w \Vdash_s B$
- $\mathcal{M}, w \Vdash_t A \vee B$ iff $\mathcal{M}, w \Vdash_t A$ or $\mathcal{M}, w \Vdash_t B$
- $\mathcal{M}, w \Vdash_s A \vee B$ iff $\mathcal{M}, w \Vdash_s A$ or $\mathcal{M}, w \Vdash_s B$
- $\mathcal{M}, w \Vdash_t \neg A$ iff for all v where $w \leq v$ we have $\mathcal{M}, v \nVdash_s A$
- $\mathcal{M}, w \Vdash_s \neg A$ iff for all v where $w \leq v$ we have $\mathcal{M}, v \nVdash_t A$
- $\mathcal{M}, w \Vdash_t A \rightarrow B$ iff for all v where $w \leq v$ we have either $\mathcal{M}, v \nVdash_s A$ or $\mathcal{M}, v \Vdash_t B$
- $\mathcal{M}, w \Vdash_s A \rightarrow B$ iff for all v where $w \leq v$ we have either $\mathcal{M}, v \nVdash_t A$ or $\mathcal{M}, v \Vdash_s B$

It's helpful in looking at the above clauses to note that, letting $\overline{s} = t$ and $\overline{t} = s$, the t- and s-satisfaction clauses for '$\rightarrow$' and '$\neg$' have a common form, namely:

- $\mathcal{M}, w \Vdash_{\mathsf{x}} \neg A$ iff for all v where $w \leq v$ we have $\mathcal{M}, v \nVdash_{\overline{\mathsf{x}}} A$
- $\mathcal{M}, w \Vdash_{\mathsf{x}} A \rightarrow B$ iff for all v where $w \leq v$ we have either $\mathcal{M}, v \nVdash_{\overline{\mathsf{x}}} A$ or $\mathcal{M}, v \Vdash_{\mathsf{x}} B$

We will exploit this uniformity below in our soundness & completeness theorems.

As with the Kripke semantics for intuitionistic logic it is easy to show that the above semantic clauses allow us to lift the conditions of PERSISTENCE and CONTAINMENT give above from atomic sentences to the full language.

Lemma 14.1 (Persistence) *For all formulas A, and worlds w, w' we have for $\mathsf{x} \in \{t, s\}$:*

$$\text{If } \mathcal{M}, w \Vdash_{\mathsf{x}} A \text{ and } w \leq w', \text{ then } \mathcal{M}, w' \Vdash_{\mathsf{x}} A$$

[5] In [2] is it pointed out that the appropriate analogue of the principle of factivity in this setting is actually the classically equivalent $\mathsf{K}A \rightarrow \neg\neg A$, the addition of which to IEL^- results in the logic IEL.

Proof By induction on the complexity of A. Basis case follows from T- PERSISTENCE or S- PERSISTENCE. The only interesting case in the induction step are those for '$\rightarrow$' and '$\neg$'. We give the case for $\rightarrow$ (that for $\neg$ is similar). Suppose, then, that (i) $\mathcal{M}, w \Vdash_{\mathsf{x}} A \rightarrow B$, that (ii) $w \leq w'$ but that (iii) $\mathcal{M}, w' \nVdash_{\mathsf{x}} A \rightarrow B$. So by (iii) there is a w'' where (iv) $w' \leq w''$ and (v) $\mathcal{M}, w'' \Vdash_{\overline{\mathsf{x}}} A$ but (vi) $\mathcal{M}, w'' \nVdash_{\mathsf{x}} B$. By (ii) and (iv) we have $w \leq w''$, which from (i) gives us that either $w'' \nVdash_{\overline{\mathsf{x}}} A$ or $w'' \Vdash_{\mathsf{x}} B$. But from (v) it follows that we must have $w'' \Vdash_{\mathsf{x}} B$, contradicting (vi). So the result follows by reductio. □

Lemma 14.2 (Containment) *For all formulas A and worlds w we have that:*

$$\mathcal{M}, w \Vdash_s A \text{ only if } \mathcal{M}, w \Vdash_t A$$

Proof By induction on the complexity of A. The basis case follows from CONTAINMENT. The only interesting case in the induction step are those for $\rightarrow$ and $\neg$. We give the case for $\rightarrow$ (that for $\neg$ is similar). Suppose that $\mathcal{M}, w \Vdash_s A \rightarrow B$, but $\mathcal{M}, w \nVdash_t A \rightarrow B$. Then we have a w' where $w \leq w'$ and $w' \Vdash_s A$ and $w' \nVdash_t B$. By the induction hypothesis we have $w' \Vdash_t A$. By the fact that $\mathcal{M}, w \Vdash_s A \rightarrow B$ and $w \leq w'$ we have that if $w' \Vdash_t A$ then $w' \Vdash_s B$. So by the induction hypothesis again we have that $w' \Vdash_t B$ giving us a contradiction. □

As mentioned above, we will be using the above model theory to characterize nonreflexive and nontransitive logics. We will do this by defining a variety of mixed notions of consequence over these models as follows.

Definition 14.2 Suppose that $\mathsf{x}, \mathsf{y} \in \{s, t\}$. Then given a sequent $[\Gamma \succ \Delta]$ say that it xy-*holds at w in* $\mathcal{M}$ iff either $\mathcal{M}, w \nVdash_{\mathsf{x}} \gamma$ for some $\gamma \in \Gamma$, or if $\mathcal{M}, w \Vdash_{\mathsf{y}} \delta$ for some $\delta \in \Delta$, and that it xy-*holds throughout* $\mathcal{M}$ iff it xy-holds at all worlds in $\mathcal{M}$. Finally we will say that a sequent is xy-valid iff it xy-holds in all models.

This definition gives rise to the following two 'mixed' notions of validity.[6] Say that a sequent is ILts valid iff it is ts valid, and ILst-valid iff it is st-valid.

Defining validity for metasequents in this setting presents us with choices which are not present in the valuational setting adopted above and in [8] and elsewhere in the literature. In the literature on metasequent validity it is typical to draw a distinction between a *local* and *global* notion of metasequent validity where:

- A metasequent $\Sigma \Rightarrow \sigma$ is *metasequentially locally valid* iff for all models m, if every member of Σ holds in m, then σ holds in m.

[6] It is worth briefly noting that we also get intuitionistic analogues of the Logic of Paradox and Strong Kleene logic out of this definition as the set of all tt and ss-valid sequents respectively. These logics are not substructural, like the mixed notions, but do have a number of interesting constructive features. I investigate some curious features of the 'Constructive Logic of Paradox' in [9], building off work in the present paper along with previous work on the constructive analogues of Parry's logic of analytic implication in [5], and Halldén's logic of nonsense in [22] (on which also see [1]).

- A metasequent $\Sigma \Rightarrow \sigma$ is *metasequentially globally valid* iff if every member of Σ holds in all models, then σ holds in all models.

As is done in [8] we will work with local metasequential validity (although see [19] and [12] for arguments concerning whether to favour one notion over the other). In broadly modal settings like the present one there are two reasonable looking notions of local metasequential validity, determined by two different ways of regarding a sequent as holding in a model, namely the following:

- A metasequent $\Sigma \Rightarrow \sigma$ is *local-local* xy-*valid* iff for all models $\mathcal{M} = \langle W, \leq, V_s, V_t \rangle$ and every $w \in W$, if every member of Σ xy-holds at w, then σ xy-holds at w.
- A metasequent $\Sigma \Rightarrow \sigma$ is *local-global* xy-*valid* iff for all models $\mathcal{M} = \langle W, \leq, V_s, V_t \rangle$, if every member of Σ xy-holds throughout $\mathcal{M}$, then σ xy-holds throughout $\mathcal{M}$.

Is there any way to make a principled choice between these two different accounts of validity in this setting? One prima facie desirable feature for an account of validity for a nonreflexive logician is that it allows one to execute the recapture strategy outlined in [7]. There it is argued that a nonreflexive logician can account for the correctness of classically valid inferences by counting them as involving suppressed metainferential premises. For example while $[\neg\neg\neg p \succ \neg p]$ is not TS-valid, the metainference

$$[p \succ p] \Rightarrow [\neg\neg\neg p \succ \neg p]$$

is TS-valid, allowing the nonreflexive logician to explain the correctness of the inference $[\neg\neg\neg p \succ \neg p]$ by pointing out that the above metainference is logically valid, and we are often correct in suppressing the atomic metainferential premise $[p \succ p]$ in non-paradoxical contexts. Explanations for the correctness of intuitionistically valid inferences like triple-negation elimination are unavailable to a nonreflexive logician who uses local-local validity.[7] As we can see, this metainference is invalid according to local-local ts-validity. Consider the following model $\mathcal{M} = \langle W, \leq, V_s, V_t \rangle$:

- $W = \{w, v\}$
- $\leq = \{\langle w, w \rangle, \langle w, v \rangle, \langle v, v \rangle\}$
- $V_s(p) = \emptyset$, $V_t(p) = \{v\}$.

Given that $w \nVdash_t p$ we have $[p \succ p]$ts-holds at w. But it is also clear that $v \Vdash_t \neg p$, as for no $\leq$-successor u of v do we have $u \Vdash_s p$, and so all $\leq$-successors u of w are such that $u \nVdash_s \neg\neg p$, from which we have $w \Vdash_t \neg\neg\neg p$. At the same time $w \nVdash_s \neg p$ as we have $v \Vdash_t p$. So $[\neg\neg\neg p \succ \neg p]$ fails to ts-hold at w. So this metainference has a local-local counterexample at w. By contrast, it is easy to see that the above metainference is local-global valid—the above counterexample being ruled out by

[7] Thank you to Pablo Cobreros for correspondence and conversations which helped me more clearly think through this point.

the fact that $[p \succ p]$ does not ts-hold at v, this feature being crucial to the local-local counterexample to this metainference.

Given this we will in the rest of this paper adopt local-global metainferential validity, and whenever we talk of a metasequent being xy-valid what we mean is that the metasequent in question is local-global xy-valid.

One immediate use of the above machinary which we can make at this point is to show that ILts and ILst are respectively $[Id]$- and $[Cut]$-companions of IL. We will treat the case of ILts in detail below, the case of ILst following similarly. First we will want to make use of the following Lemma.

Lemma 14.3 *A model $\langle W, \leq, V_s, V_t \rangle$ ts-validates all instances of $[Id]$ iff $V_s = V_t$.*

Proof For the 'only if' direction we prove the contrapositive. Suppose that $\mathcal{M} = \langle W, \leq, V_s, V_t \rangle$ and $V_s \neq V_t$, so for some p_i we have $w \in V_t(p_i)$ but $w \notin V_s(p_i)$ (as by CONTAINMENT if we have $w \in V_s(p_i)$ we also have $w \in V_t(p_i)$). From this it follows that $w \Vdash_t p_i$ but $w \nVdash_s p_i$ and so it follows directly that we do not have $\mathcal{M} \models_{ts} [p_i \succ p_i]$, and so that instance of $[Id]$ is not validated in $\mathcal{M}$.

For the 'if' direction suppose $V_s = V_t$. We prove by induction on the complexity of A that all instances of $[A \succ A]$ hold in $\mathcal{M}$. For the basis case suppose that $w \Vdash_t p_i$, and so $w \in V_t(p_i)$. As, by hypothesis, $V_t(p_i) = V_s(p_i)$ it follows that $w \in V_s(p_i)$ and so $w \Vdash_s p_i$. But w was arbitrary so at every world $w \in W$ if $w \Vdash_t p_i$ then $w \Vdash_s p_i$, and so $\mathcal{M} \models_{ts} [p_i \succ p_i]$. For the induction hypothesis suppose, then, that for all formulas B of complexity less than n we have that $\models_{ts} [B \succ B]$. Suppose, then, that $w \Vdash_t \neg B$. Then for all $w \leq v$ we have $v \nVdash_s B$. But we have $\models_{ts} [B \succ B]$, so this means that at all such v we must have $v \nVdash_t B$, and so it follows that we have $w \Vdash_s \neg B$, as desired. □

Theorem 14.2 *Intuitionistic Logic is the $[Id]$-companion of* ILts, *and the $[Cut]$-companion of* ILst.

Proof As stated above we'll treat only the $[Id]$-case here. Recall that what we need to demonstrate to show that Intuitionistic logic is the $[Id]$-companion of ILts is (a) that $[Id]$ is not valid in ILts, and (b) that the addition of all instances of $[Id]$ to ILts results in intuitionistic logic. It is easy to show that not all instances of $[Id]$ are valid in ILts. To see this consider the following simple model $\mathcal{M} = \langle W, \leq, V_s, V_t \rangle$.

- $W = \{w\}$
- $\leq = \{\langle w, w \rangle\}$
- $V_s(p_i) = \emptyset$
- $V_t(p_i) = W$

In this model we have $w \Vdash_t p_i$, but $w \nVdash_s p_i$ for all $w \in W$, from which it follows directly that $[p_i \succ p_i]$ does not ts-hold throughout $\mathcal{M}$, and so the metainference $\Rightarrow [p_i \succ p_i]$ is not ts-valid.

Now we show that restricting ourself to models which validate all instances of $[Id]$ results in intuitionistic logic. From Lemma 14.3 we have that the class of all models which validate all instances of $[Id]$ are those models of the form $\langle W, \leq, V, V\rangle$, each of which are trivially isomorphic to Kripke models for intuitionistic logic, which from [13] we know determine intuitionistic logic. □

14.3 Proof-Theory

14.3.1 A Metasequent Calculus

Here we give a proof system which we label SubIL, which is a general calculus for determining the validity of metasequents over the models defined in the previous section. This system is a metasequent calculus in the style of [8], built of Maehara's multiple-conclusion sequent calculus for intuitionistic logic, but with the addition of labels in the style of [4]. All formulas in our sequents will be labelled with truth-statuses—being of the form $t : A$ or $s : A$. In the following rules, recall that $x \in \{s, t\}$ and $\overline{s} = t$ and $\overline{t} = s$. Given a sequent $[\Gamma \succ \Delta]$ let $[\Gamma \succ \Delta]^{\mathsf{xy}}$ be the sequent $[\Gamma^{\mathsf{x}} \succ \Delta^{\mathsf{y}}]$ where $\Gamma^{\mathsf{x}} = \{x : A | A \in \Gamma\}$ and $\Delta^{\mathsf{y}} = \{y : A | A \in \Delta\}$. Due to the addition of labels this metasequent calculus gives us a general calculus for demonstrating that a metainference $\sigma_1, \ldots, \sigma_n \Rightarrow \sigma$ is xy-valid by producing a derivation of the labelled metasequent $\sigma_1^{\mathsf{xy}}, \ldots, \sigma_n^{\mathsf{xy}} \Rightarrow \sigma^{\mathsf{xy}}$.

Our calculus will consist of four different families of rules: outer structural rules (governing the structure of metasequents), inner structural rules (governing the structure of sequents), premise operational rules (governing the introduction of connectives in sequents in our metasequent premises), and conclusion operational rules (governing the introduction of connectives our metasequents conclusion sequent). We describe each of these rules in order below.

OUTER STRUCTURAL RULES

$$\frac{}{\Sigma, \sigma \Rightarrow \sigma}\,[ID] \qquad \frac{\Sigma \Rightarrow \sigma' \quad \Sigma', \sigma' \Rightarrow \sigma}{\Sigma, \Sigma' \Rightarrow \sigma}\,[CUT]$$

INNER STRUCTURAL RULES

$$\frac{}{\Sigma \Rightarrow [s : A \succ t : A]}\,[id_{st}] \qquad \frac{}{\Sigma \Rightarrow [s : A \succ s : A]}\,[id_{ss}] \qquad \frac{}{\Sigma \Rightarrow [t : A \succ t : A]}\,[id_{tt}]$$

$$\frac{\Sigma \Rightarrow [\Gamma \succ \Delta, x : A] \quad \Sigma \Rightarrow [x : A, \Gamma' \succ \Delta']}{\Sigma \Rightarrow [\Gamma, \Gamma' \succ \Delta, \Delta']}\,[cut] \qquad \frac{\Sigma \Rightarrow [\Gamma \succ \Delta]}{\Sigma \Rightarrow [\Gamma, \Gamma' \succ \Delta, \Delta']}\,[dil]$$

CONCLUSION OPERATIONAL RULES

$$\frac{\Sigma \Rightarrow [\Gamma \succ \Delta, x : A] \quad \Sigma \Rightarrow [\Gamma \succ \Delta, x : B]}{\Sigma \Rightarrow [\Gamma \succ \Delta, x : A \wedge B]}\,[\wedge R^x_c] \qquad \frac{\Sigma \Rightarrow [\Gamma, x : A, x : B \succ \Delta]}{\Sigma \Rightarrow [\Gamma, x : A \wedge B \succ \Delta]}\,[\wedge L^x_c]$$

$$\frac{\Sigma \Rightarrow [\Gamma \succ \Delta, x : A, x : B]}{\Sigma \Rightarrow [\Gamma \succ \Delta, x : A \vee B]}\,[\vee R^x_c] \qquad \frac{\Sigma \Rightarrow [\Gamma, x : A \succ \Delta] \quad \Sigma \Rightarrow [\Gamma, x : B \succ \Delta]}{\Sigma \Rightarrow [\Gamma, x : A \vee B \succ \Delta]}\,[\vee L^x_c]$$

$$\frac{\Sigma \Rightarrow [\Gamma, \overline{x} : A \succ x : B]}{\Sigma \Rightarrow [\Gamma \succ x : A \rightarrow B]}\,[\rightarrow R^x_c] \qquad \frac{\Sigma \Rightarrow [\Gamma \succ \Delta, \overline{x} : A] \quad \Sigma \Rightarrow [x : B, \Gamma \succ \Delta]}{\Sigma \Rightarrow [\Gamma, x : A \rightarrow B \succ \Delta]}\,[\rightarrow L^x_c]$$

$$\frac{\Sigma \Rightarrow [\Gamma, \overline{x} : A \succ\ ]}{\Sigma \Rightarrow [\Gamma \succ x : \neg A]}\,[\neg R^x_c] \qquad \frac{\Sigma \Rightarrow [\Gamma \succ \Delta, \overline{x} : A]}{\Sigma \Rightarrow [\Gamma, x : \neg A \succ \Delta]}\,[\neg L^x_c]$$

PREMISE OPERATIONAL RULES

$$\frac{\Sigma, [\Gamma \succ \Delta, x : A], [\Gamma \succ \Delta, x : B] \Rightarrow \sigma}{\Sigma, [\Gamma \succ \Delta, x : A \wedge B] \Rightarrow \sigma}\,[\wedge R^x_p] \qquad \frac{\Sigma, [\Gamma, x : A, x : B \succ \Delta] \Rightarrow \sigma}{\Sigma, [\Gamma, x : A \wedge B \succ \Delta] \Rightarrow \sigma}\,[\wedge L^x_p]$$

$$\frac{\Sigma, [\Gamma \succ \Delta, x : A, x : B] \Rightarrow \sigma}{\Sigma, [\Gamma \succ \Delta, x : A \vee B] \Rightarrow \sigma}\,[\vee R^x_p] \qquad \frac{\Sigma, [\Gamma, x : A \succ \Delta], [\Gamma, x : B \succ \Delta] \Rightarrow \sigma}{\Sigma, [\Gamma, x : A \vee B \succ \Delta] \Rightarrow \sigma}\,[\vee L^x_p]$$

$$\frac{\Sigma, [\Gamma, \overline{x} : A \succ x : B] \Rightarrow \sigma}{\Sigma, [\Gamma \succ x : A \rightarrow B] \Rightarrow \sigma}\,[\rightarrow R^x_p] \qquad \frac{\Sigma, [\Gamma, x : A \rightarrow B \succ \Delta, \overline{x} : A], [x : B, \Gamma \succ \Delta] \Rightarrow \sigma}{\Sigma, [\Gamma, x : A \rightarrow B \succ \Delta] \Rightarrow \sigma}\,[\rightarrow L^x_p]$$

$$\frac{\Sigma, [\Gamma, \overline{x} : A \succ\ ] \Rightarrow \sigma}{\Sigma, [\Gamma \succ x : \neg A] \Rightarrow \sigma}\,[\neg R^x_p] \qquad \frac{\Sigma, [\Gamma, x : \neg A \succ \Delta, \overline{x} : A] \Rightarrow \sigma}{\Sigma, [\Gamma, x : \neg A \succ \Delta] \Rightarrow \sigma}\,[\neg L^x_p]$$

14.4 Soundness and Completeness

We semantically interpret our labelled calculus in the expected way, adopting the notation that $\mathcal{M} \models_w x : A$ is to mean $\mathcal{M}, w \Vdash_x A$. The definitions of what it is for a labelled sequent to hold in a model, and for a metasequent to hold in a model are the obvious elaborations of the definitions given above. A labelled sequent $[\Gamma \succ \Delta]$ *holds at w in a model $\mathcal{M}$* iff either $\mathcal{M} \not\models_w \gamma$ for some $\gamma \in \Gamma$ or $\mathcal{M} \models_w \delta$ for some

$\delta \in \Delta$. A metainference $\Sigma \Rightarrow \sigma$ of labelled sequents is *valid* iff for all models $\mathcal{M}$, if every $\sigma' \in \Sigma$ holds at all worlds $w \in W$, then σ holds at all worlds $w \in W$.

Theorem 14.3 (Soundness) *If a metasequent* $\Sigma \Rightarrow \sigma$ *is derivable in* SubIL, *then* $\Sigma \Rightarrow \sigma$ *is valid.*

Proof We proceed by induction on the structure of derivations, showing that our initial sequents are valid and our rules preserve validity. For the basis case it is easy to see that $[ID]$, $[id_{ss}]$ and $[id_{tt}]$ hold in all models, and that $[id_{st}]$ holds in all models follows from Lemma 14.2.

We give an illustrative selection of cases for the induction step below.

[cut] Suppose that the last rule applied is $[cut]$, and so our derivation ends like this:

$$\frac{\Sigma \Rightarrow [\Gamma \succ \Delta, x : A] \quad \Sigma \Rightarrow [x : A, \Gamma' \succ \Delta']}{\Sigma \Rightarrow [\Gamma, \Gamma' \succ \Delta, \Delta']}\ [cut]$$

Suppose that the conclusion sequent fails to hold at some world $w \in W$, and so we have $\models_w \Gamma \cup \Gamma'$, but $\not\models_w \delta$ for any $\delta \in \Delta \cup \Delta'$. Now either (i) $w \Vdash_x A$ or (ii) $w \nVdash_x A$. If (i) then the RHS premise fails to hold. If (ii) then the LHS premise fails to hold. So either way one of the premises will fail to hold if the conclusion fails to do so.

[$\to$ $\mathbf{R_c^x}$]: Suppose that the last rule applied is $[\to R_c^x]$, and so our derivation ends like this:

$$\frac{\Sigma \Rightarrow [\Gamma, \overline{x} : A \succ x : B]}{\Sigma \Rightarrow [\Gamma \succ x : A \to B]}\ [\to R_c^x]$$

Suppose that the conclusion sequent fails to hold in $\mathcal{M}$. Then there is some $w \in W$ s.t. $\mathcal{M} \models_w \Gamma$ but $w \nVdash_x A \to B$. So there $\exists w \leq w'$ s.t. $w' \Vdash_{\overline{x}} A$ and $w' \nVdash_x B$. By Lemma 14.1 we also have $\mathcal{M} \models_{w'} \Gamma$ and so the premise sequent fails to hold at w'.

[$\to$ $\mathbf{L_p^x}$]: Suppose that the last rule applied is $[\to L_p^x]$, and so our derivation ends like this:

$$\frac{\Sigma, [\Gamma, x : A \to B \succ \Delta, \overline{x} : A], [x : B, \Gamma \succ \Delta] \Rightarrow \sigma}{\Sigma, [\Gamma, x : A \to B \succ \Delta] \Rightarrow \sigma}\ [\to L_p^x]$$

Suppose that the conclusion sequent fails to hold. In particular this means that for all $w \in W$ we have every metasequent premise holding at w (while σ fails at some w). So suppose that we have $\mathcal{M} \models [\Gamma, x : A \to B \succ \Delta]$, and suppose that we have $\mathcal{M} \models_w \Gamma$, $\mathcal{M} \not\models_w \delta$ for all $\delta \in \Delta$. Then we must have $\mathcal{M} \not\models_w x : A \to B$ (automatically making the LHS premise sequent hold at w). So there is some $w \leq w'$ such that $w' \Vdash_{\overline{x}} A$ and $w' \nVdash_x B$. So by persistence we have $w \nVdash_x B$, and so the RHS premise holds at w.

□

Theorem 14.4 (Completeness) *If* $\Sigma \Rightarrow \sigma$ *is underivable in* SubIL, *then it is invalid.*

Proof The countermodel construction we present here is a combination of that used in [8] and that for intuitionistic logic used in [18]. We proceed in stages as outlined below.

Constructing Reduction Trees Suppose that we have an enumeration $x : A_1, x : A_2, \ldots x : A_n, \ldots$ of all labelled formulas. Let $\Sigma \Rightarrow \sigma$ be a metasequent. We will construct a reduction tree of metasequents as follows.

- The root of our reduction tree is the metasequent $\Sigma \Rightarrow \sigma$
- A branch of our tree is *closed* iff (i) its leaf metasequents are of the form $\Theta, [X \succ Y] \Rightarrow [X, X' \succ Y, Y']$ (in which case we can derive this metasequent by applying $[dil]$ to an instance of $[ID]$), (ii) its leaf metasequents are of the form $\Sigma \Rightarrow [X, s : A \succ Y, t : A]$, $\Sigma \Rightarrow [X, s : A \succ Y, s : A]$, or $\Sigma \Rightarrow [X, t : A \succ Y, t : A]$ (in which case we can derive this metasequent by applying $[dil]$ to an instance of $[id_{st}]$, $[id_{ss}]$, or $[id_{tt}]$ respectively). A branch which is not closed is *open*.
- In the first stage we process our metasequent premises. To do this we extend our open branch by applying root-first one our of PREMISE OPERATIONAL RULES to a sequent $\theta_i \in \Theta$, where $\Theta \Rightarrow \theta$ is an open leaf metasequent of our tree. So, for example, if our leaf metasequent is of the form $\Sigma, [\Gamma, x : A \rightarrow B \succ \Delta] \Rightarrow \sigma$ and we apply $[\rightarrow L^x_p]$ root-first expanding out tree to give us a new leaf metasequent of
$$\Sigma, [\Gamma, x : A \rightarrow B \succ \Delta, \overline{x} : A], [\Gamma, x : B \succ \Delta] \Rightarrow \sigma$$
- We continue doing this until this process reaches a fixed point.

At the end of this stage we will have ended up with a tree of metasequents the root of which is $\Sigma \Rightarrow \sigma$. Let Σ^+ be the result of taking the union of all the Σ_i such that $\Sigma_i \Rightarrow \sigma$ appears as a node in the tree constructed in our first stage. We will then create our reduction tree using the metasequent $\Sigma^+ \Rightarrow \sigma$.

- To create a reduction tree we proceed as follows. Suppose that $\Theta \Rightarrow [X \succ Y]$ is a leaf metasequent of an open branch. We then apply the $[\wedge L^x_c]$, $[\wedge R^x_c]$, $[\vee L^x_c]$, $[\vee R^x_c]$, $[\rightarrow L^x_c]$, $[\neg L^x_c]$ rules root-first to our leaf metasequent, guided by the order in which formulas appear in our enumeration. We perform these reductions until there are no longer any formulas in the leaf metasequents of open branches to which we are able to apply our reduction which we have not yet applied it to.
- Once there are no longer any formulas in the leaf of an open branch which we can apply our reductions to we expand that branch with two new leaf nodes $\Theta \Rightarrow [X \succ Y, x : A_i]$ and $\Theta \Rightarrow [x : A_i, X \succ Y]$ where $x : A_i$ is the next formula in our enumeration. This is our $[cut]$-reduction.
- An open branch is *completed* iff there is no longer any reduction which can be applied to it.

Let us refer to the completed reduction tree for a metasequent S as $T(S)$. It is a routine matter to see that if all of the leaf metasequents of $T(S)$ are provable in SubIL then so is S, given the manner in which the construction of our reduction tree is guided by the rules of SubIL. In particular, then, if S is unprovable then there must be some branch, call it B_S, in $T(S)$ every metasequent of which is unprovable. We will assign the metasequent which results from merging together all the metasequents on B_{S_0} (where $S_0 = \Sigma^+ \Rightarrow \sigma$) the label 0.

We now move on to the third stage of our construction. Given a metasequent $\Theta \Rightarrow [X_n \succ Y_n]$ labelled n which results from merging together all the sequents on B_{S_n}, consider for each formula $x : B \to C \in Y_n$ (and similarly for $x : \neg B \in Y_n$) the metasequents:

$$\Theta \Rightarrow [X, \overline{x} : B \succ x : C]$$

It is easy to see that this sequent cannot be SubIL provable (otherwise we could prove $\Theta \Rightarrow [X \succ Y]$ by applying $[\to R^x_c]$ and then $[dil]$). Let us label this new sequent nm, where m is the position of $x : B \to C$ in our enumeration, and let $T(S_{nm})$ be its reduction tree. Continue this procedure by selecting an open branch $B_{S_{nm}}$ and for each formula of the form $x : C \to D \in Y_{nm}$ (similarly $x : \neg C \in Y_{nm}$) considering the metasequent $\Theta \Rightarrow [X_{nm}, \overline{x} : C \succ x : D]$, calling this sequent nml where l is the position of $x : C \to D$ in our enumeration.

Continue this procedure ω many times.

Converting the Reduction Trees into a Model: We will use the collection of all of these reduction trees to define our model as follows, referring to the metasequent with label l as $\Theta \Rightarrow [X_l \succ Y_l]$. Let W be the set of all labels given to metasequents, and for $l, m \in W$ set $l \leq m$ iff l is an initial segment of m. Define the V_t and V_s as follows:

- $w \in V_t(p_i)$ iff $t : p \in X_w$.
- $w \in V_s(p_i)$ iff $s : p \in X_w$.

It is easy to see by construction that the model defined here satisfies the conditions of T- PERSISTENCE and S- PERSISTENCE. To see that it satisfies CONTAINMENT suppose that (i) $w \in V_s(p_i)$ and suppose (ii) $w \notin V_t(p_i)$. By (i) we know that $s : p_i \in X_w$, and by (ii) that $t : p_i \notin X_w$. As $\Theta \Rightarrow [X_w \succ Y_w]$ is the merge of an open branch it follows from the $[cut]$-reduction that we must have $t : p_i \in Y_w$. But this would mean that sequent w is of the form $\Theta \Rightarrow [X_w, s : p_i \succ Y_w, t : p_i]$, and hence is closed! So it follows that CONTAINMENT must be satisfied.

Showing that truth and membership coincide: Having shown that this is indeed a model, we now move on to showing that $x : A \in X_w$ iff $w \Vdash_x A$. The basis case is handled by our definition of V_s and V_t. Suppose, then, that for all formulas B with complexity $< n$ we have $x : B \in X_w$ iff $w \Vdash_x B$. We deal with the two directions for $\to$ in the inductions step, the others being similar:

- Suppose that $x : B \to C \in X_w$. Then by construction for all v such that $w \leq v$ we have $x : B \to C \in X_v$. So by the $[\to L]$-reduction we have either $\overline{x} : B \in Y_v$

or $x : C \in X_v$. So by the IH we have either $v \nVdash_{\overline{x}} B$ or $v \Vdash_x C$. As this holds for all $w \leq v$ it follows, then, that $w \Vdash_x B \rightarrow C$.

- Suppose that $x : B \rightarrow C \notin X_w$. By the $[cut]$-reduction we have that $x : B \rightarrow C \in Y_w$. So by the second stage of construction there is some v such that $w \leq v$ where $\overline{x} : B \in X_v$ and $x : C \in Y_v$. So by the IH we have $v \Vdash_{\overline{x}} B$ and $v \nVdash_x C$. So it follows that $w \nVdash_x B \rightarrow C$.

Showing that our model is a counterexample: Now we need to show that this model is such that, at our root node 0 we have $\Sigma \Rightarrow [\Gamma \succ \Delta]$ failing to hold. It is easy to see that at 0 as $\Gamma \subseteq X_0$ we have $\mathcal{M} \models_0 \Gamma$. Suppose, then, that for some $x : A \in \Delta$ we have $\mathcal{M} \models_0 x : A$. Then we must have $x : A \in X_0$, but as $\Delta \subseteq Y_0$ we must also have $x : A \in Y_0$, meaning that our branch is closed! So for all $x : A \in \Delta$ we have $\mathcal{M} \not\models_0 x : A$. Suppose, then, that $[X \succ Y] \in \Sigma$. As none of the branches B_S are closed we must have, for each metasequent $\Theta \Rightarrow [X_w \succ Y_w]$ with label w, that either $X \not\subseteq X_w$ or $Y \not\subseteq Y_w$, and so either some $x : A \in X$ where $x : A \notin X_w$, or some $x : B \in Y$ where $x : B \notin Y_w$. For the first if $x : A \notin X_w$ we must have $w \nVdash_x A$ and so the sequent holds at w. For the second if $B \notin Y_w$ then we must have $B \in X_w$ by $[cut]$-closure, and so $w \Vdash_x B$, and so the sequent holds at w. As this is the case for all w in the model, it follows that all the metasequent premises hold at all worlds, while $[\Gamma \succ \Delta]$ fails at 0, and thus the sequent is invalid. □

Our completeness result brings with it some corollaries related to our different notions of metasequent validity. For example, it follows from the above results that a metasequent $\Sigma \Rightarrow \sigma$ is ts-valid iff $\Sigma^{\mathsf{ts}} \Rightarrow \sigma^{\mathsf{ts}}$ is derivable in SubIL. We can summarize these results as follows.

Corollary 14.1 *A metasequent $\Sigma \Rightarrow \sigma$ is* xy*-valid iff $\Sigma^{\mathsf{xy}} \Rightarrow \sigma^{\mathsf{xy}}$ is* SubIL *derivable.*

14.5 Relating Classical and Constructive Substructural Logics

We close by looking at the relationship between ILts and ILst and TS and ST. It is well known, and easy to show, that classical logic is determined by the class of all intuitionistic Kripke frames whose accessibility relation is the identity relation, and that the law of excluded middle—$A \vee \neg A$—corresponds (as shown in [21, p. 234]) to this class of frames.[8]

[8] See Sect. 3 of Segerberg's tour-de-force [17] for further discussion of determination results over minimal logic for related classes of frames.

It is easy to show that a similar determination fact obtains in the case of our substructural Kripke frames. The most straightforward way to see this is through the following connection between our semantics in terms of substructural Kripke models, and the semantics for TS and ST given in terms of trivaluations in Sect. 14.1.

For convenience let $\mathcal{ID}$ be the class of all substructural Kripke models where $\leq$ is just the identity relation on W.

Proposition 14.1 *1. Let $\mathcal{M} = \langle W, \leq, V_s, V_t \rangle$ be a model in $\mathcal{ID}$, and $w \in W$, and define the trivaluation v_w as follows:*

$$v_w(A) = \begin{cases} 1 & w \Vdash_s A \\ i & w \Vdash_t A \text{ and } w \nVdash_s A \\ 0 & w \nVdash_t A \end{cases}$$

Then the trivaluation v_w is a Strong Kleene valuation.

2. Let v be a Strong Kleene valuation. Then $\langle \{w\}, \{\langle w, w \rangle\}, V_s^v, V_t^v \rangle$ is a model, where $w \in V_s^v(p_i)$ iff $v(p_i) = 1$, and $w \in V_t^v(p_i)$ iff $v(p_i) \in \{1, i\}$.

Corollary 14.2 *A metasequent $\Sigma \Rightarrow \sigma$ is* XY*-valid iff it is global-local* xy*-valid in the class of all models in $\mathcal{ID}$.*

The question then arises: what metasequents correspond to the class of $\mathcal{ID}$ models? Interestingly this is sensitive to what species of xy-validity we are using. First say that a metasequent $\Sigma \Rightarrow \sigma$ is xy-*valid on a frame* $\langle W, \leq \rangle$ iff for any model $\mathcal{M} = \langle W, \leq, V_s, V_t \rangle$ if every sequent in Σ xy-holds throughout $\mathcal{M}$, then σ holds throughout $\mathcal{M}$. This is just the standard notion of validity on a frame lifted to the present setting.

Theorem 14.5 *1. Every instance of the metasequent MEM ('Metasequent Excluded Middle')*

$$(MEM) \qquad [\Gamma, A \succ \Delta], [\Gamma, \neg A \succ \Delta] \Rightarrow [\Gamma \succ \Delta]$$

is ts*-valid on a frame $\langle W, \leq \rangle$ iff $\leq$ is the identity relation on W.*

2. Every instance of the metasequent SEM ('Sequent Excluded Middle')

$$(SEM) \qquad \Sigma \Rightarrow [\Gamma \succ \Delta, A, \neg A]$$

is st*-valid on a frame $\langle W, \leq \rangle$ iff $\leq$ is the identity relation on W.*

Proof We treat only case 1, case 2 being very similar to the standard case for intuitionistic logic.

For the 'only if' direction we prove the contrapositive. Suppose that we have a frame $\langle W, \leq \rangle$ where $w \leq w'$ and $w \neq w'$. We will show that the following instance of MEM is ts-invalid on that frame.

$$[p \succ q], [\neg p \succ q] \Rightarrow [\succ q]$$

To do this consider the model $\mathcal{M} = \langle W, \leq, V_s, V_t \rangle$ on the frame $\langle W, \leq \rangle$ where $V_t(p) = V_s(p) = V_t(q) = V_s(q) = W \setminus \{u : u \leq w\}$. In this model we have $u \Vdash_{ts} [p \succ q]$ for all worlds $u \in W$, as at all worlds u where $u \not\leq w$, we have $u \Vdash_s q$, while at worlds u where $u \leq w$ we have $u \nVdash_t p$. We also have $u \Vdash_{ts} [\neg p \succ q]$, as at all worlds u we have $u \nVdash_t \neg p$, as for all those worlds $u \leq w'$ we have $w' \Vdash_s p$ and so $u \nVdash_t \neg p$, while for all worlds $u \not\leq w'$ we have $u \leq u$ and $u \Vdash_s p$ and so $u \nVdash_t \neg p$. So both premise sequents ts-hold at all worlds in W. But it is clear that $w \nVdash_{ts} [\succ q]$ as $w \notin V_s(q)$. So if $\leq$ is not the identity relation on W, then MEM is not ts-valid on that frame.

For the 'if' direction suppose that we have a model $\mathcal{M} = \langle W, \leq, V_s, V_t \rangle$ on a frame $\langle W, \leq \rangle$ where $\leq$ is the identity relation on W. Suppose further that at some world $w \in W$ we have $w \nVdash_{ts} [\Gamma \succ \Delta]$. So we have (i) $w \Vdash_t \gamma$ for all $\gamma \in \Gamma$, (ii) $w \nVdash_s \delta$ for all $\delta \in \Delta$. Suppose then that $w \Vdash_t A$. Then by (i) and (ii) it follows that $w \nVdash_{ts} [\Gamma, A \succ \Delta]$. Suppose, then, that $w \nVdash_t A$. By containment it follows then that $w \nVdash_s A$, and so as $\leq$ is the identity relation it follows that $w \Vdash_t \neg A$. So, then, by (i) and (ii) we have $w \nVdash_{ts} [\Gamma, \neg A \succ \Delta]$. So either way we have one of $[\Gamma, A \succ \Delta]$ or $[\Gamma, \neg A \succ \Delta]$ failing to hold at w when $[\Gamma \succ \Delta]$ fails to hold at w, and so by contraposition it follows that at all worlds in this model that if $[\Gamma, A \succ \Delta]$ and $[\Gamma, \neg A \succ \Delta]$ hold at that world so does $[\Gamma \succ \Delta]$. As this model was arbitrary it then follows that MEM is valid on this frame, as desired. □

What is interesting here is that the metasequent $[\Gamma, A \succ \Delta], [\Gamma, \neg A \succ \Delta] \Rightarrow [\Gamma \succ \Delta]$ has an st-counterexample in the class of $\mathcal{ID}$ models, and similarly that $\Sigma \Rightarrow [\Gamma \succ \Delta, A, \neg A]$ has a ts-counterexample in that class.[9] This is a further chapter in a well-known cautionary tale where classically equivalent principles can come apart quite dramatically in non-classical logics. In this case we can see that in the context of nontransitive and nonreflexive logics whether a principle is expressed as a *rule* (or multi-premise metainference) or an *axiom* (or zero premise metainference) can make a significant difference, with the rule-version of excluded middle being equivalent in ST to $[Cut]$ and the axiom-version being equivalent in TS to $[Id]$.

[9] For the st-counterexample consider the model $\langle \{w\}, =, V_s, V_t \rangle$ where $V_s(p) = \emptyset$ and $V_t(p) = W$, $V_s(q) = V_t(q) = W$, $V_s(r) = V_t(r) = \emptyset$—then in this model the instance $[q, p \succ r], [q, \neg p \succ r] \Rightarrow [q \succ r]$ has a st-counterexample. For the ts-counterexample consider the model $\langle \{w\}, =, V_s, V_t \rangle$ where $V_s(p) = \emptyset$ and $V_t(p) = W$—then in this model the instance $\Rightarrow [\succ p, \neg p]$ has a ts-counterexample.

14.6 Conclusion

As stated in our title, this has been a mere invitation to constructive nonreflexive and nontransitive logics. Many puzzles remain, and there is much work still to be done. To close out this paper let me mention a few of the more salient and interesting puzzles which remain.

- PROOF- THEORY: Our brief proof-theoretic investigation of these logics required us to use a labelled metasequent proof-theory. A number of questions remain open here (aside from the obvious proof-theoretic questions about Cut elimination, etc., which are also still unaddressed for the metasequent calculi for **TS** and **ST** from [8]). The obvious one is whether unlabelled metasequent proof-theory can be given for **ILts** and **ILst** (or perhaps extensions thereof).
- SEMANTICS: In one of the few papers which I know of that deals with intuitionistic versions of substructural logics [20], we are presented with a very different semantic clause for the conditional (with negation being defined in terms of the conditional). Put in the notation of the present paper their clause is:

  - If $\mathcal{M}, w \Vdash_s A \rightarrow B$ and $\mathcal{M}, w \Vdash_t A$, then $\mathcal{M}, w \Vdash_s B$.
  - $\mathcal{M}, w \Vdash_t A \rightarrow B$ iff for all $w \leq v$, either $\mathcal{M}, v \nVdash_s A$, or $\mathcal{M}, v \Vdash_t B$

  Note that this is a *partial* clause. Using this clause they are able to prove that their semantics is sound and complete w.r.t. the derivable sequents in a minor modification of the SET- FMLA sequent calculus **G1i**. We leave a more systematic investigation of the metainferences that are validated by this semantics, and a comparison with the metainferences validated by **ILst** for future work.
- INTERPRETATION: We briefly discussed above the similarities between the current semantic framework and the framework of intuitionistic epistemic logic of [2]. What metainferences are validated in the nonreflexive logic we get if we interpret valid sequents as taking us from intuitionistically true premises to intuitionistically known conclusions, or the nontransitive logic we get if we interpret valid sequents as taking us from intuitionistically known premises to intuitionistically true conclusions?

We leave the further consideration of these, and other, questions for future work.

Acknowledgements Thanks to the audiences at the inaugural workshop for the PLEXUS project, and the 2024 Asian Workshop on Philosophical Logic for their helpful comments. Special thanks are due to Pablo Cobreros, Bogdan Dicher, Ellie Ripley, Shawn Standefer, Katsuhiko Sano, and Takuro Onishi for their helpful comments. Thanks are also due to the two reviewers for the 2024 Asian Workshop on Philosophical Logic, as well as a reviewer for the present volume, for their helpful comments!

References

1. Arenhart, J.R., Omori, H.: On Woodruff's constructive nonsense logic. Studia Logica (2024)
2. Artemov, S., Protopoescu, T.: Intuitionistic epistemic logic. Rev. Symb. Logic **9**(2), 266–298 (2016)
3. Cobreros, P., Egré, P., Ripley, E., van Rooij, R.: Tolerant, classical, strict. J. Philos. Log. **41**, 347–385 (2012)
4. Cobreros, P., Rosa, E.L., Tranchini, L.: (I can't get no) antisatisfaction. Synthese **198**, 8251–8265 (2021)
5. Ferguson, T.M.: A computational interpretation of conceptivism. J. Appl. Non-Class. Logics **24**(4), 333–367 (2014)
6. Fjellstad, A.: Naive modus ponens and failure of transitivity. J. Philos. Log. **45**(1), 65–72 (2016)
7. French, R.: Structural reflexivity and the paradoxes of self-reference. Ergo **3**(5), 113–131 (2016)
8. French, R.: Metasequents and tetravaluations. J. Philos. Log. **51**(6), 1453–1476 (2022)
9. French, R.: The Constructive Logic of Paradox: Paraconsistency on the Woodruff Plan. J. Appl. Logics 12(5), 1205–1227 (2025)
10. French, R., Ripley, E.: Valuations: bi, tri and tetra. Stud. Logica. **107**, 1313–1346 (2019)
11. Girard, J.Y.: Three-Valued Logic and Cut-Elimination: The Actual Meaning of Takeuti's Conjecture. Państwowe Wydawn. Naukowe, Warszawa (1976)
12. Golan, R.: There is no tenable notion of global metainferential validity. Analysis **81**(3), 411–420 (2021)
13. Kripke, S.A.: Semantical analysis of intuitionistic logic i. In: Dummett, M., Crossley, J.N. (eds.), Formal Systems and Recursive Functions: Proceedings of the Eighth Logic Colloquium, Oxford July 1963, pp. 92–130. North Holland (1963)
14. Ripley, E.: Paradoxes and failures of cut. Australas. J. Philos. **91**(1), 139–164 (2013)
15. Ripley, E.: Revising up: strengthening classical logic in the face of paradox. Philosoph. Imprint **13**(5), 1–13 (2013)
16. Ripley, E.: Comparing substructural theories of truth. Ergo **2**, 299–328 (2015)
17. Segerberg, K.: Propositional logics related to Heyting's and Johansson's. Theoria **34**(1), 26–61 (1968)
18. Takeuti, G.: Proof Theory. Sole Distributors for the U.S.A. and Canada, Elsevier Science Pub. Co., New York (1976)
19. Teijeiro, P.: Strength and stability. Análisis Filosófico **41**(2), 337–349 (2021)
20. Thomas, M.: A Kripke-Style Semantics for Paradox-tolerant, Nontransitive Intuitionistic Logic. Unpublished Manuscript (2014)
21. Van Benthem, J.: Correspondence theory. In: Gabbay, D., Guenthner, F. (eds.) Handbook of Philosophical Logic: Volume II: Extensions of Classical Logic, pp. 167–247. Springer, Dordrecht (1984)
22. Woodruff, P.W.: On constructive nonsense logic. In: Halldén, S., (ed.), Modality, Morality and Other Problems of Sense And Nonsense: Essays Dedicated to Sören Halldén, pp. 192–205. GWK Gleerup Bokforlag (1973)

Chapter 15
A Proof-Theoretic Approach to the Binding Problem

Yukiko Yana, Koji Mineshima, and Daisuke Bekki

Abstract In the semantics of natural language, it is a common practice to distinguish between *assertion* and *presupposition* when analyzing a sentence. A theory that assigns two levels of meaning, assertion and presupposition, to a sentence is called the "two-dimensional theory." In the analysis of sentences with indefinite subjects using the two-dimensional theory, it is known that there is no guarantee that the subject in the semantic representation of the presupposition is the same individual as that in the assertion. This problem is commonly called the "binding problem." As a solution to the binding problem, Sudo (2012, 2014) proposed combining two-dimensional theory with Dynamic Predicate Logic and Plural Predicate Logic. This approach involves analyzing the anaphoric linking between assertion and presupposition as a case of cross-sentential anaphora. This paper points out empirical problems with Sudo's proposal. We then propose an alternative approach to address the binding problem by using Dependent Type Semantics, an approach grounded in proof-theoretic semantics where assertion and presupposition are treated at a single level of semantic representation.

Keywords The binding problem · Presupposition · Dependent types

---

Y. Yana (✉) · D. Bekki
Ochanomizu University, Bunkyo, Japan
e-mail: yana.yukiko@is.ocha.ac.jp

D. Bekki
e-mail: bekki@is.ocha.ac.jp

K. Mineshima
Keio University, Minato, Japan
e-mail: minesima@abelard.flet.keio.ac.jp

K. Sano et al. (eds.), *Exploring Negation, Modality and Proof*,
Logic in Asia: Studia Logica Library,
https://doi.org/10.1007/978-981-96-8566-0_15

15.1 Introduction: The Binding Problem

In the semantics of natural language, it is common to distinguish between the content that a sentence asserts, called *assertion*, and its background information, called *presupposition*. One well-known approach to presuppositions is the so-called *two-dimensional theory* [11]. According to this theory, the sentence *Mary criticized herself* is analyzed into two dimensions of meaning as shown in (1): the assertion is (1a), while the presupposition is (1b). The entire meaning of the sentence is expressed as the conjunction of the assertion and the presupposition.

(1) Mary criticized herself.

a. Assertion: $criticize(m, m)$

b. Presupposition: $female(m)$

It has been pointed out that a two-dimensional theory produces weaker predictions than empirical interpretations for certain types of sentences [8, 11]. Consider, for instance, the well-known example in (2), where the possessive NP *his bicycle* contains a pronoun and it is bound by an indefinite NP. The assertion is (2a), while the presupposition is (2b).

(2) A man was pushing his bicycle.

a. Assertion: A man was pushing his bicycle.
$\exists x(man(x) \wedge \exists y(bicycle(y) \wedge have(x, y) \wedge push(x, y))$

b. Presupposition: A man has a bicycle.
$\exists x(man(x) \wedge \exists y(bicycle(y) \wedge have(x, y)))$

In (2a) and (2b), there is no guarantee that the existentially quantified subjects stand for the same individual, and the same is true of the existentially quantified objects, thus predicting a weak interpretation. This is known as *the binding problem*.

Sudo [17, 18] proposes a solution to the binding problem within the framework of a two-dimensional theory combined with Dynamic Predicate Logic (DPL) [7]. The central idea is that the assertion and presupposition of a quantified sentence stand in an anaphoric relation that is essentially similar to cross-sentential anaphora with a quantificational antecedent. As we will see in Sect. 15.2.3, to handle the anaphoric dependencies created by universal quantifiers, Sudo [17, 18] extends the basic two-dimensional framework with Plural Predicate Logic (PPL) [22].

In Sect. 15.2, we give an overview of Sudo's two-dimensional theory and argue that it faces empirical challenges that are difficult to address without further extensions to its framework. In Sect. 15.3, we present an alternative approach to handling the binding problem, based on a framework of proof-theoretic semantics known as *Dependent Type Semantics* (DTS) [2, 4]. In DTS, assertion and presupposition are treated at a single level of semantic representation, and presupposition resolution is formulated as an inferential process for resolving presuppositional elements encoded as *underspecified types* [2]. Additionally, in Sect. 15.4, we show that by extending DTS with the strategy of *agnostic accommodation*, our approach can correctly derive interpretations for sentences involving what Sudo [17] refers to as "non-entailed" presuppositions.

15.2 DPL and Two-Dimensional Theory

15.2.1 DPL

Sudo [17] presents an approach to the binding problem by formalizing a two-dimensional theory within DPL [7], one of the frameworks of dynamic semantics for handling anaphora.

In DPL, assignment functions are updated as each sentence is processed. This feature allows DPL to provide appropriate interpretations for both intra-sentential and cross-sentential anaphora, such as E-type anaphora [6]. Let us consider the example of this kind of anaphora.

(3) A man entered. He whistled.

a. $\exists x(man(x) \wedge enter(x)) \wedge whistle(x)$

Interpreting (3a) under first-order predicate logic, the variable x of the argument of predicate *whistle* behaves as a free variable. In the context of DPL, on the other hand, the scope of the existential quantifier includes the subexpression linked by conjunction by updating the assignment function. Therefore, the variable x of the predicate *whistle* argument is bound to the existential quantifier, guaranteeing that all x are identical entities.

In Sudo's theory, the interpretation of a sentence consists of first interpreting the assertion part and then dynamically passing it to the presupposition. Here is an example sentence where the possessive NP *his mother* triggers a presupposition.

(4) A boy loves his mother.

a. Assertion: $\exists x \exists y(boy(x) \wedge love(x, y) \wedge mother\_of(y, x))$

b. Presupposition: $mother\_of(y, x)$

c. $\exists x \exists y(boy(x) \wedge love(x, y) \wedge mother\_of(y, x)) \wedge mother\_of(y, x)$

This type of presupposition is called *entailed presupposition* because, in this analysis, the assertion entails the presupposition (but not vice versa). Under the structure in (4c), which represents the overall meaning of (4), the variable in $mother\_of(y, x)$ in the second conjunct (the presupposition part) becomes free. DPL allows for providing an empirically correct interpretation by updating the assignment function based on the interpretation of the first conjunct (the assertion part), as mentioned earlier. In other words, DPL treats the arguments of the second occurrence of *mother_of* as variables bound to the initial existential quantifiers. Thus, it is possible to obtain the correct reading, which is equivalent to $\exists x \exists y(boy(x) \wedge love(x, y) \wedge mother\_of(y, x))$.

15.2.2 Universal Quantifier

As we saw in the previous section, the two-dimensional theory can derive the correct predictions for sentences with existential quantifiers. This section shows that the two-dimensional theory with DPL fails to provide the correct interpretation for sentences with universal quantifiers.

Let us consider the example sentence with a universal quantifier in (5). By connecting (5a) and (5b) with a conjunction, we obtain the semantic representation in (5c).

(5) Every boy loves his mother.

a. Assertion: $\forall x(boy(x) \to \exists y(love(x, y) \wedge mother\_of(y, x)))$

b. Presupposition: $mother\_of(y, x)$

c. $\forall x(boy(x) \to \exists y(love(x, y) \wedge mother\_of(y, x))) \wedge mother\_of(y, x)$

In DPL, when interpreting a sentence with a universal quantifier, the pair of assignment functions $\langle g, h\rangle$ is set as $h = g$. Thus, the assignment function updated by interpreting (5a) cannot provide referents for the subsequent sentence (5b). Therefore, in (5c), the variables in the second occurrence of *mother_of* become free. In other words, it is not guaranteed that the referent of *his mother* is the same as the referent introduced by the assertion. Universal quantification is *externally static* in that it blocks anaphoric links between a term that occurs in its scope and a pronoun outside of it. The interpretation of (5c) using the assignment function results in the following:

(6) $[\![\forall x(boy(x) \to \exists y(love(x, y) \wedge mother\_of(y, x))) \wedge mother\_of(y, x)]\!]$
$= \{\langle g, g\rangle \mid \forall k : (k[x]g \Rightarrow (k(x) \in I(boy) \Rightarrow \exists j : j[y]k \;\&\; \langle j(x), j(y)\rangle \in I(mother\_of) \;\&\; \langle j(y), j(x)\rangle \in I(love)) \;\&\; \langle g(y), g(x)\rangle \in I(mother\_of)\}$

The interpretation of the assertion updates the assignment functions k and j. However, the assignment function used when interpreting the second *mother_of* is g. This means that the presupposition *mother_of*(y, x) cannot refer to the content of the assertion.

The same problem arises for sentences with the determiner *no*. The semantic representation of (7a) is (7b). As we saw earlier, the second *mother_of* arguments become free variables when interpreting (7b). Therefore, as before, the two-dimensional theory with DPL cannot derive the correct prediction.

(7) a. No boy loves his mother.

b. $\forall x(boy(x) \to \neg\exists y(love(x, y) \wedge mother\_of(y, x))) \wedge mother\_of(y, x)$

From these observations, it becomes apparent that the basic two-dimensional framework with DPL is not sufficient to handle sentences with externally static operators such as universal quantification, i.e., those sentences in which the scope of the assignment function for the assertion is closed.

15.2.3 Plural Predicate Logic

In order to resolve the issues outlined in the previous section, Sudo [17, 18] presents a two-dimensional theory with PPL [22], which extends DPL to handle plurals and generalized quantifiers. In PPL, plural individuals are treated as sets of individuals, and singular individuals as singleton sets. The basic idea of PPL is to analyze predicates like *boy* as denoting sets of sets of individuals, rather than sets of individuals. Quantifiers like *every*, which take two predicates, are then interpreted as denoting relations between sets of sets of individuals. This contrasts with the basic theory where generalized quantifiers are interpreted as denoting relations between sets of individuals.

In PPL, each determiner phrase introduces a maximal set of individuals that satisfies a predicate. Assuming that the language is extended with the subset relation, the operator $\mathbf{max}_x$ is defined as follows [18, 22], where $\phi[x/y]$ is the formula obtained from ϕ by substituting all occurrences of x in ϕ with y.

(8) $\mathbf{max}_x(\phi) \equiv \neg\exists y(x \subset y \land \phi[x/y]) \land \phi$

In words, $\mathbf{max}_x(\phi)$ means that x is a maximal set which satisfies ϕ.

Let us consider the sentence with the determiner *every* we discussed earlier. In the two-dimensional theory with PPL, the sentence (5) is analyzed as (9).[1]

(9) a. Assertion: $\exists x \exists y(\mathbf{max}_x(boy(x)) \land \mathbf{max}_y(\exists z(y \subseteq x \land love(y, z) \land mother\_of(z, y))) \land every(x, y))$

b. Presupposition: $mother\_of(z, y)$

c. $\exists x \exists y(\mathbf{max}_x(boy(x)) \land \mathbf{max}_y(\exists z(y \subseteq x \land love(y, z) \land mother\_of(z, y))) \land every(x, y)) \land mother\_of(z, y)$

In (9a), x is the maximal set of the boys and y is the maximal subset of x that satisfies predicates *love*(y, z) and *mother_of*(z, y).[2] The condition *every*(x, y) has the same meaning as in the standard generalized quantifier theory, that is, x is a subset of y. The semantic representation of the whole sentence is (9c). In PPL, existential quantifiers are externally dynamic as in DPL in that they can bind variables outside their scope.

---

[1] In the analysis originally proposed by van den Berg [22] and Sudo [18], the assertion of (5) is given as:

(i) $\exists x \exists y(\mathbf{max}_x(\delta_x(boy(x))) \land \mathbf{max}_y(\exists z(y \subseteq x \land \delta_y(love(y, z) \land mother\_of(z, y)))) \land every(x, y))$.

Here, the distributive operator δ is defined as $\delta_x(\phi) \equiv \forall y \subset x(\mathbf{singleton}(y) \rightarrow \phi[x/y])$. The δ operator is necessary for handling collective predicates. Here, it is omitted for simplicity since only distributive ones are being considered. Under this simplification, there is a unique maximal set corresponding to each predicate.

[2] Here we take each predicate to be distributive. See footnote 1. The condition $y \subseteq x$ is used to guarantee the conservativity of generalized quantifiers; see [22] for details.

Thus, the variables in the presupposition part *mother_of* (z, y) can be bound by the initial existential quantifiers. This predicts the correct reading for the sentence (5).

The same analysis can be applied to examples with the determiner *no*. In the two-dimensional theory with PPL, the semantic representation of the sentence (7a) can be provided as follows.

(10) a. Assertion: $\exists x \exists y(\mathbf{max}_x(boy(x)) \wedge \mathbf{max}_y(\exists z(y \subseteq x \wedge love(y, z) \wedge mother\_of(z, y))) \wedge no(x, y))$
b. Presupposition: $mother\_of(z, y)$
c. $\exists x \exists y(\mathbf{max}_x(boy(x)) \wedge \mathbf{max}_y(\exists z(y \subseteq x \wedge love(y, z) \wedge mother\_of(z, y))) \wedge no(x, y)) \wedge mother\_of(z, y)$

(7a) can be interpreted essentially in the same way as the example with *every*, except that *every*(x, y) is replaced by *no*(x, y). In (10a), *no*(x, y) means that x and y are disjoint, as in the standard generalized quantifier theory. Thus, for the sentence to be true, the set y introduced here must be empty. As before, existential quantifiers can bind the variables in the presupposition part *mother_of* (z, y).

15.2.4 Sentential Connectives

While the two-dimensional theory extended with PPL can handle examples with determiners *every* and *no*, it fails to derive the correct predictions for complex sentences formed by externally static operators such as sentential negation, disjunction, and conditionals. The following sentence is a typical example of sentential negation.

(11) It is not the case that every boy loves his mother.
a. Assertion: $\neg\exists x \exists y(\mathbf{max}_x(boy(x)) \wedge \mathbf{max}_y(\exists z(y \subseteq x \wedge love(y, z) \wedge mother\_of(z, y))) \wedge every(x, y))$
b. Presupposition: $mother\_of(z, y)$
c. $\neg\exists x \exists y(\mathbf{max}_x(boy(x)) \wedge \mathbf{max}_y(\exists z(y \subseteq x \wedge love(y, z) \wedge mother\_of(z, y))) \wedge every(x, y)) \wedge mother\_of(z, y)$

In (11a), unlike (9a), the assertion falls within the scope of negation. Thus, when interpreting (11c), it is not possible to interpret it with the pair of assignment functions that inherit the assignment of the assertion. Consequently, as we have seen in Sect. 15.2.2, the variables in the second *mother_of* predicate remain free.

The same problem arises for complex sentences with disjunction (S_1 or S_2) and conditional (If S_1, then S_2), as illustrated in (12) and (13), respectively.

(12) Every boy loves his mother or every girl loves her mother.

(13) If every boy loves his mother, the foundation of affection is strong.

Similar to the case of negation in (11), the assertion expressed by "Every boy loves his mother" appears in the externally static environment created by the disjunction in (12) or the conditional in (13). Consequently, the two-dimensional theory fails to derive correct predictions for these examples.

The problem identified in Sect. 15.2.2 does not stem from an incapability to handle quantifiers but rather from the failure to incorporate presuppositions within the scope of quantifier binding that is introduced by assertions. Consequently, even with the extension in PPL, a complete resolution is not achieved.

One might argue that the issues arising from sentential connectives could be addressed by extending the two-dimensional theory to incorporate a mechanism for updating *local contexts*, as developed in Heim [8]. In fact, Sudo [18, Section 7.1] suggests the possibility of extending the two-dimensional theory with a mechanism to handle local contexts for analyzing sentences like (11). However, no detailed formulation is given in [18]. Exploring this possibility would require substantial extensions to Sudo's two-dimensional theory and is therefore beyond the scope of this paper; we will not discuss it further here.

Based on these observations, we can say that the two-dimensional theory extended with PPL is still facing a significant challenge in handling the binding problem. We conclude that the two-dimensional theory, which treats the anaphoric links between assertion and presupposition in the same way as cross-sentential anaphora, requires significant extensions in order to count as a viable approach.

15.3 Dependent Type Semantics

In this section, we introduce Dependent Type Semantics (DTS) [2, 4, 26], which is a proof-theoretic semantics framework based on dependent type theory [13]. DTS provides a compositional analysis of anaphora and presupposition using dependent types.[3] In contrast to the two-dimensional theory, DTS represents anaphora and presupposition in a single level of semantic representation and formulates the process of resolving them as an inferential process of type checking.

DTS has the two essential types, Π-types and Σ-type: Π-type is a generalization of the function types in simply typed λ-calculus. When a term f of Π-type $(x : A) \to B(x)$ takes a term a of type A as an argument, the term $f(a)$ is of type $B(a)$; Σ-type is a generalization of the product type in simply typed λ-calculus. A term (t, u) of

[3] See, for example, [19] for an analysis of factive presupposition and [20] for that of paycheck pronouns. [25] gives a comparison between DTS and Discourse Representation Theory (DRT) [10].

Σ-type $(x : A) \times B(x)$, which is also written as $\begin{bmatrix} x : A \\ B(x) \end{bmatrix}$, is a pair. When t is of type A and u is of type $B(t)$, $(x : A) \times B(x)$ is a type of pair (t, u). The projection functions π_1 and π_2 extract a term from a pair having a Σ-type: we write $\pi_1(t, u) =_\beta t$ and $\pi_2(t, u) =_\beta u$.

We will revisit the example sentences in the previous section from the proof-theoretic perspective of DTS and demonstrate how it can avoid the binding problem in quantified sentences.

15.3.1 Presupposition

First, we show that DTS can correctly derive predictions for sentences with existential quantifiers. In DTS, the semantic representation of (4) is (14). The variable u represents "a boy," variable v represents "her mother" and $\mathbf{love}(\pi_1 u, \pi_1 v)$ means "the entity that is a boy ($\pi_1 u =_\beta x$) and the entity that is a mother($\pi_1 v =_\beta y$) are in a relationship denoted by **love**."

$$(14) \quad \begin{bmatrix} u : (x : \mathbf{entity}) \times \mathbf{boy}(x) \\ v@\,(y : \mathbf{entity}) \times \mathbf{mother\_of}(y, \pi_1 u) \\ \mathbf{love}(\pi_1 u, \pi_1 v) \end{bmatrix}$$

We refer to the types marked with @ as *underspecified types* [2] in this semantic representation. These types perform as triggers that call proof search during type checking, and appropriate terms replace the specified variable (v in the case of (14)) through the process of type checking/inference. Here is the type inference rule for *underspecified types* (@-rules).

$$(15) \quad \left[\!\left[\Gamma \vdash \begin{bmatrix} x@A \\ B \end{bmatrix} :? \right]\!\right] = \left\{ \begin{array}{c|c} \begin{array}{c} \Gamma \\ \mathcal{D}_B \\ B'' : \mathsf{type} \end{array} & \begin{array}{c} \vdots \in [\![\Gamma \vdash A : \mathsf{type}]\!] \\ A' : \mathsf{type} \\ \vdots \in [\![\Gamma \vdash ? : A']\!] \\ M : A' \\ B[M/x] \twoheadrightarrow_\beta B' \\ \mathcal{D}_B \in [\![\Gamma \vdash B' : \mathsf{type}]\!] \end{array} \end{array} \right\}$$

where $[\![\Gamma \vdash ? : A']\!]$ is a result of the proof search, which is a set of proof diagrams.

We assume that the context used for type checking contains the following term:

$$(16) \quad f : \left(u : \begin{bmatrix} x : \mathbf{entity} \\ \mathbf{boy}(x) \end{bmatrix} \right) \to \begin{bmatrix} y : \mathbf{entity} \\ \mathbf{mother\_of}(\pi_1 u, x) \end{bmatrix}$$

This represents world knowledge that "every boy has a mother." The semantic felicity condition for (14) requires the type checking (17).

(17) $$\Gamma, f : \left(u : \begin{bmatrix} x : \textbf{entity} \\ \textbf{boy}(x) \end{bmatrix}\right) \to \begin{bmatrix} y : \textbf{entity} \\ \textbf{mother\_of}(\pi_1 u, x) \end{bmatrix}$$
$$\vdash \begin{bmatrix} u : (x : \textbf{entity}) \times \textbf{boy}(x) \\ v@\,(y : \textbf{entity}) \times \textbf{mother\_of}(y, \pi_1 u) \\ \textbf{love}(\pi_1 u, \pi_1 v) \end{bmatrix} : \mathsf{type}$$

Here, we present the formation rules, a subset of the rules used during type checking.

(18) Formation rule for Π-type: $$\frac{\Gamma \vdash A : \mathsf{type} \quad \Gamma, x : A \vdash B : \mathsf{type}}{\Gamma \vdash (x : A) \to B : \mathsf{type}} \ (\Pi F)$$

Formation rule for Σ-type: $$\frac{\Gamma \vdash A : \mathsf{type} \quad \Gamma, x : A \vdash B : \mathsf{type}}{\begin{bmatrix} x : A \\ B \end{bmatrix} : \mathsf{type}} \ (\Sigma F)$$

These formation rules demonstrate that we can use the information of A as context on the right side tree when launching the type check for B. It allows for the proof search while referencing previously uttered content.

The type checking (17) calls for the rules (15) and (18), triggering the following type checking and the proof search.

(19) a. $\Gamma, f, u \vdash (y : \textbf{entity}) \times \textbf{mother\_of}(y, \pi_1 u) : \mathsf{type}$
b. $\Gamma, f, u \vdash ? : (y : \textbf{entity}) \times \textbf{mother\_of}(y, \pi_1 u)$

Note that $u : (x : \textbf{entity}) \times \textbf{boy}(x)$ is added to the context by the formation rule for Σ-type. Thus, the proof search for (19b) finds the proof term $f(u)$.

By substituting the proof term $f(u)$ obtained from this proof search, we can incorporate the information in the presupposition, namely **mother_of**, into the semantic representation. As a result, we obtained the following semantic representation:

(20) $$\begin{bmatrix} u : (x : \textbf{entity}) \times \textbf{boy}(x) \\ \textbf{love}(\pi_1 u, \pi_1(f(u)) \end{bmatrix}.$$

This semantic representation corresponds to the empirical reading: "A boy loves the boy's mother."

15.3.2 Universal Quantifier and Negation

In this section, we show that DTS can correctly derive predictions for sentences that include universal quantifiers and negation. In DTS, the semantic representation of (5) is (21).

(21) $$\left(u : \begin{bmatrix} x : \textbf{entity} \\ \textbf{boy}(x) \end{bmatrix}\right) \to \begin{bmatrix} v@\,(y : \textbf{entity}) \times \textbf{mother\_of}(y, \pi_1 u) \\ \textbf{love}(\pi_1 u, \pi_1 v) \end{bmatrix}$$

The semantic felicity condition for (21) requires the type checking (22). Note that $u : \begin{bmatrix} x : \textbf{entity} \\ \textbf{boy}(x) \end{bmatrix}$ is added to the context by the formation rule for Π-type.

$$(22)\quad \Gamma, f, u : \begin{bmatrix} x : \textbf{entity} \\ \textbf{boy}(x) \end{bmatrix} \vdash \begin{bmatrix} v@\,(y : \textbf{entity}) \times \textbf{mother\_of}(y, \pi_1 u) \\ \textbf{love}(\pi_1 u, \pi_1 v) \end{bmatrix} : \mathsf{type}$$

This type checking triggers the following checks.

(23) a. $\Gamma, f, u \vdash (y : \textbf{entity}) \times \textbf{mother\_of}(y, \pi_1 u) : \mathsf{type}$

b. $\Gamma, f, u \vdash ? : (y : \textbf{entity}) \times \textbf{mother\_of}(y, \pi_1 u)$

The proof search for (23b) finds a proof $f(u)$.

As a result, in this case of universal quantification, we obtained the following semantic representation by proof search:

$$(24)\quad \left(u : \begin{bmatrix} x : \textbf{entity} \\ \textbf{boy}(x) \end{bmatrix}\right) \to \textbf{love}(\pi_1 u, \pi_1(f(u)))$$

This semantic representation provides a reading in which the presupposition information, **mother_of**, is bound by the assertion. This reading is equal to the empirically correct reading.

Next, let us consider the case of sentential negation. The semantic representation of (11) is (25).

$$(25)\quad \neg\left(\left(u : \begin{bmatrix} x : \textbf{entity} \\ \textbf{boy}(x) \end{bmatrix}\right) \to \begin{bmatrix} v@\,(y : \textbf{entity}) \times \textbf{mother\_of}(y, \pi_1 u) \\ \textbf{love}(\pi_1 u, \pi_1 v) \end{bmatrix}\right)$$

The semantic felicity condition for (21) requires the type checking (22).

$$(26)\quad \Gamma \vdash \neg\left(\left(u : \begin{bmatrix} x : \textbf{entity} \\ \textbf{boy}(x) \end{bmatrix}\right) \to \begin{bmatrix} v@\,(y : \textbf{entity}) \times \textbf{mother\_of}(y, \pi_1 u) \\ \textbf{love}(\pi_1 u, \pi_1 v) \end{bmatrix}\right) : \mathsf{type}$$

We can apply the formation rule for negation to (26): $\dfrac{\Gamma \vdash A : \mathsf{type}}{\Gamma \vdash \neg A : \mathsf{type}}\,(\neg F)$.[4]

By applying rules, the type checking (26) becomes (27).

$$(27)\quad \Gamma, f, u : \begin{bmatrix} x : \textbf{entity} \\ \textbf{boy}(x) \end{bmatrix} \vdash \begin{bmatrix} v@\,(y : \textbf{entity}) \times \textbf{mother\_of}(y, \pi_1 u) \\ \textbf{love}(\pi_1 u, \pi_1 v) \end{bmatrix} : \mathsf{type}$$

This judgment is equal to (22). In the case of negation, as in the previous case of universal quantification, we eventually obtain a semantic representation where the underspecified type is replaced by the proof $f(u)$. Thus, similar to the previous case, it is capable of deriving correct predictions in which the presupposition information, **mother_of**, is bound by the assertion.

In DTS, entities introduced by presupposition triggers are directly integrated into the semantic representation. As a result, DTS allows presupposed entities to be used as arguments of the predicate **love** within the assertion.

[4] See [4] for further details of the formulation rules.

This is a major difference between DTS and the two-dimensional theory with DPL and PPL. In the latter theory, entities introduced by presuppositions do not serve as arguments for predicates within the assertion. This limitation arises due to anaphoric accessibility considerations, which treat the presupposition and the assertion as separate propositions, even though they are part of the same sentence.

15.4 Non-Entailed Presupposition

In this section, we look at another type of presupposition, where the assertion and the presupposition are independent, i.e., what is called *non-entailed presupposition* in Sudo [17]. Sudo's theory demonstrated that treating presupposition and assertion two-dimensionally correctly predicts this type of presupposition. In this section, we present an analysis in DTS for "non-entailed" presuppositions, focusing on gender presuppositions as a case study, and show how predictions equivalent to Sudo's theory can be achieved within the DTS framework.

15.4.1 Case: Two-Dimensional Theory

First, we will briefly explain how predictions for non-entailed presuppositions are derived within the two-dimensional theory. A typical example is the gender presupposition triggered by a reflexive pronoun, as discussed in Sudo [17, 18]. In particular, an example such as (28), where a non-monotonic quantifier like *exactly one* binds a pronoun, poses an interesting challenge.

(28) Exactly one student criticized herself.

The empirically correct reading for the sentence (28) is "Only one student in the class was self-critical and the student is female." Note that interpreting this sentence as an instance of "entailed" presuppositions would lead to the incorrect prediction: "There is exactly one female student who criticized herself."

The semantic representation in (29) can derive the correct reading as follows.

(29) a. Assertion: $\exists x \exists y(\mathbf{max}_x(student(x)) \wedge \mathbf{max}_y(criticize(y, y)) \wedge y \subseteq x \wedge exactly\_one(x, y))$

b. Presupposition: $female(x)$

c. $\exists x \exists y(\mathbf{max}_x(student(x)) \wedge \mathbf{max}_y(criticize(y, y)) \wedge y \subseteq x \wedge exactly\_one(x, y)) \wedge female(x)$

As one can see from (29c), the presupposition only refers to the entity introduced by the assertion without contributing its content to the assertion. Consequently, the two-dimensional theory can avoid incorrect predictions.

15.4.2 Case: DTS

We provide an analysis of the type of presuppositions independent of assertion, as in (27), by refining the mechanism for deriving conventional implicature (CI) presented in [3] through the notion of underspecified types [2]. Using underspecified types, when A is presupposed independently of the assertion, we can use the anonymous variable _ and write it as $\_@A$. At the same time, a proof term of this type will not be used in subsequent semantic representations. In other words, this term does not contribute to the assertion.

We assume that the semantic representation of *herself* is given as follows:

(30) $\lambda R \lambda x.(\_@\mathbf{female}(x) \times R(x, x))$

(30) maps a 2-place predicate to a 1-place reflexive predicate with the presupposition that x is female. The additional clause $\_@\mathbf{female}(x)$ triggers a proof search for the proposition $\mathbf{female}(x)$ but the resulting proof term does not contribute anything to the assertion.

Let us consider the example for (28). The semantic representation is as follows:

(31)
$$\begin{bmatrix} u : (x : \mathbf{entity}) \times \mathbf{student}(x) \\ \_@\mathbf{female}(\pi_1 u) \\ \mathbf{criticize}(\pi_1 u, \pi_1 u) \\ (y : \mathbf{entity}) \to \left(z : \begin{bmatrix} \mathbf{student}(y) \\ \_@\mathbf{female}(y) \\ \mathbf{criticize}(y, y) \end{bmatrix} \right) \to \pi_1 u =_e y \end{bmatrix}$$

The semantic felicity condition for (31) requires type checking (32).

(32)
$$\Gamma \vdash \begin{bmatrix} u : (x : \mathbf{entity}) \times \mathbf{student}(x) \\ \_@\mathbf{female}(\pi_1 u) \\ \mathbf{criticize}(\pi_1 u, \pi_1 u) \\ (y : \mathbf{entity}) \to \left(z : \begin{bmatrix} \mathbf{student}(y) \\ \_@\mathbf{female}(y) \\ \mathbf{criticize}(y, y) \end{bmatrix} \right) \to \pi_1 u =_e y \end{bmatrix} : \mathsf{type}$$

We may accommodate necessary entities in certain instances when resolving underspecified types in DTS [2]. In previous research, the process of accommodation is commonly employed when dealing with presuppositions that cannot be resolved solely through contextual information. There are two widely acknowledged methods for accommodating information: *global accommodation* and *local accommodation*.[5] In what follows, we demonstrate that these conventional methods fail to derive the correct interpretation in DTS and propose a new accommodation strategy.

[5] The strategy known as *intermediate accommodation* [1, 23], which involves accommodating presuppositions at an intermediate context (e.g., the antecedent of a conditional sentence and the restrictor of the quantifier), is sometimes proposed. However, it will be omitted from this discussion as it does not contribute to the conclusion.

The first strategy is *global accommodation*; this is a method for resolving the presupposition at the outermost (top-level) scope. When we resolve the two occurrences of **female** in (32) as global accommodations, the type checking of the following judgments arises.

(33) a. $\Gamma,\ u : (x : \mathbf{entity}) \times \mathbf{student}(x) \vdash \mathbf{female}(\pi_1 u) : \mathsf{type}$

b. $\Gamma,\ u : (x : \mathbf{entity}) \times \mathbf{student}(x) \vdash ? : \mathbf{female}(\pi_1 u)$

To construct the proof term for (33b), it is necessary to introduce into the context the judgment $f : \left(u : \begin{bmatrix} x : \mathbf{entity} \\ \mathbf{student}(x) \end{bmatrix}\right) \to \mathbf{female}(\pi_1 u)$, which states that every student is female. This operation is semantically identical to assigning the reading that holds true only under the context "All members of the class are female; there is no male." to the sentence (28). Since this interpretation is inconsistent with the empirical reading, it is clear that resolution through global accommodation is unsuitable.[6]

The second strategy is *local accommodation*; this is a method for resolving the presupposition in the same scope as the presupposition trigger. In DTS, replacing $\begin{bmatrix} u : (x : \mathbf{entity}) \times \mathbf{student}(x) \\ \_@\mathbf{female}(\pi_1 u) \end{bmatrix}$ with $\begin{bmatrix} u : (x : \mathbf{entity}) \times \mathbf{student}(x) \\ \mathbf{female}(\pi_1 u) \end{bmatrix}$ in (32) and subsequently launching a type check is considered an equivalent operation. When resolving the two **female** as local accommodations, the following type check arises.

$$(34)\quad \Gamma \vdash \begin{bmatrix} u : (x : \mathbf{entity}) \times \mathbf{student}(x) \\ \mathbf{female}(\pi_1 u) \\ \mathbf{criticize}(\pi_1 u, \pi_1 u) \\ (y : \mathbf{entity}) \to \left(z : \begin{bmatrix} \mathbf{student}(y) \\ \mathbf{female}(y) \\ \mathbf{criticize}(y, y) \end{bmatrix}\right) \to \pi_1 u =_e y \end{bmatrix} : \mathsf{type}$$

By resolving the presuppositions using local accommodations in this manner, the sentence (28) can be true even in scenarios where other male students are criticized themselves. This interpretation, similar to the earlier situation of global accommodation, does not conform to empirical observations. Therefore, resolution through local accommodations also fails to predict the correct interpretation.

Here, we propose *agnostic accommodation* as the third strategy. The definition of the agnostic accommodation is the following:

(35) **Definition: Agnostic accommodation**

Let $\mathcal{P}$ be $\begin{bmatrix} u@\ (x : \mathbf{entity}) \times A(x) \\ B(\pi_1 u) \end{bmatrix}$ and $\mathcal{Q}$ be $\begin{bmatrix} x : \mathbf{entity} \\ B(x) \end{bmatrix}$.

When an underspecified type with the form of $\mathcal{P}$ launches a proof search, it also returns the result of the proof search with $\mathcal{P}$ replaced by $\mathcal{Q}$.

In other words, this definition states that, when applying agnostic accommodation, the information embedded in the presupposition $A(x)$ is simply eliminated from the semantic representation.

6 The empirical reading holds true even in the situation where there are some male students.

(36) $$\Gamma \vdash \begin{bmatrix} u : (x : \mathbf{entity}) \times \mathbf{student}(x) \\ \mathbf{criticize}(\pi_1 u, \pi_1 u) \\ (y : \mathbf{entity}) \to \left(z : \begin{bmatrix} \mathbf{student}(y) \\ \mathbf{criticize}(y, y) \end{bmatrix}\right) \to \pi_1 u =_e y \end{bmatrix} : \mathsf{type}$$

As a result, the interpretation "only one student in the class was self-critical." is derived.

It should be noted that applying the agnostic accommodation rule requires that $\Gamma \nvdash \neg A(a)$. Formally, this can be expressed as the following condition:

(37) **Condition: Agnostic accommodation**
If the proof term a is found such that $\Gamma \vdash a : \mathbf{entity}$, then a must satisfy the condition $\Gamma \nvdash \neg A(a)$.

This condition ensures that although gender information such as "female" does not appear in the semantic representation, it is still used in the proof process. Consequently, it is possible to prevent deriving the reading "Exactly one student criticized himself" from (36).[7] Thus, it becomes possible to derive the correct interpretation "There is precisely one student who is both self-critical and a female." by using agnostic accommodation.

It is important to note that not every underspecified type allows all three strategies: global accommodation, local accommodation, and agnostic accommodation. Thus, we assume that every underspecified type possesses three features, each of which specifies whether the corresponding accommodation strategy is permissible or not:

$$\pm G, \pm L, \pm A$$

where the feature $+G$ (resp. $-G$) indicates that the global accommodation is (resp. is not) permissible. The features $\pm L$ and $\pm A$ specify whether local accommodation and agnostic accommodation are permissible, respectively.

A remark is in order regarding the type checking of (32) where multiple underspecified types appear. Here, the two underspecified types appearing in (32) were resolved using the same strategy of agnostic accommodation. However, there is no necessity to apply the same strategy to multiple underspecified types, as shown in [12]. Thus we can perform a type check before the underspecified type within the presupposition is copied, that is, prior to β-reduction (see [12] for more details). By employing this method, (32) can be regarded as (38): **s** represents **student**, **f** represents **female**, and **c** represents **criticize**.

(38) $$\Gamma \vdash \left(\lambda P. \begin{bmatrix} u : (x : \mathbf{entity}) \times \mathbf{s}(x) \\ P(\pi_1 u) \\ (y : \mathbf{entity}) \to \left(z : \begin{bmatrix} \mathbf{s}(y) \\ P(y) \end{bmatrix}\right) \to \pi_1 u =_e y \end{bmatrix}\right) (\lambda x.(\_@\mathbf{f}(x) \times \mathbf{c}(x, x))) : \mathsf{type}$$

[7] Although the information "female" does not appear in the semantic representation, it is ensured that a proof of $\Gamma \nvdash \neg\mathbf{female}$ cannot be constructed due to (37).

With this form, it is possible to execute the type check before the underspecified type is copied. Consequently, there is no need to consider the order in which underspecified types with the same presupposition are resolved or the specific strategy applied for each resolution. It should be noted that in this case, it is necessary to determine on a case-by-case basis regarding whether to execute a proof search prior to β-reduction.

15.5 Conclusion

This paper has presented a proof-theoretic approach to the binding problem of presuppositions within the DTS framework, contrasting it with Sudo's two-dimensional theory [17, 18]. To address the binding problem caused by quantified sentences, Sudo's two-dimensional theory treats presupposition projection as an instance of cross-sentential anaphora (cross-dimensional anaphora), where pronouns in presuppositions are anaphoric to quantifiers in assertions.

In contrast, DTS integrates assertions and presuppositions within a single semantic representation. Presupposition resolution is formalized as a proof-search process, where underspecified types corresponding to presupposition triggers are resolved. We have shown how DTS' proof-theoretic framework analyzes cases that were challenging for the two-dimensional theory with DPL, such as universal quantification, as well as complex sentences containing externally static operators such as sentential negation—problems that remain unresolved even in the PPL-extended version of the two-dimensional theory [18].

Furthermore, as a natural extension of the proof-search strategy for presupposition resolution in DTS, we introduced a new method called *agnostic accommodation*, complementing the established strategies of global and local accommodation. This new approach enables DTS to handle cases of "non-entailed" presuppositions, where assertions and presuppositions are independent, without requiring substantial modifications to the DTS framework.

It is important to note that comparisons across different theoretical frameworks are inherently challenging and must be made with caution. Sudo's two-dimensional theory relies on various assumptions that extend beyond the basic DPL framework, including mechanisms for handling plurals and generalized quantifiers in PPL, as well as additional assumptions made for compositional analysis (see Sudo [18], Section 7.1). Incorporating a mechanism for local contexts would involve additional assumptions. Evaluating whether these assumptions are mere stipulations or can be independently and naturally motivated requires case-by-case assessment within the overall framework.

The same applies to DTS. While agnostic accommodation is introduced as a natural extension of the proof-search strategy based on underspecified types, determining whether it is a mere stipulation will depend on its applicability to other phenomena such as honorifics and appositives (so-called "CIs"), which can also be viewed as "non-entailed" presuppositions.[8] Furthermore, recent literature on presuppositions has shown that various types of presupposition triggers, such as factives and change-of-state predicates, exhibit different projection behaviors [5, 15, 21]. Although a comprehensive analysis of these variations based on the proof-theoretic, inferential approach presented here poses an interesting challenge, it remains open for future research.[9] The analysis of "non-entailed" presuppositions presented in this paper provides promising initial evidence that the proof-theoretic approach to presuppositions in DTS could offer valuable insights into a broader range of presupposition phenomena.

Acknowledgements We thank the two anonymous reviewers of the proceedings of AWPL2024 and the anonymous reviewer of this post-proceedings for their helpful feedback and suggestions. We also thank Katsuhiko Sano and Ryo Hatano, the organizers of AWPL2024, for their support. This work was partially supported by JSPS KAKENHI Grant Number JP23H03452, Japan, and JST CREST Grant Number JPMJCR20D2, Japan.

References

1. Beaver, D.: Accommodating topics. In: Context-dependence in the Analysis of Linguistic Meaning, pp. 79–90. Brill (2002)
2. Bekki, D.: A proof-theoretic analysis of weak crossover. In: New Frontiers in Artificial Intelligence (JSAI-isAI 2021 Workshops, JURISIN, LENLS18, SCIDOCA, Kansei-AI, AI-BIZ, Yokohama, Japan, November 13–15, 2021, Revised Selected Papers), LNAI 13856, pp. 228–241. Springer Nature Switzerland (2023)
3. Bekki, D., McCready, E.: CI via DTS. In: New Frontiers in Artificial Intelligence (JSAI-isAI 2014 Workshops. LENLS, JURISIN, and GABA, Kanagawa, Japan, October 27–28, 2014, Revised Selected Papers), Lecture Notes in Artificial Intelligence 9067, pp. 23–36. Springer, Berlin (2015)
4. Bekki, D., Mineshima, K.: Context-passing and underspecification in dependent type semantics. In: Modern Perspectives in Type Theoretical Semantics, pp. 11–41. Springer (2017)
5. Roberts, C., Simons, M.: Preconditions and projection: Explaining non-anaphoric presupposition. Linguist. Philos. **47**(4), 703–748 (2024)
6. Evans, G.: Pronouns. Linguist. Inquiry **11**(2), 337–362 (1980)
7. Groenendijk, J., Stokhof, M.: Dynamic predicate logic. Linguist. Philos. **14**(1), 39–100 (1991)
8. Heim, I.: On the projection problem for presuppositions. In: Proceedings of WCCFL 2, pp. 114–125. Stanford University, Stanford (1983)
9. Ito, Y., Mineshima, K., Bekki, D.: Indirect answers to QUD. In: Proceedings of Integrating Approaches to Social Meaning (IASM) workshop, in the 29th European Summer School in Logic, Language and Information (ESSLLI 2017), Toulouse, France (2017)
10. Kamp, H., Reyle, U.: From Discourse to Logic. Kluwer, Springer (1993)
11. Karttunen, L., Peters, S.: Conventional implicatures. In: Syntax and Semantics 11: Presupposition, pp. 1–56. Academic Press (1979)

[8] Prior work within the DTS framework includes [24] on honorifics and [14] on appositives.

[9] A comparison between the pragmatic approaches to presuppositions based on the Question Under Discussion (QUD) framework [16, 21] and the proof-theoretic approach based on DTS can be found in [9].

12. Kubota, Y., Mineshima, K., Levine, R., Bekki, D.: Underspecification and interpretive parallelism in dependent type semantics. In: Osswald, R., Retoré, C., Sutton, P., (eds.), Proceedings of the IWCS 2019 Workshop on Computing Semantics with Types, Frames and Related Structures, pp. 1–9. Association for Computational Linguistics (2019)
13. Martin-Löf, P.: Intuitionistic Type Theory. Bibliopolis (1984)
14. Matsuoka, D., Bekki, D., Yanaka, H.: Appositive projection as implicit context extension in dependent type semantics. In: Logic and Engineering of Natural Language Semantics: 20th International Conference, pp. 224–243. Springer, Berlin (2024)
15. Potts, C.: Presupposition and implicature. In: The Handbook of Contemporary Semantic Theory, pp. 168–202. Wiley (2015)
16. Simons, M., Beaver, D., Tonhauser, J., Roberts, C.: What projects and why. Semant. Linguist. Theory **20**, 309–327 (2010)
17. Sudo, Y.: On the Semantics of Phi Features on Pronouns. PhD thesis, Massachusetts Institute of Technology, 2012
18. Sudo, Y.: Presupposition Projection in Quantified Sentences and Cross-Dimensional Anaphora. University College London, Manuscript (2014)
19. Tanaka, R., Mineshima, K., Bekki, D.: Factivity and presupposition in dependent type semantics. J. Lang. Model. **5**(2), 385–420 (2017)
20. Tanaka, R., Mineshima, K., Bekki, D.: Paychecks, presupposition, and dependent types. In: Proceedings of the Fifth Workshop on Natural Language and Computer Science (NLCS 2018), EasyChair Preprint no. 215 (2018)
21. Tonhauser, J., Beaver, D., Roberts, C., Simons, M.: Toward a taxonomy of projective content. Language **89**, 66–109 (2013)
22. van den Berg, M.: Some Aspects of the Internal Structure of Discourse: The Dynamics of Nominal Anaphora. PhD thesis, Universiteit van Amsterdam (1996)
23. van der Sandt, R.A.: Presupposition projection as anaphora resolution. J. Semant. **9**, 333–377 (1992)
24. Watanabe, N., Bekki, D.: Japanese Honorification: Compositionality and Expressivity. In: Proceedings of FAJL 7: Formal Approaches to Japanese Linguistics. MIT Working Papers in Linguistics, pp. 265–276 (2014)
25. Yana, Y., Mineshima, K., Bekki, D.: The proviso problem from a proof-theoretic perspective. In: Proceedings of Logical Aspects of Computational Linguistics (LACL) (2021)
26. Yana, Y., Mineshima, K., Bekki, D.: Variable handling and compositionality: comparing DRT and DTS. J. Logic Lang. Inform. **28**(2), 261–285 (2019)

Chapter 16
Conservative Imaging and AGM Postulates

Hayato Ito

Abstract The purpose of this paper is to investigate a variant of David Lewis' imaging and to explore their relationship with conditionalization, a basic notion of belief change in Bayesianism, in terms of AGM-style postulates. This paper gives a formulation of conservative imaging restricting the selection function of possible worlds and its probabilistic postulates as an "intermediate" type of belief change in the sense that it has both the essential features of Lewis' imaging and the significant property of conditionalization, namely, conservativity and compares it with types of them, including the clarification of non-trivial and immediate condition under which conditionalization and conservative imaging coincide within a certain convex combination.

Keywords AGM theory · Bayesianism · Lewis' imaging

16.1 Introduction

Orthodox Bayesianism has two doctrines: (B1) an agent's belief state is represented by a probability function, and (B2) an agent's belief change is represented by conditionalization. Against the orthodox view, David Lewis proposes an alternative belief change called *imaging*. This method of belief changes with accepting proposition A simulates shifting the probabilities of all possible worlds to their images, which are the closest possible worlds in overall accessibility where A is true. Lewis [10] proposes imaging as a tool for understanding the probabilities of conditionals. The essential feature of imaging is that it can deal with subjunctive suppositions, whereas conditionalization cannot. Recently, many studies have focused on probabilistic puzzles by applying imaging instead of conditionalization, including, e.g., the Monty Hall problem [3, 15], Sleeping Beauty problem [2], Commutativity dilemma [9], and Judy Benjamin problem [6].

H. Ito (✉)
Department of Philosophy, Keio University, Tokyo, Japan
e-mail: itoh.phil@gmail.com

K. Sano et al. (eds.), *Exploring Negation, Modality and Proof*,
Logic in Asia: Studia Logica Library,
https://doi.org/10.1007/978-981-96-8566-0_16

While these various specific applications have been explored, there are few studies to investigate the general properties of imaging. One notable exception is an approach based on the framework of AGM [5]. The AGM theory clarifies the properties of a concrete function by exploring the bundle of postulates it satisfies, that is, by examining types of belief change rather than concrete functions. In papers that are less accessible due to being written in Japanese, Suzuki [13, 14] explicates the AGM postulates of Lewis' imaging to express its properties. The results show that Lewis' imaging satisfies the probabilistic AGM postulates representing Katsuno and Mendelzon's belief update [8]. Independently of Suzuki's study, Ramachandran [12] provides the probabilistic AGM postulates of Katsuno and Mendelzon's belief erasure [8] and claims that the reversal belief change satisfies them. From the viewpoint of AGM, Lewis' original imaging contains a linear representation incompatible with a conservative representation satisfied by conditionalization. Roughly, the difference between linearity and conservativity is that the former keeps the ratio of probability functions during a belief change, while the latter breaks the balance. Accordingly, since both cannot be in the same probability function, the conditions under which conditionalization and imaging coincide can be specified merely in trivial cases.

On the other hand, by putting a certain restriction on the probability function of imaging, another variant of imaging has also been suggested to include a conservative representation. The central idea was suggested by Isaac Levi in private correspondence with Peter Gärdenfors; see [4, 5] for the sketch of the idea. The main contribution of this paper is to give a formulation of conservative imaging and its AGM postulates as an "intermediate" type of belief change in the sense that it has both the essential features of Lewis' imaging, e.g., connection with subjunctive suppositions, and the significant property of conditionalization, conservativity, and to compare it with types of conditionalization and Lewis' imaging. The three results obtained are as follows: (i) Theorem 16.1 provides probabilistic AGM postulates satisfied by conservative imaging, (ii) Theorem 16.2 provides probabilistic AGM postulates satisfied by its reversal, and (iii) Corollary 16.1 specifies a non-trivial and immediate condition under which conditionalization and conservative imaging coincide.

The structure of this paper is as follows. Section 16.2 introduces the definition of probability, conditionalization, and AGM. In Sects. 16.3 and 16.4, we review Lewis' imaging in terms of AGM and discuss the properties of Lewis' imaging. In Sect. 16.5, we present a formulation of conservative imaging. Then, we provide two new types of belief change, modification-type and defeasance-type, and prove that the former AGM postulates are satisfied by conservative imaging and the latter by its reversal in Sect. 16.6. Finally, we show a condition under which conditionalization and conservative imaging coincide.

16.2 Preliminaries

Consider a space $\langle W, \mathscr{F}, P \rangle$ such that:

(a) W is a non-empty set of all *possible worlds* as a primitive notion;
(b) $\mathscr{F}$ is an algebra of subsets of W closed under finite union, finite intersection, and complementation, and we call its elements *propositions*; and
(c) P is a function from $\mathscr{F}$ to a subset of non-negative real numbers, and the agent's belief state is represented by the function.

For any world i and proposition A, A is true at i if and only if i belongs to A, and we call such a world an A-world. We assume that the function P satisfies the following Kolmogorov's axioms: non-negativity, normality, and finite additivity, and call it a *probability function*. For all propositions A and B in $\mathscr{F}$,

(d) *Non-negativity:* $P(A) \geq 0$;
(e) *Normality:* $P(W) = 1$; and
(f) *Finite Additivity:* $P(A \cup B) = P(A) + P(B)$,

if $A \cap B = \emptyset$, where the symbol $\emptyset$ denotes the empty set, i.e., the *impossible* proposition and W is also called the *necessary* proposition in itself, and others, *contingent*. Furthermore, let $P_\perp$ be the *absurd* function. For any proposition A,

(g) *Absurdity:* $P_\perp(A) = 1$.

The absurd function complements the undefined belief state after accepting a proposition with zero probability. Let us denote the union of the set of all probability functions and the absurd function by **P**.

Next, we define a probability function $P^\natural_A$ with accepting A. Let us say that this new function comes from P by *conditionalizing on A*. This paper follows the traditional definition [7] using a certain ratio with prior probabilities, i.e., conditional probabilities. For all propositions A and B in $\mathscr{F}$, if $P(A)$ is positive,

$$P^\natural_A(B) =^{\mathrm{df}} P(A \cap B)/P(A), \tag{16.1}$$

and if not, let $P^\natural_A(B)$ be $P_\perp(B)$. We call it the *conditionalizing of B on A*. As the doctrine (B2) in Sect. 16.1, orthodox Bayesian considers that an agent's belief change is adequately represented by conditionalization.

Example 16.1 You know that there are 10 purple candies and 20 green ones in the bag. You also know that the purple ones taste like grapes, 5 of the greens lime, and 15 of the greens huckleberry. You take one of the candies from the bag and see that it is green. Question 1: Before you take out the candy, what is the degree to which you believe the candy taken out is lime-flavored? Question 2: After that, what is the degree to which you believe that it is lime-flavored?

Let L be the proposition that the candy is lime-flavored, and G the candy is green. The former answer is $P(L) = 1/6$. Next, we interpret the example regarding an *indicative* supposition before belief change—if the candy I take out is green, what is the probability that it is lime-flavored? Let us assume that the belief change is conditionalization. The answer is $P^{\natural}_{G}(L) = P(L \cap G)/P(G) = 1/4$. Now, we observe that the degree of belief in the proposition L is increasing in that situation. This result is intuitively correct.

Then, we introduce the AGM theory [5]. This theory *extends* Bayesianism by modifying the Bayesian doctrine (B2) in Sect. 16.1 as follows.

(B2*) An agent's belief change is represented by a *type-function* of belief change constrained by a bundle of postulates in AGM.

Each type-function from **P** and $\mathscr{F}$ to **P** is constrained by a bundle of AGM-style postulates. In this paper, the formulation of AGM-style postulates is not set-theoretic [1] but probabilistic. Note that the translation from usual belief sets in set-theoretic AGM to probability functions is only in one direction.[1] Three basic types of belief change are as follows. First, consider the following *expansion*-type-function: $+ : \mathbf{P} \times \mathscr{F} \to \mathbf{P}$. The expansion-type of belief change requires the following six postulates:

(+1) $P^{+}_{A} \in \mathbf{P}$;
(+2) $P^{+}_{A}(A) = 1$;
(+3) if $A = B$, then $P^{+}_{A} = P^{+}_{B}$;
(+4) if $P(A) = 0$, then $P^{+}_{A} = P_{\perp}$;
(+5) if $P(A) = 1$, then $P^{+}_{A} = P$; and
(+6) if $A \cap B = \emptyset$, then $P^{+}_{A\cup B} = P^{+}_{A} \underset{\alpha}{+} P^{+}_{B}$, where $\alpha = P(A)/P(A \cup B)$.

(+6) includes a convex combination using a parameter α, a real number from the unit interval. From probabilities P and P', a function $P \underset{\alpha}{+} P'$ is defined by:

$$[P \underset{\alpha}{+} P'](A) =^{\mathrm{df}} \alpha \cdot P(A) + (1 - \alpha) \cdot P'(A). \tag{16.2}$$

It is obvious that the generated function is also a probability function. For any proposition A, conditionalization on A is expansion-type with accepting A.

Proposition 16.1 (Lemma 5.1 in [5]) *Conditionalization is expansion-type.*

Second, consider the following *revision*-type-function: $* : \mathbf{P} \times \mathscr{F} \to \mathbf{P}$. The revision-type of belief change requires the following six postulates:

(*1) $P^{*}_{A} \in \mathbf{P}$;
(*2) $P^{*}_{A}(A) = 1$;
(*3) if $A = B$, then $P^{*}_{A} = P^{*}_{B}$;
(*4) if $P(A) = 0$, then $P^{*}_{A} = P_{\perp}$;

[1] See, e.g., Theorem 5.8 and Theorem 5.12 in [5].

(∗5) if $P(A) > 0$, then $P^*_A = P^+_A$; and
(∗6) if $P^*_A(B) > 0$, then $P^*_{A\cap B} = (P^*_A)^+_B$.

Third, the *reversal* function of revision-type can be described by a certain combination of a revision-type-function and the probability function before the belief change. Consider the reversal function, $- : \mathbf{P} \times \mathscr{F} \rightarrow \mathbf{P}$, generated by the following convex mixture, called *Harper identity*. Given a proposition A and probability functions P and P^*,

$$P^-_A =^{\mathrm{df}} P \underset{\alpha}{+} P^*_{\overline{A}}, \tag{16.3}$$

where $\alpha < 1$ if and only if $P(A) = 1$ and $\alpha = 1$ otherwise. This reversal function is *contraction*-type-function satisfying the following six postulates:

(–1) $P^-_A \in \mathbf{P}$;
(–2) if $P(A) \neq 1$, then $P^-_A(A) < 1$;
(–3) if $A = B$, then $P^-_A = P^-_B$;
(–4) if $P(A) < 1$, then $P^-_A = P$;
(–5) if $P(A) = 1$, then $(P^-_A)^+_A = P$; and
(–6) if $P^-_{A\cap B}(\overline{A}) > 0$, then $P^-_A(C \cap \overline{A})/P^-_A(\overline{A}) = P^-_{A\cap B}(C \cap \overline{A})/P^-_{A\cap B}(\overline{A})$.

Proposition 16.2 (Theorem 5.14 in [5]) *If a function is revision-type, then the reversal function generated by the combination (3) is contraction-type.*

As the postulate (–5) shows, a reversal function need not be *recovered* by the corresponding forward function, but it can be recovered by an expansion intended to be *minimal* belief change. See [5] for the concept of minimal change.

16.3 Lewis' Imaging

Following (B2*) in the previous section, this paper explores concrete imaging functions and their type-functions. In this section, we define Lewis imaging and its generalized variant as concrete functions. It is worth remarking that there are various formulations. Besides Lewis [10], Gärdenfors [4, 5] formulates imaging using *determiners* similar to Carnapian state-descriptions. Lewis [11] proposes another method and Suzuki [13, 14] refines it. Building on Suzuki's notation and Lewis's terminology, this paper reconstructs the latter method. In the next section, we review the type-function of Lewis' imaging in terms of AGM.

The central idea of imaging is that the imaging on A shifts the probability of all possible worlds to the closest worlds in overall accessibility where A is true. At first, we need the notions of *comparative similarity* and probability of possible worlds. The notation $j \precsim_i k$ means a universal closure of three worlds i, j, and k such that j is at least as similar to i as k is. The system of comparative similarity requires the following three conditions at least. For any world i in W,

(a) *Weak ordering:* $\precsim_i$ is transitive and strongly connected, i.e., for any worlds j and k, if $j \precsim_i k$ and $k \precsim_i l$, then $j \precsim_i l$; and either $j \precsim_i k$ or $k \precsim_i j$;
(b) *Self-accessibility:* i is accessible from i; and
(c) *Strictly* $\precsim_i$ *-minimality:* for any world j different from i, $i \precsim_i j$.

Assume that for any world i,the *closest* A-worlds to i is eventually specified by the following *set-selection function*. For any world i and proposition A,

(d) *World selection:* $f(A, i)$ is the set of all closest A-worlds to i.

In order to ensure Kolmogorov's axioms in the later formulations, this paper agrees with at least the following assumptions, apart from Lewis' commitments in several literatures. For any world i and proposition A,

(e) *Limit Assumption:* there exists a closest A-world to i; and
(f) *Centering Assumption:* if i belongs in A, then i belongs in $f(i, A)$.

Furthermore, let @ be a *distribution of probability over possible worlds*. For simplicity and convenience, assume that all distributions are *finite discrete*. We roughly say that $@(i)$ is the probability of i. For any proposition A,

$$P(A) =^{\mathrm{df}} \sum_{i \in A} @(i). \tag{16.4}$$

Let us start with the additional assumption.

(g) *Uniqueness Assumption:* $f(A, i)$ is a singleton,

and we call its element a *sharp image of* A at i. The following indicator function $\mathbb{I}_A$ is called *sharp image-selection function of* A. Given a proposition A,

$$\mathbb{I}_A(i, j) =^{\mathrm{df}} \begin{cases} 1 & \text{if } i \in f(A, j), \\ 0 & \text{otherwise.} \end{cases} \tag{16.5}$$

Let $@^{\sharp!}_A$ be a *distribution of sharp image of* A. For any world i,

$$@^{\sharp}_A(i) =^{\mathrm{df}} \sum_{j \in W} @(j) \cdot \mathbb{J}_A(i, j). \tag{16.6}$$

Let us say that $P^{\sharp!}_A$ comes from P by *imaging on* A. If $P(A)$ is positive,

$$P^{\sharp!}_A(B) =^{\mathrm{df}} \sum_{i \in B} @^{\sharp!}_A(i), \tag{16.7}$$

and if not, let $P^{\sharp!}_A(B)$ be $P_\perp(B)$. We call it the *imaging of B on A* (Fig. 16.1).

Example 16.2 Let $W = \{g, h, i, j, k, l\}$, $A = \{g, k, l\}$, and $B = \{j, k, l\}$. Suppose that $g \in f(A, g)$, $l \in f(A, h)$, $g \in (A, i)$, $l \in f(A, j)$, $k \in f(A, k)$, and $l \in f(A, l)$. After the imaging on A, the probabilities of g, k, and l are stable, the probability of h shifts to l, and the probability of i shifts to g, and the probability of j shifts to l. Hence, $@^{\sharp!}_A(g) = @(g) + @(i)$, $@^{\sharp!}_A(k) = @(k)$, and $@^{\sharp!}_A(l) = @(h) + @(j) + @(l)$. Therefore, $P^{\sharp!}_A(B) = @^{\sharp!}_A(k) + @^{\sharp!}_A(l) = @(h) + @(j) + @(k) + @(l)$.

As we have seen in the example above, in the case of belief change on A, imaging allows us to deal with the probability of $\overline{A}$ before the belief change. Therefore, imaging is essentially associated with *subjunctive* supposition. Consider the following simplified example in Katsuno & Mendelzon [8].

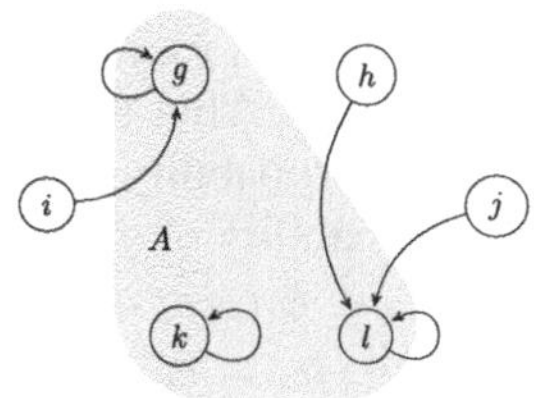

Fig. 16.1 The imaging on A

Example 16.3 There is a room with a table, a book, and a magazine. Let B be the proposition that the book is on the table and M the proposition that the magazine is on the table. You know that it is impossible for both the book and the magazine to be on the table or for both to be off the table. You command the robot to place the book on the table and, if the book is already on the table, to remain there and return. You now accept B. What is the degree to which you believe $\overline{M}$ after the belief change with accepting B?

Since fully accepting B is inconsistent with the belief state before the command, we interpret the example regarding a subjunctive supposition at the time before the belief change—suppose that B were the case; then what would be the probability of $\overline{M}$? Let us assume that the belief change is imaging. Within the imaging on B, the probability of $(M \cap \overline{B})$-worlds shifts to the $(M \cap B)$-worlds and that of $(\overline{M} \cap \overline{B})$-worlds shifts to the $(\overline{M} \cap B)$-worlds. Therefore, $P^{\sharp}_B(\overline{M}) = 1/2$, because $P^{\sharp}_B(B \cap \overline{M}) = 1/2$ and $P^{\sharp}_B(\overline{B} \cap \overline{M}) = 0$. The answer is intuitively correct.

Next, we consider a generalized variant of imaging in the sense that there is allowed to be more than one closest world. Therefore, the set-selection function $f(A, i)$ has one or more elements; we call them *blurred images of* A at i. We define the following *blurred image-selection function of* A. Note that we denote the number of elements of a proposition A by $|A|$. Given a proposition A,

$$\mathbb{J}_A(i, j) =^{\text{df}} \begin{cases} 1/|f(A, j)| & \text{if } i \in f(A, j), \\ 0 & \text{otherwise.} \end{cases} \tag{16.8}$$

Let $@^{\sharp}_A$ be a *distribution of blurred image of* A. For any world i,

$$@^{\sharp}_A(i) =^{\text{df}} \sum_{j \in W} @(j) \cdot \mathbb{J}_A(i, j). \tag{16.9}$$

Let us say that $P^{\sharp}_A$ comes from P by *general imaging on A*. If $P(A)$ is positive,

$$P^{\sharp}_A(B) =^{\mathrm{df}} \sum_{i \in B} @^{\sharp}_A(i), \qquad (16.10)$$

and if not, let $P^{\sharp}_A(B)$ be $P_{\perp}(B)$. We call it the *general imaging of B on A* (Fig. 16.2).

Example 16.4 Let $W = \{g, h, i, j, k, l\}$, $A = \{g, k, l\}$, and $B = \{j, k, l\}$. Suppose that $g \in f(A, g)$, $g, l \in f(A, h)$, $g, k \in f(A, i)$, $k, l \in f(A, j)$, $k \in f(A, k)$, and $l \in f(A, l)$. After the general imaging on A, the probabilities of g, k, and l are stable, the probability of h shifts to g and l, the probability of i shifts to g and k, and the probability of j shifts to k and l. Hence, $@^{\sharp}_A(g) = @(g) + @(h)/2 + @(i)/2$, $@^{\sharp}_A(k) = @(i)/2 + @(j)/2 + @(k)$, $@^{\sharp}_A(l) = @(h)/2 + @(j)/2 + @(l)$. Therefore, $P^{\sharp}_A(B) = @^{\sharp!}_A(k) + @^{\sharp!}_A(l) = @(h)/2 + @(i)/2 + @(j) + @(k) + @(l)$.

16.4 Properties of Lewis' Imaging

Suzuki [13, 14] explicates the AGM postulates of Lewis' imaging by translating Katsuno and Mendelzon's *belief update* [8] to probabilistic postulates. Let $\Diamond : \mathbf{P} \times \mathscr{F} \to \mathbf{P}$ be the *update-type-function* with accepting A. The update-type of belief change requires the following eight postulates:

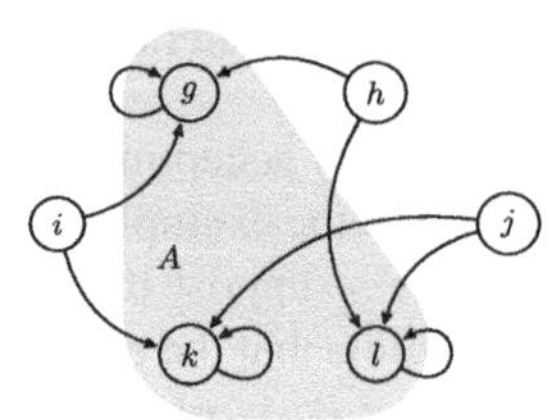

Fig. 16.2 The imaging on A

($\Diamond$1) $P^{\Diamond}_A \in \mathbf{P}$;
($\Diamond$2) $P^{\Diamond}_A(A) = 1$;
($\Diamond$3) if $A = B$, then $P^{\Diamond}_A = P^{\Diamond}_B$;
($\Diamond$4) if $P(A) = 0$, then $P^{\Diamond}_A = P_{\perp}$;
($\Diamond$5) if $P(A) = 1$, then $P^{\Diamond}_A = P$;
($\Diamond$6) if $P^{\Diamond}_{A \cap B}(C) = 1$, then $(P^{\Diamond}_A)^{+}_B(C) = 1$;
($\Diamond$7) if $P^{\Diamond}_A(B) > 0$ and $(P^{\Diamond}_A)^{+}_B(C) = 1$, then $P^{\Diamond}_{A \cap B}(C) = 1$; and
($\Diamond$8) if $P(A) > 0$, then $(P \underset{\alpha}{+} P')^{\Diamond}_A = P^{\Diamond}_A \underset{\alpha}{+} P'^{\Diamond}_A$.

Proposition 16.3 (Theorem 4 in [13] and Theorem 3 in [14]) *Lewis' imaging is update-type.*

Ramachandran [12] formulates the postulates of reversal belief change by translating Katsuno and Mendelzon's *belief erasure* [8] to probabilistic postulates and argues that the reversal function satisfies all of them. For any proposition A, let $P^{\flat}_A$

be the reversal function generated by the following combination. Given a proposition A and probability functions P and $P^\sharp$,

$$P_A^\flat =^{df} P \underset{\alpha}{+} P_{\overline{A}}^\sharp. \tag{16.11}$$

Let $\blacklozenge : \mathbf{P} \times \mathscr{F} \rightarrow \mathbf{P}$ be the *erasure*-type-function. The erasure-type of belief change requires the following seven postulates:

($\blacklozenge$1) $P_A^\blacklozenge \in \mathbf{P}$;
($\blacklozenge$2) if $P_A^\blacklozenge(B) = 1$, then $P(B) = 1$;
($\blacklozenge$3) if $A = B$, then $P_A^\blacklozenge = P_B^\blacklozenge$;
($\blacklozenge$4) if $P(A) = 0$, then $P_A^\blacklozenge = P$;
($\blacklozenge$5) if $P(A) \neq 1$, then $P_A^\blacklozenge(A) < 1$;
($\blacklozenge$6) if $P(A) = 1$, then $(P_A^\blacklozenge)_A^+ = P$; and
($\blacklozenge$7) if $P(A) > 0$, then $(P \underset{\alpha}{+} P')_A^\blacklozenge = P_A^\blacklozenge \underset{\alpha}{+} P_A'^\blacklozenge$.

Proposition 16.4 (Theorem 1 in [12]) *If a function is update-type, then the reversal function generated by the combination (11) is erasure-type.*

In the postulates of belief update and belief erasure, ($\Diamond$8) and ($\blacklozenge$7) are called linear representations. This property requires the ratio of probability functions to be equal before and after the belief change. Let $P_A^\dagger$ be a probability function after a belief change with accepting A.

(a) *Linearity:* if $P(A) > 0$, then $(P \underset{\alpha}{+} P')_A^\dagger = P_A^\dagger \underset{\alpha}{+} P_A'^\dagger$.

The linear representation is incompatible with the following conservative representation as the same probability function.

(b) *Conservativity:* if $P(A) > 0$ and $P(B) = 1$, then $P_A^\dagger(B) = 1$.

It is apparent that conditionalization has a conservative representation. If $P(A) > 0$ and $P(B) = 1$, then $P(A \cap \overline{B}) = 0$. Hence, $P_A^\natural(B) = P(A \cap B)/P(A) = P(A \cap B)/P(A \cap B) = 1$. The intuitive explanation of incompatibility is that linearity keeps the ratio of probability functions during a belief change, but conservativity breaks the balance. Gärdenfors proves it formally.

Proposition 16.5 (Theorem 2 in [4] and Theorem 5.3 in [5]) *No probability function is both linear and conservative.*

Conditionalization and Lewis' imaging often have different results, but it is possible for conditionalization and imaging to coincide. The general conditions are not well known except in trivial cases. Under the uniform probability distribution, the exception is formulated clearly by Suzuki [13, 14].

Proposition 16.6 (Theorem 3 in [13] and Theorem 2 in [14]) *Given A, if* $@(i) = 1/|W|$ *and* $f(A, i) = A$ *for any world* i, *then* $P_A^{\sharp} = P_A^{\natural}$.

16.5 Conservative Imaging

The properties of Lewis' imaging are now explicated by probabilistic AGM postulates that include a linear representation. On the other hand, it is possible to consider that there is a variant of imaging that includes a conservative representation satisfied by conditionalization. This paper considers conservative imaging and its type-function as a bundle of "intermediate" properties that have both the essential features of imaging, its connection with subjunctive suppositions, and the important property of conditionalization, that is, a conservative representation.

Let us begin with defining conservative imaging as a concrete function. The central idea of conservative imaging is to exclude the possible worlds with probability zero from the agent's images. Conservative imaging shifts the probabilities of all possible worlds to the closest worlds where A is true with certain positive probabilities. Thus, if the probability of A is zero, then no probability shifts to A-world. Under a certain formal restriction on the set-selection function, we define the following *conservative image-selection function of* A. Let $W^{\uparrow}$ be the subset of W such that $@(i)$ is positive.

$$W^{\uparrow} =^{\text{df}} \{i \in W : @(i) > 0\} \tag{16.12}$$

Given a proposition A,

$$\mathbb{K}_A(i, j) =^{\text{df}} \begin{cases} 1/|f(A \cap W^{\uparrow}, j)| & \text{if } i \in f(A \cap W^{\uparrow}, j), \\ 0 & \text{otherwise.} \end{cases} \tag{16.13}$$

Let $@_A$ be a *distribution of conservative image of* A. For any world i,

$$@_A(i) =^{\text{df}} \sum_{j \in W} @(j) \cdot \mathbb{K}_A(i, j). \tag{16.14}$$

Let us say that P_A comes from P by *conservative imaging on* A. If $P(A)$ is positive,

$$P_A(B) =^{\text{df}} \sum_{i \in B} @_A(i), \tag{16.15}$$

and if not, let $P_A(B)$ be $P_{\perp}(B)$. We call it the *conservative imaging of* B *on* A (Fig. 16.3).

Example 16.5 Let $W = \{g, h, i, j, k, l\}$, $A = \{g, k, l\}$, and $B = \{j, k, l\}$. Suppose that $g \in f(A, g), l \in f(A, h), g \in (A, i), l \in f(A, j), k \in f(A, k)$, and $l \in f(A, l)$. After the imaging on A, the probabilities of g, k, and l are stable, the probability of h shifts to l, and the probability of i shifts to g, and the probability of j shifts to l. Hence, $@^{\sharp!}_A(g) = @(g) + @(i)$, $@^{\sharp!}_A(k) = @(k)$, and $@^{\sharp!}_A(l) = @(h) + @(j) + @(l)$. Therefore, $P^{\sharp!}_A(B) = @^{\sharp!}_A(k) + @^{\sharp!}_A(l) = @(h) + @(j) + @(k) + @(l)$. Let $W=\{g, h, i, j, k, l\}$, $A=\{g, k, l\}$, $B = \{j, k, l\}$, and $W^{\uparrow} = \{g, i, k\}$. Suppose that $g \in f(A, g)$, $g \in f(A, h)$, $g, k \in f(A, i)$, $k \in f(A, j)$, $k \in f(A, k)$, $l \in f(A, l)$, and a strict relation of comparative similarity such that $l \prec_l k \prec_l g$. After the conservative imaging on A, the probabilities of g and k are stable, the probability of h shifts to g, the probability of i shifts to g and k, the probability of j shifts to k, and the probability of l shifts to k. Hence, $@_A(g) = @(g) + @(h) + @(i)/2$, $@_A(k) = @(i)/2 + @(j) + @(k) + @(l)$. Therefore, $P_A(B) = @_A(k) = @(i)/2 + @(j) + @(k) + @(l)$.

In the same way as Lewis' imaging, e.g., in Examples 2 and 4, conservative imaging on A allows us to deal with the probability of $\overline{A}$ before the belief change. Thus, conservative imaging is associated with subjunctive supposition. Let us consider the following situation, which is a modified version of Katsuno and Mendelzon's classic example.

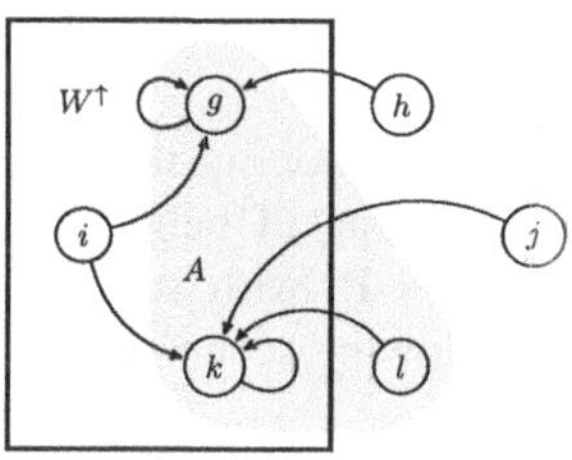

Fig. 16.3 The imaging on A

Example 16.6 There is a room with a table, a book, a magazine, and a lightbulb. Let B be the proposition that the book is on the table, M the magazine is on the table, and L the light is on. You are a confident person and believe that once both the book and the magazine are on the table, the lightbulb is switched on, and otherwise, the light is off, although it remains unclear whether your belief is true. Furthermore, you believe with equal 1/3 degrees of confidence that only a book is on the table, only a magazine is on the table, and both are on the table. You command the robot to place the book on the table and, if the book is already on the table, to remain there and return. As a result, you accept B. What is the degree to which you believe $M \cap L$ at this time?

Since fully accepting B is inconsistent with the belief state before the command, we interpret the example regarding a subjunctive supposition at the time before the belief change - suppose that B were the case; then what would be the probability of $M \cap L$? Firstly, let us assume that the belief change is Lewis' imaging. Suppose that all $(\overline{B} \cap M \cap \overline{L})$-worlds are closest to $(B \cap M \cap \overline{L})$-worlds and $(B \cap M \cap L)$-worlds. Thus, after the imaging on B, the half probability of $(\overline{B} \cap M \cap \overline{L})$-worlds shifts to the $(B \cap M \cap L)$-worlds. Therefore, $P^{\sharp}_B(M \cap L) = P(B \cap M \cap L) + P(\overline{B} \cap M \cap L) + P(\overline{B} \cap M \cap \overline{L})/2 = 1/3 + 0 + 1/6 = 1/2$. However, the intuitively correct answer is 2/3. Intuitively, after the command, the probability of $(\overline{B} \cap M \cap \overline{L})$-worlds should shift to $(B \cap M \cap L)$-worlds, not $(B \cap M \cap \overline{L})$-worlds because the agent believes that whenever the book and the magazine are on the desk, the light is on. Conservative imaging then realizes the belief change. Since $(B \cap M \cap \overline{L})$-worlds have zero-probability, the probability of $(\overline{B} \cap M \cap \overline{L})$-worlds shifts to the closest worlds with positive probability, i.e. $(B \cap M \cap L)$-worlds. Therefore, $P_B(M \cap L) = P(B \cap M \cap L) + P(\overline{B} \cap M \cap L) + P(\overline{B} \cap M \cap \overline{L}) = 1/3 + 0 + 1/3 = 2/3$. The difference is derived from conservativity. Before the belief change, the probabilities of initial states, $P((B \cap \overline{M} \cap \overline{L}) \cup (\overline{B} \cap M \cap \overline{L}) \cup (B \cap M \cap L))$, is 1. The probability of $M \cap L$ is positive. $P_{M \cap L}((B \cap \overline{M} \cap \overline{L}) \cup (\overline{B} \cap M \cap \overline{L}) \cup (B \cap M \cap L)) = 1$ after the command. Hence, this belief change is represented by conservativity. Thanks to Gärdenfors' impossibility theorem (Proposition 16.5 in Sect. 16.4), Lewis' imaging does not explain the example since $P^{\sharp}_{M \cap L}((B \cap \overline{M} \cap \overline{L}) \cup (\overline{B} \cap M \cap \overline{L}) \cup (B \cap M \cap L)) < 1$.

16.6 Properties of Conservative Imaging

Next, we define the probabilistic AGM postulates that conservative imaging satisfies. The type of belief change is defined by the *modification*-type-function, $\triangle : \mathbf{P} \times \mathscr{F} \to \mathbf{P}$, with accepting A. The modification-type of belief change requires the following seven postulates:

($\triangle$1) $P^{\triangle}_{A} \in \mathbf{P}$;
($\triangle$2) $P^{\triangle}_{A}(A) = 1$;
($\triangle$3) if $A = B$, then $P^{\triangle}_{A} = P^{\triangle}_{B}$;
($\triangle$4) if $P(A) = 0$, then $P^{\triangle}_{A} = P_{\bot}$;
($\triangle$5) if $P(A) = 1$, then $P^{\triangle}_{A} = P$;
($\triangle$6) $P^{\triangle}_{A \cap B}(C) = 1$ if and only if $(P^{\triangle}_{A})^{+}_{B}(C) = 1$; and
($\triangle$7) if $P(A) > 0$ and $P(B) = 1$, then $P^{\triangle}_{A}(B) = 1$.

($\triangle$1), ($\triangle$2), ($\triangle$3), and ($\triangle$4) correspond directly to those of belief revision. ($\triangle$1) requires that belief modification is represented by a probability function. ($\triangle$2) requires that the belief state after belief change with A accepts A. ($\triangle$3) is an equivalence rule. ($\triangle$4) is a consistency requirement in belief modification as a rationality of belief changes. ($\triangle$5) corresponds directly to (+5) and is weaker than ($*$5). ($\triangle$5) says that if an agent already accepts A before accepting with A, then the belief state is the same before and after the belief change. ($\triangle$6) is weaker than ($*$6) as a requirement of so-called minimality of belief changes. ($\triangle$7) is a conservative representation. Then, P_A satisfies all of them.

Theorem 16.1 *Conservative imaging is modification-type.*

Proof – (△1) P_A satisfies the Kolmogorov's axioms. First, it is obvious that $P_A \geq 0$. Second, if $P(A) > 0$, then $P_A(W) = \sum_{i \in W} \sum_{j \in W'} @(j) \cdot \mathbb{K}_A(i, j) = 1$. If $P(A) = 0$, then $P_A(W) = P_\perp(W) = 1$. Third, if $P(B \cap C) = 0$, then $P_A(B \cup C) = \sum_{i \in B \cup C} \sum_{j \in W} @(j) \cdot \mathbb{K}_A(i, j) = \sum_{i \in B} \sum_{j \in W} @(j) \cdot \mathbb{K}_A(i, j) + \sum_{i \in C} \sum_{j \in W} @(j) \cdot \mathbb{K}_A(i, j) = P_A(B) + P_A(C)$.

- (△2) If $P(A) > 0$, then $P_A(A) = \sum_{i \in A} \sum_{j \in W} @(j) \cdot \mathbb{K}_A(i, j) = \sum_{i \in W} @(i) = 1$. If $P(A) = 0$, then $P_A(A) = P_\perp(A) = 1$.
- (△3) and (△4) Obvious.
- (△5) Suppose that $P(A) = 1$. $P_A(B) = \sum_{i \in B} \sum_{j \in W} @(j) \cdot \mathbb{K}_A(i, j) = \sum_{i \in B} \sum_{j \in A} @(j) \cdot \mathbb{K}_A(i, j) + \sum_{i \in B} \sum_{j \in \overline{A}} @(j) \cdot \mathbb{K}_A(i, j) = \sum_{i \in B} @(i) = P(B)$.
- (△6) (⇒) First, suppose that $P_{A \cap B}(C) = 1$. From Proposition 16.1, if $P(A) > 0$ and $P(B) > 0$, then $(P_A)^\natural_B(C) = \dfrac{\sum_{i \in B \cap C} \sum_{j \in W} @(j) \cdot \mathbb{K}_A(i, j)}{\sum_{i \in B} \sum_{j \in W} @(j) \cdot \mathbb{K}_A(i, j)} =$

$$\frac{\sum_{i \in B \cap C} \sum_{j \in W} @(j) \cdot \mathbb{K}_A(i, j)}{\sum_{i \in B \cap C} \sum_{j \in W} @(j) \cdot \mathbb{K}_A(i, j) + \sum_{i \in B \cap \overline{C}} \sum_{j \in W} \cdot @(j) + \mathbb{K}_A(i, j)}.$$

Since $P_{A \cap B}(C) = 1$, $\sum_{i \in B \cap \overline{C}} \sum_{j \in W} @(j) \cdot \mathbb{K}_A(i, j) = 0$. Therefore, $(P_A)^\natural_B(C) = 1$.

If $P(A) = 0$ or $P(B) = 0$, then $(P_A)^\natural_B(C) = P_\perp(C) = 1$.

(⇐) Second, suppose that $(P_A)^\natural_B(C) = 1$. If $P(A \cap B) > 0$, then $\dfrac{\sum_{i \in B \cap C} \sum_{j \in W} @(j) \cdot \mathbb{K}_A(i, j)}{\sum_{i \in B} \sum_{j \in W} @(j) \cdot \mathbb{K}_A(i, j)} = 1$, from the above. We have $A \cap B \subseteq C$. Therefore, $P_{A \cap B}(C) = \sum_{i \in C} \sum_{j \in W} @(j) \cdot \mathbb{K}_{A \cap B}(i, j) = 1$. If $P(A \cap B) = 0$, then p.

- (△7) Suppose that $P(A) > 0$ and $P(B) = 1$. Since there is no $(A \cap \overline{B})$-worlds with positive probability and $P(B) = 1$, $P_A(B) = \sum_{i \in B} \sum_{j \in W} @(j) \cdot \mathbb{K}_A(i, j) = \sum_{i \in W} @(i) \cdot \dfrac{|B \cap f(A \cap \{h \colon @(h) > 0\}, i)|}{|f(A \cap \{h \colon @(h) > 0\}, i)|} = 1$.

In the same way as the belief revision and update, we consider the reversal type of belief change. Let P_A be the reversal function generated by the following convex combination. Given a proposition A and probability functions P and P ,

$$P_A =^{\mathrm{df}} P \underset{\alpha}{+} P_{\overline{A}}. \tag{16.16}$$

Let $\blacktriangle : \mathbf{P} \times \mathscr{F} \rightarrow \mathbf{P}$ be the *defeasance*-type-function. The defeasance-type of belief change requires the following six postulates:

($\blacktriangle$1) $P_A^{\blacktriangle} \in \mathbf{P}$;
($\blacktriangle$2) if $P(A) \neq 1$, then $P_A^{\blacktriangle}(A) < 1$;
($\blacktriangle$3) if $A = B$, then $P_A^{\blacktriangle} = P_B^{\blacktriangle}$;
($\blacktriangle$4) if $P(A) = 0$, then $P_A^{\blacktriangle} = P$;
($\blacktriangle$5) if $P(A) = 1$, then $(P_A^{\blacktriangle})_A^{+} = P$; and
($\blacktriangle$6) if $P(\overline{A}) > 0$ and $P(B) = 1$, then $P_A^{\blacktriangle}(B) = 1$.

($\blacktriangle$1), ($\blacktriangle$2), ($\blacktriangle$3), and ($\blacktriangle$5) correspond directly to those of belief contraction, ($\blacktriangle$4) is weaker than (–4), and ($\blacktriangle$6) is a conservative representation. Then, P_A satisfies all of them.

Theorem 16.2 *If a function is modification-type, then the reversal function generated by the combination (16) is defeasance-type.*

Proof – ($\blacktriangle$1) Since $P \in \mathbf{P}$ and $P_A \in \mathbf{P}$, it is obvious that $P_A \in \mathbf{P}$.
– ($\blacktriangle$2) Suppose that $P(A) \neq 1$. Hence, $P(\overline{A}) \neq 0$. $P_A(A) = a \cdot P(A) + (1 - a) \cdot P_{\overline{A}}(A) = a \cdot P(A)$. Thus, $P_A(A) < 1$.
– ($\blacktriangle$3) Obvious.
– ($\blacktriangle$4) Suppose that $P(A) = 0$. Hence, $P(\overline{A}) = 1$. $P_A(B) = a \cdot P(B) + (1 - a) \cdot P_{\overline{A}}(B) = a \cdot P(B) + (1 - a) \cdot P(B) = P(B)$, by ($\triangle$5) of Theorem 16.1.
– ($\blacktriangle$5) Suppose that $P(A) = 1$. $(P_A)_A^{\natural}(B) = P_A(A \cap B)/P_A(A)$. From the supposition, $P_{\overline{A}}(A) = P_{\overline{A}}(A \cap B) = 0$. Since $P(A) = 1$, $(P_A)_A^{\natural}(B) = P(A \cap B)/P(A) = P(B)$
– ($\blacktriangle$6) Suppose that $P(\overline{A}) > 0$ and $P(B) = 1$. $P_A(B) = a \cdot P(B) + (1 - a) \cdot P_{\overline{A}}(B)$. From ($\triangle$7) of Theorem 16.1, $P_{\overline{A}}(B) = P(B) = 1$. We have $P_A(B) = 1$.

Since conditionalization and conservative imaging share conservativity, we can specify a non-trivial condition under which conservative imaging and conditionalization coincide in the sense that there are no unusual restrictions on the distribution or comparative similarity over possible worlds. The condition comes naturally from Proposition 16.1. Consider that after the conservative imaging on A, if $P(A \cap B) = 0$, then $(A \cap B)$-worlds are ignored from our images. From the observation, using the convex mixture, it is sufficient to specify the ratio of the probability functions to be exactly conditionalization. Therefore, the following corollary is derived from Proposition 16.1 and Theorem 16.1.

Corollary 16.1 *If* $P(A \cap B) = 0$*, then* $P^{\triangle}_{A \cup B} = P^{\triangle}_{A} \underset{\alpha}{+} P^{\triangle}_{B}$*, where* $\alpha = P(A)/P(A \cup B)$*, if and only if for any* A*,* $P^{\triangle}_{A} = P^{+}_{A}$*.*

Proof ($\Rightarrow$) Suppose that $P(A \cap B) = 0$. From (2), $P^{\triangle}_{A \cup \overline{A}}(A \cap B) = P^{\triangle}_{A}(A \cap B) \cdot P(A) + P^{\triangle}_{\overline{A}}(A \cap B) \cdot P(\overline{A})$ for all B. From ($\triangle$5) and $P(A \cup \overline{A}) = 1$, $P^{\triangle}_{A \cup \overline{A}}(A \cap B) = P(A \cap B)$. If $P(A) > 0$, then $P^{\triangle}_{\overline{A}}(A \cap B) = 0$ because of ($\triangle$2). Therefore, $P(A \cap B) = P(A) \cdot P^{\triangle}_{A}(A \cap B)$. It is obvious that $P^{\triangle}_{A}(A \cap B) = P^{\triangle}_{A}(B)$ from ($\triangle$2). Thus, $P^{\triangle}_{A}(B) = P(A \cap B)/P(A) = P^{+}_{A}(B)$ for all B. If $P(A) = 0$, then $P^{\triangle}_{A} = P_{\perp} = P^{+}_{A}$. The 'if' direction is obvious.

16.7 Conclusion

Our formal results are as follows: (i) we have shown the formulation of conservative imaging that excludes the possible world with probability zero from the agent's images and its AGM postulates that include a conservative representation, (ii) in the same way as the belief revision and update, we have shown its reversal function and AGM postulates, and (iii) we then have shown the condition under which conservative imaging and conditionalization coincide. The following table summarizes the discussions and results of this paper. Note that the same row indicates the same properties in the entry of postulates (Table 16.1).

An important issue open for future investigation is as follows. In terms of decision theory, Lewis' imaging explains causal aspects of our decisions [11]. However, especially in the situation of Corollary 16.1, conservative imaging also seems evidential because it entails conditionalization that explains evidential aspects of our decisions [7]. We need to investigate what the operation of removing zero-probability worlds from our images corresponds to in the distinction between causal and evidential decision theories.

Acknowledgements I would like to acknowledge the helpful suggestions made by Satoru Suzuki on the draft. I am truly grateful to Koji Mineshima and anonymous referees for their comments and the Japan Science and Technology Agency for the support. I emphasize that I am solely responsible for the shortcomings.

References

1. Alchourrón, C. E., Gärdenfors. P. & Makinson, D.: On the logic of theory change: partial meet contraction and revision functions. J. Symb. Logic **50**(2): 510–530 (1985)
2. Cozic, M.: Imaging and sleeping beauty: a case for double-halfers. Int. J. Approx. Reason. **52**(2), 137–143 (2011)
3. Cross, C.: A characterization of imaging in terms of popper functions. Philos. Sci. **67**(2), 316–338 (2000)

Table 16.1 Summary of this paper about properties of various belief changes

Type of belief change	Expansion	Revision	Update	Modification
Function	Proposition 16.1		Proposition 16.3	Theorem 16.1
Example	Example 16.1		Example 16.3	Example 16.6
Postulates	+1	$*1$	$\Diamond 1$	$\Delta 1$
	+2	$*2$	$\Diamond 2$	$\Delta 2$
	+3	$*3$	$\Diamond 3$	$\Delta 3$
	+4	$*4$	$\Diamond 4$	$\Delta 4$
	+5		$\Diamond 5$	$\Delta 5$
	+6			
		$*5$		
		$*6$		
			$\Diamond 6$ & $\Diamond 7$	$\Delta 6$ (stronger)
			$\Diamond 8$ (linearity)	
				$\Delta 7$ (conservativity)
Reversal type		Contraction Proposition 16.2	Erasure Proposition 16.4	Defeasance Theorem 16.2

4. Gärdenfors, P.: Imaging and conditionalization. J. Philos. **79**(12), 747–760 (1982)
5. Gärdenfors, P.: Knowledge in Flux: Modeling the Dynamics of Epistemic States. The MIT Press (1988)
6. Günther, M.: Learning conditional information by Jeffrey imaging on Stalnaker conditionals. J. Philos. Log. **47**(5), 851–876 (2018)
7. Jeffrey, R.: The Logic of Decision, 2nd edn. University of Chicago Press (1983)
8. Katsuno, H., Mendelzon, A.: On the difference between updating a knowledge base and revising it. In: Gärdenfors, P. (ed.), Belief Revision, pp. 183–203. Cambridge University Press (1992)
9. Leitgeb, H.: Imaging all the people. Episteme **14**(4), 463–479 (2017)
10. Lewis, D.: Probabilities of conditionals and conditional probabilities. Philos. Rev. **85**(3), 297–315 (1976)
11. Lewis, D.: Causal decision theory. Australas. J. Philos. **59**(1), 5–30 (1981)
12. Ramachandran, R.: Belief erasure using partial imaging. In: Li, J. (eds.), AI 2010: Advances in Artificial Intelligence, pp. 52–61. Springer (2010)
13. Suzuki, S.: On imaging (in Japanese). Ronshu **17**, 24–40 (1998)
14. Suzuki, S.: Imaging reconsidered (in Japanese). Ronshu **18**, 79–102 (1999)
15. Suzuki, S.: Prolegomena to general-imaging-based probabilistic dynamic epistemic logic. In: Washio, T., Satoh, K., Takeda, H., Inokuchi, A. (eds.), New Frontiers in Artificial Intelligence, JSAI 2006, pp. 118–132. Springer (2007)

Chapter 17
Unified Inductive Logic: From Formal Learning to Statistical Inference to Supervised Learning

Hanti Lin

Abstract While the traditional conception of inductive logic is Carnapian, I develop a Peircean alternative and use it to unify formal learning theory, statistics, and a significant part of machine learning: supervised learning. Some crucial standards for evaluating non-deductive inferences have been assumed separately in those areas but can actually be justified by a unifying principle.

Keywords Induction · Peirce · Formal learning theory · Machine learning · Statistics

17.1 Introduction

According to the Carnapian/Bayesian view, inductive logic is a matter of securing a "high proportion" of worlds that make the conclusion true within the domain of the worlds that make the premises true (Carnap [1]). An alternative view of inductive logic is developed here: it is a matter of being guaranteed that, once "enough" premises (data or evidence) are given, a true conclusion would be obtained.

This idea can be traced back to C. S. Peirce (1902/[12]: CP 2.780-1), and I propose that it be systematically developed along the following line of thought: We should be given a certain kind of guarantee *at least* when the amount of evidence is arbitrarily large. What kind of guarantee? A natural idea is to seek (i) a guarantee to *actually* get exactly the true answer to the question posed in one's context of inquiry. If that is unachievable, we should seek (ii) a guarantee to have a *high physical objective probability* of getting exactly the true answer. If even that is unachievable, we should seek, or settle with, (iii) a guarantee to have a high probability of getting *close* to the true answer. Note that those three guarantees, (i)–(iii), form a sequence of increasingly lower standards:

H. Lin (✉)
University of California, Davis, CA, USA
e-mail: ika@ucdavis.edu

K. Sano et al. (eds.), *Exploring Negation, Modality and Proof*,
Logic in Asia: Studia Logica Library,
https://doi.org/10.1007/978-981-96-8566-0_17

(i) a guarantee to *actually* get exactly the true answer,
↓ weaken

(ii) a guarantee to *have a high chance* of getting *exactly* the true answer,
↓ weaken again

(iii) a guarantee to have a high chance of getting *close to* the true answer.

When we tackle an empirical problem, we ought to strive for the highest achievable of such standards—achievable relative to the empirical problem undertaken. In a slogan: *strive for the highest achievable!* This is a principle that, I claim, unifies multiple areas that study non-deductive inferences. In particular, many disciplines—including formal learning theory, hypothesis testing and estimation in classical statistics and supervised learning in machine learning—appear to *assume* distinct standards to evaluate the different types of inference methods that they each study. But, if I am right, those evaluative standards need not be *assumed*, let alone assumed *separately*; instead, they can be justified by the unifying principle I propose.

To make my claim precise, many familiar concepts (such as empirical problems and various modes of convergences) need to be reformulated in a uniform setting. This will be addressed in the first half of this paper (through Sect. 17.3). Then, in the second half (beginning in Sect. 17.4), the main mathematical results can be stated, and their philosophical significance will be explained. I will conclude by looking into both the history and the future: revisiting Peirce's original ideas and exploring the potential for further unification, including a version of Bayesianism.

17.2 Varieties of Empirical Problems

Examples first:

Example 1 (*Enumerative Induction*) The **easy raven problem** poses a question: Are all ravens black? Two potential answers: `Yes` versus `No`, which form the hypothesis space $\mathsf{H} = \{\texttt{Yes}, \texttt{No}\}$. Evidence is gathered by observing ravens one by one, and noting their colors as either black (`1`) or nonblack (`0`). So the space of the possible evidential states, E, is the tree of all finite binary sequences. A possible world for this problem takes the form:

$$w = (e_1 e_2 e_3 \cdots, h),$$

where $e_1 e_2 e_3 \cdots$ is an infinite binary sequence, and h is the competing hypothesis true in that world w. It is assumed in the background that either all ravens are black or, if not, a counterexample would be observed sooner or later if the evidence were to accumulate indefinitely. This assumption rules out only one possible world, $(\texttt{111}\cdots, \texttt{No})$, in which the true answer is `No` (not all ravens are black) and we would still always only observe black ravens $\texttt{111}\cdots$. This background assumption is formalized by a set W of possible worlds—the set of all worlds of the form $(e_1 e_2 e_3 \cdots, h)$ except for $(\texttt{111}\cdots, \texttt{No})$.

If the above background assumption is relaxed to include the possible world $(\texttt{111}\cdots, \texttt{No})$, we obtain the *hard* raven problem, which was studied in formal learning theory only quite recently (Lin [9]).

The above example suggests that, in general, an empirical problem has at least three components: (i) competing hypotheses, (ii) data sequences as possible evidence, (iii) a background assumption. Indeed, these three components also figure in another classic empirical problem:

Example 2 (*Statistical Testing*) The **fair coin problem** poses a question: Is the coin fair? So the hypothesis space H is $\{\texttt{Fair}, \texttt{Unfair}\}$. Evidence is obtained by tossing the coin repeatedly and observing the results, either landing heads ($\texttt{1}$) or landing tails ($\texttt{0}$). So the evidence space E for this problem is the tree of binary sequences (as in the easy raven problem). Assumed in the background is the standard IID assumption in statistics: that the bias θ of the coin, i.e., the probability of landing heads, stays constant through time and coin tosses are independent. Under the IID assumption, each possible bias θ in the unit interval $[0, 1]$ determines a probability function $\mathbb{P}_\theta$ defined over E (which is the binary tree). A possible world for this problem takes this form:

$$w = (e_1 e_2 e_3 \cdots, \theta, \mathbb{P}_\theta).$$

In this world, θ is the true bias of the coin, $\mathbb{P}_\theta$ is the true probability function that represents the data-generation mechanism, and it turns out to generate the data sequence $e_1 e_2 e_3 \cdots$. The background assumption is represented by a set of possible worlds, W, defined as the set of all worlds of the above form. The hypothesis $\texttt{Fair}$ asserts that $\theta = 0.5$, which is true in the worlds in which $\theta = 0.5$. Similarly for the hypothesis $\texttt{Unfair}$, which asserts the negation $\theta \neq 0.5$.

Some clarifications are in order. First, the background assumption is very weak. For example, it is logically compatible with this possible world:

$$w = (\texttt{1010}\cdots, 0.5, \mathbb{P}_{0.5}),$$

in which the coin is (still) fair and alternates between landing heads ($\texttt{1}$) and tails ($\texttt{0}$). It is also logically compatible with this possible world:

$$w = (\texttt{1111}\cdots, 0.5, \mathbb{P}_{0.5}),$$

in which the coin is (still) fair and it turns out to always land heads—yes, this is logically possible. In fact, the background assumption is even compatible with any worlds of the following form, where $e_1 e_2 e_3 \cdots$ is an arbitrary binary sequence:

$$w = (e_1 e_2 e_3 \cdots, 0.5, \mathbb{P}_{0.5}),$$

in which the coin is fair and it turns out to land heads or tails according to the pattern $e_1 e_2 e_3 \cdots$. All those worlds are ruled in, following the practice of classical statistics.

Second, note that probabilities are assigned to the data sequences in the evidence space. That is, each probability function $\mathbb{P}_\theta$ is defined over the evidence space, which represents a stochastic mechanism for generating data or evidence and conforms to the use of *objective physical probabilities*, or *chances*, in classical statistics. In contrast, Bayesian statistics, and Bayesian epistemology in general, also allow probabilities to be assigned to possible worlds, but those probabilities represent subjective degrees of belief. A comparison of the present work with Bayesianism will be provided in the concluding section.

The fair coin problem presupposes that there exist probabilities, so probabilities have to figure in the possible worlds in use. In contrast, for the easy raven problem, it suffices to use worlds of simpler forms without probabilities. So there should be flexibility in our design of possible worlds:

Definition 1 (*Possible World*) A possible world is an ordered pair or tuple of the form:

$$w = (e_1 e_2 \cdots, \mathsf{other}),$$

where the first component $e_1 e_2 \cdots$ is an infinite data sequence, understood as the one produced in that world, and the second component other specifies the other relevant elements of that world.

Many different things can go into other, such as a distinguished statement true in world w, or the distribution of physical probabilities that exist objectively in that world.

I have mentioned three components of an empirical problem: (hypothesis, evidence, and assumption). There is a fourth component, to be motivated below:

Example 3 (*Statistical Estimation*) The **coin bias problem** is basically the same as the fair coin problem except that it poses a more fine-grained question: What is the bias of the coin? So the hypothesis space is $\mathsf{H} = [0, 1]$, the unit interval. The evidence space E is the same; so is the background assumption W. The possible worlds in use are the same, too, taking the form $w = (e_1 e_2 e_3 \cdots, \theta, \mathbb{P}_\theta)$. But our guess of the bias, as a hypothesis $h \in \mathsf{H}$, can be more or less accurate; the loss of accuracy can be measured by the difference between the guess h and the true bias. That is, the loss of accuracy of guessing $h \in [0, 1]$ in a world w, written $\mathsf{Loss}(h, w)$, can be defined as the difference between the guess h and the true bias in w.

So we need this additional definition:

Definition 2 (*Loss Function*) A loss function is a function $\mathsf{Loss} : \mathsf{H} \times \mathsf{W} \to \mathbb{R}$ that maps any hypothesis h in H together with any world in W to a non-negative real number, denoted by $\mathsf{Loss}(h, w)$, under the constraint that, in each world $w \in \mathsf{W}$, there is a unique hypothesis $h \in \mathsf{H}$ that attains zero loss of accuracy: $\mathsf{Loss}(h, w) = 0$.

The uniqueness constraint is adopted in this paper only for the sake of simplicity. It allows us to talk about convergence to *the truth* in a world w: the unique hypothesis that attains zero loss of accuracy in that world w. It is not hard to generalize, referring to *a* truth or truth*s* in a world, but let's opt for simplicity here in order to focus on more important issues.

When we were thinking about the first two examples (the easy raven problem and the fair coin problem), we were not compelled to think about the loss function only because the role of the loss function is already played by the talk of truth and falsity. That is, the loss of accuracy has only two values: 0 for getting the truth, 1 for getting a falsehood. More precisely, in the easy raven and fair coin problems, the loss function is simple: $\mathsf{Loss}(h, w)$ is equal to 1 minus the truth value of hypothesis h in world w.

Thus, for the sake of uniformity, every empirical problem is required to include a loss function as a fourth component. In fact, as will become clear in the next section, this fourth component is even a necessity in order to produce a uniform treatment of various standards for evaluating inference methods. Hence the following definition:

Definition 3 (*Empirical Problem*) An empirical problem is a quadruple (H, E, W, Loss) with the following interpretations and requirements:

- H is a set of hypotheses, understood as the hypotheses under consideration.
- E is a tree, called the evidence tree, in which the nodes are finite data sequences $e_1 \cdots e_n$ ordered by sequence extension, and every branch is infinite.
- W is a set of possible worlds such that the infinite data sequences that appear in the worlds therein are exactly the branches of the evidence tree E. This set W is meant to contain all and only the possible worlds compatible with one's background assumption.
- Loss is a loss function on $\mathsf{H} \times \mathsf{W}$.

This definition is also general enough to cover problems in supervised learning, from binary classification to nonparametric regression, to be discussed in section 17.6.

The components of an empirical problem have their own roles to play. The evidence space E and the hypothesis space H are used to define inference methods as functions from the former to the latter (with a minor refinement to be formally stated below). The other two components are used to define various standards for assessing inference methods, as we will see below. Those standards examine each inference method by considering how that method performs for truth seeking across a range of the possible worlds—the worlds in W. So, W represents the background assumption against which inference methods are evaluated. These ideas will be fleshed out in the formal definitions in the next section.

17.3 Modes of Convergence as Evaluative Standards

Here are the objects of evaluation:

Definition 4 (*Inference Method*) An inference method for an empirical problem (H, E, W, Loss) is a function $M : \mathsf{E} \to \mathsf{H} \cup \{?\}$; that is, M can receive any finite data sequence that figures as a node of the evidence tree E, and then output one of the hypotheses in H or a question mark ? that represents judgment suspension.

Only the first two components of a problem, H and E, are needed to define the inference methods for that problem. The remaining two components, W and Loss, are used to define standards for assessing inference methods. But preliminaries first:

Definition 5 Here are some notation conventions:

- Let $\mathbb{P}_w$ denote the probability measure true in w if w contains such a thing.
- Let $h^{M,n,w}$ denote the hypothesis or output of method M at stage n in world w, that is, $M(e_1 e_2 \cdots e_n)$, where $e_1 e_2 \cdots e_n$ is the sequence of the first n data points received in world w.
- Similarly, let $\hat{h}^{M,n}$ be the random variable that maps each possible data sequence of length n to the output of M on that sequence; so $\hat{h}^{M,n}$ can be intuitively understood to denote the hypothesis that method M outputs at stage n (leaving worlds and data sequences unspecified), following the standard use of random variables in statistics.

The random variable notation $\hat{h}^{M,n}$ is particularly convenient for expressing probabilities like the following:

$$\mathbb{P}_w\Big[\,\mathsf{Loss}\big(\hat{h}^{M,n}, w\big) < \epsilon\Big] =_{\mathrm{df}} \mathbb{P}_w\Big\{e_1 e_2 \cdots e_n : \mathsf{Loss}\big(M(e_1 e_2 \cdots e_n), w\big) < \epsilon\Big\}.$$

The right hand side means the probability, in world w, of observing a data sequence $e_1 e_2 \cdots e_n$ of length n such that the loss of M's output is below the threshold ϵ. This phrase can be more concisely expressed using the notation on the left-hand side: it denotes the probability, in world w, for the loss of M's output to be below ϵ given sample size n.

Here is the thing:

Definition 6 (*Three Basic Modes of Convergence*) An inference method M for a problem (H, E, W, Loss) can be said to achieve one or another mode of convergence to the truth. The following defines three modes.

- **Convergence with Nonstochastic Identification:**
  for any world $w \in \mathsf{W}$,
  there exists sample size N such that,
  for all $n \geq N$,

$\text{Loss}(h^{M,n,w}, w) = 0$;
that is, given sample size n or more, M outputs exactly the truth in w.

- **Convergence with Stochastic Identification:**
  for any probability threshold $1 - \delta < 1$,
  for any world $w \in \mathsf{W}$,
  there exists sample size N such that,
  for all $n \geq N$,

  $\mathbb{P}_w$ exists, and
  $\mathbb{P}_w\big[\,\text{Loss}(\hat{h}^{M,n}, w) \;=\; 0\,\big] \;>\; 1 - \delta$;
  that is, given sample size n or more, the probability for M to output exactly the truth is high (greater than $1 - \delta$) in w.

- **Convergence with Stochastic Approximation:**
  for any upper bound on loss $\epsilon > 0$,
  for any probability threshold $1 - \delta < 1$,
  for any world $w \in \mathsf{W}$,
  there exists sample size N such that,
  for all $n \geq N$,

  $\mathbb{P}_w$ exists, and
  $\mathbb{P}_w\big[\,\text{Loss}(\hat{h}^{M,n}, w) \;<\; \epsilon\,\big] \;>\; 1 - \delta$;
  that is, given sample size n or more, the probability that M outputs a hypothesis ϵ-close to the truth is high (greater than $1 - \delta$) in w.

These three basic modes of convergence have been mostly developed and studied separately in different areas, couched in very different languages, and employed to talk about apparently different subjects. The formalism developed here provides a uniform reformulation of those modes. Moreover, the plain English glosses accompanying the above definitions (the clauses following 'that is') make it clear that the old, familiar modes of convergence are covered:

(i) **Convergence w/ Nonstochastic Identification**: It goes by the name *identification/decidability in the limit* in formal learning theory (Kelly [7]: Chap. 3), originally designed for theory choice in a deterministic setting.
(ii) **Convergence w/ Stochastic Identification**: This captures the conjunction of *consistency in significance level* and *consistency in power* as studied in statistical hypothesis testing (Lehmann [3]: Chap. 3). It also captures the so-called *model selection consistency* in statistical model selection (Claeskens et al. [2]: Chap. 4).
(iii) **Convergence w/ Stochastic Approximation**: This captures so-called *estimation consistency* in the statistical theory of estimation (Lehmann [3]: Chap. 2). As you will see below (section 17.6), it also captures the consistency of learning algorithms in supervised learning, which includes classification (Shalev-Shwartz et al. [15]: Part I) and nonparametric regression (Györfi et al. [6]: Chap. 1).

Table 17.1 The Hierarchy of the Three Basic Modes of Convergence

(i)	Convergence w/ Nonstochastic Identification
	|
(ii)	Convergence w/ Stochastic Identification
	|
(iii)	Convergence w/ Stochastic Approximation

It is not hard to develop variants of the above modes of convergence. Let me briefly outline some notable ones. For each of the three modes defined above, exchanging the quantifiers "for any world" and "there exists sample size" gives us a higher standard, a mode of *uniform* convergence. There are other variants, which can be obtained by thinking about: *rates* of convergence (Györfi et al. [6]: Chap. 1), *monotonic* convergence or somewhat *stable* convergence if not perfectly monotonic (Lin [9]), and convergence for *almost all* worlds in W—almost all in a topological sense (Lin [8]). Those modes of convergence and their combinations have been studied in one or another area. From an epistemological point of view, they correspond to higher or lower standards for evaluating inference methods. However, I will focus on the three basic modes just defined, which suffice for making the philosophical point I want to make below.

17.4 Towards Unification

The three basic modes of convergence have been reformulated in a uniform notation to clarify their close connection. Although they are developed in distinct areas, they set standards for assessing *all* inference methods in *any* empirical problems. The standards range from high to low—from actually getting the truth, to probably getting exactly the truth, to probably getting close to the truth. Hence, the hierarchy presented in Table 17.1.

We need just one more definition before the first result can be stated:

Definition 7 (*Achievability*) A mode of convergence is said to be achievable for an empirical problem if some inference method for that problem satisfies that mode of convergence.

Then we have:

Proposition 1 *Consider the hierarchy of the three modes of convergence in Table 17.1.*

- *For the easy raven problem, the highest achievable is mode* (i).
- *For the fair coin problem, it is mode* (ii).
- *For the coin bias problem, it is mode* (iii).

See the appendix for the proof. The novelty of this result consists in what is *not* achievable. The claims of achievability are already obtained with greater generality in formal learning theory and statistics. But I still provide elementary proofs of those achievability claims, considering that readers familiar with one of the two areas might not be familiar with the other.

Since the novelty is in the claims of *un*achievability, let me explain the proof strategy. Mode (i) is unachievable for the two statistical problems due to a specific kind of *underdetermination by data*: those two problems allow that one and the same infinite data sequence can be generated under different hypotheses. Mode (ii) is unachievable for the coin bias problem because of considerations about *cardinality*: the number of competing hypotheses (all the real numbers in the unit interval) is strictly greater than the number of possible evidential inputs (all finite binary sequences). So, for any inference method, there is at least one hypothesis that is doomed never to be an output.

I suspect that the proof strategy just sketched allows us to generalize the result to cover wide classes of empirical problems. However, I will focus on the three paradigm problems—easy raven, fair coin, and coin bias—to streamline the discussion and emphasize more pressing points: a picture of different areas unified into a cohesive whole.

Although statistics and formal learning theory may appear to be very different, they operate with the same guiding principle in the unifying picture I propose:

Strive for the Highest Achievable!

Statisticians design inference methods with a minimum qualification in mind, which is standardly called consistency in statistics but is just the stochastic mode (ii) or (iii) defined above. Statisticians do not aim at the higher mode (i). The *cause* underlying statisticians' practices may lie in how their textbooks are written and their PhD programs are structured. But the *reason* that justifies what statisticians do, I propose, is based on this fact: the empirical problems addressed by statisticians are too hard to make the higher mode (i) achievable. Similarly, formal learning theorists do not aim at the lower modes (ii) and (iii), because they should not: since the empirical problems they study make the higher mode (i) achievable, they should use that higher standard to evaluate inference methods.

So, statistics and formal learning theory need not be regarded as two separate areas that *assume* different standards to evaluate different kinds of inference methods. Instead, there is a unifying principle: strive for the highest achievable. The use of different standards in different areas—stochastic versus nonstochastic—need not be assumed and can be *justified* by the proposed principle.

Similarly, hypothesis testing and parameter estimation need not be viewed as distinct subfields of statistics with separate evaluative standards. The use of different standards—identification versus approximation—need not be assumed and can be *justified* by the proposed principle.

17.5 Deterministic Versus Stochastic?

Although formal learning theory is often presented in a way that gives the impression that it only concerns deterministic hypotheses (as in Kelly [7]), let's be more careful. The hypothesis "All ravens are black" is neutral regarding whether the actual world is deterministic or indeterministic. This hypothesis can be true in a deterministic world, but it can also be true in an indeterministic world—a world in which all ravens turn out to be black by chance. We have represented this world without probabilities by:

$$w = (\texttt{111}\cdots, \texttt{Yes}).$$

Strictly speaking, this w only represents a very coarse-grained world, for it says nothing whatsoever about whether determinism is true or false. The coarse-grained world w can be realized by two different, more fine-grained worlds:

$$\begin{aligned} u &= (\texttt{111}\cdots, \texttt{Yes}, \mathbb{P}_{\text{trivial}}), \\ u' &= \big(\texttt{111}\cdots, \texttt{Yes}, \mathbb{P}'\big), \end{aligned}$$

where, $\mathbb{P}_{\text{trivial}}$ is a probability function that only assigns trivial probabilities 0 and 1—in particular, it assigns probability 1 to the constant sequence $\texttt{111}\cdots$. But $\mathbb{P}'$ is a nontrivial probability function, assigning (say) .7, to the constant sequence $\texttt{111}\cdots$. In both of those fine-grained worlds, it is true that all ravens are black. However, it is true deterministically in the first world u, and true only by chance in the second world u'. Hence the following definition:

Definition 8 (*Probabilistic Extension*) Suppose that an evidence tree E is given. Consider a world $w = (e_1 e_2 \cdots, h)$ that contains no probability measure. This world w is said to have another world w' as a probabilistic extension if w' takes the form $w' = (e_1 e_2 \cdots, h, \mathbb{P})$, which differs from w only by adding a (trivial or nontrivial) probability measure $\mathbb{P}$ over the evidence tree E.

Definition 9 (*Fine-Grained Version*) Fine-grained versions of a problem (H, E, W, Loss) are problems of the form: (H, E, W', Loss'), with the same H, the same E, but a different W' and a different Loss', satisfying the following constraints:

- W' can be obtained from the original W by, first, removing each world w that contains no probability measure and, then, replacing w with one or multiple probabilistic extensions of w.
- $\mathsf{Loss}'(h, w') = \mathsf{Loss}(h, w)$ if w' is a probabilistic extension of w.

Then we have:

Proposition 2 *Consider the three modes of convergence in Table* 17.1. *If the highest mode* (i) *is achievable for a problem* $\mathcal{P}$, *then the three modes* (i)–(iii) *are all achievable for any fine-grained version of* $\mathcal{P}$ *that only involve countably additive probability measures.*

See the appendix for the proof.

Corollary 1 *Mode* (i) *implies mode* (ii) *in the weak sense stated in the preceding proposition. Mode* (ii) *implies mode* (iii) *in the standard sense.*

The easy raven problem fails to make (ii) and (iii) achievable only because of its somewhat misleading mathematical representation. The fine-grained versions of that problem do a better representational job, being explicitly neutral between determinism and indeterminism, making all the three modes provably achievable.

So it is misleading to say that formal learning theory differs from statistics in that the former presupposes determinism. This claim is incorrect; there is no such presupposition. The distinction between these two fields is best understood as a division of labor guided by a common principle: strive for the highest achievable.

17.6 Supervised Learning

The present setting also covers problems studied in supervised learning, whose simplest examples concern binary classification:

Example 4 (*Binary Classification*) Suppose that we want to determine whether a given object is a cat by examining a (pixelated) image of it. Or suppose that we want to determine whether a given watermelon is tasty by examining properties such as its skin color distribution and the sound it produces when tapped. More generally, suppose that we want to classify an object of a certain kind into category 0 or category 1, and do it on the basis of the feature that it has in a countable set of mutually exclusive features: $\mathsf{X} = \{x_1, x_2, x_3 \ldots\}$. A **(binary) classifier** is an indicator function $h : \mathsf{X} \to \{0, 1\}$. Suppose, further, that we are given a set H of (some or all) such classifiers. We would like to pick a good classifier from H in light of examples. An **example** is an object with a specified feature $x \in \mathsf{X}$ and a specified category 0 or 1; so an example is formally an ordered pair $(x, 0)$ or $(x, 1)$. Such examples form an **example space**: $\mathsf{X} \times \{0, 1\}$. A data sequence $(e_1 \cdots e_n)$ is a finite sequence of such ordered pairs. All such data sequences form an evidence tree $\mathsf{E} = (\mathsf{X} \times \{0, 1\})^{<\infty}$. A **learning algorithm** is just an inference method $M : \mathsf{E} \to \mathsf{H} \cup \{?\}$, although people in machine learning typically only consider $M : \mathsf{E} \to \mathsf{H}$. To sum up: a **task of binary classification** can be formally identified as an ordered pair (X, H) consisting of two elements:

- a countable feature space $\mathsf{X} = \{x_1, x_2, x_3 \ldots\}$,
- a set $\mathsf{H} \subseteq \{0, 1\}^{\mathsf{X}}$ as a candidate pool of classifiers.

Those two elements suffice to determine the other elements of such a task, including classifiers, examples, an evidence tree E, and learning algorithms. *But what counts as a good classifier—good for predictive purposes?* The quality of a classifier depends on the state of the world. Assume that examples are generated by a probabilistic mechanism, represented by a probability distribution D over the example space $\mathsf{X} \times \{0, 1\}$, where $D(x, y)$ denotes the probability that the next example has feature x and belongs to category y. The **predictive risk** of a classifier h with respect to the (true but unknown) distribution D is given by the probability of *misclassification*:

$$\mathsf{Risk}(h, D) \;=\; D\big\{(x, y) : h(x) \neq y\big\} \;=\; \sum_{(x,y):\, h(x) \neq y} D(x, y)\,.$$

The smaller, the better. All this points to an empirical problem $(\mathsf{H}, \mathsf{E}, \mathsf{W}, \mathsf{Loss})$, called a **(binary) classification problem**:

- Question: Which classifier is the best in class (in H) for predictive purposes? So the hypothesis space is H, the given candidate pool of classifiers.
- E, the evidence tree, is the tree of sequences of examples (x, y) taken from the example space $\mathsf{X} \times \{0, 1\}$.
- W, which represents the background assumption, is the set of all possible worlds of the form $w = (e_1 e_2 \cdots, D, \mathbb{P}_D)$, where D is an arbitrary probability distribution over the example space, and $\mathbb{P}_D$ is the IID probability measure (on E) generated from distribution D.
- $\mathsf{Loss}(h, w) = \mathsf{Risk}(h, D_w) - \min_{h' \in \mathcal{H}} \mathsf{Risk}(h', D_w)$, where D_w denotes the distribution in world w

This concludes the last of the four examples examined in this paper. It is no coincidence that textbooks in machine learning use the symbol h to denote classifiers, referring to them as hypotheses (Shalev-Shwartz et al. [15]).

Once we see what binary classification is, generalizations are straightforward. When the category set $\mathsf{Y} = \{0, 1\}$ is generalized to a set of finite categories, $\mathsf{Y} = \{1, 2, \ldots, k\}$, we obtain problems of *multiclass classification*. When both the feature space X and the category space Y become continuous, being Euclidean spaces, we have problems of *nonparametric regression*. These generalizations encompass nearly the entire area of supervised learning.

In supervised learning, the minimum qualification for good learning algorithms is called *consistency*, which is essentially the mode of convergence (iii) as defined in this paper—convergence with stochastic approximation. However, consistency does not need to be *assumed* as a minimum qualification in supervised learning. It can be justified as follows. By the principle of striving for the highest achievable, the achievability of at least mode (iii) in the hierarchy in Table 17.1 implies that mode (iii) or a higher mode has to be achieved, which in turn implies, thanks to corollary 1, that mode (iii) must be achieved at the very least—as a minimum qualification. The same argument also justifies mode (iii) as a minimum qualification for good estimators in statistical estimation.

17.7 Closing

This paper reformulates three basic modes of convergence to the truth using a uniform notation and explores their potential to create a unified picture of inductive logic. Additional modes of convergence (as mentioned at the end of section 17.3) should be investigated to enrich the simple hierarchy in Table 17.1 and to determine whether more fields can be incorporated into the unified framework. In this closing note, let me explain how we arrived at this point and where we might go from here.

In hindsight, Peirce already did a lot for us over a century ago. As pointed out in Lin's [10] discussion of Peirce's convergentism, Peirce sought to justify enumerative induction by appeal to the mode of convergence with nonstochastic identification (Peirce [12]: CP 2.775, 7.125), and he *also* studied statistical estimation using the mode of convergence with stochastic approximation (Peirce [12]: CP 2.669-93). Unfortunately, after Peirce, those two modes of convergence took diverging paths. The stochastic one helped establish the estimation theory in statistics (Fisher [4]), and the nonstochastic one helped create formal learning theory (Putnam [13], Gold [5])—but those two fields have long been regarded as largely unrelated. I propose that we return to Peirce's idea and reinforce it with the principle developed above "Strive for the Highest Achievable!"

The result is a conception of inductive logic that unifies formal learning theory, classical statistics, and a significant part of machine learning: supervised learning. But can it be extended to cover other areas of machine learning, such as reinforcement learning and unsupervised learning? I am optimistic for reinforcement learning, for it has been given a largely uniform foundation as supervised learning (Mohri et al. [11]). Of course, a detailed argument is needed nonetheless, but that has to be left to future work.

However, the prospect for incorporating unsupervised learning into the unified picture remains unclear. In fact, even theorists of machine learning have difficulty evaluating algorithms of unsupervised learning by a rigorous standard—let alone a standard defined as a mode of convergence. To talk about convergence to the truth, there needs to be a truth to begin with, but such a truth is typically missing in unsupervised learning, as observed in a standard textbook on the theoretical foundation of machine learning: "[A] basic problem is the lack of "ground truth" for clustering, which is a common problem in unsupervised learning" (Shalev-Shwartz & Ben-David [15]: 308).

The most influential approach to inductive logic to date in philosophy is the Bayesian one. I suspect that it can be developed in a manner that fits into the unified picture I have painted. On the Bayesian approach, inductive logic should be founded on the idea that probabilities, as rational degrees of belief, ought to be updated by conditionalization on the new evidence (Carnap [1]). But is it that all priors—initial assignments of probabilistic degrees of belief—are equally permissible? The answer is "yes" according to the *subjectivists* in Bayesian epistemology. But the answer is "no" according to anti-subjectivists, including the *objectivists*, who maintain that permissible priors must be "flat", conforming to the principle of indifference or a variant of it. But the objectivists are not the only people who would like to constraint the candidate pool of permissible priors by something more than the axioms of probability. For a certain group of Bayesians who are less known to the philosophical community but work in a branch of statistics called *Bayesian nonparametric statistics*, it is customary to rule out some prior assignments of degrees of belief by appeal to considerations about convergence (see Rousseau [14] for a review). The idea is simple: a Bayesian prior is permissible only if updating it successively via conditionalization ensures a sequence of posteriors that achieves a specific mode of convergence to the truth, commonly called *Bayesian consistency*. Thus, even Bayesians may find their place within the unifying framework I have outlined for inductive logic; however, the details remain to be addressed in future work.

I hope this offers some initial reason for optimism about the prospect of a unified inductive logic.

Acknowledgements I thank Konstantin Genin for his patient and inspiring discussions. I also thank two anonymous reviewers for their detailed and constructive comments on earlier versions of this paper.

17.8 Appendix: Proofs

17.8.1 Proof of Proposition 1

To establish the first part for the easy raven problem, it suffices to verify that mode (i) is achieved in the easy raven problem by the inference method that outputs `Yes` exactly when the input contains no `0`, and outputs `No` otherwise.

Now, establish the second part for the fair coin problem as follows. Mode (i), convergence with nonstochastic identification, is unachievable because the fair coin problem features this kind of underdetermination by data: The background assumption W in the fair coin problem is so weak that it contains a pair of worlds in W sharing the *same* infinite data sequence (say, heads, tails, heads, tails, and so on), while one world makes `Fair` true and the other makes `Unfair` true. So, to converge to the truth in one of those two worlds is to converge to a falsehood in the other. Mode (i) is thus unachievable. As to the achievability of mode (ii) for the fair coin problem, it can be proved by Bernoulli's law of large numbers as follows. Let M be the inference method that outputs `Fair` exactly when the observed frequency $\bar{X}_n$ of heads in the sample is *close enough* to 0.5 in the sense that $|\bar{X}_n - 0.5| < 1/\sqrt[4]{n}$, where n is the sample size. Now recall this version of Bernoulli's law of large numbers:

$$\mathbb{P}_\theta(|\bar{X}_n - \theta| < \epsilon) \;\geq\; 1 - \frac{1}{4n\epsilon^2}\,.$$

In any world in which the truth is `Fair`, the probability for M to get the truth is obtained by letting $\theta = 0.5$ and $\epsilon = 1/\sqrt[4]{n}$ in Bernoulli's law of large numbers:

$$\begin{aligned}
&\mathbb{P}_{.5}\left(|\bar{X}_n - 0.5| < 1/\sqrt[4]{n}\right)\\
&\geq 1 - \frac{1}{4n(1/\sqrt[4]{n})^2}\\
&= 1 - \frac{1}{4\sqrt{n}},
\end{aligned}$$

which converges to 1 as n tends to infinity. Moreover, in any world where the truth is `Unfair` with true bias $\theta \neq 0.5$, when the sample size n is large enough to ensure that $1/\sqrt[4]{n} < \frac{1}{2}|\theta - 0.5|$, the probability for M to get the truth is

$$\begin{aligned}
&\mathbb{P}_{\theta}\left(|\bar{X}_n - 0.5| \geq 1/\sqrt[4]{n}\right)\\
&\geq \mathbb{P}_{\theta}\left(|\bar{X}_n - \theta| < 1/\sqrt[4]{n}\right)\\
&\geq 1 - \frac{1}{4\sqrt{n}},
\end{aligned}$$

which converges to 1 as n tends to infinity. So M achieves mode (ii).

Now, establish the third part for the coin bias problem as follows. Mode (i) is unachievable for the same reason as in the fair coin problem. As for mode (ii), note that the set of all possible (evidential) inputs is the set of all nodes in the binary tree. So there are countably many possible inputs. But there are uncountably many hypotheses. Let M be an arbitrary inference method. It follows that there is some hypothesis h_M that M never outputs in any world. So, the probability for M to output (exactly) the truth always stays zero in the worlds in which the truth is h_M. So M fails to achieve mode (ii). But mode (iii) is achievable, for a well-known reason: the inference method that always outputs the observed frequency of heads achieves mode (iii), convergence with stochastic approximation, thanks to Bernoulli's law of large numbers (again). Q.E.D.

17.8.2 Proof of Proposition 2

Suppose that mode (i) is achievable for a problem $(\mathsf{H}, \mathsf{E}, \mathsf{W}, \mathsf{Loss})$, which has a refined version $(\mathsf{H}, \mathsf{E}, \mathsf{W}', \mathsf{Loss}')$. Let M be an inference method for the original problem that achieves mode (i), which implies that there is a one-to-one correspondence between the worlds in W and the infinite branches of the evidence tree E:

$$w = (e_1 e_2 \cdots, -) \quad \mapsto \quad e_1 e_2 \cdots \in \mathsf{E}.$$

So, for the sake of convenience, identify those worlds with the infinite branches of the evidence tree: $\mathsf{W} = \mathsf{E}$. Let $\mathsf{Success}(M, n)$ be the set of the worlds w in W, or equivalently the branches b in E, such that, by stage n, M has output the truth and would never drop it in b. By definition, $\mathsf{Success}$ is monotonic in this sense:

whenever $n \leq n'$,

$$\mathsf{Success}(M, n) \subseteq \mathsf{Success}(M, n') .$$

Since M achieves mode of convergence (i), it follows that

$$\bigcup_{n \in \mathbb{N}} \mathsf{Success}(M, n) = \mathsf{E} .$$

Let w' be an arbitrary world in W', and let $\mathbb{P}_{w'}$ denote be the probability measure over E in w'. Then, by countable additivity, we have:

$$\begin{aligned} & \lim_{n \to \infty} \mathbb{P}_{w'}\big(\mathsf{Success}(n)\big) \\ = \; & \mathbb{P}_{w'}\big(\bigcup_{n \in \mathbb{N}} \mathsf{Success}(n)\big) \\ = \; & \mathbb{P}_{w'}(\mathsf{E}) \\ = \; & 1 \end{aligned}$$

Now, note that the probability for M to get the truth by stage n in world w' is greater than or equal to $\mathbb{P}_{w'}\big(\mathsf{Success}(n)\big)$, which approaches 1 as $n \to \infty$ thanks to the above. So M achieves mode (ii), convergence with stochastic identification. Achieving mode (ii) immediately implies achieving mode (iii). Q.E.D.

References

1. Carnap, R.: On inductive logic. Philos. Sci. **12**(2), 72–97 (1945)
2. Claeskens, G., Hjort, N.L.: Model Selection and Model Averaging. Cambridge University Press (2008)
3. Lehmann, E.L. (ed.): Elements of Large-Sample Theory. Springer, New York (1999)
4. Fisher, R.A.: Statistical Methods for Research Workers. Oliver & Boyd (1925)
5. Gold, E.M.: Language identification in the limit. Inf. Control **10**(5), 447–474 (1967)
6. Györfi, L., Kohler, M., Krzyzak, A., Walk, H.: A Distribution-free Theory of Nonparametric Regression. Springer, New York (2002)
7. Kelly, K.T.: The Logic of Reliable Inquiry. Oxford University Press (1996)
8. Lin, H.: The hard problem of theory choice: a case study on causal inference and its faithfulness assumption. Philos. Sci. **86**(5), 967–980 (2019)
9. Lin, H.: Modes of convergence to the truth: steps toward a better epistemology of induction. Rev. Symb. Logic **15**(2), 277–310 (2022)
10. Lin, H.: "Convergence to the Truth". In: Sylvan, K., Sosa, E., Dancy, J., Steup, M. (eds.), The Blackwell Companion to Epistemology, 3rd edn. Wiley Blackwell (2025)
11. Mohri, M., Rostamizadeh, A., Talwalkar, A.: Foundations of Machine Learning. MIT Press (2018)
12. Peirce, C.S.: Collected Papers of Charles Sanders Peirce (Volumes I–VIII). InteLex Corp (1994)
13. Putnam, H.: Trial and error predicates and a solution to a problem of Mostowski. J. Symb. Logic **30**(1), 49–57 (1965)
14. Rousseau, J.: On the frequentist properties of Bayesian nonparametric methods. Ann. Rev. Stat. Its Appl. **3**, 211–231 (2016)
15. Shalev-Shwartz, S., Ben-David, S.: Understanding Machine Learning: From Theory to Algorithms. Cambridge University Press (2014)

Chapter 18
Composition and Plural Identity

Byeong-uk Yi

Abstract This article argues that the thesis of composition as identity (CAI) is incoherent because its proponents assume the existence of composite objects (i.e., objects with proper parts) to propose the thesis to cast light on their nature. To do so, the article proves that CAI is logically equivalent to the (part-whole) triviality thesis, the thesis that everything is a part of itself but of nothing else. In proving this, the article distinguishes three formulations of CAI resulting from different analyses of the composition relation, and proves that all three formulations of the thesis are logically equivalent to the triviality thesis. It also argues that CAI is logically equivalent to the triviality thesis on any adequate analysis of composition.

Keywords Mereology · Composition · Plural identity · Plural logic · Composition as identity (CAI) · Mereological sum · Fusion · Part/whole · Part-whole triviality · Mereological nihilism · Plato

18.1 Introduction: Two Theses on Composition and Identity

In the *Theaetetus*, Plato argues against a thesis about knowledge of simple and complex objects: "the elements are unknowable and the complexes knowable" [17, 201c–206b]. To do so, he formulates a metaphysical thesis on parts and wholes. The thesis is a precursor of what modern philosophers call the *thesis of composition as identity* (in short, CAI).

To argue against the epistemological thesis, which we might call the *knowability thesis*, he distinguishes two views of *complexes*, i.e., "things composed of elements" [*ibid.*, 202b], while using the first syllable of the name "Socrates" (SO) as a repre-

My work for this article was supported in part by a SSHRC Insight Grant [Grant No. 435-2022-0190], which is hereby gratefully acknowledged.

B. Yi (✉)
Department of Philosophy, University of Toronto, Toronto, Canada
e-mail: b.yi@utoronto.ca

K. Sano et al. (eds.), *Exploring Negation, Modality and Proof*,
Logic in Asia: Studia Logica Library,
https://doi.org/10.1007/978-981-96-8566-0_18

sentative example of a complex. The first of the views holds that the syllable SO, for example, is just the letters in it: "the two letters, S and O, ... are the first syllable of" the name [*ibid.*, 203c].[1] The second view holds that the syllable, for example, is "a whole" or "a *single* form resulting from the combination of the several elements [e.g., the letters] when they fit together" [*ibid.*, 204a, my italics]. Then he argues that both views conflict with the knowability thesis: one can have knowledge of complexes but not of elements (i.e., objects without *proper* parts).[2] To show that the thesis conflicts with the second view,[3] he argues that the view implies that the syllable SO, for example, "must have no parts" because "when a thing has parts, the whole is necessarily all the parts" [*ibid.*, 204a], where he uses "part" in a narrow sense for *proper* parts.[4]

This argument rests on a metaphysical thesis on parts and wholes. To formulate the thesis, say that some things (taken together) *compose* something, if the latter is a *whole composed of* the former; and that something is a *composite* (or *complex*) *object*, if it is a whole composed of proper parts. Then we can formulate the thesis as follows:

C1. *Plato's Thesis*: If something (i.e., a single thing) has a proper part, then its proper parts (taken together) compose it and (thus) it is the same thing as the proper parts (taken together).[5]

On this thesis, the syllable SO is the same thing as the letters S and O (taken together) although it is not the same thing as either of them. Assuming the thesis (C1), Plato concludes that the syllable, for example, cannot have a proper part, for C1 implies, as he assumes, that if something (e.g., SO) has a proper part, then it cannot have a proper part. Thus, he says that a composite object (e.g., SO), on the second view, must be "an absolutely single form, indivisible into parts" [*ibid.*, 205c]. If so, on the knowability thesis, wholes or composites must also be elements and thus unknowable.

While Plato takes C1 for granted in this argument,[6] it is not clear why one must accept it. It is a substantial metaphysical thesis with significance reaching far beyond the epistemology of composites.[7]

---

[1] On this view, a complex (e.g., SO) is not one object but many objects (e.g., two letters, S and O) and apparent singular terms for complexes (e.g., "the syllable SO") is *semantically plural* (albeit *syntactically singular*), that is, they refer to many things (e.g., S and O) taken together. For the present purpose, however, it is not necessary to discuss the view.

[2] A *proper* part of an object is a part of the object not identical with it. An *improper* part of an object is a part identical with it.

[3] He argues that the first view conflicts with the thesis because the letters S and O are knowable if and only if they both are (i.e., S is knowable and so is O).

[4] In this article, I follow the modern practice to use "part" in a broad sense for both proper and improper parts, unless noted otherwise.

[5] Or, equivalently, the proper parts (taken together) are the same things as the (single) thing.

[6] Note also that Plato assumes C1 in the *Parmenides* [18, 129c–d] (see Yi [28, 170]).

[7] Note that it is not plausible at all to hold C1 on syllables, words, etc. Both "so" (or SO) and "os" (or OS) are composed of S and O, but these two letters (taken together) cannot be the same things as both SO and OS, for the two syllables are not the same thing. To avoid this problem, defenders of C1

Modern philosophers have proposed and had debates over its close cousins. Most of the recent debates are influenced by David Lewis's statement of what he calls "the Thesis of *Composition as Identity*" in *Parts of Classes* [10, 82, original italics]. Using "fusion" for wholes composed of parts (proper or improper), he states the thesis as follows:

> I say that composition . . . the many-one relation of many parts to their fusion . . . is like identity. The 'are' of composition is, so to speak, the plural form of the 'is' of identity. Call this the Thesis of *Composition as Identity*. [10, 82, original italics]

In this passage, he states a thesis about the predicate "compose": the relation of composition that the "compose" in, e.g., "S and O *compose* SO" refers to is a *plural cousin* of the standard *singular identity* relation that the "is" of identity refers to, a kind of identity relation, *plural identity*, that (unlike the standard, *singular* identity) can pertain to *many* things (e.g., S and O) taken together. On this thesis, one can replace the "compose" in, e.g., "S and O *compose* SO" with the plural form "are" of the "is" for singular identity to obtain, e.g., "S and O *are* (the same things as) SO" (and *vice versa*). Thus the thesis implies a cousin of C1:

C2. *Composition as Identity* (CAI): Some things (taken together) *compose* something if and only if the former things (taken together) *are* (the same things as) the latter thing.

This thesis, which is usually called the *composition as identity thesis* (in short, CAI), results from replacing with "compose" the italicized "are (the same things as)" in a logical truth:

> Some things (taken together) *are* (the same things as) something if and only if the former things (taken together) *are* (the same things as) the latter thing.

While inspiring the above formulation of the thesis that composition is (plural) identity, Lewis ends up rejecting the thesis that *identifies* composition with plural identity (which implies C2) to settle with a weaker thesis that takes it to be *similar* in certain respects to plural identity.[8] The latter, unlike the former, does not imply C2.[9] Still, some philosophers inspired by his suggestion of the stronger thesis defend C2. Moreover, some philosophers (e.g., Plato [18, 129c–d], Frege [4, 28]) assume the thesis without giving an explicit statement thereof.[10] But I think the thesis has a serious problem.

---

might restrict the thesis to a special kind of composites, those for which different arrangements (e.g., order) of their components do not yield different composites (e.g., spatial regions). The subsequent discussion of C1 applies to the thesis as restricted to such composites as well. The same holds for C2, which is presented below.

[8] See Lewis [10, 84–87]. Note that the weaker thesis does not imply C2. For more on the difference between the strong and weak theses, see Yi [24, 145f], which calls them the *strong* and *weak composition theses*. See also, e.g., Yi [28, 170f] and [29, S4471f].

[9] One might take the passage quoted above to state the weaker thesis. In this article, however, I consider only the stronger thesis, which implies C2, and henceforth means C2 by "CAI".

[10] See, e.g., Yi [28, 170] and [28, 187] for discussions of the assumption of CAI in Plato's *Parmenides* [18, 129c–d] and Frege's *Foundations of Arithmetic* [4, 28], respectively.

We can see that CAI (i.e., C2) implies an implausible thesis about parts and wholes that proponents of the thesis cannot accept:

T1. *Weak part-whole triviality* (*mereological nihilism*): Nothing has a proper part.[11]

The same holds for C1, the precursor of C2 that Plato assumes in his argument against the knowability thesis. In this argument, we have seen, he takes C1 to imply a logical equivalent of T1:

T1′. If something (e.g., SO) has a proper part, it cannot have a proper part.

He applies T1′ to conclude that the syllable SO, for example, "must have no parts" [17, 204a]. While he uses this consequence of C1 to reject the knowability thesis, I think the consequence gives a good reason to reject C1, not the knowability thesis. The same holds for C2, which also implies T1.

In this article, I develop this argument by proving that both C1 and C2 imply T1. While they both imply T1, we can see that C2 is stronger than C1. C1 is logically equivalent to T1, but C2 to a stronger thesis:

T2. *Part-whole triviality*: Something is a part of something if and only if the former is identical with the latter.[12]

The equivalence between C1 and T1 and that between C2 and T2 yield complete characterizations of C1 and C2. And the characterizations show that it is incoherent for proponents of the theses to hold them, for they *assume* that some things (e.g., SO, water molecules, spatial areas) have proper parts.[13]

To prove that C1 and C2 are logically equivalent to T1 and T2, respectively, it is necessary to use an extension of elementary logic, *plural logic*, that can capture logical relations pertaining to plural (as well as singular) constructions of natural languages (e.g., those involved in C1 and C2). I present the basics of plural logic in Sect. 18.2. Then I prove the logical equivalence between C1 and T1 in Sect. 18.3, and that between C2 and T2 in Sect. 18.4. In Sect. 18.5, I conclude by arguing that the proofs show that those who hold CAI (i.e., C2) are incoherent.

Now, it would be useful to have an overview of earlier works related to the argument against CAI sketched above. Note that the thesis that composition is identity, strictly speaking, is stronger than C2. While the former is a thesis about the *relations* of composition and plural identity, one that identifies the former with the latter, C2 is just concerned with the *extensions* of the relations. But it is usual to focus on C2 (or the like) in debates over the thesis and to call C2 (or the like) the *composition as*

[11] Sider calls T1 "mereological nihilism" [19, 20], while taking it to be the thesis called *nihilism* by van Inwagen, who officially defines it as the thesis that some things compose something if and only if they are one [23, 72f]. See also Calosi [1] and Loss [12].

[12] See Sect. 18.2.1 for more on the difference between T1 and T2.

[13] Sider, who calls T1 "mereological nihilism", *contrasts* T1 with CAI; he says, "Those willing to accept mereological nihilism . . . have no need for composition as identity" because "it is natural [for them] to say that there really is no such relation as parthood" [19, 77]. In a later article [20], he defends T1 but still does not relate the thesis to CAI. See also note 47. For a discussion of Sider's views on CAI and mereology, see, e.g., Yi [29, S4493ff].

identity thesis (or *CAI*), for the relations cannot be identical if C2 is false. And most of the debates focus on a key part of C2. Note that C2 can be divided into two parts:

C2a. *CAI-A*: If some things (taken together) *compose* something, the former things (taken together) *are* (the same things as) the latter thing.
C2b. *CAI-B*: Some things (taken together) *compose* something, if the former things (taken together) *are* (the same things as) the latter thing.

And CAI-B is fairly plausible.[14] So most arguments against C2 are aimed at CAI-A, and this thesis, instead of C2, is often called the *composition as identity thesis* (or *CAI*).

Yi [24, 146–149], who presents one of the earliest arguments against CAI, argues that CAI-A conflicts with the assumption that there is a composite (e.g., Genie) composed of two disjoint objects (e.g., Tom and Jerry). We can basically apply the argument in [24] to show that CAI-A conflicts with the weaker assumption that there is a composite object (i.e., an object with a proper part), which means that CAI-A *trivializes* the part-whole relation because it implies T1, the weak part-whole triviality thesis. And Yi [28, 183–184] proves that CAI (viz., C2) implies T1. To prove this, the article uses *plural logic*, one of the usual *definitions* of the predicate for composition, and the part-whole *reflexivity* (T3: $x < x$). This almost shows that CAI *logically* implies T1 (and T2),[15] but falls a step short by relying on a non-logical thesis (viz., T3). (The article argues, however, that T3 is "an analytic truth any theory of part and whole must include as a theorem" [28, 184].) Yi [29] makes two improvements. The article proves that CAI-B logically implies T2, and thereby proves that CAI (viz., C2) *logically* implies T2 and *vice versa*. But the article rests on the same definition of the composition predicate used in Yi [28] (see Definition 18.6(a) in Sect. 18.4.1). Although it is the definition most proponents of CAI assume in proposing the thesis, the question arises whether they can avoid the trivialization of the part-whole relation by adopting *alternative* definitions of the predicate (e.g., Definition 18.6(b) or 18.6(c) in Sect. 18.4.1). In this article, I establish the negative answer to this question. I prove that CAI is logically equivalent to T1 on *all* the usual definitions of the predicate (Sect. 18.4.3). Moreover, I clarify the *constraint* that adequate definitions of the predicate must conform to, and use the constraint to prove that CAI is logically equivalent to T1 on *any* adequate definition of the predicate (Sect. 18.4.3).[16]

---

[14] CAI-B follows from the thesis that everything compose itself, which follows from the thesis that everything is a part of itself (T3 in Sect. 18.2.1) on usual definitions of the predicate "compose".

[15] Note that T1, together with T3, implies T2 (see Sect. 18.2.1).

[16] Sider [21], who holds CAI (or CAI-A), proves that CAI-A implies a thesis weaker than T2 (see Yi [29, S4486ff]). Using his proof, Calosi [1] proves that CAI-A is equivalent to T1, and Loss [12] that CAI-A implies T1. While using plural logic, all of these works formulate the proofs in systems that include non-logical axioms of mereology (the standard theory of part and whole), including T3 among others: Sider [21, 212] and Calosi [1, 221] assume most of the axioms of mereology, and Loss [12], who uses Sider's proof of a thesis weaker than T2, inherits the axioms assumed by Sider and holds that it follows from the definition of the predicate for composition (e.g., Definition 18.6(a) in Sect. 4.1)) that everything is composed by a single object (viz., itself) [12, 371], while it is necessary to use T3 to show this (see note 44). Thus, none of them addresses the question whether

18.2 Plural Logic

Consider the characterizations of C1 and C2 in terms of T1 and T2:

1. C1 is logically equivalent to T1.
2. C2 is logically equivalent to T2.[17]

To prove these characterizations, it is necessary to invoke the logic of plural constructions of natural languages, including those involved in C1 and C2. To do so, it is useful to use a system of logic, *plural logic*, recently developed by, e.g., Yi [26, 27], McKay [13], and Oliver and Smiley [14] to give a systematic account of logical features of plural constructions, (e.g., C1, C2), including their relations to singular constructions (e.g., T1, T2). To account for the logic of plural constructions, plural logic extends elementary logic to regimented languages we can get by extending them by adding symbolic counterparts of basic plural constructions of natural languages: "they", "some things (are such that)", "are (the same things as)", "compose", etc. It is useful to have an overview of the basics of plural logic to prepare for proofs of characterizations of C1 and C2.[18]

18.2.1 Plural Languages

To characterize the logic of C1 and C2, it is useful to formulate them in (*regimented*) *plural languages*, extensions of elementary languages that include symbolic counterparts of plural constructions of natural languages. We can formulate them in basic plural language resulting from extending elementary languages by adding *plural cousins* of elementary language variables, quantifiers, and predicates:

a. plural variables: xs, ys, etc.
b. plural quantifiers: Π (the universal), Σ (the existential)
c. plural predicates: H (**is.one.of**), $\approx$ (**be.the.same.as**), C (**compose**), etc.

(Plural languages include elementary language variables, quantifiers, and predicates as *singular* variables, quantifiers, and predicates.) Plural variables are symbolic counterparts of plural pronouns (e.g., "they") used anaphorically, as "they" are in "Students do well in classes *they* enjoy".[19] The plural universal and existential quantifiers amount to the determiners "any" and "some" that combine with plural common nouns, as in "*Some* students, if they work together, can solve the problem" and "*Any* students who can collaborate to solve this problem must be very good at collaboration". Plural predicates have one or more *plural argument places*, argument places

---

CAI (or CAI-A) *logically* implies T1, and defenders of CAI might respond to them by rejecting some of the non-logical theses used in their proofs.

[17] See Theorems 18.1 and 18.2 in Sects. 18.3–18.4.

[18] The overview given below is based on Yi [26, 27].

[19] Plural variables are primitive plural terms.

that *can* admit plural terms (e.g., plural variables).[20] For example, the second argument place of "H", both argument places of "$\approx$", and the first argument place of "C" are plural, and "$H(x, xs)$ (it *is one of* them)", "$xs \approx ys$ (they *are* (*the same things as*) they)", and "$C(xs, y)$ (they *compose* it)" are well-formed.[21] Now, plural argument places might admit singular terms (e.g., the singular variable "x") as well. Such argument places are called *neutral* while the others are called *exclusively plural*. All plural argument places of predicates we will consider are *neutral* ones. For example, "$H(x, y)$ (it *is one of* it)", "$xs \approx y$ (they *are* (*the same things as*) it)", and "$C(x, y)$ (it *composes* it)" are also well-formed.

Like the singular identity predicate "=", the plural predicate "H (**is.one.of**)" is a basic logical predicate. And we can use it to define other logical predicates. The plural identity predicate "$\approx$ (**be.the.same.as**)" can be defined as follows:

Definition 18.1 (*Plural identity*) $xs \approx ys \equiv: \forall z(zHxs \leftrightarrow zHys)$.

(I use, e.g., "$zHxs$" as a short for "$H(z, xs)$".)[22]

While basic plural languages do not have complex terms amounting to *plural definite descriptions* (e.g., "the parts of Socrates"), we can introduce them via contextual definitions (see Yi [26, 244f] and [27, Sect. 5.1]). Suffice it, for the present purpose, to note the notation I use for one kind of plural definite descriptions: $[x : \varphi(x)]$, where $\varphi(x)$ is a formula. For example, "$[x : x < s]$", where "$<$" is a singular predicate for the part-whole relation and "s" a singular term for Socrates, is a plural language counterpart of "the (things that are) parts of Socrates".

Some might object that the symbolic predicate "H" is not a proper counterpart of the English "is one of" because the second argument of this predicate, they hold, cannot admit a singular term.[23] If this is correct, however, we can use the predicate to define its *neutral* cousin, "is.one.of", as follows:

x is.one.of $ys \equiv:$ anything z is such that x is one of $[ys$ and $z]$.

---

[20] By contrast, elementary language predicates (e.g., "=", "$<$ (is a part of)") are *singular* predicates, those whose argument places can admit *only* singular terms (e.g., the singular variable "x").

[21] The first argument place of "H" and the second one of "C" are *singular* argument places, i.e., argument places that can admit only singular terms. All the three examples given above are 2-place plural predicates, but plural languages may include plural predicates with any finite number of argument places.

[22] Using this definition, we can show that plural identity is an equivalence relation. That is, the following are logical truths of plural languages:

a. *Plural reflexivity*: $xs \approx xs$.
b. *Plural symmetry*: $xs \approx ys \rightarrow ys \approx xs$.
c. *Plural transitivity*: $xs \approx ys \wedge ys \approx zs \rightarrow xs \approx zs$.

[23] For a discussion of this objection and the response given below, see, e.g., Yi [24, 179f]. See also the discussion of PL2 in Sect. 18.2.2.

Replacing the plural variable "ys" in "[ys and z]" with a singular term (e.g., "y") yields a complex plural term (e.g., "[y and z]"), which can be admitted into the second argument place of "is one of".[24] Similarly, we can define "H" as a predicate with a *neutral* argument place in terms of its cousin with an *exclusively plural* one, "H'" (**is-one-of**), that matches the English "is one of". To do so, we can consider plural languages that include "H'" as a primitive predicate and "@ (**and**)" as a symbolic counterpart of the English "and" as used to combine terms to yield complex plural terms, as in "Socrates *and* Plato", "Socrates's parts *and* Plato", "Socrates *and* Plato's teacher", etc. In such languages, we can define "H" as a cousin of "H'" as follows:

Definition 18.2 $H(x, ys) \equiv: \forall z\, H'(x, [ys@z])$.

Then the second argument place of "H" is neutral; "$H(x, y)$" (viz., "$\forall z\, H'(x, [y@z])$") is well-formed because "$[y@z]$" is a plural term.[25]

Now, we can formulate C1 and C2 in plural languages with mereological predicates. Consider plural languages that, in addition to logical expressions of plural languages (e.g., "xs", "Π", "H"), include the two-place plural predicate "C (**compose**)" while extending elementary languages that include singular predicates for mereological relations: "$<$ (is a part of)", "O (overlaps)", "$\lneq$ (is a proper part of)", etc. In such languages, we can formulate C1 and C2 as follows:

C1. *Plato's Thesis*: $\exists y\, y \lneq x \rightarrow C([z : z \lneq x], x) \wedge x \approx [z : z \lneq x]$. (If something has a proper part, then its proper parts compose it and it is the same thing as the proper parts.)

C2. CAI: $C(xs, y) \leftrightarrow xs \approx y$. (Some things *compose* something if and only if the former (taken together) *are* (the same things as) the latter.)[26]

In the same languages (which include elementary languages with singular mereological predicates), we can also formulate T1 and T2:

T1. *Weak part-whole triviality*: $\sim\ y \lneq x$. (Or, $\sim \exists x\, \exists y\, y \lneq x$. Nothing has a proper part.)

T2. *Part-whole triviality*: $x < y \leftrightarrow x = y$. (Something is a part of something if and only if the former is identical with the latter.)

Note that mereological predicates can be reduced to one of them (e.g., "$<$", "O"). It is usual to define the proper part predicate "$\lneq$" as follows:

---

[24] So "Anything is such that *Cicero* is one of [*Tully* and it (= the thing)]", for example, is well-formed. Incidentally, this sentence is true if and only if *Cicero* is *Tully* (see PL2 in Sect. 18.2.2).

[25] And we can see that 'xHy' and '$x = y$' are logically equivalent, and so are "$xHys$" and "$xH'ys$" (see the discussion of PL2 in Sect. 18.2.2).

[26] In giving the symbolic formulation of C1 and C2, I follow the usual practice of omitting frontal universal quantifiers (singular or plural), which we can take to be implicit to bind the free variables in the explicit formulas. (The same hold for T1 and T2 below.) With the universal quantifiers made explicit, C2, which involves a plural variable, is as follows:

C2. CAI: $\Pi xs\, \forall y (C(xs, y) \leftrightarrow xs \approx y)$.

Definition 18.3 (*Proper part*) $x \lneq y \equiv: x < y \wedge x \neq y$.

And we can define the composition predicate "C" in terms of "<" or "O" in plural languages (see Definition 18.6 in Sect. 4.1).

Clearly, T1 is logically equivalent to a thesis Plato takes to follow from C1:

T1′. $x \lneq y \rightarrow \sim x \lneq y$. (If something is a proper part of something, the former is not a proper part of the latter.)

And it is straightforward to use Definition 18.3 to see that T1 is logically equivalent to the following:

T1″. $x < y \rightarrow x = y$. (If something is a part of something, the former is identical with the latter.)

So T2, which implies T1″, implies T1. And the converse of T1′ is logically equivalent to the reflexivity of the part-whole relation:

T3. *Part-whole reflexivity*: $x < x$.

Thus, T2 is logically equivalent to the conjunction of T1 and T3. And this means, as we can see, that C2 (i.e., CAI) is logically equivalent to the conjunction of C1 and T3.[27] So we might take C1 to be weaker than C2. But this does not mean that there is a great division between the theses. For T3 might be considered a *logical* (or *analytic*) truth resulting from a suitable definition of the wide notion of *part*.[28] For example, one might take the narrow notion of *part* (Plato's) to be more basic than the wide notion to replace Definition 18.3 with a definition of "<" in terms of "$\lneq$":

Definition 18.4 (*Alternative to Definition* 18.3) $x < y \equiv: x \lneq y \vee x = y$.

This definition yields T3 as a logical (or analytic) truth (see also Definition 18.5(b) in Sect. 18.4.1). So Plato's Thesis, C1, might be considered a formulation of the composition as identity thesis (CAI).

18.2.2 Plural Logic

Unlike elementary logic, the logic of plural languages is not completely axiomatizable.[29] But we can give strong systems of plural logic capturing much of their logic that are sufficient for most purposes. To prove the characterizations of C1 and C2, we can use the system of plural logic given in Yi [25, 179ff] and [27]. Presented below are basic theorems of the system used in proving the characterizations:

---

[27] For C1 and C2 are logically equivalent to T1 and T2, respectively (Theorems 18.1 and 18.2).

[28] See, e.g., Yi [28, 183f].

[29] See Yi [25, 181] and [27, 257–262]. And see, e.g., Yi [27, 256–262] for a model-theoretic characterization of plural logic.

PL1. $\exists x\, xHxs$.
PL2. $xHy \leftrightarrow x = y$.
PL3. $H(x, [y : \varphi(y)]) \leftrightarrow \varphi(x)$.
PL4. $\Sigma xs\, (xs \approx [x : \varphi(x)]) \leftrightarrow \exists x\varphi(x)$.
PL5. $xs \approx ys \leftrightarrow \forall z(zHxs \leftrightarrow zHys)$.[30]
PL6. *Substitutivity of Plural Identity* ($\approx E$): $xs \approx ys \wedge \varphi(xs) \rightarrow \varphi(ys)$.

PL1–PL6 include implicit singular or plural universal quantifiers ($\forall$, Π) binding free variables in the explicit formulas. With the quantifiers made explicit, PL6, for example, is:

PL6. $\Pi xs \Pi ys(xs \approx ys \wedge \varphi(xs) \rightarrow \varphi(ys))$.

And the system yields (a) the standard singular universal and existential quantifier introduction and elimination rules, such as $\forall$-elimination ($\forall E$), and (b) their cousins for plural universal and existential quantifiers, such as Π-elimination (ΠE). For example, PL6 logically implies its proper instances by Π-elimination (ΠE), which result from properly substituting the plural variables in the explicit formulas with suitable terms (singular or plural).

Two remarks about proper instances of PL1–PL6 are in order. First, we can get instances of *plural quantifications* with *singular* as well as plural terms, for the plural quantifiers Π and Σ do not amount to "some two or more" and "any two or more", but to "some one or more" and "any one or more".[31] Thus proper instances of, e.g., PL6 include "$xs \approx y \wedge H(s, xs) \rightarrow H(s, y)$", where the singular "$y$" replaces the plural "$ys$" in the theorem. Second, the proper application of ΠE with defined plural terms (e.g, the plural definite description "$[x : x < s]$") requires that they are not empty.[32] To show this, it is necessary to prove the matching plural existential quantifications (e.g., "$\Sigma xs\, xs \approx [x : x < s]$"), which one can prove by, e.g., applying PL4. With, e.g., "$\Sigma xs\, xs \approx [x : x < s]$" proved, we can apply ΠE to PL5 to get "$[x : x < s] \approx y \wedge H(s, [x : x < s]) \rightarrow H(s, y)$".

Now, some might object to PL2. To do so, they might hold that the formula is ill-formed because "xHy", which amounts to, roughly, "it is one of *it*", is ill-formed (see Sect. 18.2.1). Or they might argue that PL2 is false because "xHy", if well-formed, must be taken to amount to "$x < y$" (and Cicero, for example, has a proper part). For example, Payton [15, S4588f] and [16, 8067, note 31] rejects PL2 to defend CAI and related theses (e.g., the many-one identity thesis).[33] While arguing that defenders of

[30] This results from the definition of $\approx$ (Definition 18.1). Note, however, that PL6 yields one direction of the biconditional: $xs \approx ys \rightarrow \forall z(zHxs \leftrightarrow zHys)$.

[31] By contrast, we cannot get instances of singular quantifications (e.g., PL2) by substituting singular variables with plural terms.

[32] Note also that predications involving any terms, including complex terms (e.g., "$C([x : x < s], y)$"), imply the existential propositions featuring them (e.g., "$\Sigma xs\, (xs \approx [x : x < s])$").

[33] The many-one identity thesis holds that there is a single thing that is identical with many things (taken together). See, e.g., Yi [28, 179–182], where I argue that the thesis is logically inconsistent.

the theses must reject PL2 by rejecting the axiom that (together with PL1) yields the thesis in usual axiomatizations of plural logic,[34] he takes the thesis (and the axiom), it seems, to be devoid of any justification or motivation. But there are strong reasons to consider PL2 a logical truth.[35]

Suppose that the English "is one of" (as the objectors assume) can admit only plural terms in its second argument place. This does not mean that the symbolic formula "xHy" must be ill-formed. For the predicate "H", as explained above (Sect. 18.2.1), can be taken to be introduced via a definition in languages that include a symbolic counterpart of the English "is one of" whose second argument place is *exclusively plural*: "H'" (see Definition 18.2). Using the definition, as noted above (Sect. 18.2.1), we can see that the second argument place of "H" is *neutral*. Moreover, we can prove that PL2 follows from axioms or theorems pertaining to "H'" (and "@ (**and**)") in systems of plural logic for languages that include the predicate, such as the following:

PL2a. $H'(x, [y@z]) \leftrightarrow [x = y \vee x = z]$. (Something is one of this and that if and only if it is identical with either the former or the latter.)
PL2b. $H'(x, [ys@z]) \leftrightarrow [H'(x, ys) \vee x = z]$. (Something is one of these and that if and only if it is one of them or identical with it.)

With Definition 18.2, PL2a implies PL2. And we can use the definition (and PL1) to prove that PL2b implies the equivalence between "$H(x, ys)$" and "$H'(x, ys)$".[36]

And it is straightforward to see that PL2a is a logical truth. Consider (a)–(d):

a. Cicero is a philosopher (Pc).
b. Seneca is a philosopher (Ps).
c. Cicero and Seneca are philosophers.
d. Any *one of* Cicero and Seneca is a philosopher ($\forall x(H'(x, [c@s]) \rightarrow Px)$).

---

[34] The axiom is one direction of PL2: $xHy \rightarrow x = y$. Payton's argument against the axiom [15, S4588f] assumes both CAI and the *unrestricted composition thesis*, the thesis that any things whatsoever compose a single thing (while assuming that some things have proper parts). Note that the two theses, taken together, imply Eleatic monism, the view that there is something identical with everything, as I argue in [28, 184] and [29, S4481f] (the argument uses a part of Theorem 18.2). Incidentally, Carrara and Lando [2, S4220] suggest that those who hold the two theses would reject plural logic, but does not specify which of PL1–PL6 can be plausibly rejected.

[35] The replies given below to the objections to PL2 are essentially the same as the responses to objections to PL2 (and PL2a given) given in Yi [24, 146–8] (see discussions of the first and third objections). See also Yi [28, 178f].

[36] Thus, theorems of systems that include PL2a–PL2b (and Definition 18.2) also include their cousins for "H":

PL2a′. $H(x, [y@z]) \leftrightarrow [x = y \vee x = z]$.
PL2b′. $H(x, [ys@z]) \leftrightarrow [H(x, ys) \vee x = z]$.

Note that PL2 follows from PL2a′ as well.

(a) and (b) logically imply (c) and thus (d), of which (c) can be considered a paraphrase, and both (c) and (d) logically imply both (a) and (b). These logical relations cannot hold unless PL2a is a logical truth.[37] By contrast, those who take "x < c", for example, to imply "xHc" (while rejecting T1) would have to deny that (a) and (b) logically imply (c) and (d), for Cicero has proper parts (e.g., his nose) that are not philosophers. (Assume that "$x < c$" implies "xHc", and let n be Cicero's nose. Then "$H'(n, [c@s])$" holds, while "$P(n)$" does not. So (d) fails, while both (a) and (b) hold.)[38]

18.3 Characterization of Plato's Thesis

While stating a precursor of CAI, we have seen (Sect. 18.1), Plato assumes that the thesis, C1, implies T1 (or, equivalently, T1$'$). We can use plural logic to show that he was right to assume this. It is straightforward to prove the characterization of C1 in terms of T1:

Theorem 18.1 *C1 is logically equivalent to T1.*

Clearly, T1 implies C1. The converse follows from the following:

Lemma 18.1 *C1 implies the following:*

a. $y \lneq x \to [z : z \lneq x] \approx x.$
b. $[z : z \lneq x] \approx x \to \forall y(H(y, [z : z \lneq x]) \leftrightarrow H(y, x)).$
c. $[z : z \lneq x] \approx x \to \forall y(y \lneq x \leftrightarrow y = x).$
d. $y \lneq x \to \exists y\,(y \lneq x \wedge y = x).$
e. $\sim y \lneq x.$

So we can complete a proof of Theorem 18.1 by proving Lemma 18.1:

Proof Assume C1. Clearly, (a) follows from C1. (b) is a logical truth following from PL6 ($\approx E$). (c) follows from (b) by PL3 and PL2. (d) follows from (a) and (c). Finally, (e) follows from (d) by the definition of "$\lneq$" (Definition 18.3).

18.4 Characterization of CAI

We can now prove the characterization of CAI (C2) in terms of T2:

Theorem 18.2 *C2 (i.e., CAI) is logically equivalent to T2.*

---

37 We can give essentially the same reason for taking PL2b to be a logical truth.

38 For more on the theorems given above, including PL2, see, e.g., Yi [28, 177ff] and [29, S4475ff].

To do so, it is necessary to use an analysis of the *composition* relation that relates it to other mereological attributes (i.e., properties or relations): atomicity, part-whole, overlap, disjointedness, etc. Systems of *mereology* (deriving from the Greek *meros* "part") formulate analyses of composition in terms of simple mereological relations as definitions of the predicate "compose (or C)", and proponents of CAI assume the definitions in formulating the thesis.[39] We can use them to prove Theorem 18.2. In addition to *definitions* of mereological predicates, systems of mereology include proper *axioms*. But it is not necessary to use any such axiom to prove the theorem. Thus it is a theorem of plural logic and this shows that CAI and T2 are *logically* equivalent.

18.4.1 Mereology and Analyses of Composition

Classical mereology (CM) is a standard theory of part and whole that stems from the works of Leśniewski [8, 9], Tarski [22], and Leonard and Goodman [7], who present different but logically equivalent systems of axioms of mereology. Systems of classical mereology involve analyses of various mereological attributes: atomicity, part-whole, overlap, disjointedness, composition, etc. They take one of them (e.g., part-whole, overlap) as basic to analyze other mereological attributes in terms of them, and formulate the analyses as definitions of predicates for the attributes. Many of the systems take the *part-whole relation* as basic and use a predicate for the relation (e.g., "$<$") as a primitive while giving definitions for other mereological predicates. But some might take other mereological predicates as primitives. Leonard and Goodman [7] use the predicate for *disjointedness* as a primitive, and Goodman [5, 46–56] takes as a primitive the predicate for its complement, "*overlaps* (or O)", to reformulate their system, including their analysis of composition.

Now, systems of classical mereology give three major analyses of composition. Two of them use the part-whole relation as basic and the other might be taken to use the overlap relation as basic. The analyses can be formulated as definitions of the composition predicate "C (**compose**)" in plural languages.

Let $\mathscr{L}_<$ and $\mathscr{L}_o$ be plural languages that include "$<$ (is a part of)" and "O (overlaps)", respectively, as primitive singular two-place predicates. Then we can define "O" and "$<$" in $\mathscr{L}_<$ and $\mathscr{L}_o$, respectively:

Definition 18.5 (*Overlap and Part-whole*)

a. $xOy \equiv: \exists z(z < x \wedge z < y)$. $[\mathscr{L}_<]$
b. $x < y \equiv: \forall z(zOx \rightarrow zOy)$. $[\mathscr{L}_o]$

And we can formulate three definitions of the composition predicate:

Definition 18.6 (*Composition*)

---

[39] They usually call wholes composed of some things their *fusions* or (*mereological*) *sums*.

a. $C(xs, y) \equiv: \forall x(xHxs \rightarrow x < y) \wedge \forall z(z < y \rightarrow \exists x[xHxs \wedge xOz])$. $[\mathscr{L}_<]$
b. $C(xs, y) \equiv: \forall x(xHxs \rightarrow x < y) \wedge \forall z(\forall x[xHxs \rightarrow x < z] \rightarrow y < z)$. $[\mathscr{L}_<]$
c. $C(xs, y) \equiv: \forall x(xOy \leftrightarrow \exists z[zHxs \wedge xOz])$. $[\mathscr{L}_o]$

The first two definitions, (a)–(b), are given in $\mathscr{L}_<$ and the third, (c), in $\mathscr{L}_o$.[40]

The three definitions are not logically equivalent, although they are equivalent in some systems of mereology, which include proper axioms governing "<" and "O".[41] But it is not necessary for our purpose to decide in favor of one or another of the definitions, for all of them yield Theorem 18.2. That is, we can use any of the definitions to prove the theorem.

18.4.2 Three Theses on Composition

We can take Definitions 18.6(a)–18.6(c) to involve only the primitives in relevant languages. In 18.6(a), "xOz" can be taken to abbreviate "$\exists w(w < x \wedge w < z)$" (Definition 18.5(a)). Similarly, T1 and T2 can be considered formulas in $\mathscr{L}_o$ involving abbreviations by Definition 18.5(b). But it is useful for convenience of exposition to consider *definition systems* (i.e., systems of definitions) that include definitions for extending $\mathscr{L}_<$ and $\mathscr{L}_o$ to languages that include defined expressions as well. Let $S_1^<$ be a definition system for extending $\mathscr{L}_<$ that includes Definitions 18.5(a) and 18.6(a), $S_2^<$ one for $\mathscr{L}_<$ that includes 18.5(a) and 18.6(b), and S^o one for $\mathscr{L}_o$ that includes Definitions 18.5(b) and 18.6(c). Then we can refer to these systems to formulate Theorem 18.2 as specified with definitions of the composition predicate:

Theorem 18.3 (Formulations of Theorem 18.2 with specification of definitions)

a. C2 is equivalent to T2 in $S_1^<$.
b. C2 is equivalent to T2 in $S_2^<$.
c. C2 is equivalent to T2 in S^o.

With the definitions specified by the systems, we can use plural logic to prove this theorem. This shows that C2 and T2 as, e.g., formulas of $\mathscr{L}_<$ involving abbreviations by Definitions 18.5(a) and 18.6(a) are *logically equivalent* on plural logic.

Theorem 18.3(a) is proved in Yi [29].[42] We can follow the same strategy for the proof of 18.3(a) to prove 18.3(b)–(c). Moreover, we can improve on the proof by attending to some theses on composition verified by all the three definitions:

Comp-1. $\forall x\, x < x \rightarrow C(y, y)$.

---

[40] See Tarski [22] and Lewis [10] for Definition 18.6(a), Link [11, 17 and 24] for 18.6(b), and Leonard and Goodman [7] and Goodman [5, 52] for versions of 18.6(c).

[41] See, e.g., Hovda [6] for a discussion of the three definitions. Note that he takes all the definitions (including 18.6(c)) to be given in $\mathscr{L}_<$, but I think this fails to do justice to the elegance and power of Definition 18.6(c).

[42] See also Yi [28, 183f], Calosi [1], and Loss [12] (see note 16).

Comp-2. $\forall x\, x < x \wedge yHxs \wedge \forall z(zHxs \rightarrow z < y) \rightarrow C(xs, y)$.
Comp-3. $C(xs, y) \wedge zHxs \rightarrow z < y$.

(Note that Comp-2 implies Comp-1.) It is straightforward to prove that all these theses hold in any of the three definition systems:

Lemma 18.2

a. *Comp-1, Comp-2, and Comp-3 hold in* $S_1^<$.
b. *Comp-1, Comp-2, and Comp-3 hold in* $S_2^<$.
c. *Comp-1, Comp-2, and Comp-3 hold in* S^o.[43]

So we can prove Theorem 18.3 by proving the following:

Theorem 18.4 *Suppose that Comp-2 and Comp-3 hold. Then C2 is equivalent to T2.*

18.4.3 Proof of Characterization of C2

To prove Theorem 18.4, it is useful to divide C2 into two parts:

C2a. CAI-A: $C(xs, y) \rightarrow xs \approx y$.

C2b. CAI-B: $xs \approx y \rightarrow C(xs, y)$.

Now, notice that the theorem follows from the following:

Lemma 18.3

a. *C2b implies* $\forall x\, C(x, x)$.
b. *C2b and Comp-3 imply* $\forall x\, x < x$ *(T3)*.
c. *Comp-2 and T3 imply* $C([z : z < y], y)$.
d. *C2a and* $C([z : z < y], y)$ *imply T2.*
e. *Comp-1, Comp-3, and T2 imply C2.*

So it suffices to prove Lemma 18.3. To do so, let "$< y >$" abbreviate the plural definite description "$[z : z < y]$". Then T3 implies "$\exists xs\, (xs \approx < y >)$" (by PL4), and "$H(z, < y >)$" and "$z < y$" are logically equivalent (by PL3). Now, we can prove the lemma:

Proof [Lemma 18.3(a)] Consider an instance of C2b: $x \approx x \rightarrow C(x, x)$. This implies $C(x, x)$, for $x \approx x$ is a logical truth (by PL5). □

Proof [Lemma 18.3(b)] Consider an instance of Comp-3: $C(x, x) \wedge xHx \rightarrow x < x$. This implies $C(x, x) \rightarrow x < x$, for xHx is a logical truth (by PL2). So C2b and Comp-3 imply T3 (by Lemma 18.3(a)). □

---

[43] The proof of this lemma is left as an exercise.

Proof [Lemma 18.3(c)] T3 implies $\exists xs\, xs \approx [z : z < y]$ (by PL4). So we can use the complex plural term $< y >$ (i.e., $[z : z < y]$) to get an instance of Comp-2:

$$\forall x\, x < x \wedge H(y, < y >) \wedge \forall z(H(z, < y >) \rightarrow z < y) \rightarrow C(< y >, y).$$

And T3 implies the antecedent of this conditional (by PL3). So T3 and Comp-2 imply the consequent: $C(< y >, y)$. □

Proof [Lemma 18.3(d)] Consider:

i. $C(< y >, y) \rightarrow < y > \approx y$.
ii. $< y > \approx y$
iii. $H(x, < y >) \leftrightarrow xHy$.
iv. $x < y \leftrightarrow x = y$ (T2).

Assume C2a and $C(< y >, y)$, which implies $\exists xs\, xs \approx < y >$ (see note 32). Then C2a implies (i) (by ΠE). And $C(< y >, y)$ and (i) imply (ii), which implies (iii) (by PL6). Finally, (iii) implies (iv) (by PL2–PL3). □

Proof [Lemma 18.3(e)] Assume Comp-1, Comp-3, and T2. Then the following are equivalent:

i. $C(xs, y)$.
ii. $\forall z(zHxs \leftrightarrow z = y)$.
iii. $\forall z(zHxs \leftrightarrow zHy)$.
iv. $xs \approx y$.

(ii) implies (iii) (by PL2). (iii) implies (vi) (by PL4). And with T2 and Comp-1, (iv) implies (i). For T2 (which implies T3) and Comp-1 imply $C(y, y)$, and $C(y, y)$ and (iv) imply (i) (by PL6). Finally, with T2 and Comp-3, (i) implies (ii):

A. With Comp-3, $C(xs, y)$ and $zHxs$ imply $z < y$. And $z < y$ with T2 implies $z = y$. So (with T2 and Comp-3) $C(xs, y)$ and $zHxs$ imply $z = y$.

B. To show that (with T2 and Comp-3) $C(xs, y)$ and $z = y$ imply $zHxs$, it suffices to show that (with T2 and Comp-3) $C(xs, y)$ implies $yHxs$. Now, with T2 and Comp-3, $C(xs, y)$ and $xHxs$ imply $x = y$ and thus $yHxs$, for $C(xs, y)$ and $xHxs$ imply $x < y$ (by Comp-3) and $x < y$ implies $x = y$ (by T2). So $C(xs, y)$ and $\exists x\, xHxs$ imply $yHxs$. And $\exists x\, xHxs$ is a logical truth (by PL1). So (with T2 and Comp-3) $C(xs, y)$ imply $yHxs$.

So (i) and (iv), i.e., $C(xs, y)$ and $xs \approx y$, are equivalent. □

This completes the proof of Theorem 18.4. And the theorem, together with Lemma 18.2, implies Theorem 18.3. Moreover, I think Theorem 18.4 yields Theorem 18.2.

18.4.4 Constraints on Analyses of Composition

To derive Theorem 18.2 from Theorem 18.4, it is necessary to supplement C2 with a suitable definition of the composition predicate "C" that specifies the notion of *composition* underlying the thesis. Lewis assumes Definition 18.6(a) [10, 73] in stating what he calls "the Thesis of *Composition as Identity*" [10, 82, original italics], which inspires the formulation of C2. And Theorem 18.2 holds for C2 as supplemented with this definition (Theorem 18.3(a)).

Proponents of C2 might base the thesis on other definitions of "C" to avoid commitment to T2. But I do not think one can make C2 precise with a suitable definition of "C" without yielding a thesis that, like T2, trivializes the part-whole relation. Assuming other definitions of "C" given in systems of mereology (Definitions 18.6(b) and 18.6(c)), too, makes C2 logically equivalent to T2 (Theorem 18.3(b)–(c)). Now, proponents of C2 might opt for yet another definition to make a precise formulation of the thesis. But it is hard to see a good reason for them to reject all the three standard definitions and, moreover, to deny the equivalences they yield. The definitions are equivalent in standard systems of mereology, which are accepted by proponents of C2, and the equivalences (e.g., $C(xs, y) \leftrightarrow \forall x(xHxs \rightarrow x < y) \wedge \forall z(z < y \rightarrow \exists x[xHxs \wedge xOz]))$ are theorems of the systems.

Moreover, Theorem 18.4 shows that Theorem 18.2 holds for C2 with *any* definition that verifies Comp-2 and Comp-3. I do not think a definition of "C" supplementing C2 can be considered *adequate* unless the definition yields both theses. Comp-3 directly results from the informal notion of *a whole composed of parts*, which is meant to be captured by a proper analysis of composition. On this notion, if some things compose something, the latter is a *whole* of which the former are *parts*. Now, while this condition is necessary for composition, it is not sufficient: some parts of a whole (e.g., the end points of a line), taken together, fail to compose the whole (e.g., the line). For parts composing a whole must be sufficient to, so to speak, *cover* any part of the whole. Comp-1 and Comp-2 result from analyses of composition meant to capture this condition. The idea is that anything by itself is sufficient to cover any part of it (Comp-1), and that given this, adding additional parts to it does not undermine the coverage already provided by it (Comp-2). The mereological definitions of the composition predicate must capture this informal talk of "coverage" with formal talk of "part ($<$)", "overlap (O)", etc.[44] So we can impose the following *constraint* on an adequate definition of "C":

> *Constraint on definition of the predicate for composition*: An adequate definition of the composition predicate "C" must verify Comp-2 and Comp-3.

[44] This requires taking the part-whole relation to be reflexive: $\forall x\, x < x$ (T3). In some systems of mereology, T3 results from the definition of "$<$" (e.g., Definition 18.5(b)). The same does not hold in systems that include $<$-based definition systems (e.g., $S_1^<$, $S_2^<$)—hence the need to add T3 in the antecedents of Comp-1 and Comp-2. In such systems, however, the definition of "C" is meant to be supplemented with axioms implying T3, for T3 can be considered analytic. Note also that CAI-B (i.e., C2b) and Comp-3 imply T3 (Lemma 18.3(b)). So proponents of CAI cannot reject T3.

With this constraint, Theorem 18.4 implies Theorem 18.2 as elaborated on as follows:

Theorem 18.5 (Elaborate formulation of Theorem 18.2) *C2 is equivalent to T2 (on any adequate definition of the composition predicate "C").*

18.5 Concluding Remarks

The composition as identity thesis (CAI) trivializes the part-whole relation. Its formal version, C2, we have seen (Sect. 18.3), is logically equivalent to the *part-whole triviality thesis* (T2): $x < y \leftrightarrow x = y$ (Theorem 18.5). On this thesis, nothing has a proper part (T1) while everything is an improper part of itself (T3). In the *Theaetetus* [17, 202–204], we have seen (Sect. 18.1), Plato in a way anticipates that CAI trivializes the part-whole relation. He states a cousin of CAI, viz., Plato's Thesis (C1), and takes the thesis to imply a logical equivalent of T1 (viz., T'_1), which trivializes the proper part relation. Although he does not prove this assumption, we have seen that he is right to assume it. C1 is logically equivalent to T1 (Theorem 18.1). And the gap between C1 and CAI (i.e., C2) is negligible, for T3 is arguably an analytic truth.[45]

Theorems 18.1 and 18.2 (or 18.5), which characterize C1 and C2, show that they trivialize the part-whole relation. This means that they are *logically* consistent. Both are implied by, e.g., Eleatic monism combined with T3: there is something identical with everything and it is a part of itself. But this does not mean that it is coherent to hold the theses. Although some (e.g., Parmenides) might end up accepting C1 or C2 because they hold T1 or T2 for an independent reason, proponents of C1 or C2 do not hold them for this reason. Rather, they *assume* the existence of composite objects (i.e., wholes with proper parts), and take the theses to formulate an insight into their nature. But the theses contradict the underlying assumption.

To argue against the knowability thesis, we have seen, Plato gives the syllable SO as a prime example of a composite, one with two proper parts (S and O). Then he concludes that the syllable "must have no [proper] parts" [17, 204a] by noting that C1, which he assumes, implies a logical equivalent of T1: If something (e.g., SO) has a proper part, it has no proper part (T_1'). C1, which indeed implies this thesis, implies that SO, like anything else (e.g., a spatial region), has no proper part. But this gives no reason to conclude that SO has no proper part, for there is no good reason to accept C1, which Plato takes for granted without giving any reason to accept. Taking C1 to imply T1 (or T_1'), I think, he should have taken the initial assumption that SO has proper parts to be far more plausible than C1. Thus, Theorem 18.1, which shows that he was right to take C1 to imply T1 (or T_1'), gives a good reason to reject C1.

---

45 See, e.g., Definition 18.4 in Sect. 18.2.1 and Definition 18.5(b) in Sect. 18.4.1. See also footnote 44.

Modern proponents of CAI have the same problem. They also assume the existence of composites. In proposing CAI (or its weak cousin), we have seen (Sect. 18.1), Lewis considers composition "the *many*-one relation of *many* parts to their fusion" [10, 82, my italics].[46] This assumes that there is a whole with proper parts. Similarly, proponents of CAI assume the existence of composites to take the thesis to cast light on their relation to their parts.[47] Unlike Plato, who takes its cousin (C1) to trivialize the part-whole relation, they fail to see that the same holds for CAI. Still, CAI also trivializes the part-whole relation, for the thesis implies T2 (Theorem 18.2). And this shows that they are inconsistent in holding CAI.

Acknowledgements I presented an earlier version of this article at Peking University and in *The 6th Asian Workshop in Philosophical Logic* (AWPL24) and the 2024 annual meeting of Korean Society for Philosophy of Science. I wish to thank the audiences, including Yifeng Ding and Chunghyoung Lee, for useful comments and discussions. I would also like to thank Chungsik Bak, Joe Hung, and anonymous referees for AWPL24 for useful comments on earlier versions. Needless yet important to say, I am solely responsible for any remaining errors and infelicities.

References

1. Calosi, C.: Composition is identity and mereological nihilism. Philos. Quart. **66**, 219–235 (2016)
2. Carrara, M., Lando, G.: Mereology and identity. Synthese **198** (Suppl 18), 4205–4227 (2020/2021)
3. Cotnoir, A.J., Baxter, D. (eds.): Composition as Identity. Oxford University Press, Oxford (2014)
4. Frege, G.: Foundations of Arithmetic. Translated by J. L. Austin. Blackwell, Oxford (1980)
5. Goodman, N.: The Structure of Appearance, 3rd edn. Reidel, Dordrecht (1977)
6. Hovda, P.: What is classical mereology? J. Philos. Logic **38**, 55–82 (2009)
7. Leonard, H.S., Goodman, N.: The calculus of individuals and its uses. J. Symb. Log. **5**, 45–55 (1940)
8. Leśniewski, S.: Podstawy og lnej teoryi mnogości I. Prace Polskiego Koła Naukowego w Moskwie, Sekcya matematyczno-przyrodnicza, Moskow (1916). English translation by D. I.

[46] This cannot be taken to mean that the relation holds only for wholes with many parts, for the relation is also taken to hold between anything and itself.

[47] For example, Sider formulates CAI as follows:

> Composition as identity is the . . . strangely compelling doctrine that . . . a *single* whole is genuinely identical to its *many* parts taken together—identical in the very same sense . . . in which I am identical to myself and 2+2 is identical to 4. [21, 211, my italics]

And Calosi takes CAI to hold that "*many* things (the *proper parts* of something) can be strictly identical to just one thing (their mereological fusion)" [1, 220, my italics]. In both formulations, CAI includes the assumption of the existence of composites (or proper parts), and is logically inconsistent. (Note, however, that their formal formulations of CAI do not imply the assumption, and follow from T2.)

Barnett as 'Foundations of the General Theory of Sets I'. In: Surma, S.J., Srzednicki, Jan J.T., Barnett, D.I., Rickey, V.F. (eds.) Stanislaw Leśniewski: Collected Works, vol. 1, pp. 129–173. Kluwer, Dordrecht (1992)

9. Leśniewski, S.: O podstawach matematyki. Przegląd Filozoficzny **30**: 164–206; **31**: 261–291; **32**: 60–101; **33**: 77–105; **34**: 142–170 (1927–1931). English translation by D. I. Barnett as 'On the Foundations of Mathematics'. In: Surma, S. J., Srzednicki, Jan J.T., Barnett, D.I., Rickey, V.F. (eds.) Stanislaw Leśniewski: Collected Works, vol. 1, pp. 174–382. Kluwer, Dordrecht (1992)
10. Lewis, D.: Parts of Classes. Basil Blackwell, Oxford (1991)
11. Link, G.: The logical analysis of plural and mass terms: A lattice-theoretical approach. In: Bäuerle, R. et al. (eds.) Meaning, use, and interpretation of language, pp. 303–323. de Gruyter, Berlin (1983). Reprinted in G. Link, Algebraic Semantics in Language and Philosophy, pp. 11–34. CSLI, Stanford (1998)
12. Loss, R.: A sudden collapse to nihilism. Philos. Quart. **68**, 370–375 (2018)
13. McKay, T.: Plural Predication. Oxford University Press, Oxford (2006)
14. Oliver, A., Smiley, T.: Plural Logic, 2nd edn. Oxford University Press, Oxford (2016)
15. Payton, J. D.: How to identify wholes with their parts. Synthese **198** (Suppl 18): 4571–4593 (2019/2021)
16. Payton, J.D.: Composition as identity, now with all the pluralities you could want. Synthese **199**, 8047–8068 (2021)
17. Plato: The Theaetetus of Plato. Translation by M.J. Levett. Hackett, Indianapolis (1992)
18. Plato: Parmenides. Translation by M.L. Gill & P. Ryan. Hackett, Indianapolis (1996)
19. Sider, T.: Parthood. Philos. Rev. **116**, 51–91 (2007)
20. Sider, T.: Against parthood. In: Bennett, K., Zimmerman, D.W. (eds.) Oxford Studies in Metaphysics, vol. 8, pp. 237–293. Oxford University Press, Oxford (2013)
21. Sider, T.: Consequences of collapse. In: Cotnoir and Baxter [3], pp. 211–221 (2014)
22. Tarski, A.: Les fondaments de la géométrie des corps. Księga Pamiątkowa Pierwszego Polskiego Zjazdu Matematycznego (supplement to Annales de Société Polonaise de Mathématique) **7**, 29–33 (1929). Translated as 'Foundations of the geometry of solids'. In: Tarski, A., Logic, Semantics, Metamathematics, 2nd edn., pp. 24–29. Hackett, Indianapolis, IN (1983)
23. van Inwagen, P.: Material Beings. Cornell University Press, Ithaca (1990)
24. Yi, B.-U.: Is mereology ontologically innocent? Philos. Stud. **93**, 141–160 (1999)
25. Yi, B.-U.: Is two a property? J. Philos. **96**, 163–190 (1999)
26. Yi, B.-U.: The logic and meaning of plurals: Part I. J. Philos. Logic **34**, 459–506 (2005)
27. Yi, B.-U.: The logic and meaning of plurals: Part II. J. Philos. Logic **35**, 239–288 (2006)
28. Yi, B.-U.: Is there a plural object? In: Cotnoir and Baxter [3], pp. 169–191 (2014)
29. Yi, B.-U.: Is composition identity? Synthese **198** (Suppl 18), S4467–S4501 (2018/2021)

The manufacturer's authorised representative in the EU is Springer Nature Customer Service Centre GmbH, Europaplatz 3, 69115 Heidelberg, Germany. If you have any concerns regarding our products, please contact ProductSafety@springernature.com

Printed and bound by CPI Group (UK) Ltd, Croydon, CR0 4YY
22/09/2026
02220985-0003